Large eddy simulation (LES) is one of the most powerful computational tools available today for studying complex turbulent flows.

This book is the first to offer a comprehensive review of LES – the history, state of the art, and promising directions for future research. Among topics covered are fundamentals of LES; LES of incompressible, compressible, and reacting flows; LES of atmospheric, oceanic, and environmental flows; and LES and massively parallel computing.

The book grew out of an international workshop that, for the first time, brought together leading researchers in engineering and geophysics to discuss developments and applications of LES models in their respective fields.

All those whose work involves turbulence modeling will find this book an invaluable source of information.

T0297803

# Large Eddy Simulation of Complex Engineering and Geophysical Flows

# Large Eddy Simulation of Complex Engineering and Geophysical Flows

Edited by

**BORIS GALPERIN**
*University of South Florida*

**STEVEN A. ORSZAG**
*Princeton University*

CAMBRIDGE UNIVERSITY PRESS
Cambridge, New York, Melbourne, Madrid, Cape Town, Singapore,
São Paulo, Delhi, Dubai, Tokyo

Cambridge University Press
The Edinburgh Building, Cambridge CB2 8RU, UK

Published in the United States of America by Cambridge University Press, New York

www.cambridge.org
Information on this title: www.cambridge.org/9780521131339

© Cambridge University Press 1993

First published 1993
This digitally printed version 2010

A catalogue record for this publication is available from the British Library

Library of Congress Cataloguing in Publication data
Large eddy simulation of complex engineering and geophysical flows /
edited by Boris Galperin. Steven A. Orszag.
     p.     cm.
Includes index.
ISBN 0-521-43009-7 (hc)
1. Turbulence – Mathematical models.   2. Eddies – Mathematical
     models.   I. Galperin, Boris.   II. Orszag, Steven A.
TA357.5.T87L37   1993
620.1'064 – dc20                                                   92-27775
                                                                        CIP

ISBN 978-0-521-43009-8 Hardback
ISBN 978-0-521-13133-9 Paperback

Additional resources for this publication at www.cambridge.org/9780521131339

# Contents

Contents                                                          vii

# Contributors

Alvaro A. Aldama
*Department of Civil Engineering and Operations Research*
*Water Resources Program*
*Princeton University*
*Princeton, New Jersey*

Keith W. Bedford
*Great Lakes Forecasting Program*
*Department of Civil Engineering*
*Ohio State University*
*Columbus, Ohio*

William Cabot
*Center for Turbulence Research*
*NASA–Ames Research Center*
*Moffett Field, California*

Mark A. Cane
*Lamont–Doherty Geological Observatory*
*Columbia University*
*Palisades, New York*

Keeley R. Costigan
*Department of Atmospheric Science*
*Colorado State University*
*Fort Collins, Colorado*

*Contributors*

William R. Cotton
*Department of Atmospheric Science*
*Colorado State University*
*Fort Collins, Colorado*

William P. Dannevik
*A-Division and Center for Compressible Turbulence*
*Lawrence Livermore National Laboratory*
*Livermore, California*

Gordon Erlebacher
*Institute for Computer Applications in Science and Engineering*
*NASA–Langley Research Center*
*Hampton, Virginia*

Joel H. Ferziger
*Department of Mechanical Engineering*
*Stanford University*
*Stanford, California*

Piotr J. Flatau
*Department of Atmospheric Science*
*Colorado State University*
*Fort Collins, Colorado*

Patrick C. Gallacher
*NOARL*
*Stennis Space Center*
*Mississippi 39529*

Peyman Givi
*State University of New York at Buffalo*
*Department of Mechanical and Aerospace Engineering*
*Amherst, New York*

Jackson R. Herring
*National Center for Atmospheric Research*
*Boulder, Colorado*

Greg Holloway
*Institute of Ocean Sciences*
*Sidney, British Columbia*

Kiyosi Horiuti
*Institute of Industrial Science*
*University of Tokyo*
*Minato-Ku, Tokyo*

Mohammed Y. Hussaini
*Institute for Computer Applications in Science and Engineering*
*NASA–Langley Research Center*
*Hampton, Virginia*

George E. Karniadakis
*Department of Mechanical and Aerospace Engineering*
*Princeton University*
*Princeton, New Jersey*

Robert M. Kerr
*National Center for Atmospheric Research*
*Boulder, Colorado*

Alan R. Kerstein
*Combustion Research Facility*
*Sandia National Laboratories*
*Livermore, California*

Cecil E. Leith
*Center for Compressible Turbulence*
*Lawrence Livermore National Laboratory*
*Livermore, California*

Marcel Lesieur
*Institut de Mécanique de Grenoble*
*Unité Mixte de Recherche CNRS*
*Institut National Polytechnique de Grenoble*
*Université Joseph Fourier*
*Grenoble-Cedex, France*

Cyrus K. Madnia
*State University of New York at Buffalo*
*Department of Mechanical and Aerospace Engineering*
*Amherst, New York*

Patrick A. McMurtry
*Department of Mechanical Engineering*
*University of Utah*
*Salt Lake City, Utah*

James C. McWilliams
*National Center for Atmospheric Research*
*Boulder, Colorado*

Suresh Menon
*School of Aerospace Engineering*
*Georgia Institute of Technology*
*Atlanta, Georgia*

Olivier Métais
*Institut de Mécanique de Grenoble*
*Unité Mixte de Recherche CNRS*
*Institut National Polytechnique de Grenoble*
*Université Joseph Fourier*
*Grenoble-Cedex, France*

Chin-Hoh Moeng
*National Center for Atmospheric Research*
*Boulder, Colorado*

Parviz Moin
*Department of Mechanical Engineering*
*Stanford University*
*Stanford, California*

Peter Müller
*Department of Oceanography*
*School of Ocean and Earth Science and Technology*
*University of Hawaii*
*Honolulu, Hawaii*

Xavier Normand
*Institut de Mécanique de Grenoble*
*Unité Mixte de Recherche CNRS*
*Institut National Polytechnique de Grenoble*
*Université Joseph Fourier*
*Grenoble-Cedex, France*

Steven A. Orszag
*Program in Applied and Computational Mathematics*
*Princeton University*
*Princeton, New Jersey*

Roger A. Pielke
*Department of Atmospheric Science*
*Colorado State University*
*Fort Collins, Colorado*

Ugo Piomelli
*Department of Mechanical Engineering*
*University of Maryland*
*College Park, Maryland*

Saad A. Ragab
*Department of Engineering Sciences and Mechanics*
*Virginia Polytechnic Institute and State University*
*Blacksburg, Virginia*

Ulrich Schumann
*DLR, Institute of Atmospheric Physics*
*Oberpfaffenhofen, Germany*

Shaw-Ching Sheen
*Department of Engineering Sciences and Mechanics*
*Virginia Polytechnic Institute and State University*
*Blacksburg, Virginia*

Aristeu Silveira-Neto
*Institut de Mécanique de Grenoble*
*Unité Mixte de Recherche CNRS*
*Institut National Polytechnique de Grenoble*
*Université Joseph Fourier*
*Grenoble-Cedex, France*

Joseph Smagorinsky
*Department of Geological and Geophysical Sciences*
*Princeton University*
*Princeton, New Jersey*

Ilya Staroselsky
*Program in Applied and Computational Mathematics*
*Princeton University*
*Princeton, New Jersey*

Robert L. Walko
*Department of Atmospheric Science*
*Colorado State University*
*Fort Collins, Colorado*

John C. Wyngaard
*Department of Meteorology*
*Pennsylvania State University*
*University Park, Pennsylvania*

Victor Yakhot
*Program in Applied and Computational Mathematics*
*Princeton University*
*Princeton, New Jersey*

Woon K. Yeo
*Department of Civil Engineering*
*Myong Ji University*
*Seodarmoon–Ku, Seoul, Korea*

Thomas A. Zang
*NASA–Langley Research Center*
*Hampton, Virginia*

# Preface

Large eddy simulation (LES) is one of the most powerful computational tools available today for the calculation of turbulent flows. The name of this method reflects its very essence: whereas large-scale flow structures are calculated explicitly, small-scale processes, taking place below the limits of numerical resolution, are parameterized using models of various degrees of complexity. Usually such parameterizations are done in the form of eddy viscosity, but characteristic of LES has been an extensive use of nonlinear eddy viscosities introduced in the early 1960s. The first to use a nonlinear viscosity in numerical models of global atmospheric circulation was Joseph Smagorinsky, but in Chapter 1 of this volume he traces even earlier utilizations of such viscosities in simulations of compressible flows with shocks. Truly large eddy simulations in a form not much different from those in use today were performed by James Deardorff in the early 1970s on convective and channel flow turbulence. Because, by today's standards, computer resources of 20 years ago were quite primitive, extensive experimentation with nonlinear viscosities in geophysical models was not feasible. Besides, these models seemed to have problems that were much more acute than subtleties in the choice of eddy viscosity, and soon the center of gravity in geophysical research shifted elsewhere.

An opposite tendency was developing in engineering, where efforts were concentrated on direct simulation of the three-dimensional Navier–Stokes equation in simple geometries resolving all scales up to the viscous, or Kolmogorov, scale (the method known as direct numerical simulation, or DNS). Rapid progress in this field was stimulated by growing computer power, the introduction of vector computer architecture, and the intensive development of spectral methods. Of course, DNS could be applied only to low Reynolds number flows with relatively small spatial scales, but those are often sufficient in practical engineering problems. At the same time, DNS codes were used in the LES mode, whereas resolved scales were considerably larger and molecular viscosity was replaced by nonlinear eddy viscosity of the Smagorinsky type to account for the unresolved small-scale mixing. This led to a revitalization of interest in LES in the engineering community.

In the early 1980s meteorologists revived their interest in LES, but it was used mostly in relatively small scale calculations of various features of planetary boundary layers. For some ten years the exploration of the potential of LES in engineering and in geophysics has been going on by almost parallel routes, without much interaction. This weak communication between the fields can be attributed to different objectives in respective LES applications: while engineering flows are relatively small scale and neutrally stratified, geophysical flows occupy large domains and are usually subjected to a host of external factors, such as density stratification, rotation, curvature, and differential rotation.

Recently, however, the overlap between LES applications in geophysics and engineering has been rapidly increasing due to the widening variety of problems made tractable by this method. Among problems of equal importance in both fields are flows in the vicinity of solid walls, flows with open surfaces, rotating flows, flows with chemical reactions and stochastic backscatter. Additional common ground may be flows with two-dimensionalization and relaminarization due to external factors (extra strains in Peter Bradshaw's terminology). In such flows, of particular interest to large-scale geophysics, the inherent backscatter of energy from three-dimensional to two-dimensional regions in the wave number space may become an important energy source for large-scale dynamics, due to the inverse energy cascade.

An international workshop entitled "Large Eddy Simulation: Where Do We Stand?", the first of its kind, took place on December 19–21, 1990, at St. Petersburg Beach, Florida. It was intended to bring together engineers and geophysicists actively engaged in the development and application of LES models. Its objectives were the assessment of the state of the art of LES, facilitation of the cross-disciplinary exchange of information and experience, cross-fertilization of the fields that are or may be benefiting from the application of LES, and an outline of the most promising directions for the future research. A total of 25 presentations by international experts were clustered in five sessions, each of which was devoted to a particular area of engineering or geophysical sciences. The information presented and deliberations that followed were summarized in two panel discussions. The third panel discussion was devoted to assessing the state of the art of the rapidly developing application of massively parallel computers in LES.

This book is the result of the workshop and it contains most of the information presented there. However, the chapters have been updated, and since LES is a rapidly developing field, some contributions made after the workshop have been included. The book therefore actually reflects the status of LES and of applications of massively parallel computers as of the beginning of 1993.

The book consists of four major sections, Fundamentals of LES, LES in Engineering, LES in Geophysics, and LES and Massively Parallel Computing. The first section covers major historical highlights as well as important recent contributions, such as applications of two-point closures, renormalization group theory of turbulence, dynamic eddy viscosity, and formulations of the stochastic backscatter. The second section describes applications of LES for simulations of incompressible, compressible, and reacting flows important in engineering. The third section details applications of LES for geophysical sciences, such as meteorology, physical oceanography, and environmental engineering. The sections on LES in Engineering and LES in Geophysics begin with broad overview chapters which should give the reader a global picture of those respective areas and general guidelines about the rest of these sections' presentations. Finally, updated transcripts of the panel discussions should help the reader to summarize and digest the information presented. The fourth section is based on the panel discussion on LES and massively parallel computing, but it has been substantially extended and upgraded to include important recent developments.

We hope that this book will be useful to a wide community of scientists, engineers, researchers, and graduate students working in different areas of fluid mechanics and interested in the computation of turbulent flows. We also hope that it will promote the continued flourishing of large eddy simulation.

In conclusion, we acknowledge sponsorship of the workshop by the University of South Florida, Office of Naval Research, NASA–Langley Research Center, and Army Research Office, and by the computer companies IBM, Intel, and Silicon Graphics. We also acknowledge the great help of Ms. L. Ellenburg and her staff at the Division of Conferences and Institutes, University of South Florida, Ms. M. Kastrenakes from CommuniK Advertising, Inc., Ms. K. Birchett, and Ms. E. Bedell, Ms. L. Niswander, and Ms. L. Kelbaugh from the Department of Marine Science, University of South Florida, in organizing and running the workshop, and Mr. J. Chad Edmisten from the Marine Science Graphics Department, University of South Florida, for his great help in the layout and design of the artwork. We also appreciate the assistance, patience, and understanding of the authors who have helped us to put this book together in camera-ready form.

*Steven A. Orszag*

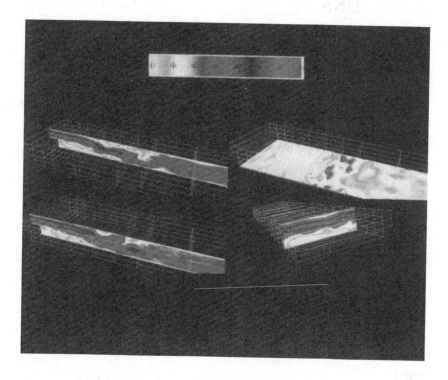

Plate 1   Color contours of instantaneous streamwise velocity along the span, on a horizontal plane, and on a vertical plane downstream of the step [$Re = 8,888$] (see Chapter 8).

Plate 2   Plot of subgrid product concentration contours for the pseudoin-compressible case at $t^* = 0.549$ (see Chapter 15).

Plate 3   Plot of subgrid unmixedness contours for the pseudoincompressible case at $t^* = 0.549$ (see Chapter 15).

A colour version of these plates is available for download from
www.cambridge.org/9780521131339

DNS                                    Beta

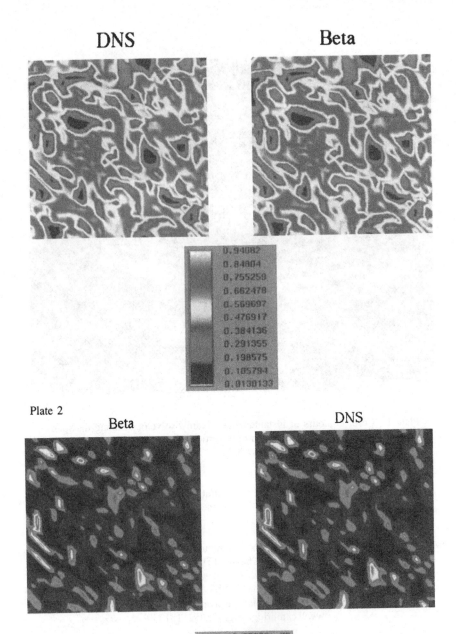

0.94082
0.84804
0.755259
0.662478
0.569697
0.476917
0.384136
0.291355
0.198575
0.105794
0.0130133

Plate 2

Beta                                    DNS

3.25963e-09
-0.00381074
-0.00762148
-0.0114322
0.015243
-0.0190537
0.0228644
0.0266752
0.0304859
0.0342967
-0.0381074

Plate 3

# PART ONE

# FUNDAMENTALS OF LARGE EDDY SIMULATION

# 1

## Some Historical Remarks on the Use of Nonlinear Viscosities

JOSEPH SMAGORINSKY

### 1.1 Introductory Remarks

This chapter reviews the early experiences of the application of computational methods to the hydrodynamics of large-scale atmospheric motions: the beginnings of numerical weather prediction (see, for example, Smagorinsky 1983).

In the early 1950s, the state of the art was a quasi-geostrophic, quasi-two-dimensional model of large-scale atmospheric dynamics. Of course, the models were time-dependent and nonlinear. We had just emerged from the era when we dealt exclusively with barotropic models that simulated atmospheric evolutions on time scales of one or, at the most, two days. In these models total kinetic energy was assumed to be conserved. Models were also being constructed for the next hierarchical step, that is, baroclinic models which allowed potential–kinetic energy conversions, thus including the processes related to storm development (baroclinic instability) and thereby extending the validity of the models by another day or two. The small, but nontrivial, large-scale vertical component of motion was implied, even quasi-geostrophically.

The crucial achievement in the mid-1950s was the successful construction by Norman Phillips (1956) of an energetically self-sustaining model which emulated the prime external radiative energy source and the viscous energy sink. This model was capable of being integrated for extended intervals and exhibited many of the atmosphere's nonlinear characteristics such as the fundamental energy or index cycle. Phillips had used a conventional linear viscosity and conductivity with the coefficient

$0.2L^{4/3}$ (in CGS units), determined in accordance with Richardson's (1926) empirical law, $L$ being identified with the grid size. What was being observed in these longer-term integrations was the development of extended vortex lines, and at the Institute for Advanced Study the phenomenon was aptly dubbed "noodling."

John von Neumann, who oversaw Charney's group at the Institute, recalled some of his and Richtmyer's earlier experience (1950) with hydrodynamic shocks. In particular, they were dealing with one-dimensional flows, and were seeking a means "to introduce (artificial) dissipative terms into the equations so as to give the shocks a thickness comparable to (but preferably somewhat larger than) the spacing of the points of the network. Then the differential equations (more accurately, the corresponding difference equations) may be used for the entire calculation, just as though there were no shocks at all." Von Neumann and Richtmyer were guided by earlier physical insights on the roles of dissipative mechanisms, that is, viscosity and heat conduction, in the behavior of shocks. The form of the dissipation was derived heuristically and was "introduced for purely mathematical reasons." The viscosity was taken to be proportional to the magnitude of the divergence, which, in a one-dimensional flow, is indistinguishable from the magnitude of the deformation.

A second reference has to do with a study by Phillips (1959) on the occurrence of nonlinear computational instability in the numerical integration of a barotropic, nondivergent atmospheric model. It was motivated by his experience, a few years earlier (Phillips 1956), with the unprecedented long integrations of a quasi-geostrophic, baroclinic, two-level "general circulation model," the first of this genre. After several weeks of simulated time, the appearance of large truncation errors caused an almost explosive increase of the total energy of the system. This instability at the smallest resolvable scales could not be suppressed by reducing the time step, as is customary in the case of linear Courant–Friedrich–Lewy (CFL) instability. However, by periodically eliminating all components with wave lengths smaller than 4 times the grid size, that is, an artificial "smoothing," he was able to suppress the nonlinear instability.

About 1960, J. Charney and N. Phillips conveyed by personal communication that they had successfully used a two-dimensional version of the von Neumann–Richtmyer nonlinear viscosity, proportional to the deformation, to control grid-scale filamentation in trial numerical integrations. Presumably these were for quasi-geostrophic flows.

In a recently published interview with Charney, conducted in August 1980 (Lindzen, Lorenz and Platzman 1990, p. 61), Charney said that after using the nonlinear viscosity in 1955 or 1956 he tried a linear viscosity and "found the results were just about as good and, therefore, gave up the idea of using the non-linear viscosity, ... other techniques are probably preferable filtering techniques. Because the one problem with these viscosities is that in the long run they can give you unrealistic transports of momentum."

The present account is based on an attempt in the early 1960s to rationalize the derivation of a nonlinear viscosity based upon the principles of modern turbulence theory, particularly those related to Heisenberg–Kolmogorov similarity in the inertial subrange of three-dimensional, isotropic turbulence. What I tried to do at the time was to particularize the classical results from the theory of turbulence to three-dimensional horizontally isotropic turbulence, in which the vertical component is quasi-hydrostatically constrained. The formulation, however, in general is applicable to the primitive equations of motion, that is, where gravity waves are admissible modes.

The notion of an inertial subrange of fluid motions implies a regime free from sources and sinks of kinetic energy and separated in wave number space from the processes responsible for molecular viscous decay. That is, an equilibrium should exist such that the total kinetic energy of the subrange is conserved, and that the flow of kinetic energy cascades from low to higher wave numbers purely by the nonlinear interactions of the inertial forces.

One purpose for seeking such a subrange in the atmosphere was that, in the numerical integration of the equations of motion, an artificial threshold was created by the grid size. For longer waves, the dynamics of the motions were dealt with explicitly, whereas for the shorter waves, the motions had to be dealt with statistically, that is, parametrically, or ignored entirely. Generally, unless the statistical dynamics of the turbulence are understood well enough, the explicit dynamics cannot adequately communicate with the implied viscous subrange. Fortunately, the results of similarity arguments (Taylor 1935; von Kármán and Howarth 1938; Kolmogorov 1941a,b; Onsager 1945, 1949; Batchelor 1946; von Weizsäcker 1948; Heisenberg 1948a,b; Bass 1949) provided some insight into the statistical properties of isotropic turbulent transfer in the inertial subrange which, as will be shown, appears to be formally applicable to atmospheric motions.

The present account includes a somewhat shortened version of an unpublished manuscript (Smagorinsky 1962). The latter was written at a time when a nonlinear viscosity was being used in an atmospheric primitive equation general circulation model (Smagorinsky 1963). However, the model was constrained to filter out external gravity wave modes. The essential differences between the derivations in this chapter and the early draft are in the justification of the quasi-hydrostatic impact on the viscosity coefficients; in the choice of, and the parametric relationships between, the numerical coefficients occurring in the turbulence formulation; in correcting the numerical coefficient from 6 to $\frac{2}{3}$ in (39) below; and in a rationale for the conditions governing the choice of the single nondimensional "constant" in the turbulence exchange coefficient.

Much has been learned since, both theoretically and experimentally. The present perspective reflects on some of the many developments of the past 28 years. Numerous applications of a nonlinear viscosity have enlarged the base of experience in meteorology, oceanography and a variety of other fluid dynamical problem areas.

One may infer the location of a relative inertial subrange for atmospheric motions, at least for those motions sufficiently removed from the lower boundary and the equator (Figure 1) (Smagorinsky 1974). It was already well-established in the early 1950s (Onsager 1949; Fjørtoft 1953) that because of the vorticity conservation by quasi-horizontal motions, the long barotropic waves (zonal or east–west wave numbers 1 to 3) have associated with them a net transfer of kinetic energy to lower wave numbers. This property of geostrophic turbulence is responsible for the $-3$ power falloff of spectral energy density for the largest atmospheric scales (Charney 1971). It is the range where enstrophy (half the squared vorticity) is cascaded to smaller scales, but where, in the net, kinetic energy decascades to longer scales, ultimately maintaining the jet stream against dissipation. Furthermore, the seasonal monsoon resulting from the planetary-scale continents provides a quasi-stationary thermal forcing with a zonal wave number of about 2. Also, the baroclinic instability process (Charney 1947; Eady 1949; Fjørtoft 1951) provides a major source of energy in zonal wave numbers 4 to 7 (see, for example, Holton 1979, pp. 5, 36). Large-scale condensation processes provide an additional source at somewhat higher wave numbers. The atmospheric mesoscale, nominally from about 5 to 500 km in horizontal dimensions, harbors many intense, but generally intermittent and sparse, phenomena in the extratropics, such as mesoscale convective complexes and fronts which may not appear in a spectrum taken at any one time. Proceed-

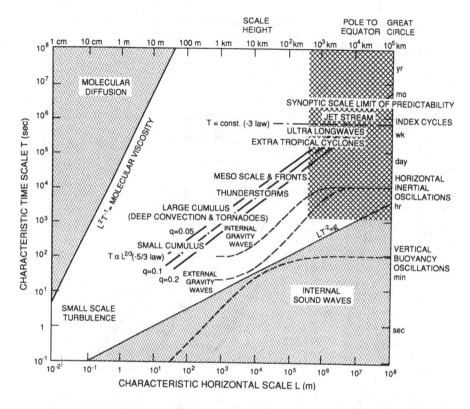

Fig. 1  A space–time domain for characteristic atmospheric phenomena. The unstippled region encompasses most of the kinetic-energy-containing phenomena, with the predominance of extratropical cyclones, ultralong waves and the jet stream. The cross-hatched area denotes scales and phenomena typically resolved by general circulation models, that is, the macroscale. The central heavy diagonal lines are $q = L^{2/3} T^{-1} =$ constant in units of $m^{2/3}$ $s^{-1}$. (After Smagorinsky 1974, based on a figure by H. Fortag as modified by K. Ooyama.)

ing somewhat further downscale, we encounter convective phenomena of various types. However, it has been pointed out by Emanuel (1984) that "only slant-wise convection is capable of directly generating kinetic energy in the mesoscale and occurs only intermittently in strongly baroclinic situations." The mesoscale range has proven to be particularly

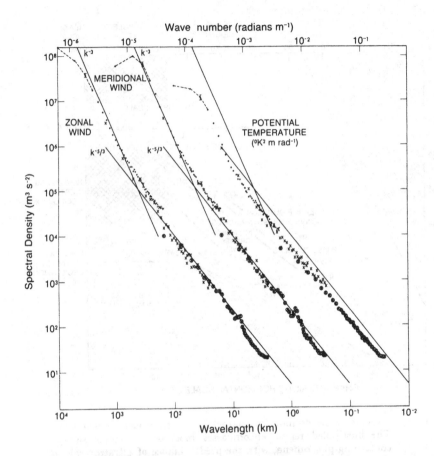

Fig. 2   Variance power spectra of wind and potential temperature near
the tropopause from GASP aircraft data. The spectra for meridional wind and temperature are shifted one and two decades to the
right, respectively; lines with slopes −3 and −$\frac{5}{3}$ are entered at the
same relative coordinates for each variable for comparison. (After
Nastrom and Gage, 1985.)

useful for setting the grid size in studying large-scale motions by numerical simulations.

Recent definitive measurements of atmospheric energy spectra in the
span of horizontal scales 2 to 10,000 km (Figure 2) (Nastrom and Gage
1985) reaffirm that while a −3 slope fits the spectrum for scales larger
than 300–400 km in the "geostrophic turbulence" range, −$\frac{5}{3}$ fits best for
the mesoscales, that is, for wavelengths between 2.6 km and 300–400
km.

## 1.2 The Mean Equations of Motion

If we average the Navier–Stokes equations over a wave number interval $k \leq k' \leq \infty$, we find

$$\frac{D\mathbf{v}}{Dt} + 2\mathbf{\Omega} \times \mathbf{v} = -\frac{1}{\rho}\nabla p^* + \frac{1}{\rho}\nabla \cdot \mathbf{\Gamma}^* - \mathbf{g}. \tag{1}$$

Here, $D()/Dt$ is the individual time rate of change on a particle, $\nabla$ is the three-dimensional gradient vector operator, $\mathbf{v}$ is the average three-dimensional relative velocity vector, $p^*$ is the average molecular pressure, $\mathbf{\Gamma}^*$ is the Reynolds stress dyadic, $\mathbf{g}$ is the gravitational acceleration vector, $\rho$ is the density (its Reynolds fluctuations having been neglected) and $\mathbf{\Omega}$ is the earth's angular velocity vector. We have assumed that the wave number, $k$, is sufficiently small that the Reynolds stresses are large compared with the molecular viscous stresses, which have therefore been omitted.

$\mathbf{\Gamma}^*$ may be written as the sum of its average value in all directions and a deviation, $\mathbf{\Gamma}$,

$$\mathbf{\Gamma}^* = \frac{|\mathbf{\Gamma}^*|}{3}\mathbf{\Upsilon} + \mathbf{\Gamma} \qquad \text{and} \qquad |\mathbf{\Gamma}| = 0, \tag{2}$$

where $\mathbf{\Upsilon}$ is the idemfactor and $|\mathbf{\Upsilon}| = 3$.

By analogy with the kinetic theory of gases, we interpret $|\mathbf{\Gamma}^*|/3$ as the pressure due to the eddy kinetic energy. Hence, the macropressure is (Hinze 1959, p. 15)

$$p = p^* - |\mathbf{\Gamma}^*|/3, \tag{3}$$

and (1) becomes

$$\frac{D\mathbf{v}}{Dt} + 2\mathbf{\Omega} \times \mathbf{v} = -\frac{1}{\rho}\nabla p + \frac{1}{\rho}\nabla \cdot \mathbf{\Gamma} - \mathbf{g}, \tag{4}$$

in which $(1/\rho)\nabla \cdot \mathbf{\Gamma} \equiv \mathbf{F}$ is the frictional force vector per unit mass.

## 1.3 Application of Elasticity Theory

The general theory of viscous fluids conventionally suggests that the strains of individual fluid elements give rise to stresses analogous to those in elastic media. However, in fluids, the pressures are proportional to the rate of strain, since a fluid yields to a shearing stress. One can, therefore, adapt the formalisms developed in the theory of elasticity provided that the fluid velocity, $\mathbf{v}$, is used instead of the elastic displacements.

Therefore, we assume, for small deformations, that $\mathbf{\Gamma}/\rho$ is proportional to the rate of strain dyadic, $\mathbf{S} = \nabla\mathbf{v}$, through a tensor, $\mathbf{C}$, such that

$$\mathbf{\Gamma} = \rho|\mathbf{C} \cdot \mathbf{S}|, \tag{5}$$

in analogy to the elastic function in elasticity. In general, $\mathbf{C}$ is a tetradic, which has $3^4$ ($= 81$) components.

The kinetic energy equation may be formed by the scalar multiplication of $\mathbf{v}$ and (4) and then integrating over the entire volume. Assuming closed boundaries, one finds

$$\frac{\partial}{\partial t} \iiint \frac{\rho v^2}{2} \, dV = \iiint p \nabla \cdot \mathbf{v} \, dV - \iiint \rho \mathbf{v} \cdot \mathbf{g} \, dV ,$$

$$- \iint_{z=0} (\mathbf{v} \cdot \boldsymbol{\Gamma}) \cdot d\mathbf{A} - \iiint \rho \epsilon \, dV, \qquad (6)$$

where

$$\epsilon \equiv |\mathbf{S} \cdot \boldsymbol{\Gamma}| / \rho \qquad (7)$$

is the local energy dissipation or decay, $\iiint \rho \epsilon \, dV$ is the energy transformation function between the explicit and eddy flow, $dV$ is a volume element, $d\mathbf{A}$ is a horizontal area element, $\iiint (p \nabla \cdot \mathbf{v} - \rho \mathbf{v} \cdot \mathbf{g}) \, dV$ is the potential–kinetic energy conversion, and $\iint_{z=0} (\mathbf{v} \cdot \boldsymbol{\Gamma}) \cdot d\mathbf{A}$ is the energy dissipation due to the eddy stresses tangential to the lower boundary at $z = 0$, $z$ being the vertical coordinate, opposite to the direction of $\mathbf{g}$.

Hence, if we write the matrices of the tensor components symbolically as

$$\mathbf{S} \to S_{rs}, \quad \boldsymbol{\Gamma} \to \tau_{mn}, \quad \mathbf{C} \to C_{mnrs}, \qquad (8)$$

then the equations for the stress components are

$$\tau_{mn} / \rho = \sum_{rs} C_{mnrs} S_{rs} \qquad (9)$$

and

$$\epsilon = \sum_{mnrs} C_{mnrs} S_{mn} S_{rs}. \qquad (10)$$

It follows from energy considerations (Sokolnikoff 1946, p. 63) that there exists a reflective symmetry along any arbitrary axis, so that

$$S_{rs} = S_{sr}, \quad \tau_{mn} = \tau_{nm}, \quad C_{mnrs} = C_{rsmn} = C_{nmrs} = C_{mnsr}, \quad (11)$$

in which case there are only 6 independent $S_{rs}$ and $\tau_{mn}$, and 21 independent $C_{mnrs}$.

## 1.4 Axial Symmetry

We will now confine our attention to a scale of motion sufficiently small that these motions can be regarded as horizontally isotropic, that is, symmetric with respect to the vertical axis. This definitely excludes the barotropic long-wave regime and probably the energy-producing baro-clinic instability regime.

The axial symmetry of $\tau_{mn}$ and $S_{rs}$ contracts their matrices to 6 × 6, resulting in 9 nonzero $C_{mnrs}$, of which only 3 are independent when (2) is taken into account:

$$\sum_m \tau_{mm} = 0, \qquad (12)$$

$$
\begin{aligned}
\alpha &\equiv -\tfrac{1}{2}C_{1133} = -\tfrac{1}{2}C_{2233} = \tfrac{1}{4}C_{3333} , \\
\beta &\equiv \tfrac{1}{2}(C_{1111} - C_{1122}) = \tfrac{1}{2}(C_{2222} - C_{1122}) = C_{1212} , \\
\gamma &\equiv C_{1313} = C_{2323} .
\end{aligned}
\qquad (13)
$$

The stress components may now be written

$$
\begin{aligned}
\tau_{11}/\rho &= \beta(S_{11} - S_{22}) + \alpha(\nabla \cdot \mathbf{v} - 3S_{33}) , \\
\tau_{22}/\rho &= -\beta(S_{11} - S_{22}) + \alpha(\nabla \cdot \mathbf{v} - 3S_{33}) , \\
\tau_{33}/\rho &= -2\alpha(\nabla \cdot \mathbf{v} - 3S_{33}) , \\
\tau_{23}/\rho &= \tau_{32}/\rho = 2\gamma S_{23} , \\
\tau_{13}/\rho &= \tau_{31}/\rho = 2\gamma S_{13} , \\
\tau_{12}/\rho &= \tau_{21}/\rho = 2\beta S_{12} ,
\end{aligned}
\qquad (14)
$$

where

$$\nabla \cdot \mathbf{v} = S_{11} + S_{22} + S_{33} = -\frac{d}{dt}\ln \rho \qquad (15)$$

by the equation of continuity.

Also, in the case of axial symmetry, one may calculate $\epsilon$ from (10) and (14):

$$\epsilon = \beta D^2 + \alpha M^2 + \gamma P^2, \qquad (16)$$

where

$$
\begin{aligned}
D^2 &\equiv (S_{11} - S_{22})^2 + (2S_{12})^2 , \\
M^2 &\equiv (\nabla \cdot \mathbf{v} - 3S_{33})^2 , \\
P^2 &\equiv (2S_{13})^2 + (2S_{23})^2 .
\end{aligned}
\qquad (17)
$$

It is easy to see that for $\alpha, \beta, \gamma \geq 0$, $\epsilon$ is positive-definite, thus ensuring positive energy dissipation.

## 1.5 Spherical Curvilinear Coordinates

It will be useful to record the forms of the viscous terms in spherical curvilinear coordinates. We denote the longitude, $\lambda$ (in the direction of the unit vector $\mathbf{i_1}$), the latitude, $\phi$ ($\mathbf{i_2}$), and the radial distance from the center of the earth, $r = a + z$ ($\mathbf{i_3}$), where $a$ is the radius of mean sea level and $z$ is the elevation above it. The spatial curvilinear differentials are

$$\delta x = r \cos \phi \delta \lambda, \quad \delta y = r \delta \phi, \quad \delta z = \delta r, \tag{18}$$

and the curvilinear relative velocity components are

$$u = r \cos \phi \frac{d\lambda}{dt}, \quad v = r \frac{d\phi}{dt}, \quad w = \frac{dr}{dt}. \tag{19}$$

The strains are (Morse and Feshbach 1953, p. 117)

$$
\begin{aligned}
S_{11} &= \frac{\partial u}{\partial x} + \frac{w}{r} - \frac{v \tan \phi}{r}, \\[4pt]
S_{22} &= \frac{\partial v}{\partial y} + \frac{w}{r}, \\[4pt]
S_{33} &= \frac{\partial w}{\partial z}, \\[4pt]
S_{13} &= \frac{1}{2} \left[ \frac{\partial u}{\partial z} + \frac{\partial w}{\partial x} \right], \\[4pt]
S_{23} &= \frac{1}{2} \left[ r \frac{\partial (v/r)}{\partial z} + \frac{\partial w}{\partial y} \right], \\[4pt]
S_{12} &= \frac{1}{2} \left[ \cos \phi \frac{\partial (u/\cos \phi)}{\partial y} + \frac{\partial v}{\partial x} \right].
\end{aligned}
\tag{20}
$$

The frictional force vector,

$$\rho \mathbf{F} = \nabla \cdot \mathbf{\Gamma}, \tag{21}$$

in curvilinear component form is

$$
\begin{aligned}
\rho F_x &= \frac{\partial \tau_{11}}{\partial x} + \frac{1}{\cos^2 \phi} \frac{\partial (\cos^2 \phi \, \tau_{12})}{\partial y} + r^{-3} \frac{\partial (r^3 \tau_{13})}{\partial z}, \\[4pt]
\rho F_y &= \frac{\partial \tau_{12}}{\partial x} + \frac{1}{\cos^2 \phi} \frac{\partial (\cos^2 \phi \, \tau_{22})}{\partial y} + r^{-3} \frac{\partial (r^3 \tau_{23})}{\partial z} - \frac{\tan \phi}{r} \tau_{33}, \\[4pt]
\rho F_z &= \frac{\partial \tau_{13}}{\partial x} + \frac{1}{\cos \phi} \frac{\partial (\cos \phi \, \tau_{23})}{\partial y} + r^{-3} \frac{\partial (r^3 \tau_{33})}{\partial z}.
\end{aligned}
\tag{22}
$$

## 1.6 The Quasi-Static Constraint

In large-scale meteorological applications, it is customarily useful to assume that the motions are subject to the quasi-hydrostatic constraint. This is clearly justified by a scale analysis (see, for example, Holton 1979, p. 39). The approximation is valid well into the atmospheric mesoscale.

In order to assess further the relative roles of the three viscous coefficients, $\alpha, \beta$ and $\gamma$, which were distilled from imposing symmetry and horizontal isotropy (vertical axial symmetry) constraints, it will be instructive to examine the energy transformations. We have already derived the energy equation for the three-dimensional domain (6) and the energy dissipation $\epsilon$, Eqs. (16) and (17), consistent with the symmetries we have assumed.

In the aforementioned unpublished manuscript (Smagorinsky 1962), the approach was to employ a "hand-waving" scale analysis to argue the consequences of the hydrostatic approximation on the turbulence formulation. At present, we offer a somewhat more elegant energetic argument based upon the preservation of positive-definiteness for the dissipation.

It is generally established that, for a quasi-static constraint, it is necessary to make certain "fixes" on the terms of the equations of motion in order to restore consistency in the energy equation. In particular, a fundamental property to preserve is that the individual change of kinetic energy should depend *only* on the work done by the pressure gradient and external forces. Upon making the quasi-static approximation,

$$0 = -\frac{1}{\rho}\frac{\partial p}{\partial r} - g, \tag{23}$$

the energy equation appropriate to the hydrodynamic system assumes, perforce, the horizontally two-dimensional form that no longer preserves the properties of work done by the pressure gradient and external forces. A correct kinetic energy equation consistent with the quasi-static approximation can be derived if certain changes are made in the inertial terms of the two horizontal equations of motion (see, for example, Smagorinsky 1958):

• the terms $[(2u/r\cos\phi) + 2\Omega]$ and $2vw/r$ are dropped from the inertial terms of the first and second equations of motion, respectively, and
• where $r$ appears undifferentiated in the horizontal equation of motion, it is replaced by $a$, the mean radius of the earth.

A problem also arises with the viscous terms. To examine this problem, let us form the energy dissipation term in the energy equations corresponding to the horizontal and the vertical separately:

$$\epsilon_H = \beta D^2 + \alpha M\Pi + \gamma N, \tag{24}$$

$$\epsilon_V = -\alpha M(\Pi - M) + \gamma(P^2 - N), \tag{25}$$

where

$$\Pi \equiv S_{11} + S_{22} \tag{26}$$

is the horizontal divergence, and

$$N \equiv 2S_{13}\frac{\partial u}{\partial z} + 2S_{23}\frac{\partial v}{\partial z} \ . \qquad (27)$$

Note that $\epsilon_H + \epsilon_V = \epsilon$, when (24) and (25) are compared with (16). For nonstatic motions, the viscous transformations between the horizontal and vertical coordinates are $\alpha M\Pi + \gamma N$. We see from (16) that $\epsilon$ is positive-definite, so that according to (6) it can be associated with local kinetic energy dissipation. On the other hand, for quasi-static motions, (24) is the operative energy dissipation function, but there is no assurance that either $\alpha M\Pi$ or $\gamma N$ is dissipative. To ensure consistency between the energy dissipation function and the quasi-static constraint, we first of all require that in (13)

$$\alpha \equiv -\tfrac{1}{2}C_{1133} = -\tfrac{1}{2}C_{2233} = \tfrac{1}{4}C_{3333} = 0. \qquad (28)$$

The other troublesome term, $\gamma N$, is easier to deal with. Using (20) we find that the stresses in (27) are

$$2S_{13} = \frac{\partial u}{\partial z} + \frac{\partial w}{\partial x} \ ,$$
$$2S_{23} = \frac{\partial v}{\partial z} + \frac{\partial w}{\partial y} \ , \qquad (29)$$

where $r$, when a factor, has been replaced by $a$. A scale analysis for large-scale atmospheric motions (Holton 1979, p. 36) indicates that the first terms are several orders of magnitude larger than the second terms involving $w$, so that

$$2S_{13} \approx \frac{\partial u}{\partial z} \ ,$$
$$2S_{23} \approx \frac{\partial v}{\partial z} \ . \qquad (30)$$

From (27) and (17) we then find that

$$N \approx (2S_{13})^2 + (2S_{23})^2$$
$$\approx \left(\frac{\partial u}{\partial z}\right)^2 + \left(\frac{\partial v}{\partial z}\right)^2 \qquad (31)$$
$$\approx P^2,$$

which is positive-definite. Using (28) and (31), it is easy to see that (24) is positive-definite,

$$\epsilon \approx \epsilon_H \approx \beta D^2 + \gamma P^2. \qquad (32)$$

We have therefore concluded that, for quasi-static motions, the transformations of kinetic energy of the horizontal motions associated with the $\alpha$-term in (24) must be ignored and the significant stresses are due only to the "elastic" functions $\beta$ and $\gamma$, provided that the strain components, $2S_{13}$ and $2S_{23}$, are calculated according to (30). The stresses in (14) then become

$$\tau_{11}/\rho = \beta(S_{11} - S_{22}) \equiv \beta D_T ,$$
$$\tau_{22}/\rho = -\beta(S_{11} - S_{22}) \equiv -\beta D_T ,$$
$$\tau_{33}/\rho = 0 ,$$
$$\tau_{23}/\rho = \tau_{32}/\rho = 2\gamma S_{23} , \tag{33}$$
$$\tau_{13}/\rho = \tau_{31}/\rho = 2\gamma S_{13} ,$$
$$\tau_{12}/\rho = \tau_{21}/\rho = 2\beta S_{12} \equiv \beta D_S ,$$

where

$$D_T \equiv S_{11} - S_{22} \quad \text{and} \quad D_S \equiv 2S_{12} \tag{34}$$

are the horizontal tension and shearing strains, respectively, and $2S_{13}$ and $2S_{23}$ are the two horizontal components of the vertical shearing strain.

We may now write the appropriate viscous force components from (22):

$$\rho_H F_1 \equiv \rho F_x = \frac{\partial(\rho\beta D_T)}{\partial x} + \frac{1}{\cos^2 \phi}\frac{\partial(\rho\beta D_S \cos^2 \phi)}{\partial y} + \frac{\partial(2\rho\gamma S_{13})}{\partial z} ,$$

$$\rho_H F_2 \equiv \rho F_y = \frac{\partial(\rho\beta D_S)}{\partial x} - \frac{1}{\cos^2 \phi}\frac{\partial(\rho\beta D_T \cos^2 \phi)}{\partial y} + \frac{\partial(2\rho\gamma S_{23})}{\partial z} , \tag{35}$$

$$\rho_V F \equiv \rho F_z = \frac{\partial(2\rho\gamma S_{13})}{\partial x} + \frac{1}{\cos \phi}\frac{\partial(2\rho\gamma S_{23} \cos \phi)}{\partial y} = 0 .$$

## 1.7 Similarity Hypothesis for Three-Dimensionally Isotropic Turbulence

Heisenberg (1948b) extended Kolmogorov's similarity theory of three-dimensionally isotropic turbulence by assuming that the action of the small eddies on the large eddies is analytically equivalent to a turbulent friction such that the energy decay, $\epsilon$, is related to the magnitude of the wave number, $k$, by (Batchelor 1953, pp. 125–132)

$$\epsilon = [f + K(k)]2H(k), \tag{36}$$

where

$$H(k) = \int_0^k (k'')^2 E(k'') \, dk'', \tag{37}$$

$f$ is the molecular kinematic viscosity and $K(k)$ has been known as the eddy viscosity. Here, $E(k)$ is the energy density or spectrum function of turbulence and is related to the energy decay by $\epsilon = d(kE)/dt$. Furthermore, $2H(k)$ is, physically, the mean square rate of strain. Assuming similarity in the contribution to $K(k)$ over all wave numbers from $k$ to $\infty$, dimensional arguments, along with the requirement that $K(k)$ depends on $k$ and $E(k)$ only, give

$$K(k) = A \int_k^\infty (k')^{-3/2} [E(k')]^{1/2} \, dk'. \tag{38}$$

Here, the nondimensional constant, $A$, which may be referred to as the Heisenberg constant, is to be evaluated experimentally. The dummy variables $k'$ and $k''$ have been used according to the convention $0 \le k'' \le k \le k' \le \infty$.

Bass (1949) pointed out (see also Proudman 1951 and Batchelor 1953) that (36) and (38) together constitute an integral equation for $E(k)$ with a boundary condition that $K(k) \to 0$ as $k \to \infty$, the solution of which is

$$E(k) = \tfrac{2}{3} k_0^{-3} \epsilon^{2/3} l^{5/3} (1 + V^4)^{-4/3}. \tag{39}$$

It is easy to find then that

$$2H(k) = \epsilon^{2/3} l^{-4/3} (1 + V^4)^{-1/3}, \tag{40}$$

$$f + K(k) = \epsilon^{1/3} l^{4/3} (1 + V^4)^{1/3}, \tag{41}$$

where the following abbreviations have been introduced:

$$l \equiv k_0/k, \tag{42}$$

$$V \equiv (f^3/\epsilon)^{1/4} l^{-1}, \tag{43}$$

$$k_0^4 \equiv \tfrac{3}{8} A^2. \tag{44}$$

The length $(f^3/\epsilon)^{1/4}$ is a measure of the scale of motions where molecular viscous energy decay is taking place and is known as the Kolmogorov scale. If $k_0$ is of the order of 0.1–1.0, then in the inertial subrange $V^4 \ll 1$ and $E(k)$ and $H(k)$ are independent of $f$:

$$E(k) \approx \tfrac{2}{3} k_0^{-3} \epsilon^{2/3} l^{5/3} = C_K \epsilon^{2/3} k^{-5/3} , \tag{45}$$

$$2H(k) \approx \epsilon^{2/3} l^{-4/3} ,$$

$$K(k) \approx \epsilon^{1/3} l^{4/3} \propto \epsilon^{1/3} k^{-4/3}. \tag{46}$$

The spectrum (45) and the wave number dependence of eddy viscosity (46) were established by Kolmogorov (1941a) based on dimensional considerations. The coefficient,

$$C_K \equiv \frac{2}{3} k_0^{-4/3} = \left( \frac{8}{9A} \right)^{2/3} , \tag{47}$$

has been called the Kolmogorov constant (usually elsewhere denoted by $\alpha$). From (46) we see that $l = (K^3/\epsilon)^{1/4}$ is analogous to the Kolmogorov dissipation scale, $(f^3/\epsilon)^{1/4}$. Since, in the inertial subrange, $f \ll K(k)$, we find from (43) that $V^4 = (f/K)^3 \ll 1$, thus justifying the previously made assumption.

For equilibrium, that is, $E(k) \propto k^{-5/3}$, in the inertial subrange, (45) requires that $\epsilon$ be independent of $k$. Therefore, it not only represents the energy sink for large $k$ in the viscous subrange, but also must be the energy source in the low wave number range. As we shall see later, a weaker condition on $\epsilon$ will be sufficient to maintain an inertial subrange.

An early determination of $A$ by Heisenberg (1948a) was 0.85. Proudman (1951) later obtained $0.45 \pm 0.05$.

From (42), (45) and (46), we find that in the inertial subrange

$$\epsilon = K(k)2H(k) = (k_0/k)^2[2H(k)]^{3/2}, \tag{48}$$

$$K(k) = (k_0/k)^2[2H(k)]^{1/2}. \tag{49}$$

In the viscous subrange, $V^4 \gg 1$, so that

$$E(k) \approx (A\epsilon/2f^2)^2 k^{-7} ,$$
$$2H(k) \approx \epsilon/f , \tag{50}$$
$$K(k) \approx 0 .$$

## 1.8 Horizontally Isotropic Three-Dimensional Turbulence

The results of the preceding section are not immediately useful for our purpose since they were derived for three-dimensionally isotropic turbulence, while we are concerned with the range of atmospheric turbulence that is three-dimensional, on one hand, but is two-dimensionally isotropic and quasi-static, on the other. It is therefore necessary to consider the kinematic consequences of the dimensionality. This has been investigated by Ogura (1952), Lee (1951) and Reid (1959), who have shown that, although the functional forms of the energy spectrum for two- and three-dimensionally isotropic turbulence are different, they may be used interchangeably with small error.

In retrospect, it appears that in the early 1960s there was little support for the idea of applying three-dimensionally isotropic theory to the quasi-two-dimensional large-scale atmospheric circulation. Even as recently as 1985, Deardorff (1985, p. 340) wrote that the "sub-grid-scale formulation has found more use in three-dimensional large-eddy turbulence modeling than in the originally intended global-circulation applications."

However, Kraichnan (1971) concluded that, in the energy cascade range, the functional form for $E(k)$ in (45) is equally valid for two-dimensional as well as for three-dimensional turbulence. Only the value of the numerical coefficient must be different. We will therefore assume that the conclusions drawn from Heisenberg's theory for three-dimensionally isotropic turbulence are applicable to three-dimensional turbulence which is horizontally isotropic. Although we assume isotropy in the horizontal, we do not impose the constraint of two-dimensionality on the flow in general.

Since the Heisenberg hypothesis and the experimental results are effectively valid for homogeneous turbulence, we will assume that (32) is applicable locally, per unit mass, as well as in the large. We assume, furthermore, that the energy transformations due to the $\beta$ and $\gamma$ terms may be partitioned so that the similarity arguments may be applied to each partition separately; that is, one can write

$$\epsilon \approx \epsilon_H = {}_H\epsilon_H + {}_V\epsilon_H, \tag{51}$$

where

$$_H\epsilon_H \approx \beta D^2 \tag{52}$$

and

$$_V\epsilon_H \approx \gamma P^2 \tag{53}$$

represent the quasi-static energy transformations due to the stresses in the horizontal and the vertical, respectively. Comparing (52) and (53) with (48), and remembering that $2H(k)$ is the mean square rate of strain identifiable with $D^2$ and $P^2$, while $K(k)$ is the exchange coefficient identifiable with $\beta$ and $\gamma$, one finds in the inertial subrange, using (49), that

$$\beta = (k_0/_Hk)^2|D|, \tag{54}$$

$$\gamma = (k_0/_Vk)^2|P|, \tag{55}$$

and, with (52) and (53), that

$$_H\epsilon_H = (k_0/_Hk)^2|D|^3, \tag{56}$$

$$_V\epsilon_H = (k_0/_Vk)^2|P|^3. \tag{57}$$

We have assumed $k_0$ to be spherically isotropic.

We note an implied violation of the stress–strain relations of elasticity upon which (5) is based, namely, that the strains are small so that the stress coefficients are independent of the strains. Nevertheless, we will continue on.

In what follows, we will drop the posterior subscript, $H$. It will be understood that the anterior subscript, $H$, refers to horizontal variations of the horizontal velocity components and that the anterior subscript, $V$, refers to vertical variations of the horizontal velocity components.

## 1.9 Interpretation of Finite-Difference Grids

The 1962 draft manuscript offered a rationale for determining the numerical value of the Kármán-like coefficient in (49). It is included here intact for historical perspective. It has proven to be incorrect, and is followed by what is a more defensible approach. However, the comments on the fuzziness of the spectral cutoff with the method of finite differences used, I think, are still valid.

In the similarity formulation, $k$ is the reference wave number in the inertial subrange across which the energy cascade takes place. For a discrete spectral representation of (34) and (35), $_H k 2^{-1/2} = k_x = k_y$ and $_V k = k_z$ are taken to correspond to the one-dimensional wave number one greater than the maximum resolvable by the explicit dynamics; that is, $k_x$ and $k_z$ are the smallest wave numbers within the Reynolds stress regime. In finite differences the cutoff depends on the method of differencing and the grid size, and is not sharp. One would expect that a fully consistent formulation of the eddy transfers should depend upon the difference form of the nonlinear terms, since this will determine the spectral properties of the cascade process in the neighborhood of the smallest resolvable scales. However, for our present purposes it seems reasonable to assume that the maximum resolvable one-dimensional wave number is the local earth grid size, $\Delta s$ in the horizontal and $\Delta z$ in the vertical. Hence

$$_H l = k_0/_H k \approx 0.28\Delta s, \tag{58}$$

$$_V l = k_0/_V k \approx 0.40\Delta z. \tag{59}$$

The numerical coefficient in (59) was taken as the rounded value of the Kármán constant and in (58) was $2^{-1/2}$ times that value. In the case of the lateral viscosity (58), this amounts to a value of 1.29 for $A$.

If (59) is applied to the boundary layer, then it reduces to the Prandtl theory with the mixing length, $k_0/_V k$, while $\Delta z$ is the distance to the wall.

The above rationalization for choosing the numerical value of the exchange coefficient is what was followed in the early applications of the nonlinear viscosity (Smagorinsky 1963; Smagorinsky, Manabe and Holloway, 1965). The discussion which now follows reflects a more correct line of reasoning and goes back to some of the earliest thinking of Lilly (1967) and Deardorff (1971).

In (54) and (55), $k_0/k = l$ is proportional to the integral scale of turbulence corresponding to a truncated spectrum, as is the case of a finite-difference grid. For a given grid size, $\Delta$, Deardorff (1985, p. 337) defines an integral scale,

$$L = 2\Delta, \qquad (60)$$

such that motions on scales greater than two times the grid interval are resolved explicitly. The corresponding orbital wave number (Heisenberg 1948b) is

$$k = 2\pi/L = \pi/\Delta, \qquad (61)$$

which coincides with Lilly's (1967, p. 204) conclusion as to "the largest wave number unambiguously representable on a finite difference mesh."

The horizontal and vertical wave numbers are

$$_Hk = 2\pi/_HL = \pi/\Delta s \qquad (62)$$

and

$$_Vk = 2\pi/_VL = \pi/\Delta z. \qquad (63)$$

Using (42), (54) and (62), and identifying $\beta$ with $_HK$, one can derive in the horizontal

$$_Hl = (k_0/\pi)\Delta s, \qquad (64)$$

$$_HK = (k_0/\pi)^2(\Delta s)^2|D| = 4c_s(\Delta s)^2|D|, \qquad (65)$$

where

$$c_s \equiv (k_0/2\pi)^2. \qquad (66)$$

This corresponds to the definition of $c_s$ used by Yakhot and Orszag (1986). Somewhat different definitions of $c_s$ are used by other authors. Similarly, in the vertical, one finds

$$_Vl = (k_0/\pi)\Delta z, \qquad (67)$$

$$_VK = (k_0/\pi)^2(\Delta z)^2|P| = 4c_s(\Delta z)^2|P|. \qquad (68)$$

Note that in boundary layer turbulence, if $l$ is taken to be analogous to the Prandtl mixing length and $k^{-1} = L/2\pi = \Delta/\pi$ is the distance from the wall, then in the present instance, $k_0$ plays the role of the Kármán constant in boundary layer turbulence. Thus turbulence at the grid size appears to be analogous to boundary layer turbulence, while the distance from the wall and the grid size each define an integral scale of turbulence.

From this we see that the error made in (58) and (59) was that the coefficients should have been $k_0/\pi$ instead of $0.40/2^{1/2}$ in (58) and $k_0/\pi$ instead of 0.40 in (59); that is, the horizontal exchange coefficient should have been about 6 times smaller and the vertical about 11 times smaller.

Strictly speaking, the smallest resolvable scale will vary horizontally depending on the local value of the map-scale factor of the conformal map projection being used.

The derivation closes to within one numerical parameter, $A$, or alternatively, by (44), to

$$k_0 = (3/8)^{1/4} A^{1/2} \tag{69}$$

or to

$$l/L = (3/2^7)^{1/4} A^{1/2}/\pi, \tag{70}$$

or, by (66), to $c_s$ or, by (47), to $C_K$, one of which needs to be evaluated empirically.

Although $_HL$, $_vL$, $_Hl$ and $_vl$ are each functions of the specified grid size, the ratios

$$(_Hl/_HL) = (_vl/_vL) = (l/L) \tag{71}$$

are not and, together with $k_0$, are isotropic.

Early on, Lilly (1967) pointed out that the Kolmogorov spectrum requires that the numerical coefficient in (65), $4c_s$, be deducible from the Kolmogorov constant, $C_K$. And as we noted earlier, Lilly also defines the largest wave number representable in the finite-difference mesh as $\pi/\Delta$, which corresponds to Deardorff's definition of the integral scale (60). From (47) and (66), one finds

$$c_s = (3C_K/2)^{-3/2}/(2\pi)^2 = 0.0138 C_K^{-3/2}. \tag{72}$$

Lilly's (1967) $k$ (not to be confused with our $k$, the wave number) is the same as Deardorff's $c$, such that

$$c = 2^{5/4} c_s^{1/2}. \tag{73}$$

Lilly gets, in his notation, $k \approx (3\alpha/2)^{-3/4}/\pi \approx 0.23\alpha^{-3/4}$, which translates to $c_s = (0.23)^2 2^{-5/2} C_K^{-3/2} = 0.00935 C_K^{-3/2}$ in our present notation.

## 1.10 Comparison of Some Results

Much of the research activity on the subject of large eddy simulation in the past 28 years has been a search for the "right" value of the elusive nondimensional coefficient.

We will now discuss the results of several authors (this review does not aim to be exhaustive) in experimentally or theoretically determining the arbitrary coefficient, either the Kolmogorov constant or one related to it. As we have already noted, there is a variety of definitions in the literature of the numerical coefficient in the nonlinear viscosity. Some of the interrelationships are summarized in (74) from (47), (66), (69), (70), (72) and (73):

$$
C_K = \frac{2}{3}k_0^{-4/3} = \left(\frac{8}{9A}\right)^{2/3} = \frac{2^{-1/3}}{3}\left(\frac{\pi l}{L}\right)^{-4/3},
$$

$$
c_s = \left(\frac{k_0}{2\pi}\right)^2 = \left[\frac{(3/8)^{1/2}}{(2\pi)^2}\right]A = \left(\frac{l}{L}\right)^2 = 2^{-5/2}c^2 .
$$

$$(74)$$

The reported and derived values are summarized in Table 1.

We, first of all, note the obvious. As a result of assuming isotropy for the coefficient $k_0$, not only is $l/L$ also isotropic, but so are $c$, $c_s$, $C_K$ and $A$.

We have already referred to Proudman's (1951) experimental determination of $A \approx 0.40$–$0.50$. This yields by (69) and (70): $k_0 = 0.495$–$0.553$ and $l/L = 0.0788$–$0.0881$.

Von Neumann and Richtmyer (1950) reported that the equivalent of $c_s = 0.25$ yielded good results in practice for the representation of one-dimensional shocks. On the other hand, J. Charney and N. Phillips (personal communication) found that the coefficient $c_s = 0.016$ gave reasonable results in quasi-geostrophic integrations.

It has also already been remarked that Smagorinsky (1963) used the present formulation of a nonlinear lateral viscosity with $A = 1.29$ ($c_s = 0.020$) in a numerical atmospheric general circulation experiment using the primitive equations of motion employing a two-level model that admits internal gravity waves. The resulting subgrid-scale lateral transfer of momentum was found to be too intense compared with that due to the explicitly resolved motions.

In a subsequent paper, Smagorinsky, Manabe and Holloway (1965) presented results from long time integrations with a nine-level, hemispheric, primitive equation, general circulation model (admitting both internal and external gravity waves). Several different values of the lat-

Table 1. *Parameter values from several investigations*

- <u>underlined values</u> given by author(s)
- all others are derived

| Source | $A$ | $c$ | $C_K$ | $c_s$ | $k_0$ | $l/L$ |
|---|---|---|---|---|---|---|
| *Smagorinsky (1962, 1963), Smagorinsky et al. (1965)* | 1.29 | 0.336 | 0.780 | <u>0.020</u> | 0.889 | 0.141 |
| *Lilly (1963) 3D* | 1.24 | 0.330 | <u>0.800</u> | 0.0193 | 0.872 | 0.139 |
| *Charney & Phillips (about 1960)* | 1.03 | 0.301 | 0.906 | <u>0.016</u> | 0.795 | 0.126 |
| *Heisenberg (1948a) 3D* | <u>0.85</u> | 0.273 | 1.030 | 0.0132 | 0.721 | 0.115 |
| *Kraichnan (1971) 3D* | 0.537 | 0.217 | <u>1.400</u> | 0.00832 | 0.573 | 0.0912 |
| *Deardorff (1971) 3D re: Lilly convection* | 0.503 | <u>0.210</u> | 1.46 | 0.00780 | 0.555 | 0.0883 |
| *Proudman (1951) 3D upper bound A* | <u>0.500</u> | 0.209 | 1.47 | 0.00776 | 0.553 | 0.0881 |
| *Smagorinsky (present)* | 0.478 | 0.205 | 1.51 | 0.00741 | <u>0.541</u> | 0.0861 |
| *Yakhot & Orszag (1986) 3D* | 0.432 | 0.195 | <u>1.617</u> | 0.00671 | 0.515 | 0.0819 |
| *Yakhot & Orszag (1986) RNG* | | | <u>1.617</u> | <u>0.0062</u> | | |
| *Proudman (1951) 3D lower bound A* | <u>0.400</u> | 0.187 | 1.70 | 0.00620 | 0.495 | 0.0788 |
| *Rosati & Miyakoda (1988)* | 0.226 | 0.141 | 2.49 | <u>0.00350</u> | 0.372 | 0.0592 |
| *Deardorff (1971) mean shear* | 0.193 | <u>0.130</u> | 2.77 | 0.00299 | 0.343 | 0.0547 |
| *Deardorff (1971)* | 0.114 | <u>0.100</u> | 3.93 | 0.00177 | 0.264 | 0.0420 |
| *Kraichnan (1971) 2D* | 0.0514 | 0.0671 | <u>6.69</u> | 0.000797 | 0.177 | 0.0282 |
| *Lilly (1963) 2D* | 0.0480 | 0.0649 | <u>7.00</u> | 0.000744 | 0.171 | 0.0273 |

eral exchange coefficient were tried, but $c_s = 0.020$ still seemed to give
the best results.

Deardorff (1971) considered the performance of different values of $c$ for
different kinds of flows, but seemed most satisfied with 0.10. He noted,
however, that Lilly (1962) had success with 0.21 in thermal convection
integrations. Deardorff expressed the view that the presence of large
shear would require a smaller value, 0.13. In a later paper, Deardorff
(1985) revived the question of the influence of shear and held that "the
$K$ proportionality constant $2^{-1/2}c^2$ needs to be reduced by a factor of
at least two in flows dominated by mean shear."

We refer to a rather remarkable paper by Yakhot and Orszag (1986)
which frees turbulence theory from empiricism and determines the basic
inertial subrange parameters through a "renormalization group (RNG)
analysis." We will not attempt a critique of the methodology and will
mention only some of their results. From first principles, they calculate the Kolmogorov constant, $C_K = 1.617$, and the coefficient in
(65), $(k_0/\pi)^2 \equiv 4c_s = 0.0248$. Our calculation of $c_s$ corresponding to
$C_K = 1.617$ is 0.00671. It should be noted that Yakhot and Orszag defined a somewhat different integral scale, $L_Y$, than did Lilly and Deardorff. They suggest that it should be viewed as corresponding to the
largest fluctuating scale within the inertial range of the system. It appears that $L = 2^{5/4}L_Y$.

Mahlman and Umscheid (1987) report that numerical integrations of
a tropo-spheric-stratospheric-mesospheric model (SKIHI – a global 40-
level model extending to 0.01 mb) using finite differences and a nonlinear
viscosity gave much poorer results in the smallest scales for horizontal
grid lengths of 5° latitude (555 km) than for 3° and especially 1°. This
was attributed in part to the fact that the smallest resolvable scale lay in
the –3 region for the coarsest grid. The use of a passive tracer in the integrations revealed that noisy filamentation was occurring at the smallest
resolvable scales. An abundance of gravity waves was also evident.

In a recent oceanographic application, Rosati and Miyakoda (1988)
concluded, "Certainly the utility of [sic] non-linear viscosity permitting smaller and more realistic values of $A_M$ [the viscosity coefficient]
in the open ocean without causing false computational spatial oscillations, which would occur when using a constant value of $A_M$, has been
demonstrated." They used $c_s = 0.0035$. This is the same value used by
Mahlman and his co-workers in their SKIHI integrations.

## 1.11 Two-dimensional and Three-dimensional Regimes in the $-\frac{5}{3}$ Region

The measure of merit in the choice of the exchange coefficient constant, $k_0$, of course, must be that the inertial energy cascade in the explicit regime is just balanced by the energy removal at grid scale, with no energy accumulation at grid scale as a result of the nonlinear cascade in the explicit domain. Moreover, the viscous decay should not be so large locally that the smallest resolvable scales are excessively damped. In the net, the shape of the energy density spectrum in the inertial subrange, $E(k) \propto k^{-5/3}$ in (45), should be preserved at the smallest resolvable scales.

We shall now proceed with a rationale for the choice of $k_0$ that is designed to preserve the empirical spectral properties of the inertial range. We first of all note that (51), together with (56), (57), (45) and (47), gives

$$\left[\frac{3}{2}Ek^{5/3}\right]^{3/2} = \epsilon k_0^{-2} \equiv \epsilon_{00}$$

$$= |D|^3/_H k^2 + |P|^3/_V k^2 . \tag{75}$$

Since $\epsilon = d(kE)/dt$, and for equilibrium of the spectrum $d(Ek^{5/3})/dt = 0$, we obtain from (75)

$$\frac{3}{2}\epsilon_{00}^{1/3}k_0^2 = \frac{3}{2}(\epsilon k_0^4)^{1/3} = \frac{dk^{-2/3}}{dt}. \tag{76}$$

In Figure 1 we can see that atmospheric phenomena with horizontal scales of $10^2$ to $10^5$ m are arrayed along a line for which the characteristic time, $T$, must be scaled against characteristic length, $L$, as $qT = L^{2/3}$, where $q$ is a constant. It corresponds to a $-\frac{5}{3}$ energy density falloff. Taking $L$ as the integral scale, $2\pi/k$, we find

$$\frac{3}{2}\epsilon_{00}^{1/3}k_0^2 = \frac{dk^{-2/3}}{dt} = \frac{q}{(2\pi)^{2/3}}. \tag{77}$$

Figure 1 shows that $q$ is about 0.1 m$^{2/3}$ s$^{-1}$. Furthermore, the atmospheric spectra obtained by Nastrom and Gage (1985) (Figure 2) can be utilized to determine that $\epsilon_{00} \approx 3 \times 10^{-4}$ m$^2$ s$^{-3}$ in the inertial subrange (see below).

Inserting these two empirical values into (73), one concludes that the value

$$k_0 = 0.541 \tag{78}$$

is necessary to preserve the climatological properties of the atmospheric energy spectrum and of the zoo of atmospheric mesoscale and smaller

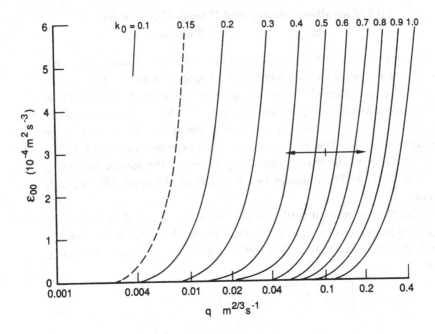

Fig. 3  $k_0$ as a function of $q$ and $\epsilon_{00} = \epsilon k_0^{-2}$ according to equation (77).

phenomena in the inertial range. Referring to Table 1, we find that our value of $k_0$ is close to the values obtained by other investigators using various means, theoretical and empirical, as applied to three-dimensional turbulence in the inertial range. However, (78) should be appropriate to the average value of $k_0$, that is, zeroth order, over the entire inertial range. The error of estimating $q$ from Figure 1 is quite large. We can see from Figure 3 that an error span of a factor of $\frac{1}{2}$ to 2 ($q = 0.05$ to $0.2$ $m^{2/3}$ $s^{-1}$) gives a span of uncertainty in $k_0$ of 0.38 to 0.77. However, $k_0$ is less sensitive to the error in estimating $\epsilon_{00}$ for $\epsilon_{00} > 2 \times 10^{-4} m^2$ $s^{-3}$. The present determination of $k_0$ is probably significant to only one figure.

As indicated earlier, Kraichnan (1971) concluded that the form of $E(k)$ in (45) was equally valid in the inertial range for two-dimensional and three-dimensional turbulence; only the value of $C_K$ ($C$ in Kraichnan's notation) should be increased by a factor of 4.78 for the two-dimensional case ($C = C_{2D} = 6.69$) compared with that of the three-dimensional case ($C = C_{3D} = 1.40$). Lilly (1983) came to a similar conclusion, but his choice for the two-dimensional coefficient ($\alpha$ in his notation) was 7 and for the three-dimensional coefficient was 0.8, a ratio of 9.

Lilly (1983) tried to match the atmospheric energy spectra in the two-dimensional and three-dimensional portions of the $-\frac{5}{3}$ range (roughly at a horizontal wave length of 200 km) by estimating that $\epsilon$ in the two-dimensional region is $10^{-5} \mathrm{m^2 \ s^{-3}}$ and in the downscale three-dimensional region is $10^{-4}$ to $10^{-3} \mathrm{m^2 \ s^{-3}}$. This seems to result in an ambiguity in that both $C_K$ and $\epsilon$ vary with $k$ in (45). However, this can be reconciled by noting that the inertial $-\frac{5}{3}$ range requires only that the product $C_K \epsilon^{2/3}$ be invariant with $k$ in (45), or, according to (47), that $\epsilon k_0^{-2} \equiv \epsilon_{00}$ be independent of $k$. On the other hand, from (77), it follows that, to zeroth order, both $\epsilon$ and $k_0$ must be constant in the inertial range.

Kraichnan's and Lilly's conclusions may be related to Deardorff's experience (1971, 1985) that a smaller coefficient is required for shear-dominated flows. This could mean that generally the larger three-dimensional coefficient is appropriate, but that in the vicinity of strong shear, for example near the jet stream, a smaller value, tending toward two-dimensionality, should be used.

The definitive atmospheric spectra for the tropopause region (the level of jet stream maximum) obtained by Nastrom and Gage (1985) (Figure 2) afford the opportunity to calculate $\epsilon k_0^{-2}$ by applying (75). We first of all note that in the $-\frac{5}{3}$ region, to very good approximation, there is equipartition of energy density in the zonal and meridional spectral components. Furthermore, Lilly (1983) has reckoned the bias due to using the horizontal wind components and the two components of wave number as 0.54 in estimating the three-dimensional spectrum. Therefore, we find that

$$\left[ \frac{3}{2} E k^{5/3} \right]^{3/2} = \left[ \frac{1}{0.54} \left( \frac{3}{2} \right) {}_H E_H k^{5/3} \right]^{3/2}$$
$$= \left\{ \frac{2}{0.54} \left( \frac{3}{2} \right) \left[ {}_H E(U)_H k^{5/3} \right] \right\}^{3/2} = \epsilon k_0^{-2} , \tag{79}$$

where ${}_H E(U)$ is the Nastrom and Gage zonal wind energy density spectrum as a function of the horizontal wave number, ${}_H k$.

The average of several points on the Nastrom and Gage $-\frac{5}{3}$ spectrum yields, through (75), $\epsilon k_0^{-2} \approx 3 \times 10^{-4} \mathrm{m^2 \ s^{-3}}$, which should be invariant over the entire range of the inertial spectrum.

According to Kraichnan's (1971) theoretical estimates of $C_K = 6.69$ and 1.40 appropriate to two-dimensional and three-dimensional turbulence (see Table 1), one finds that $k_0$ assumes values of 0.18 and 0.57, respectively, while $\epsilon$ ranges from $10^{-5}$ to $10^{-4} \text{m}^2 \text{ s}^{-3}$, respectively. Similarly, Lilly's (1983) values of $C_K$ in the two- and three-dimensional regions, 7.0 and 0.8, or $k_0 = 0.17$ and 0.87, give $\epsilon$ ranging from $0.9 \times 10^{-5}$ to $2.3 \times 10^{-4} \text{m}^2 \text{ s}^{-3}$, respectively. These ranges are in good agreement with the range cited by Lilly (1983) based on other methods of estimating the dissipation rate.

Lilly found that he had to take what, he thought, was a very low level for the dissipation rate in the two-dimensional region, $10^{-5} \text{ m}^2 \text{ s}^{-3}$, in order to match up with the smaller-scale three-dimensional spectrum. However, his choice is supported by our independent result. Note, however, that strictly speaking two-dimensional turbulence does not dissipate energy because the energy does not cascade to the small scales.

We have, therefore, to deal with the idea that the inertial dissipation rate itself may vary smoothly with scale within the inertial range, becoming smaller with increasing wavelength. These results also imply that there is a first order decrease in $k_0$ with increasing wavelength in the inertial range, but that the determination of the functional relationship, given $\Delta s$ and $\Delta z$, does not close by our present argument. That is, one is still lacking a relationship in terms of the aspect ratio, the ratio of the resolved characteristic vertical to horizontal scales as reflected by the choice of $\Delta z$ and $\Delta s$. Asymptotically, this ratio is 0 and 1 for two- and three-dimensional turbulence, respectively. It should depend on the static stability (Lilly 1983).

It thus appears that the $-\frac{5}{3}$ range not only is relevant to three-dimensional turbulence but also is hospitable in its largest scales to a spillover of two-dimensional turbulence from the $-3$ range. This condition seems to be supported by the observations of Nastrom and Gage (1985). To get a $-\frac{5}{3}$ spectrum in two-dimensional turbulence one needs an energy source at large $k$. Lilly (1983) suggested this source to be convective storms. We have already noted that Emanuel (1984) thinks that only slantwise convection can do it, though intermittently. In a recent paper, Lilly (1989) made use of turbulence closure models to effectively interpolate the spectrum between the $-3$ and $-\frac{5}{3}$ domains without invoking a special dissipation mechanism in the transition region between wavelengths 200 and 600 km.

The results of Deardorff, Kraichnan and Lilly lead one to speculate that neither Kolmogorov–Heisenberg three-dimensional turbulence nor the Onsager–Fjørtoft two-dimensional geostrophic turbulence is appropriate to the mesoscale, quasi-two-dimensional, quasi-static range addressed here, the range resolved by horizontal grids typically of the order of 100 km in mesh size. An intermediate aspect ratio, one which comes closer to relating to Deardorff's experience, may be in order. Table 1 gives the values of the various popular parameters corresponding to $c_s = 0.0035$, a recent value found empirically to give good results both in the ocean (Rosati and Miyakoda 1988) and in the atmosphere (J. Mahlman, personal communication). As one can see, it comes close in parameter-space to Deardorff's results with shear.

The notion of an intermediate range of validity for the Heisenberg–Kolmogorov spectrum seems to differ from conventional wisdom. This leaves open the question as to the physical basis for the broad range of validity of the $-\frac{5}{3}$ spectrum in (45) and the functional dependence of $k_0$ (or alternatively one of the other related nondimensional parameters) on dimensionality (that is, the aspect ratio) and horizontal scale. The argument that $\epsilon k_0^{-2}$ is invariant suggests that the variation of $k_0$ with $k$ in the inertial range depends on the variation in $\epsilon(k)$.

A clue comes from Lesieur (1990, pp. 135–136). He considers forcing in a narrow spectral band (he actually considers a line source) at some low wave number characteristic of "large energy containing eddies." For equilibrium of the energy spectrum, the kinetic energy flux, $\epsilon$, through $k_i$ undergoes a discontinuous jump at $k_i$ from zero to the kinetic energy injection rate, $\epsilon_0$.

In the atmospheric case, however, the low wave number energy forcing, $I(k)$, occurs in a rather broad finite wave number band through conversion of potential energy to kinetic (see Kung and Tanaka 1983). This band, concentrated mainly between zonal wave numbers 1 and 10, is identifiable with baroclinic instability and with global-scale, continental-ocean thermal forcing. In such a configuration, one should expect kinetic energy injection at the rate $\epsilon$ in the transitional range of wave numbers surrounding $k_i$.

For spectral equilibrium to exist, a balance between the spectral energy transfer, $-\partial \epsilon(k)/\partial k$, the energy forcing, $I(k)$, and the molecular dissipation, $J(k)$, must be satisfied at each $k$ (Lilly 1989) such that

$$\frac{\partial E}{\partial t} = 0 = -\frac{\partial \epsilon(k)}{\partial k} + I(k) - J(k). \tag{80}$$

Integrating over the entire spectral range with $\epsilon = 0$ at the termini gives

$$\int_0^\infty \left[ I(k) - J(k) \right] dk = 0, \tag{81}$$

which reflects an overall balance between forcing at very low $k \sim k_i$ and molecular dissipation at very high $k \sim k_d$, the Kolmogorov wave number, which corresponds to a dissipation scale of about 1 mm in the atmosphere (Lesieur 1990).

For wave numbers much smaller than $k_d$, one finds using (80)

$$\epsilon(k) = \int_0^k I(k)\ dk, \quad k \ll k_d. \tag{82}$$

Assuming that input into $I(k)$ comes from the band between $k_i - \delta$ and $k_i + \delta$, one can write

$$\epsilon(k) = 0, \quad 0 \le k \le k_i - \delta, \tag{83}$$

$$\epsilon(k) = \int_{k_i - \delta}^k I(k)\ dk, \quad k_i - \delta \le k \le k_i + \delta, \tag{84}$$

$$\epsilon_0 \equiv \int_{k_i - \delta}^{k_i + \delta} I(k)\ dk \tag{85}$$

and

$$\epsilon(k) = \epsilon_0 = \text{constant}, \quad k_i + \delta \le k \ll k_d. \tag{86}$$

Equation (84) constitutes a continuous change of $\epsilon(k)$ from zero to $\epsilon_0$ over a finite forcing interval (Figure 4).

We found that in the inertial range

$$\epsilon_{00} \equiv \epsilon(k)\,[k_0(k)]^{-2} = \text{constant}. \tag{87}$$

From (84) and (86) it follows that

$$k_0^2(k) = \frac{1}{\epsilon_{00}} \int_{k_i - \delta}^k I(k)\ dk = \frac{\epsilon(k)}{\epsilon_{00}}, \quad k_i - \delta \le k \le k_i + \delta, \tag{88}$$

and

$$k_0^2(k) = \epsilon_0/\epsilon_{00} = \text{constant}, \quad k_i + \delta \le k \ll k_d. \tag{89}$$

The value of $\epsilon_{00}$ has been determined empirically from the Nastrom–Gage spectrum. The kinetic energy injection rate, $\epsilon_0$, can be estimated from the potential–kinetic energy conversion spectra of Kung and Tanaka (1983) with an appropriate correspondence between the zonal wave number and $k$.

We thus have deduced a functional shape of $k_0(k)$ which depends on $I(k)$. The resulting $k_0(k)$ in the inertial range should be consistent with the zeroth order determination of $k_0 \approx 0.5$ in (78). Presumably, the details of $k_0(k)$ in the inertial range may be somewhat different for the oceans than for the atmosphere.

The glaring inconsistency to be reconciled is suggested by Figure 4: the transitional increase of $\epsilon$ with wave number lies in the $-3$ region, not the $-\frac{5}{3}$. Perhaps backscatter considerations, which are receiving a great deal of attention, can come to the rescue.

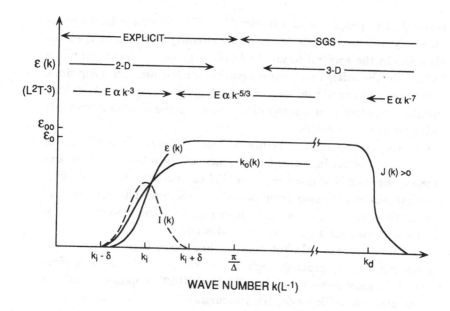

Fig. 4   Schematic variation of the equilibrium spectral energy flux, $\epsilon(k)$, as the result of forcing, $I(k)$. The balancing molecular dissipation, $J(k)$, occurs at $k_d < k$. Shown also is the variation of the nondimensional coefficient $k_0(k)$ with wave number. The grid integral scale wave number, $\pi/\Delta$, in this example was taken to be larger than $k_i + \delta$.

Strictly speaking the performance of this inertial range nonlinear viscosity formulation must be judged with both the vertical and horizontal stresses operating. Not many applications have taken this step despite the ubiquitous occurrence of strong vertical shear in connection with the jet stream in the atmosphere and the western boundary currents in the ocean.

## 1.12 Expectations

In the early 1970s, spectral representation in models of the large-scale atmosphere became popular. The cost of the computational overhead of the nonlinear viscosity became an issue, both to calculate the square root in $|D|$ and to carry some of the hydrodynamic calculations into the finite-difference form for the numerical integration. As a result, the scale-selective properties of a linear $\nabla^4$ or $\nabla^8$ viscosity gained appeal, especially because the calculation of the viscosity term can be done analytically in the spectral form. And this, very largely, is the situation today. On the other hand, oceanographic models are still using finite differences because of convenience in dealing with irregular lateral continental boundaries, and many of the users of the nonlinear viscosity today are ocean modelers.

However, there may be a return to finite differences and the use of nonlinear viscosities in atmospheric models as resolution gets finer and greater efficiency is affected with parallel computer architecture. Further impetus would come from attaining a more faithful rendering of columnwise physical processes, such as radiation and convection, and the use of semi-Lagrangian advective schemes.

Finally, one cannot help but wonder whether the first solution to the nonlinear instability problem might not be the simplest, the most effective and the most universal, that is, Phillips' (1959) proposal to truncate the smallest resolvable mode. He demonstrated this procedure by eliminating all finite-difference components with wavelengths smaller than four times the grid size. In the case of spectral methods, the course is even more direct: a complete truncation of the highest harmonic. This would attack the symptoms directly, removing energy at a rate just matching the cascade rate empirically so as to preserve the shape of the spectrum at the limit of resolution. Such an approach might well deserve testing in a variety of flows. One such attempt (called to my attention by Boris Galperin) by Browning and Kreiss (1989) gave negative results. They found that when the viscosity term is deleted from the two-dimensional incompressible Navier–Stokes equations and "the energy in high wave numbers [is] removed by setting the amplitudes of all wave numbers above a certain point in the spectrum to zero, the 'chopped' solution differs considerably from the convergent solution, even at early times."

## Acknowledgments

My thanks to Jerry Mahlman for bringing me up to date on recent experience and practices in meteorology and oceanography. I am also indebted to Boris Galperin for his very helpful editorial comments.

# References

BASS, J. (1949) Sur les bases mathématiques de la théorie de la turbulence d'Heisenberg. *C. R. Acad. Sci., Paris* **228** (part 1), 228.

BATCHELOR, G.K. (1946) The theory of axisymmetric turbulence. *Proc. Roy. Soc. A* **186**, 480–502.

BATCHELOR, G.K. (1953) *The Theory of Homogeneous Turbulence.* Cambridge University Press, 197 pp.

BROWNING, G.L. AND KREISS, H.-O. (1989) Comparison of numerical methods for the calculation of two-dimensional turbulence. *Math. Comput.* **52**, 369–388.

CHARNEY, J.G. (1947) The dynamics of long waves in a baroclinic westerly current. *J. Meteorol.* **4**, 135–162.

CHARNEY, J.G. (1971) Geostrophic turbulence. *J. Atmos. Sci.* **28**, 1087–1095.

DEARDORFF, J.W. (1971) On the magnitude of the subgrid-scale eddy coefficient. *J. Comput. Phys.* **7**, 120–133.

DEARDORFF, J.W. (1985) Sub-grid-scale turbulence modeling. In *Issues in Atmospheric and Oceanic Modeling, Part B. Weather Dynamics, Advances in Geophysics.* Ed. Barry Saltzman and Syukuro Manabe, pp. 337–343. Academic Press.

EADY, E.T. (1949) Long waves and cyclone waves. *Tellus* **1**, 35–52.

EMANUEL, K. (1984) What does "mesoscale" mean? In *Dynamics of Mesoscale Weather Systems.* Ed. J.B. Klemp, pp. 3–12. NCAR Summer Colloquium, Boulder, CO.

FJØRTOFT, R. (1951) Stability properties of large-scale disturbances. In *Compendium of Meteorology.* Ed. T.F. Malone, pp. 454–463. American Meteorological Society.

FJØRTOFT, R. (1953) On the changes in the spectral distribution of kinetic energy for two-dimensional non-divergent flow. *Tellus* **5**, 225–230.

HEISENBERG, W. (1948a) On the theory of statistical and isotropic turbulence. *Proc. Roy. Soc. A* **195**, 402–406.

HEISENBERG, W. (1948b) Zur Statistischen Theorie der Turbulenz. *Z. Phys.* **124**, 628–657.

HINZE, J. O. (1959) *Turbulence: An Introduction to Its Mechanism and Theory.* McGraw-Hill, 585 pp.

HOLTON, J.R. (1979) *An Introduction to Dynamic Meteorology.* International Geophysics Series, 23, (2nd edition). Academic Press, 391 pp.

KÁRMÁN, T. VON AND HOWARTH, L. (1938) On the statistical theory of isotropic turbulence. *Proc. Roy. Soc. A* **164**, 192–215.

KOLMOGOROV, A.N. (1941a) The local structure of turbulence in an incompressible viscous fluid for very large Reynolds numbers. *C. R. Acad. Sci. URSS* **30**, 301–305.

KOLMOGOROV, A.N. (1941b) Dissipation of energy in locally isotropic turbulence. *C. R. Acad. Sci. URSS* **32**, 16–18.

KRAICHNAN, R.H. (1971) Inertial-range transfer in two- and three-dimensional turbulence. *J. Fluid Mech.* **47**, 525–535.

KUNG, E.C. AND TANAKA, H. (1983) Energetic analysis of the global circulation during the special observing periods of FGGE. *J. Atmos. Sci.* **40**, 2575–2592.

LEE, T.D. (1951) Difference between turbulence in a two-dimensional fluid and in a three-dimensional fluid. *J. Appl. Phys.* **22**, 524.

LESIEUR, M. (1990) *Turbulence in Fluids* (2nd revised edition). Kluwer, 412 pp.

LILLY, D.K. (1962) On the numerical simulation of buoyant convection. *Tellus* **14**, 148–172.

LILLY, D.K. (1967) The representation of small-scale turbulence in numerical simulation experiments. In: *Proceedings of IBM Scientific Symposium on Environmental Sciences.* IBM Form No. 320-1951, pp. 195–210.

LILLY, D.K. (1983) Stratified turbulence and the mesoscale variability of the atmosphere. *J. Atmos. Sci.* **40**, 749–761.

LILLY, D.K. (1989) Two-dimensional turbulence generated by energy sources at two scales. *J. Atmos. Sci.* **46**, 2026–2030.

LINDZEN, R.S., LORENZ, E.N. AND PLATZMAN, G.W. (1990) *The Atmosphere – A Challenge: The Science of Jule Gregory Charney.* American Meteorological Society, 321 pp.

MAHLMAN, J.D. AND UMSCHEID, L.J. (1987) Comprehensive modeling of the middle atmosphere: The influence of horizontal resolution. In *Transport Processes in the Middle Atmosphere.* Ed. G. Visconti and R. Garcia, pp. 251–266. Reidel.

MORSE, P.M. AND FESHBACH, H. (1953) *Methods of Theoretical Physics.* McGraw-Hill, New York, 2 vols., 1978 pp.

NASTROM, G.D. AND GAGE, K.S. (1985) A climatology of atmospheric wave number spectra of wind and temperature observed by commercial aircraft. *J. Atmos. Sci.* **42**, 950–960.

NEUMANN, J. VON AND RICHTMYER, R.D. (1950) A method for the calculation of hydrodynamic shocks. *J. Appl. Phys.* **21**, 232–237.

OGURA, Y. (1952) The structure of two-dimensionally isotropic turbulence. *J. Meteorol. Soc. Japan* **30**, 27–32.

ONSAGER, L. (1945) The distribution of energy by turbulence. *Phys. Rev.* **68**, 286.

ONSAGER, L. (1949) Statistical hydrodynamics. *Nuovo Cimento*, Suppl. AL **VI**, Series IX, 279–287.

PHILLIPS, N.A. (1956) The general circulation of the atmosphere: A numerical experiment. *Q. J. Roy. Soc.* **82**, 123–164.

PHILLIPS, N.A. (1959) An example of non-linear computational instability. In *The Atmosphere and the Sea in Motion*. Ed. B. Bolin, pp. 501–504. Rockefeller Institute Press.

PROUDMAN, I. (1951) A comparison of Heisenberg's spectrum of turbulence with experiment. *Proc. Camb. Phil. Soc.* **47**, 158–176.

REID, W. H. (1959) The dynamics of isotropic turbulence in two dimensions. Tech. Rep. 23, Brown University, Div. of Applied Math.

RICHARDSON, L.F. (1926) Atmospheric diffusion shown on a distance-neighbour graph. *Proc. Roy. Soc.* A **110**, 709–737.

ROSATI, A. AND MIYAKODA, K. (1988) A general circulation model for upper ocean simulation. *J. Phys. Oceanogr.* **18**, 1601–1626.

SMAGORINSKY, J. (1958) On the numerical integration of the primitive equations of motion for baroclinic flow in a closed region. *Mon. Wea. Rev.* **86**, 457–466.

SMAGORINSKY, J. (1962) The formulation of eddy transport processes for the quasi-static inertial sub range of atmospheric motions. Unpublished manuscript.

SMAGORINSKY, J. (1963) General circulation experiments with the primitive equations, Part I: The basic experiment. *Mon. Wea. Rev.* **91**, 99–152.

SMAGORINSKY, J. (1974) Global atmospheric modeling and the numerical simulation of climate. In *Weather and Climate Modification*. Ed. W. N. Hess, pp. 633–686. Wiley.

SMAGORINSKY, J. (1983) The beginnings of numerical weather prediction and general circulation modeling: Early recollections. In *Advances in Geophysics*, Vol. 25. Ed. B. Saltzman, pp. 3–37. Academic Press.

SMAGORINSKY, J., MANABE, S. AND HOLLOWAY, J.L. (1965) Numerical results from a nine-level general circulation model of the atmosphere. *Mon. Wea. Rev.* **93**, 727–768.

SOKOLNIKOFF, I.S. (1946) *Mathematical Theory of Elasticity*. McGraw-Hill, 373 pp.

*Joseph Smagorinsky*

TAYLOR, G.I. (1935) Statistical theory of turbulence. *Proc. Roy. Soc. A*
   **151**, 421–478.

WEIZSÄCKER, C.F. VON (1948) Das spektrum der Turbulenz bei grossen
   Reynoldsschen Zahlen. *Z. Phys.* **124**, 614–627.

YAKHOT, V. AND ORSZAG, S.A. (1986) Renormalization group analysis of
   turbulence, Part I: Basic theory. *J. Sci. Comp.* **1**, 3–51.

# 2

---

# Subgrid-Scale Modeling

## JOEL H. FERZIGER

## 2.1 Introduction

Although large eddy simulation (LES) of turbulent flows has been practiced for nearly 30 years and the name has been used for close to 20 years, some confusion about what constitutes a large eddy simulation remains and a definition may be useful. We shall make this more precise below, but for now, we define a large eddy simulation as any simulation of a turbulent flow in which the large-scale motions are explicitly resolved while the small-scale motions are represented approximately by a model (in engineering nomenclature) or parameterization (in the geosciences).

The most straightforward means of defining the large-scale field is via filtering (Leonard, 1974). When the Navier–Stokes equations are filtered, the resulting equations for the large-scale component of the velocity contain terms representing the effect of the small scales on the large ones; these subgrid-scale (SGS) Reynolds stresses must be modeled or parameterized. A large body of evidence from completed simulations demonstrates that, when the SGS Reynolds stress is a small part of the total (ensemble- or time-averaged) turbulence-produced Reynolds stress, the results produced by LES are relatively insensitive to the quality of the model; this is as expected. Under these circumstances, typically encountered in the simulation of so-called building-block flows, the choice of model and values of the parameters are of only moderate importance.

On the other hand, when LES is applied to complex and/or high Reynolds number flows (sometimes called very large eddy simulation, or VLES), much of the Reynolds stress lies in the unresolved scales and model quality becomes much more important; this is also the case for Reynolds-averaged models. A number of people have tried to apply LES methods and models which were successful for simple flows to complex flows; in those cases in which the quality could be measured, the results were rather disappointing. While all the blame cannot be laid on the SGS model, the latter is almost surely a major contributor to the lack of success.

It is important to note that in engineering turbulent flows, the eddies are strongly three-dimensional and unsteady even at the largest scales, so LES must also possess these characteristics. In the geosciences, the largest scales are nearly two-dimensional (although the small scales are always three-dimensional) and the definition of LES may be not as clear. These differences are important and the models employed in the two fields may need to be different. Some people have done two-dimensional simulations of engineering flows and call them LES; in the author's opinion these computations are not LES, and if the results are to have validity, the models employed need to be investigated carefully.

If enough resolution can be employed, any turbulent flow can be simulated accurately by LES. In fact, given sufficiently fine resolution, LES becomes direct numerical simulation (DNS), whose accuracy is unquestioned. Unfortunately, for flows of practical importance, the computational and memory requirements of DNS or fine-grid LES render such simulations unfeasible. Even with the expected advances in computers coming from the introduction of massively parallel machines and larger memory chips, the cost of such simulations will remain out of reach except for a limited range of well-chosen flows. If LES or VLES is to become a practical tool, construction of better SGS models will be a pacing item.

In this chapter, we shall make reference to simulations based on ensemble- or time-average models. These will be referred to collectively as Reynolds-averaged models or parameterizations; the Reynolds-averaged Navier–Stokes (RANS) equations together with a turbulence model for closure are the basis for most engineering and geophysical "predictions" of turbulent flows. The quotation marks around the word "prediction" are meant to indicate that, when this type of modeling is employed, "postdiction" is a more accurate descriptor of what one is doing.

In the geosciences, it is difficult to see how to apply Reynolds averaging in the sense defined above. Instead, one often uses a scale separation argument, that is, that there are scales between the ones that are explicitly simulated and those represented parametrically that contain very little energy. This argument is not strictly correct, but it is often the case that the represented scales are two-dimensional while the approximated ones are three-dimensional. Nonetheless, one needs to be careful about the parameterizations employed; it is doubtful that one could take an engineering model without modification.

In the next section, a short review of the fundamentals of LES relevant to the remainder of the chapter is given. This is followed by a brief description of current SGS models. Section 2.4 discusses some recent developments. Section 2.5 describes the models used to treat the viscous sublayer near solid boundaries. Section 2.6 discusses geophysical and other applications of LES. We close with a discussion of what is needed for the future.

It is important to make another point about the differences between engineering and the geosciences. What is considered acceptable accuracy in one field may be totally unacceptable in another. In aeronautical and some mechanical engineering problems, accuracy of a few percent is demanded, whereas civil engineers and geoscientists are sometimes satisfied with results that differ from experiments and/or field measurements by as much as a factor of 2. Thus, what one group calls a good model may be completely unsatisfactory to the other.

Finally, the reader is referred to Chapter 6 by Piomelli, this volume, for a different viewpoint on the issues covered here.

## 2.2 Fundamentals

In the author's opinion, it is important to begin with a definition of what is to be computed and what must be modeled. So doing requires separation of the filtering from the numerical approximations, which complicates the passage from the Navier–Stokes equations to a computer program. The extra effort is worthwhile only if significant advantages accrue. There are two benefits. The first is extra clarity, which facilitates accurate comparison of results with experiments and/or DNS results. The second is that it provides a route to theoretical development of models and allows direct testing of the models.

The need to make precise what is to be computed in LES led Leonard (1974) to propose filtering as a means of defining the large-scale velocity field, $\overline{u}_i$,

$$\overline{u}_i = \int G(x, x'; \Delta) u_i(x') \, dx' \qquad (1)$$

where $\Delta$ is a length scale that selects the minimum resolved eddy size. In a homogeneous field, $G(x, x') \to G(|x - x'|)$; the Fourier transform of Eq. (1) is then

$$\hat{\bar{u}}_i(k) = \hat{G}(k)\hat{u}_i(k) \tag{2}$$

This equation plays an important role both in LES and in theoretical developments of models. Relations similar to Eq. (2) can be used to define filters used in conjunction with spectral methods based on function sets other than the Fourier basis.

When the filter (1) is applied to the Navier–Stokes equations, the result is precisely the Navier–Stokes equations for $\bar{u}_i$ with the additional term

$$\frac{\partial \tau_{ij}}{\partial x_j} = \frac{\partial}{\partial x_j}(\overline{u_i u_j} - \bar{u}_i \bar{u}_j) \tag{3}$$

on its left-hand side. The nomenclature "subgrid-scale Reynolds stress tensor" is applied to $\tau_{ij}$ as well as certain components of it that arise when the decomposition

$$u_i = \bar{u}_i + u'_i \tag{4}$$

is substituted into Eq. (3).

The decomposition (4) has some advantages; the individual terms have physical significance which allows them to be modeled separately. However, it has been pointed out by a number of authors, most recently by Germano (1992), that some of the terms resulting from the decomposition are not Galilean invariant and this important invariance property may be lost if care is not taken. Separate treatment of the terms resulting from decomposition of the Reynolds stress also increases the cost of doing a simulation; unless significant advantages are realized by doing so, it may be better to model $\tau_{ij}$ as a whole. The present author has been an advocate of separate treatment of the various terms; recent results show that slightly greater accuracy results from doing so (Piomelli et al., 1987) but the cost/benefit ratio may not be large enough to justify continuation of this practice. The combination of simplicity, invariance, and greater computational speed has led an increasing number of authors to prefer the simpler approach to SGS modeling.

## 2.3 SGS Models

### 2.3.1 Smagorinsky Model

Eddy viscosity models, whose origins can be traced to Boussinesq, are by far the most commonly used turbulence models in either RANS or

LES. So when he needed a model for the small scales in a global me-
terological simulation, it was natural that Smagorinsky (1963) turned
to an eddy viscosity model. The model that now bears Smagorinsky's
name (and its RANS ancestors) can be derived in many ways: heuristically, as Boussinesq, Prandtl, and Smagorinsky did; from a production
equals dissipation argument applied to the dynamic equations for the
SGS Reynolds stress; from various turbulence theories (for example, see
Lilly, 1965), the most recent being renormalization group (RNG) theory
(Yakhot and Orszag, 1986).

All of these approaches yield the same linear relationship between the
SGS Reynolds stress (one can argue whether it is $\tau_{ij}$ of Eq. (3) or one of
the components resulting from its decomposition that is being modeled)
to the strain rate of the large or resolved eddies,

$$\tau_{ij} = -2\nu_T S_{ij} = -\nu_T \left( \frac{\partial \overline{u}_i}{\partial x_j} + \frac{\partial \overline{u}_j}{\partial x_i} \right) \tag{5}$$

where $\nu_T$ is the SGS eddy viscosity. The theories also provide a value
for the parameter which cannot be obtained by heuristic arguments.

As in RANS modeling, $\nu_T$ is generally assumed to be a scalar quantity,
although there is no reason it should not be a fourth rank tensor; some
suggestions about the form of the fourth rank tensor have been made in
the context of RANS modeling. The derivations of Eq. (5) listed above
suggest that the eddy viscosity be expressed as

$$\nu_T = (C_S d)^2 (S_{ij} S_{ij})^{1/2} S_{ij} = (C_S d)^2 |S| S_{ij} \tag{6}$$

where $d$ is the length scale associated with a typical SGS eddy and $C_S$
is the single model parameter. It is intuitively logical that the length
scale $d$ be the length scale $\Delta$ used in the definition of the filter; see Eq.
(1).

A number of arguments based on more fundamental considerations
suggest that a better choice for the length scale is

$$d = (L\Delta^2)^{1/3} \tag{7}$$

where $L$ is the integral length scale of the turbulence. However, the
difficulty of computing the integral scale and therefore of applying Eq.
(7) has led to the use of $\Delta$ for the length scale $d$ in place of expression (7). Although the arguments leading to this approximation suggest
that the "constant" $C_S$ should become a function of $\Delta/L$ and thus of
Reynolds number, it appears to have been successful in all cases to which
it has been applied; this may be a consequence of the relatively small

Reynolds number range (and consequently the small range of $L/\Delta$) in the simulations.

For isotropic filters and/or simple flows, the definitions of length scales just given are unique, so no problems are associated with their application. In flows with strong anisotropy (such as highly sheared flows), the flow becomes highly anisotropic and the appropriate choice of length scale is less clear. A particularly perplexing issue has been the need to use different values of the model parameter in shear flows and isotropic turbulence; this has been observed by numerous authors. This question may be related to the length scale issue but it may equally well be a consequence of a fundamental difference between the physics of shear flows and isotropic turbulence.

Some theories suggest that the model be augmented by a random force with a $k^4$ spectrum. This force is intended to represent the backscatter, or return of energy from the small scales to the large ones. The necessity of including this term in simulations is not yet clear. Leith (1990) has shown that a random force can cause a flow to develop behavior typical of turbulent flows, but his is a transitional flow so the issue cannot be regarded as settled.

Several problems related to the Smagorinsky model have received attention over the years. The value of the parameter $C_S$ predicted by the theories (about 0.2) does a good job for isotropic turbulence; this is no surprise since the theories are based on isotropic turbulence. As noted above, for inhomogeneous shear flows, many authors have found that this value of $C_S$ must be reduced by half or more. This discrepancy has not been completely explained, although it appears to be related to an increase in the backscatter in flows with mean strain or shear (McMillan et al., 1980).

Nor has the question of the choice of which length scale to use for $\Delta$ in the model been adequately resolved for flows in which the turbulence and/or the filter are anisotropic. The most common choice is

$$\Delta = (\Delta x_1 \Delta x_2 \Delta x_3)^{1/3} \tag{8}$$

where $\Delta x_i$ is the filter width in the $i$th direction. Other suggestions have been made but have not won much support. Note that, as long as the ratios of the three length scales do not vary much within the flow, the differences can be resolved by adjusting the parameter $C_S$. This is emphatically not the case in wall-bounded flows.

## 2.3.2 Scale Similarity Model

A recent development appears to provide a way of circumventing many of the issues just raised, thereby eliminating the corresponding difficulties. We shall describe and discuss this dynamic model in Section 2.4. Before doing so, we present the scale similarity model proposed by Bardina et al. (1981); the basis for this model contains the ideas on which the dynamic model is based and may be thought of as an oversimplified or heuristic version of RNG theory. It assumes that the principal scales that require modeling, the ones whose scales lie just below the filter size, are similar in structure to the smallest resolved scales. If that is so, one can argue that the SGS stress should take the following form (for further details, see the reference):

$$\tau_{ij} = \overline{u}_i \overline{u}_j - \overline{\overline{u}}_i \overline{\overline{u}}_j \tag{9}$$

Here, $\overline{\overline{u}}_i$ is the twice-filtered velocity field; in contrast to RANS modeling, $\overline{\overline{u}}_i \neq \overline{u}_i$.

The second filter need not be identical to the first, although in most simulations that have used this model, the two have been the same. Originally, Bardina et al. included a model constant in Eq. (9) and showed that it had a value close to unity. More recent work has shown that it must be unity to ensure Galilean invariance.

This model, although it reproduces the structure of the SGS stresses rather well, does not dissipate sufficient energy and it must be combined with the Smagorinsky model in order to have all of the desired properties; the result is usually called the mixed model. In the mixed model the major function of the scale similarity component is to transfer energy from the smaller resolved scales to the larger ones.

## 2.3.3 Other Issues

Whether other models are improvements upon the Smagorinsky and mixed models is an open question. A number of such models have been proposed, a few of which will be reviewed very briefly here.

It is not difficult to derive dynamic equations for the SGS Reynolds stress tensor. Using these as the basis for a model is an idea closely related to trends in RANS modeling. It was tried early on by Deardorff (1973) without notable success. His lack of success was probably due to significant differences between the behavior of the large scales and that of the subgrid scales, especially with regard to the pressure and the terms which contain it. It is very likely that the latter need to be modeled differently (at least with different constants) than their RANS model counterparts. Deardorff did not have sufficient data to allow him to take these differences into account. Direct simulation results may allow such investigations, but no one has yet attempted this.

Replacement of the second order viscous term of the Smagorinsky model with a higher order viscous term, for example, a term proportional to fourth or higher derivatives of the velocity, is an interesting idea. Fourth order viscosities have been used in numerical calculations of compressible flow to smooth shocks without damaging the remaining flow, and there is theoretical justification for using them in LES. They are implicit in the simulations made by Kuwahara and his colleagues (see Kawamura and Kuwahara, 1985); in these, a fourth order viscosity arises as the truncation error of the third order upwind difference method applied to the advective terms. (It should be noted that the majority of the flows they simulate are transitional rather than fully turbulent. Thus they have laminar inflow conditions, which may be important.) Although using the truncation error of a numerical scheme as a model is an old idea, the model is then dependent on the grid and numerical approximations and may have little relation to the physics of the flow simulated.

Use of the eddy-damped quasi-normal Markovianized (EDQNM) theory of turbulence to compute the properties of the SGS turbulence while the large scales are treated by LES was suggested by Aupoix (1987). He obtained impressive results and was able to compute at high Reynolds numbers but only for isotropic turbulence. Moreover, replacing the Smagorinsky model with EDQNM increases the cost by an order of magnitude. If this method is to be a contender as an SGS model, especially for anisotropic turbulence and complex geometries (to which spectral ideas are difficult to apply), the EDQNM portion of the calculation will need to be simplified so that it is much less expensive.

An approach that combines some features of the methods described in the preceding two paragraphs begins by using EDQNM to compute the energy spectrum; from the results, one computes a spectral eddy viscosity which allows LES to reproduce exactly the spectrum up to the cutoff; for a review of this work see the book by Lesieur (1990). With this method, one can simulate flows with inertial $k^{-5/3}$ behavior at the cutoff wave number and thus, in principle, flows at any Reynolds number. This approach corresponds to using a more complex higher order eddy viscosity than those described above. A version of this method which can be used in physical space, called the structure function method, has been reported by P. Comte and M. Lesieur (private communication); it appears promising but has not yet been tested sufficiently for definitive evaluation.

Finally, the use of coarse grid direct numerical simulation has been proposed. This is similar to the method which uses upwind difference approximations in that it uses numerical error to model the scales that are not resolved. Some of the results obtained with this approach appear reasonable but it is difficult to say what is being computed, making it hard to convince readers and potential users of the value of the approach.

### 2.3.4 Complex and High Reynolds Number Flows

As noted above, the aim of any type of turbulence simulation is to predict certain parameters of high Reynolds number flows in complex geometries reliably. This is a tall order but the economic value of such a method would be enormous. The very existence of this volume is testimony to the fact that such a method does not now exist except, perhaps, for a limited range of flows.

It has been hoped for many years that RANS models would fill the need and, indeed, they have had numerous successes. The best general-purpose RANS models can be counted on to be accurate to within about 25%; it is best not to try to make this value precise, as it is flow dependent and affected by many parameters. Naturally, applications which require the greatest accuracy provide RANS models with their greatest challenges. Many people have searched for the ideal or universal RANS model but without success. It is the author's opinion that, unfortunately, this endeavor is unlikely to meet with success because the assumption that low order moments of velocity fluctuations can be expressed as functionals of higher order moments, which is fundamental to this type of modeling, is incorrect. However, no better means of computing mean flows has been devised. It is probably best to regard these models as sophisticated engineering correlations rather than scientific laws; one can then modify or tune the models for particular applications, provided care is taken.

Some people have suggested that VLES might replace RANS modeling as the method of choice. It does indeed have a chance of success but it will come at considerable cost because LES is necessarily three-dimensional and time-dependent and RANS calculations are often two-dimensional and steady. Thus the best opportunities for early application of LES are probably to flows that are inherently unsteady and three-dimensional. This is the case in weather forecasting and other geophysical applications where LES is essentially an operational tool. In engineering, some of the flows offering the best chances for early success are engine cylinder flows, unsteady flows over bluff bodies, and gas turbine flows.

There are two further issues. First, what is one to do for the majority of engineering flows for which LES is likely to remain too expensive for day-to-day use in the near future? In such cases, one can use RANS modeling for routine calculations but periodically update or tune the model parameters with the aid of data obtained from LES or laboratory experiments. Such a *zonal* model will be particular to the flows it is designed to simulate and care will be needed to ensure that it is more than a one-flow model. The large eddy simulations used in this type of model improvement can be either of building-block flows or of simplified versions of the complex flow. Methods of this kind are already in use in some industrial settings.

Second, if VLES is used for industrially important flows, in which much of the turbulence resides in the small scales, the quality of the SGS model becomes important. The grid size may become comparable to the integral scale of the turbulence. If this occurs, the latter can no longer be used as the length scale in the model because it would give an eddy visosity larger than the RANS eddy viscosity and the computed flow would be completely stable; that is, a Reynolds-averaged flow would be computed. It would then become necessary to introduce a means of predicting a model length scale and LES would then inherit many of the problems in RANS modeling.

## 2.4 Self-Consistent Parameter Determination

As noted above, an issue that has plagued LES from its inception is the choice of the model parameters. Both $C_S$ and $\Delta$ need to be prescribed but, despite considerable effort, except for simple flows, no agreed-upon procedures for doing so have been produced. A recent suggestion by Germano (1992) has produced hope of breaking this deadlock. We shall give only a brief introduction to his model, which has been termed the *dynamic model*; see Chapter 6 by Piomelli, this volume.

In Germano's approach, the definition (3) of the SGS stress is adopted and the LES equations of motion are filtered a second time to obtain equations for the doubly filtered velocity field $\tilde{\bar{u}}_i$. The tilde represents the second filter; a new symbol is used to emphasize that the second filter need not be the same as the first one. The equations obtained are similar to those produced by the first filtration except that $\tau_{ij}$ is replaced by a new SGS Reynolds stress:

$$T_{ij} = \overline{\widetilde{u_i u_j}} - \overline{\widetilde{u}}_i \overline{\widetilde{u}}_j \qquad (10)$$

The length scale associated with double filtration will be denoted by $\overline{\widetilde{\Delta}}$.

If the Smagorinsky model is applied to $T_{ij}$ as well as to $\tau_{ij}$ (with the length scale and strain rate appropriate to the double filter but the same model parameter $C_S$) it is not hard to show that

$$(T_{ij} - \tilde{\tau}_{ij})\overline{\widetilde{S}}_{ij} = -2C_S \left( \overline{\widetilde{\Delta}}^2 |\overline{\widetilde{S}}| \overline{\widetilde{S}}_{ij}\overline{\widetilde{S}}_{ij} - \overline{\Delta}^2 |\widetilde{\overline{S}}| \widetilde{\overline{S}}_{ij}\overline{\widetilde{S}}_{ij} \right) \qquad (11)$$

Here the absolute value of a tensor quantity is defined as the square root of the trace of the square of the tensor, that is, $|V| = \sqrt{V_{ij}V_{ji}}$. If one is doing LES based on the equations for $\overline{u}_i$, every quantity in Eq. (11) is computable and one can solve for the parameter $C_S$.

In principle, $C_S$ can be different at every point of the flow at every instant. However, because the quantity multiplying $C_S$ in Eq. (11) can be of either sign or zero and fluctuates considerably in both time and space, use of this equation can result in numerical instability and/or nonrealizable fields. For this reason, Germano et al. (1990) suggested that Eq. (11) be averaged; they used planar averaging for channel flow but other choices (such as time averaging) may be more appropriate in other flows. This model is relatively new and a number of issues need to be worked out.

As noted above, the second filter can be chosen arbitrarily, but if it is too narrow, Eq. (11) contains little information and is not reliable, and if it is too wide, important local information is averaged away. Germano et al. found $\overline{\widetilde{\Delta}}/\overline{\Delta} = 2$ to be optimum for channel flow including a transitional case.

Because changing the value of the parameter $C_S$ can compensate for an incorrect choice of the length scale $\Delta$, this model does away with the need to prescribe the length scale accurately. Furthermore, variation of the model parameter with nondimensional strain rate, Reynolds number, and other properties of the flow is accounted for automatically in this model. Thus many of the issues that have perplexed LES practitioners over the past decade may be rendered moot. However, further testing is needed before these issues can be laid to rest, though it is not clear whether the problems with complex high Reynolds number flow simulations will be resolved by using the dynamic model.

## 2.5 Wall Models

Another issue of great importance in the simulation of high Reynolds number and/or complex flows is modeling of the flow in the vicinity of a wall. This question receives less attention than SGS modeling because it is not as amenable to theoretical treatment, but the subject is at least as important. Before discussing the wall models, we shall review some findings obtained from experiments and direct and large eddy simulations made with no-slip wall boundary conditions.

It is well-known that shear flows near solid boundaries contain alternating streaks of high- and low-speed fluid. The streaks are thin, and if they are not adequately resolved, the turbulence energy production in the vicinity of the wall (which is a large fraction of the total energy production) is underpredicted (Kim and Moin, 1986). Generally, underprediction of the turbulence production results in reduction of the Reynolds stress and, thus, the skin friction. Consequently, some of the overall parameters of the flow will not be predicted correctly.

Simulations suggest that wall-region turbulence and turbulence far from the wall are relatively loosely coupled. Chapman and Kuhn (1986) showed that a simulation with an artificial boundary condition imposed at the top of the buffer layer (at approximately $y^+ = 100$) displayed most of the characteristics of the wall layer found in a simulation in which the entire flow is computed. Thus accurate prediction of the flow near the wall does not require accurate simulation of the outer flow.

On the other hand, Piomelli et al. (1987) and others have shown that, by using relatively crude lower boundary conditions to represent the effect of the wall region, one can accurately simulate the central part of the flow in a channel. Thus details of the flow in the wall region need not be known in order to simulate the outer region. We conclude that either region can be well-simulated if it is given the correct shear stress and a reasonable approximation of the fluctuations at the interface between it and the other zone.

These results suggest that useful simulations can be done without resolving the entire flow. This is important because a very fine grid in all directions is required to resolve the wall region, so if it can be represented via a model, huge savings can result. The earliest simulations of channel flow did this out of necessity and yielded reasonably good results. For rough walls, one has little choice but to use a model to represent the wall region.

Deardorff's (1970) original model contained weaknesses that were soon remedied by Schumann (1973); the latter's model, with some modifications, is still widely used. It assumes that the instantaneous velocity at the grid point nearest a wall is exactly correlated with the shear stress at the wall point directly below it:

$$\tau_w(x, z) = \frac{\overline{u}_1(x, y_1, z)}{U_1(y_1)} < \tau_w > \tag{12}$$

where $y_1$ is the height of the first grid point, $< \tau_w >$ is the mean wall shear stress, and $U_1(y_1)$ is the mean velocity at the first grid point. With the aid of this model, Schumann obtained considerably improved results.

Still another model was proposed by Mason and Callen (1986). They assumed that the logarithmic profile, which is known to represent the mean velocity profile in a region relatively far from the wall, also holds locally and instantaneously. This assumption is, of course, incorrect but their boundary condition is often used by meteorologists, who are quite satisfied with the results it produces. Piomelli et al. (1987) tested this model along with others and found it to be inadequate for engineering applications. This provides an excellent example of how the differing needs of two fields can lead to opposite conclusions relative to the effectiveness of a model.

Piomelli et al. (1987) used direct simulation results to test models for the wall layer; these included Schumann's model and two new models they constructed. The latter were based on insights into the physics of boundary layers gained from experiments and direct simulations.

The first model is based on the idea that Reynolds stress-producing events that originate near the wall do not move vertically away from the wall but, rather, at a small angle to the wall. This leads to the so-called shifted model, an improvement on Eq. (12),

$$\tau_w(x, z) = \frac{\overline{u}_1(x + \Delta_s, y_1, z)}{U_1(y_1)} < \tau_w > \tag{13}$$

where $\Delta_s = y_1 / \cos 8°$ is a spatial shift, $8°$ being the observed mean angle of event trajectories.

The second model is based on the observation that events containing significant Reynolds stress involve vertical movement, so that it is the vertical component of the velocity rather than the horizontal one that should be correlated with the wall shear stress:

$$\tau_w(x, z) = < \tau_w > - C u_r \overline{v}(x + \Delta_s, y_1, z) \tag{14}$$

Both of these models give improved agreement with experiments and direct simulations for channel flow, including cases with transpiration from the walls and high Reynolds number flows.

These models have been applied only to flows over flat walls with very mild pressure gradients. They are almost certainly inadequate for flows which separate and/or reattach or flows over complex-shaped walls. They may work in three-dimensional boundary layers because the flow angle changes slowly with distance from the wall in the lower part of the flow. In the author's opinion, no reliable simulations of fully turbulent flows in complex geometry have yet been made. Because experimental data are scarce and lack detail, the development of trustworthy methods for simulating these flows will probably require simulations with no-slip conditions. Simulations of this kind are currently under way at Stanford and other institutions.

## 2.6 Effects of Other Strains

Nearly everything in this chapter is directed toward engineering applications of LES in general and high-technology applications in particular. Furthermore, almost all the results cited are derived from theory and/or simulations of incompressible flows without "extra strains" such as stratification, rotation, compressibility and curvature. Obviously, there are flows in which extra strains are important and other fields in which LES can and does play an important role, including civil engineering and the geosciences. Indeed, there is such a wide variety of applications that it is impossible to cover them all here.

In these applications, the accuracy demanded of a calculation may be much lower than in mechanical and aeronautical engineering. This is an advantage but there are additional complexities that more than compensate for it. In the geosciences, the ratio of horizontal to vertical length scales is so large that it is impossible to use a grid fine enough to capture the length scales on which three-dimensional motions are important in global simulations. Consequently, the equations solved often depend on the scale treated (microscale, mesoscale, or global scale); the choice of governing equations on the largest scales is still something of an open question. These flows also are determined in an important way by extra strains, including the spherical shape of the planet, rotation, buoyancy, and phase change and the need to accurately predict a single realization of the flow, for example, the weather. It is clear that the issues are very different from those in engineering. Finally, the measured data are often incomplete and/or inaccurate, and laboratory experiments are often impossible.

Meteorologists and oceanographers (and, to a smaller extent, space scientists) concerned with the prediction of global circulation generally use equations that are essentially two-dimensional with a Smagorinsky eddy viscosity to parameterize the unaccounted-for motions. At smaller scales, truly three-dimensional equations may be used. Demanding that a model (with a single value of the parameter) account for the wide range of phenomena encountered in the geosciences may be asking too much. It would not be surprising if the model needed to be modified for different situations or if different parameters were needed on different scales. It appears that, for the most part, the models have been accepted without stringent testing; one could hardly have done more. When these items are taken into consideration, one should regard the results that have been achieved as nothing short of amazing.

It is the author's belief (he is largely an outsider in the geosciences and may be a bit naive) that a systematic approach may help to build a firmer foundation for modeling. The task is enormously difficult and progress will come slowly. An interesting possibility is the following. One can perform simulations of small-scale phenomena, for example, the planetary boundary layer or the mixed upper layer of the ocean (see Chapter 22 by Cane, this volume) and use the results to construct the parameterizations employed in simulations on the next larger scale. Many cases need to be run to ensure that the full range of parameters and all significant phenomena are included in the database. By bootstrapping in this way, and allowing two-way interaction between simulations at different scales, it may be possible to construct computer codes (called models in the geosciences) that allow the full gamut of phenomena on all scales to be predicted. There are many difficulties in this scenario for which solutions are not available; for example, it is not clear how large storms can be included in this procedure.

Some progress has been made in simulating flows that include extra strains. It appears that the strains can be roughly divided into two classes. Some strains, such as rotation, curvature, and stratification, affect the large scales much more strongly than the small scales. For flows governed by these strains, it is quite likely that the SGS models used for incompressible flows without extra strains can be used without major modification. For example, large eddy simulations of a stable planetary boundary layer made by Mason and Derbyshire (1990) agree very well with both direct simulations of the similar flows by Coleman et al. (1989) and field data.

On the other hand, for "strains" whose action is principally in the small scales, it is not yet clear what needs to be done. Compressible turbulence at low Mach numbers can be treated with incompressible models. At higher Mach numbers, small shock waves ("eddy shocklets") develop, and the flow behavior is quite different (Blaisdell et al., 1991). RANS models have been constructed for these flows by Zeman (1990), and it appears that similar ideas will be needed for SGS models if the flows are to be accurately simulated using LES.

In combusting flows, flames are normally thin with respect to even the smallest scales of the turbulence and LES is again very difficult. There are two possible types of LES. The first is applicable only when the chemistry is simple enough to be characterized by one or two constants. In this case, one can modify the constants (reaction rate and diffusivity) in a manner that increases the thickness of the flame while maintaining the flame speed constant. In the second approach, the flame is idealized as an infinitely thin sheet in the LES. The flame is then treated as a thin, flat sheet and the function of the SGS model is to account for the "wrinkles" that occur on the scales that are not resolved. Such a suggestion has been made by Ashurst et al. (1988).

## 2.7 Conclusions and Recommendations

After years of being regarded as a method of second choice relative to direct simulation, LES is receiving increased attention. The principal reasons are dissatisfaction with the current RANS turbulence models on the one hand and the inherent limitations of direct simulation on the other.

Improved models for both the small-scale turbulence and the wall layer are also needed if LES is to become a useful engineering tool. The recently proposed dynamic model offers promise of removing many of the difficulties that have plagued LES and give it an important advantage with respect to RANS modeling. Improved models for the wall region, especially for separating and reattaching flows, are needed just as badly and are an important subject for future research.

Finally, some "extra strains," namely those that mainly affect the large scales, appear to be relatively easy to incorporate into large eddy simulations; little if any modification of the SGS models is required. Others, which act on scales smaller than the Kolmogoroff scale, may require significant changes in the SGS model.

## Acknowledgments

The author wishes to thank the many students and colleagues who have contributed their hard work and insights to this work. In particular, colleagues Professor W. C. Reynolds and Professor P. Moin, former students Professor U. Piomelli and Dr. K. S. Yang, and present students M. Bohnert and K. Shah have made valuable contributions to the author's knowledge of this subject.

## References

ASHURST, W.T., SIVASHINSKY, G.I., AND YAKHOT, V. (1988) Flame front propagation in nonsteady hydrodynamic fields. *Comb. Sci. Tech.*, **62**, 273.

AUPOIX, B. (1987) Application de modeles dans l'espace spectral a d'autres niveaux de fermature en turbulence homogene. Dissertation, University of Claude Bernard-Lyon I.

BARDINA, J., FERZIGER, J.H., AND REYNOLDS, W.C. (1980) Improved subgrid models for large eddy simulation. AIAA Paper 80-1357.

BLAISDELL, G.A., MANSOUR, N.N., AND REYNOLDS, W.C. (1991) Numerical Simulations of Compressible Homogeneous Turbulence. Report TF-50, Stanford University, Dept. of Mechanical Engineering.

CHAPMAN, D.R. AND KUHN, G.D. (1986) The limiting behavior of turbulence near a wall. *J. Fluid Mech.*, **70**, 265.

COLEMAN, G.N., FERZIGER, J.H., AND SPALART, P.R. (1990) A numerical study of the turbulent Ekman layer. Report TF-48, Stanford University, Dept. of Mechanical Engineering.

DEARDORFF, J.W. (1970) A numerical study of three-dimensional turbulent channel flow at large Reynolds number. *J. Fluid Mech.*, **41**, 452.

DEARDORFF, J.W. (1973) Three dimensional numerical study of turbulence in an entrained mixing layer. In *AMS Workshop in Meteorology*, American Meteorological Society, p. 271.

GERMANO, M. (1992) Turbulence: the filtering approach. *J. Fluid Mech.*, **238**, 325.

GERMANO, M., PIOMELLI, U., CABOT, W.H., AND MOIN, P. (1991) A dynamic subgrid scale eddy viscosity model. *Phys. Fluids A*, **3**, 1760.

KAWAMURA, T., AND KUWAHARA, K. (1985) Direct simulation of a turbulent inner flow by a finite difference method. AIAA Paper 85-0376.

KIM, J.J., AND MOIN, P. (1986) The structure of the vorticity field in turbulent channel flow, Part 2: Study of ensemble averaged fields. *J. Fluid Mech.*, **162**, 339.

LEITH, C.E. (1990) Stochastic backscatter in a subgrid scale model–plane shear mixing layer. *Phys. Fluids A*, **2**, 297.

LEONARD, A. (1974) Energy cascade in large eddy simulations of turbulent fluid flows. *Adv. Geophys.*, **18A**, 237.

LESIEUR, M. (1990) *Turbulence in Fluids*, 2nd ed. Kluwer.

LILLY, D.K. (1965) On the computational stability of numerical solutions of time-dependent, nonlinear, geophysical fluid dynamic problems. *Mon. Wea. Rev.*, **93**, 11.

MASON, P.J., AND CALLEN, N.S. (1986) On the magnitude of the subgrid scale eddy-coefficient in large- eddy simulation of turbulent channel flow. *J. Fluid Mech.*, **162**, 439.

MASON, P.J., AND DERBYSHIRE, S.H. (1990) Large eddy simulation of the stably stratified atmospheric boundary layer. *Bound.- Layer Meteorol.*, **53**, 117.

McMILLAN, O.J., FERZIGER, J.H., AND ROGALLO, R.S. (1980) Test of new subgrid scale models in strained turbulence. AIAA Paper 80-1339.

PIOMELLI, U., FERZIGER, J.H., AND MOIN, P. (1987) Models for large eddy simulations of turbulent channel flows including transpiration. Report TF-31, Stanford University, Dept. of Mechanical Engineering.

SCHUMANN, U. (1973) Ein Untersuchung ueber der Berechnung der Turbulent Stroemungen im Platten- und Ringspalt-Kanelen. Dissertation, Karlsruhe University.

SMAGORINSKY, J. (1963) General circulation experiments with the primitive equations, Part I: The basic experiment. *Mon. Wea. Rev.*, **91**, 99.

YAKHOT, V., AND ORSZAG, S.O. (1986) Renormalization group methods in turbulence. *J. Sci. Comp.*, **1**, 3.

ZEMAN, O. (1990) Dilatation dissipation: The concept and application in modeling compressible mixing layers. *Phys. Fluids A*, **2**, 178.

# 3

## Some Basic Challenges
## for Large Eddy Simulation Research

STEVEN A. ORSZAG, ILYA STAROSELSKY,

AND VICTOR YAKHOT

### 3.1 Introduction

The Navier–Stokes equations give a coarse-grained continuum description of the many-particle dynamics of a fluid governed by Newton's equations of motion. In principle, it is possible to determine the behavior
of the fluid by simply solving Newton's equations for each of the fluid
molecules and then averaging the results over small volumes of space.
However, for realistic fluids such an approach is out of reach; furthermore, even if it could be done, most of the information would not be
of interest. The Navier–Stokes equations embody a qualitatively different understanding of the underlying physics than simply simulating
the molecular dynamics of the fluid. They provide a simplified, coarse-
grained description of the fluid flow which, despite its enormous simplicity compared with the molecular dynamics, contains all the relevant
physics of the fluid at scales much larger than the mean free path of
the molecules. The fundamental reason for the success of the Navier–
Stokes equations is the existence of a spectral gap between the scales of
molecular motions and the scales of the smallest excited eddies in fluid
flows.

The problem of obtaining a continuum description of turbulence is
different. It is commonly thought that large eddy simulation (LES) of
turbulence is a compromise between direct numerical simulation (DNS)
and Reynolds-averaged Navier–Stokes (RANS) solution of turbulence
transport models. In a RANS solution, all dynamical degrees of freedom smaller than the size of the largest (energy-containing) eddies are

averaged over, so there is no dynamical information about smaller scales. On the other hand, in DNS, all eddies down to the dissipation scale must be simulated with accuracy. LES seems to lie between the two extremes of DNS and RANS. In LES, a fine grid (with grid size $\Delta$) is used to calculate a system of modified Navier–Stokes equations in which eddies of size less than $O(\Delta)$ are removed from the dynamics. Thus, in LES eddies significantly larger than $\Delta$ are calculated in detail so their statistical properties (like correlation functions, structure functions and spectra) are computable. Eddies smaller than $\Delta$ are treated by turbulence transport modeling techniques so that the information available about them includes only the single-point quantities like the kinetic energy at subgrid scales and the dissipation at subgrid scales.

The key step in both LES and RANS is the derivation of the underlying dynamical equations averaged over small scales. The only difference between LES and RANS is the definition of a small scale; in LES, small scales are smaller than the grid size $\Delta$, while in RANS small scales are smaller than the largest eddies, of size $L$. In the equilibrium statistical mechanics of fluids, the distinction between small and large scales is clear – there is a spectral gap. In turbulent fluids, this distinction is no longer clear-cut because turbulence has broad spectral excitation. From this point of view, LES and RANS are equivalent, except for the location of the spectral filter. Indeed, if the grid size of an LES simulation is taken larger and larger, self-consistency requires that LES results approach the RANS results.

The absence of a spectral gap in turbulence gives stringent self-consistency requirements for LES and RANS. However, this is only part of the problem. Similar features may be found in the theory of critical phenomena and in high-energy physics where renormalization group and related nonlinear methods have been successful in describing experimental data. In a turbulent flow, large-scale structures are responsible for the very existence of turbulence since they participate in turbulence production. The problem is that there is no spectral gap between these structures and the secondary turbulent eddies produced by them. Thus, in the development of a subgrid or transport model, extreme care must be taken to design a procedure sensitive enough not to eliminate or blur these vital production mechanisms. A typical energy spectrum in a complex flow (representative of, for example, turbulent flow past a cylinder) has a sharp spike at some small frequency $\omega_0$. In the case of flow past a cylinder, such a spike would correspond to massive separation of large-scale vortices (vortex shedding), while the broad spectrum at higher frequencies describes the small-scale features of the flow. In order to describe the physics of this flow properly, neither LES nor RANS models should filter out or substantially blur the peak at $\omega_0$ and, hence, the essential physics of the flow. It is in fact a stringent requirement that the models be constructed in such a way that the low-frequency part of the spectra (including the peak at $\omega_0$) not be affected by the procedure used to remove unwanted degrees of freedom from the dynamical description of the flow.

It is both reasonable and realistic to expect that the removal of small-scale and high-frequency components of the dynamics of a flow be described in terms of modified molecular transport coefficients like an eddy viscosity. However, it may *not* be realistic to describe the low-frequency, large-scale features (like those associated with the peak at $\omega_0$) in terms of modified transport coefficients. Indeed, a spectral peak like that at $\omega_0$ reflects the large-scale production process itself. Turbulence theory, like that embodied in the renormalization group (RNG) analysis of turbulence (to be described below), can be effective in the removal of small-scale, high-frequency components from the dynamics, leading to dynamical equations for LES and RANS. Indeed, RNG turbulence theory is effective in the description of small scale-dynamics, energy cascade and related scaling behavior (like that associated with the Kolmogorov spectrum). However, it is not yet clear whether this approach is sufficient to construct models for production processes, like those embodied in the spectral peak at $\omega_0$. Other approaches based on theories like secondary instability and hyperscale instability may be necessary. To date, there has been little work on the development of such models of production processes. At the present time, it is still thought to be necessary to resolve the production processes in both LES and RANS. This can lead to stringent resolution requirements for both LES and RANS at large Reynolds numbers. If one wants to reduce these resolution requirements (by underresolving flow features responsible for production), it is necessary to construct theoretical models of these production processes, a challenge for the future. However, recent work suggests that RNG turbulence models may retain enough of the basic physics to enable us to describe these production processes (and coherent events).

In this chapter, we review some of the basic features of LES and the techniques used to derive the LES equations. In Section 3.2, we discuss the conditions for self-consistency of LES. Then, in Section 3.3, we review the RNG theory of the Navier–Stokes equations and the RNG-based derivation of a subgrid model for LES. In Section 3.4, we discuss the effect of rigid walls on LES. In Section 3.5, we illustrate the problematics of the LES modeling of complex flows on the example of a subgrid-scale model for turbulent combustion.

## 3.2 Self-Consistency Requirements of LES

The dynamical equations for LES are derived by applying a spatial filter or smoothing operator to coarse-grain the velocity field described by the Navier–Stokes equations. If we define $\mathcal{F}_\lambda$ to be a filter operator with length scale $\lambda$, then $\mathbf{v}_\lambda = \mathcal{F}_\lambda \mathbf{v}$ is the $\lambda$-filtered velocity (usually called the supergrid velocity if $\lambda = \Delta$, the grid scale). The object of the theory of LES is to derive a closed system of dynamical equations for $\mathbf{v}_\lambda$. In the next section we will show how this is done using RNG theory.

In the LES equations, the subgrid-scale velocity $\mathbf{v}_\lambda^{\text{sg}} = \mathbf{v} - \mathbf{v}_\lambda$ is not available explicitly and is, therefore, modeled. Since $\mathbf{v}_\lambda^{\text{sg}}$ is treated by a turbulence model, it is possible to evaluate its effects only in an average sense. One of the major self-consistency requirements for the subgrid filter $\mathcal{F}_\lambda$ is the independence of the results of the filter width $\lambda$. Furthermore, if $\lambda < \mu$ are two distinct filter widths, then the LES dynamical equations for $\mathbf{v}_\lambda$ should be such that $\mathcal{F}_\mu \mathbf{v}_\lambda$ has nearly the same statistical properties as $\mathbf{v}_\mu$. The latter test is particularly difficult to apply in practice and has yet to be done.

These self-consistency tests on LES have some deep implications. First, if the large eddy dynamical equations are to be integrated accurately, we should choose the grid size $\Delta$ much smaller than $\lambda$ so that differencing errors in the LES dynamical equations are small. In DNS it is necessary to choose the grid size $\Delta$ smaller than $O(\ell_d)$, where $\ell_d$ is the Kolmogorov dissipation scale. This requirement ensures that the smallest excited inertial-range eddies are adequately resolved by the grid. In RANS it is typical to choose the grid size $\Delta$ much larger than $\ell_d$ but small compared with the filter width $\lambda$, which is chosen to be an integral scale $L$ (characterizing the largest eddies in the flow). On the other hand, in LES, it is common to choose the filter width of order two to three times the grid size $\Delta$, which itself is much larger than $\ell_d$, but much smaller than $L$.

The self-consistency requirements discussed above come into play if we realize that the filter width $\lambda$ may be chosen anywhere in the range from $\ell_d$ to $L$. For $\lambda < O(\ell_d)$ DNS is obtained, $\lambda = O(L)$ results in a RANS model, while an intermediate $\lambda$ produces LES. The fundamental question of the self-consistency of LES modeling is the independence of the predicted results of $\lambda$ as it varies over this range. If the results were truly independent of $\lambda$, one could confidently calculate the solution of RANS equations to predict the large-scale properties of turbulence. If there is no range of $\lambda$ where results are $\lambda$-independent, one must calculate DNS solutions to get a valid description of turbulence. The basis for current interest in LES is the hope that there is indeed a broad range of $\lambda$ across which there exists statistical independence of the properties of $\mathbf{v}_\lambda$. The verification of this assumption remains a challenge for the future.

## 3.3 RNG Derivation of LES Dynamics

Here we give a brief review of the renormalization group theory of the forced Navier–Stokes equations (Yakhot and Orszag, 1986a, 1986b, 1987). We will show how this theory can be used to eliminate small scales from the dynamics, thus leading to both LES and RANS models. We shall also show that the resulting dynamical equations for LES must be solved with sufficient resolution to resolve turbulence production mechanisms, in particular wall streaks in turbulent channel flows. At high Reynolds numbers $R$, it is well known that DNS requires $O(R^{9/4})$ storage and $O(R^3)$ work. LES *does* reduce these storage and work requirements somewhat. However, if wall streaks must be resolved, these work requirements are reduced only to $O(R^2)$. In order to calculate high Reynolds number flows with more modest storage and work requirements, it will be necessary to model the production mechanisms themselves.

### 3.3.1 Introduction to RNG Methods for Turbulence

The RNG model is derived for a Newtonian incompressible fluid in an infinite domain, stirred by a Gaussian random force $\mathbf{f}$,

$$\frac{\partial v_i}{\partial t} + v_j \nabla_j v_i = -\nabla_i p + \nu_o \nabla_j \nabla_j v_i + f_i,$$
$$\nabla_i v_i = 0, \tag{1}$$

where $\nu_o$ is the molecular viscosity; the density, $\rho$, has been absorbed into the pressure, $p$. The stirring force is defined by its two-point correlation function in wavevector and frequency space,

$$\langle \hat{f}_i(\mathbf{k},\omega)\hat{f}_j(\mathbf{k}',\omega')\rangle = \begin{cases} 2D_o(2\pi)^{d+1}k^{-y}P_{ij}(\mathbf{k})\delta(\mathbf{k}+\mathbf{k}')\delta(\omega+\omega'), \\ \qquad\qquad \Lambda_L < k < \infty, \\ 0, \qquad\quad 0 < k \le \Lambda_L, \end{cases} \tag{2}$$

where $\mathbf{k}$ is the wavevector, $\omega$ is the frequency, $k = |\mathbf{k}|$ and $d$ is the dimension of space. The parameter $y$ is chosen to give the Kolmogorov form of the energy spectrum and in three dimensions $y = d = 3$. Statistical homogeneity in space and time is guaranteed through $\delta(\mathbf{k}+\mathbf{k}')\delta(\omega+\omega')$. The projection operator $P_{ij}(\mathbf{k}) = \delta_{ij} - k_i k_j/k^2$ renders the force statistically isotropic and divergence-free. In the limit of an infinite Reynolds number, the integral scale $\Lambda_L \to 0$.

Small-scale fluctutations are eliminated from explicit appearance in (1) by averaging over the force field at small scales. The force given by (2) is postulated to represent the average effect of the large-scale features of the flow, including the effect of initial and boundary conditions. After the elimination of small scales is completed, the force is dropped from the resulting equations of motion for the large scales, and initial and boundary conditions are restored. Statistical properties of turbulence obtained from numerical solution of Eq. (1) subject to the forcing defined by Eq. (2) are in good agreement with experimental data (Panda et al., 1989). This shows that (1)–(2) is a reasonable model of turbulence.

There are also strong fundamental considerations which support the correspondence principle of Yakhot and Orszag (1986a). In a recent work, A. Migdal, S. Orszag and V. Yakhot (unpublished manuscript) showed that the renormalization procedure and elimination of infrared divergencies generate naturally the intrinsic stirring force with algebraic spectrum (2), if one starts with a random stirring force acting only at the largest scales of the flow. In fact, they used the Gel'fand representation of the three-dimensional delta function,

$$\delta(\mathbf{k}) = \lim_{\epsilon \to 4} \frac{(4-\epsilon)k^{1-\epsilon}}{4\pi},$$

to show the equivalence of force at $k \approx 0$ to an algebraic force of the form (2).

The Fourier transform of (1) leads to

$$\hat{v}_i(\hat{k}) = G^o(\hat{k})\hat{f}_i(\hat{k}) - \frac{i\lambda_o}{2}G^o(\hat{k})P_{imn}(\mathbf{k})\int \hat{v}_m(\hat{q})\hat{v}_n(\hat{k}-\hat{q})\frac{d\hat{q}}{(2\pi)^{d+1}}, \tag{3}$$

where $\hat{k} = (\mathbf{k}, \omega)$, $G^o(\hat{k}) = (-i\omega + \nu_o k^2)^{-1}$, $\lambda_o = 1$, $P_{imn}(\mathbf{k}) = k_m P_{in}(\mathbf{k}) + k_n P_{im}(\mathbf{k})$ and $d\hat{q}$ denotes the $(d + 1)$-dimensional integral over the wavevector and frequency components of $\hat{q}$. Equation (3) is used for $0 < k < \Lambda_o$, where $\Lambda_o$ is an ultraviolet cutoff beyond which the viscosity coincides with the molecular viscosity $\nu_o$.

The details of the RNG scale-elimination procedure have been described in Yakhot and Orszag (1986a). The velocity modes with wave numbers in a narrow band below the cutoff are expanded in a perturbation series in powers of the local Reynolds number at the cutoff. This perturbation expansion is substituted into the equations for the velocity modes $\hat{v}^<(\hat{k})$ which correspond to large scales $k \to \Lambda_L \to 0$ and long times $\omega \to 0$. Next, an ensemble average over the force field in the high wave number band eliminates the small scales from the equations for the large scales. The large-scale variables are assumed to be independent of the ensemble average over the force at high wave numbers, and this condition is nearly satisfied for $\hat{k} \to 0$. Modifications to the long-time, large-scale equations include a correction to the molecular viscosity.

The procedures of scale separation, substitution and averaging are iterated in order to remove more and more small scales from the equations for the large scales. Corrections to the molecular viscosity accumulate with each iteration, and in this way, a cutoff-dependent eddy viscosity $\nu(\Lambda)$ is generated.

The scale-elimination procedure leads to exact results for large scales and long times in the limit $\epsilon \to 0$, where $\epsilon = 4 + y - d$. In this case, the equations for the large-scale velocity components $\mathbf{v}^<$ are found at lowest order in the expansion in powers of $\epsilon$,

$$\frac{\partial v_i^<}{\partial t} + v_j^< \nabla_j v_i^< = -\nabla_i p^< + \nu(\Lambda)\nabla_j \nabla_j v_i^< + F_i^< + f_i^<, \qquad (4)$$

where wave numbers above $\Lambda$ have been removed, $\nu(\Lambda)$ is the effective viscosity and $\mathbf{F}$ is an induced force acting at large scales.

The differential equation for the effective viscosity $\nu(\Lambda)$ is found from repeated removal of infinitesimal shells. At lowest order in $\epsilon$,

$$\frac{d\nu(\Lambda)}{d\Lambda} = -\frac{A_d^0}{2} \frac{\nu(\Lambda)\bar{\lambda}^2(\Lambda)}{\Lambda}, \qquad (5)$$

$$\bar{\lambda}^2(\Lambda) = \frac{2D_o S_d}{(2\pi)^d} \frac{1}{\nu^3(\Lambda)\Lambda^\epsilon}, \qquad (6)$$

where $A_d^0 = (d^2 - d)/[2d(d + 2)]$, $S_d$ is the area of a unit sphere in $d$-dimensions and $\bar{\lambda}(\Lambda)$ is the effective Reynolds number of the flow at the

wave number $\Lambda$, based on the modified viscosity $\nu(\Lambda)$. The solution to
(5) with initial condition $\nu(\Lambda_o) = \nu_o$ is

$$\nu(\Lambda) = \nu_o\left(1 + \frac{3A_d^0}{\epsilon\nu_o^3}\frac{D_oS_d}{(2\pi)^d}(\Lambda^{-\epsilon} - \Lambda_o^{-\epsilon})\right)^{1/3}. \tag{7}$$

Substituting (7) into (6) one finds

$$\bar{\lambda}^2 = \frac{\epsilon}{3A_d^0} \tag{8}$$

for $\Lambda \ll \Lambda_o$. Since the effective Reynolds number is proportional to $\epsilon^{1/2}$
at lowest order in $\epsilon$, all perturbation expansions used in the theory are
valid at lowest order in $\epsilon$ for $\epsilon \to 0$.

Also at lowest order in $\epsilon$, the induced force $\mathbf{F}$ is Gaussian and is given
by its two-point correlation function $(k \ll \Lambda)$

$$\langle \hat{F}_i(\hat{k})\hat{F}_j(\hat{k}')\rangle = D'2D_o(2\pi)^{d+1}k^2P_{ij}(\mathbf{k})\delta(\hat{k} + \hat{k}'), \tag{9a}$$

$$\frac{dD'(\Lambda)}{d\Lambda} = -\frac{d^2 - 2}{4d(d+2)}\frac{\bar{\lambda}^2(\Lambda)}{\Lambda^{\epsilon-1+d}}. \tag{9b}$$

The solution to (9b) is found using (5) and (6), and will be discussed
later.

The induced force $\mathbf{F}$ with the correlation function $\langle \hat{F}_i\hat{F}_j\rangle \propto k^2$ is negligible compared with the bare force $\mathbf{f}$ in the range $\Lambda_L k < \Lambda_o$ where the
bare force exists. However, $\mathbf{F}$ is not negligible in the final equations for
the large scales $\mathbf{v}^<$ when $\mathbf{f}$ is dropped and initial and boundary conditions are restored. Thus $\mathbf{F}$ will be important for large eddy simulations.

The unknown amplitude $D_o$ was found by Dannevik et al. (1987)
from overall energy conservation by the nonlinear terms. They found the
equation for the energy spectrum tensor $E_{ij}(k)$ from the nondimensional
form of Eq. (4) with the cutoff $k = \Lambda$. In that case the Eq. (4) has an
effective Reynolds number $\bar{\lambda} \propto \epsilon^{1/2}$. At second order in the $\epsilon$-expansion
they recovered the EDQNM equations introduced by Orszag (1970).
These equations express the mean dissipation rate $\bar{\mathcal{E}}$ in terms of the
lowest-order energy spectrum $E(k)$ and the effective viscosity $\nu(k)$,

$$\bar{\mathcal{E}} = \bar{\mathcal{E}}\Big(\nu(k), E(k)\Big). \tag{10}$$

Combining (9) and (10), they derived the relation

$$1.59\bar{\mathcal{E}} = \frac{2D_oS_d}{(2\pi)^d}. \tag{11}$$

Using relation (11), one finds for $\Lambda_L < k$

$$E(k) = 1.61\bar{\mathcal{E}}^{2/3}k^{-5/3}, \tag{12}$$

$$\nu(k) = 0.49\bar{\mathcal{E}}^{1/3}k^{-4/3}. \tag{13}$$

The coefficients of (12) and (13) are based on the lowest order of expansion in powers of $\epsilon = 4$, and agree well with the experimentally known parameters.

### 3.3.2 The Subgrid Model

The RNG subgrid-scale model is found from the full solution (7) for $\nu(\Lambda)$ and the induced force $\mathbf{F}$ given by (9). Defining the filter width $\Delta = 2\pi/\Lambda$ and using (11) for $D_o$ in (7), the expression for subgrid-scale eddy viscosity can be written as

$$\nu(\Delta) = \nu_o \left[ 1 + H \left( \frac{0.12\bar{\mathcal{E}}}{\nu_o^3 (2\pi)^4} \Delta^4 - C \right) \right]^{1/3}. \tag{14}$$

Here, the Heaviside function $H(x)$ is defined as

$$H(x) = \begin{cases} x, & x > 0, \\ 0, & x < 0, \end{cases} \tag{15}$$

and

$$C = \frac{\bar{\mathcal{E}}(2\pi)^d}{2D_o S_d} \frac{3A_d^0}{8} \gamma^{-4}, \tag{16}$$

$$\Lambda_o = \gamma \left( \frac{\bar{\mathcal{E}}}{\nu_o^3} \right)^{1/4}. \tag{17}$$

The value $\gamma = 0.2$ is taken on the basis of experimental data for grid turbulence and, together with (13) for $\bar{\mathcal{E}}/D_o$, yields $C = 73.5$.

There is a jump discontinuity in $\nu(\Delta)$ at the grid scale $\Delta = 2\pi/\Lambda_o$, represented by the Heaviside function, because the spectrum $E(k)$ was assumed to change abruptly from the Kolmogorov form for $k \leq \Lambda_o$ to $E(k) = 0$ for $k > \Lambda_o$. The advantage of (14) is its ability to turn on in regions of high turbulence activity, where the argument of the Heaviside function is positive. When the model turns off and $\nu(\Delta) = \nu_o$, the simulation becomes a DNS and the exact dynamical equations are solved. This feature of the model is very useful for the simulation of transitional flows.

Equation (14) can be further simplified with an approximate expression for $\bar{\mathcal{E}}$ in terms of the large-scale flow

$$\bar{\mathcal{E}} \equiv \frac{\nu_o}{2} < (\nabla_j v_i + \nabla_i v_j)^2 > \approx \frac{\nu}{2} (\nabla_j v_i^< + \nabla_i v_j^<)^2. \tag{18}$$

Substitution of (18) into (14) yields an implicit expression for $\nu(\Delta)$,

$$\nu(\Delta) = \nu_o \left[ 1 + H \left( \frac{0.12\Delta^4}{2\nu_o^3 (2\pi)^4} \nu(\Delta)(\nabla_j v_i^< + \nabla_i v_j^<)^2 - C \right) \right]^{1/3}. \tag{19}$$

In the regions of a high Reynolds number flow, where the filter width is much larger than the smallest scales of the order of the Kolmogorov length scale, the Smagorinsky formula is recovered,

$$\nu(\Delta) = c_s \Delta^2 S, \tag{20}$$

$$S_{ij} = \tfrac{1}{2}(\nabla_j v_i^< + \nabla_i v_j^<); \quad S = (S_{ij} S_{ij})^{1/2}, \tag{21}$$

where $c_s = 0.003$.

Solution of (19) in the form of a cubic equation for $\nu$,

$$\nu^3 = \nu_o^3 + H\left(\frac{0.12\Delta^4}{2(2\pi)^4}\nu(\nabla_j v_i^< + \nabla_i v_j^<)^2 - C\right), \tag{22}$$

is problematic due to the unphysical roots of (22). An alternative formulation uses, instead of (18), the approximation for $\bar{\mathcal{E}}$,

$$\bar{\mathcal{E}} \approx \frac{\tau_e^2}{\nu}, \tag{23}$$

where $\tau_e$ is an eddy subgrid-scale stress,

$$\tau_e = -\nu S. \tag{24}$$

Substituting (24) into (14) results in the quartic equation for $\nu$,

$$\nu^4 = \nu_o^3 \nu + H\left(\frac{0.12\Delta^4}{2(2\pi)^4}\tau_e^2 - C\right), \tag{25}$$

which has only one real root.

It is now generally recognized as important to account for energy transfer from small to large scales in LES, especially in wall regions and in transitional flows (Zang and Piomelli, Chapter 11, this volume; Piomelli and Zang, 1991). As discussed above, the removal of small scales $\Lambda < k < \Lambda_o$ results in the induced force $\mathbf{F}$,

$$F_i(\mathbf{x}, t) = \int \hat{F}_i(\hat{k})e^{i\mathbf{k}\cdot\mathbf{x} - i\omega t}\frac{d\hat{k}}{(2\pi)^{d+1}}, \tag{26}$$

and the two-point correlation in wavevector-frequency space is given by (9). The force $\mathbf{F}$, along with a constant eddy viscosity, yields an energy spectrum proportional to $k^2$ in the energy-containing range. The existence of a $k^2$ spectrum at low wave numbers is likely to occur in nature, because a uniform distribution of energy among Fourier modes leads to a $k^2$ spectrum (Kraichnan, 1959). Thus any uniform distribution of energy at low $k$ due to initial conditions gives rise to a $k^2$ spectrum. Furthermore, a $k^2$ spectrum will persist for all time (in a decreasing wave number region above $k = 0$), since the transfer of energy to low wave numbers from the remaining supergrid wave numbers scales as $k^4$ (Proudman and Reid, 1954).

The amplitude $A_F(k, \Delta, \Delta t)$ of the force $\hat{\mathbf{F}}(\hat{k})$ is obtained from (9),

$$A_F(k, \Delta, \Delta t) = \left( D' 2 D_o (2\pi)^{d+1} k^2 \Delta \Delta t \right)^{1/2}, \qquad (27)$$

where $\Delta t$ is the numerical time step. The realizations of $\hat{\mathbf{F}}(\hat{k})$ must render the force random, isotropic and divergence free.

To find the coefficient $D'$ for both high and low Reynolds number flow regions, the differential equation (9b) is integrated using (5), (6) and the initial condition $\nu(\Lambda_o) = \nu_o$. Using Eq. (11) for $D_o$ and changing variables

$$dz = \frac{1}{\nu_o} \frac{d\nu}{d\Lambda} \, d\Lambda, \qquad (28)$$

one finds for $D'$

$$D' = 0.07 \frac{\bar{\mathcal{E}}}{\nu_o^3} \left( \frac{\nu_o^3}{0.12\bar{\mathcal{E}}} \right)^{9/4} \int_1^{\nu/\nu_o} \frac{(z^3 + C - 1)^{5/4}}{z} \, dz, \qquad (29)$$

where $C = 74.5$. Using the expression (23) for $\bar{\mathcal{E}}$ it is possible to rewrite Eq. (29) in terms of $\nu_o$, $\nu$ and $v_i^<$.

In summary, the RNG subgrid model consists of the quartic equation for $\nu(\Delta)$ given by (25) and the induced force $\mathbf{F}$ with the amplitude given by (23), (27) and (29). The effective viscosity $\nu(\Delta)$ goes abruptly to $\nu_o$ for $\Delta$ in the dissipation range because the dissipation-range energy spectrum was assumed to be zero. This feature ensures that $\nu(\Delta) = \nu_o$ at walls and enables the model to capture the spotty nature of wall turbulence by returning the exact dynamical equations in the case when $\Delta < 2\pi/\Lambda_o$. The induced force accounts for energy transfer from small to large scales. The induced energy spectrum proportional to $k^2$ implies uniform distribution of energy at large scales, which is likely to be found in nature.

### 3.3.3 LES of Channel Flow with the RNG Subgrid Model

In order to test the RNG LES model, A. Yakhot et al. (1989) performed several simulations of turbulent channel flow for different Reynolds numbers. Here we will briefly describe the results reported in A. Yakhot et al. (1989). Three runs were performed.

**Run 1.** $R_* = 125$ (DNS), with the computational box $l_x/R_* \times l_y/R_* \times l_z/R_* = 5 \times 5 \times 2$; Fourier–Chebyshev modes: $2N \times 2M \times P = 16 \times 64 \times 32$; grid spacing in streamwise and spanwise directions: $\Delta x = 39$, $\Delta y = 10$; initial conditions: perturbed Poiseuille laminar profile, $R_*^{(0)} = 125$, integration time: $T_* = 2000$, time step $\Delta t_* = 0.05$.

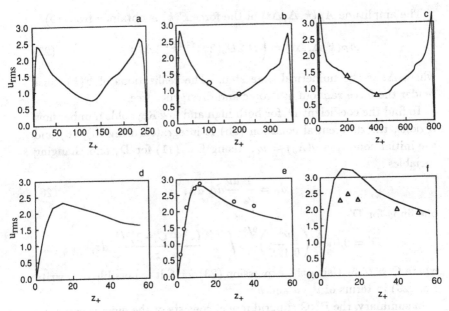

Fig. 1   Turbulent intensities $u_{rms}$ in channel flow: (a,d) $R_* = 125$; (b,e) $R_* = 185$; (c,f) $R_* = 400$. ○ , data from Kreplin and Eckelman (1979) for $R_c = 3850$; △ , data from Hussain and Reynolds (1975) for $R_c = 13,800$.

**Run 2.** $R_* = 185$ (LES), with the computational box $l_x/R_* \times l_y/R_* \times l_z/R_* = 5 \times 7 \times 2$; Fourier–Chebyshev modes: $2N \times 2M \times P = 16 \times 64 \times 64$; grid spacing in streamwise and spanwise directions: $\Delta x = 60$, $\Delta y = 20$; initial conditions: perturbed Poiseille laminar profile, $R_*^{(0)} = 185$; integration time: $T_* = 2600$, time step $\Delta t_* = 0.05$.

**Run 3.** $R_* = 400$ (LES), with the computational box $l_x/R_* \times l_y/R_* \times l_z/R_* = 5 \times 5 \times 2$; Fourier–Chebyshev modes: $2N \times 2M \times P = 16 \times 64 \times 64$; grid spacing in streamwise and spanwise directions: $\Delta x = 125$, $\Delta y = 31$; initial conditions: data of $R_* = 185$ at the steady state; integration time: $T_* = 2000$, time step $\Delta t_* = 0.05$.

The LES Runs 2–3 had been tested for dependence on the initial data. No differences had been found in the statistically steady velocity field starting from a laminar profile or from the DNS Run 1.

Case 2 was run for the same conditions as the direct simulation reported by Kim et al. (1987). This latter calculation was done using $192 \times 160 \times 129$ grid points, and $\Delta x = 12$, $\Delta y = 7$. In the LES simulations, about 60 times fewer grid points have been used.

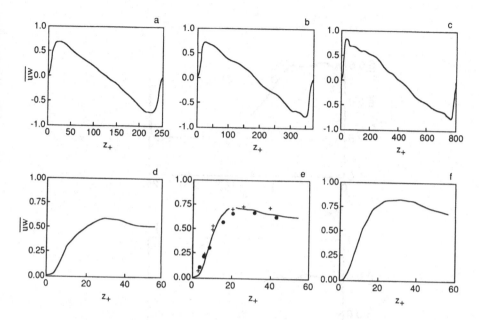

Fig. 2   Reynolds stess $\overline{uw}$ in channel flow: (a,d) $R_* = 125$; (b,e) $R_* = 185$;
(c,f) $R_* = 400$. •, +, data from Eckelman (1974) for $R_* = 142$ and
205, respectively.

Some of the key results are as follows:

**a. Mean Properties.** The profiles of the mean velocity normalized
by the wall-shear velocity demonstrate a very well pronounced logarithmic layer in LES simulations. The friction coefficient $C_f$, bulk mean
velocity $U_m$ and centerline velocity $U_c$ normalized by the wall-shear velocity $u_*$ had been compared with the experimental correlations proposed by Dean (1978). The agreement with the experimental data was
good.

**b.   Turbulent Intensities and Reynolds Shear Stress.** Turbulence intensities normalized by the friction velocity $u_*$ are shown in
Fig. 1, a–c. The subscript rms indicates root-mean-square averaging
taken over the horizontal $x, y$-plane. The Reynolds stress is plotted in
Fig. 2, a–c, and the correlation coefficient $\overline{uw}/u_{rms}w_{rms}$ is plotted in
Fig. 3. Note that at $R_* = 185$ (Run 2) the intensities are in good
agreement with available experimental data.

**c. Near-Wall Behavior of Turbulent Intensities and Reynolds
Stress.** The turbulent intensities in the near-wall region are shown in
Fig. 1, d–f and the Reynolds stress is plotted in Fig. 2, d–f. One can

*Steven A. Orszag et al.*

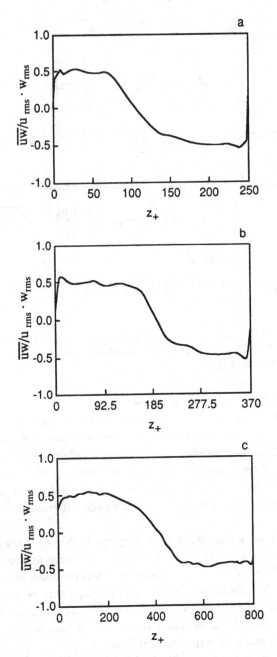

Fig. 3  Correlation coefficient $\overline{uw}/u_{rms}w_{rms}$ in channel flow: (a) $R_* = 125$;
(b) $R_* = 185$; (c) $R_* = 400$.

Table 1. *Near-wall behavior of turbulent intensities and Reynolds stress*

| $z_+$ | $u/z_+$ | $v/z_+$ | $w/(z_+^2 \times 10^3)$ | $\overline{uw}/(z_+^3 \times 10^3)$ | $\overline{uw}/u_{rms}w_{rms}$ |
|-------|---------|---------|-------------------------|--------------------------------------|-------------------------------|
| | | | $R_* = 125$, DNS | | |
| 2.402 | 0.346 | 0.156 | 7.588 | 11.92 | 0.454 |
| 5.383 | 0.315 | 0.116 | 5.514 | 8.37 | 0.482 |
| 9.515 | 0.234 | 0.082 | 3.507 | 4.48 | 0.546 |
| | | | $R_* = 185$, LES | | |
| 0.891 | 0.405 | 0.235 | 9.460 | 13.34 | 0.348 |
| 2.002 | 0.406 | 0.208 | 10.722 | 13.25 | 0.307 |
| 3.555 | 0.403 | 0.178 | 5.890 | 11.54 | 0.486 |
| 5.544 | 0.379 | 0.149 | 4.904 | 10.00 | 0.538 |
| 7.966 | 0.325 | 0.122 | 3.914 | 6.99 | 0.550 |
| | | | $R_* = 400$, LES | | |
| 1.926 | 0.451 | 0.209 | 4.905 | 10.15 | 0.46 |
| 4.329 | 0.445 | 0.172 | 5.695 | 11.89 | 0.47 |
| 7.686 | 0.377 | 0.137 | 3.009 | 5.80 | 0.51 |

see that the locations of the $u$-intensity and the Reynolds stress maxima are at about 15 and 20–30, respectively, in good agreement with known experimental data. From Fig. 2 it could be expected that $u \approx a_1 z_+$, $v \approx b_1 z_+$, $w \approx c_1 z_+^2$, $\overline{uw} \propto d_1 z_+^3$. The calculated values of $a_1, b_1, c_1, d_1$ are presented in Table 1.

Although the agreement with experimental data is rather good for all Reynolds numbers considered, it was mentioned that the resolution problems had begun to appear in the $R_* = 400$ run. Indeed, runs at $R_* = 1000$ were not successful unless the resolution was increased. To see the effects of resolving the wall streaks, three simulations were performed at $R_* = 585$ using different spanwise resolutions, $N_y = 32, 64, 128$. The skin friction coefficient was measured for each of these three runs and compared with the observed experimental value $C_f^{(exp)}$. In the high-resolution case, $N_y = 128$, $C_f$ approaches the experimental value to within about 2%. On the other hand, when $N_y = 64$, the skin friction is underpredicted by about 20% and when $N_y = 32$ the skin friction is underpredicted by about 40%. Inadequate spanwise resolution leads to a poor description of wall streaks and, hence, turbulence production.

Since the scale of a wall streak was $O(100)$ spanwise and $O(1000)$ streamwise in wall units, it was necessary to use $O(R)$ grid points spaced in both streamwise and spanwise directions. In the direction transverse to the wall, the streaks were of size $O(10)$ wall units, but in this direction it is common to use a highly nonuniform grid, so there are only weak resolution requirements in the direction normal to the wall. We conclude that resolution of wall streaks requires storage and work resources of order $R^2$, somewhat better than for DNS, but still prohibitive at large $R$. Further reduction in these computational requirements would rely on a model for the turbulence production process. The absence of such models limits the ability of the LES approach to compute high Reynolds number flows.

### 3.4 Wall Functions

Since the first LES of a turbulent channel flow, conducted by Deardorff (1970) more than twenty years ago, there has been an effort to develop improved subgrid models. In his simulations, Deardorff (1970) used the Smagorinsky (1963) subgrid model which is based on the following assumptions:

1. The flow is homogeneous and isotropic.
2. The Kolmogorov hypotheses are valid and the energy spectrum of the velocity fluctuations in the inertial range is given by

$$E(k) = C_K \bar{\mathcal{E}}^{2/3} k^{-5/3} \tag{30}$$

for $1/L < k < l_d$, and the integral and dissipation scales approach the limits $L \to \infty$ and $l_d \to 0$. The mean dissipation rate of energy is $\bar{\mathcal{E}}$ and the Kolmogorov constant is $C_K$.

3. The mean dissipation rate of energy $\bar{\mathcal{E}}$ is given by

$$\bar{\mathcal{E}}(\Delta) = \nu(\Delta) S_{ij} S_{ij}, \tag{31}$$

where $S_{ij}$ is the rate of strain tensor averaged over the scales $l < \Delta$ and the effective viscosity $\nu(\Delta)$ reflects the effect of the small-scale fluctuations with $l < \Delta$ on the large-scale components of the velocity field.

Using assumptions 1–3 and evaluating $\bar{\mathcal{E}}$ according to

$$\bar{\mathcal{E}}(\Delta) = 2\nu(\Delta) \int_0^{2\pi/\Delta} k^2 E(k) dk = \frac{3}{2} \nu(\Delta) C_K \bar{\mathcal{E}}^{2/3}(\Delta) \left(\frac{2\pi}{\Delta}\right)^{4/3}, \tag{32}$$

one finds that

$$\nu(\Delta) = c_s \Delta^2 S, \qquad (33)$$

where $S = (S_{ij}S_{ij})^{1/2}$ and the Smagorinsky constant $c_s = (2\pi)^{-2}$ $[2/(3C_K)]^{3/2}$. The fact that $c_s$ is a constant independent of $\Delta$ is a consequence of the Kolmogorov spectrum (30). Departures from the Kolmogorov spectrum, for example in the dissipation range, would render $c_s$ dependent on $\Delta$.

Large eddy simulations with the Smagorinsky subgrid model (33) do not represent wall turbulence accurately. This is easily understood in terms of the differences in the behavior of the true turbulence stress tensor, $\overline{v_i v_j}$, and the model subgrid-scale stress tensor, $\tau_{ij} - \delta_{ij}\tau_{ll}/3 = -\nu(\Delta)S_{ij}$. The normal component $v$ of the fluctuating velocity approaches zero at the wall as $y^2$, whereas the components $u$ and $w$, parallel to the wall, scale with $y$, the distance from the wall. Thus, eddies near the wall "flatten" towards the horizontal plane $(x, z)$, so that $\overline{uv} \sim y^3$. The model subgrid-scale stress does not reflect this wall anisotropy and, in general, does not scale correctly with the distance from the wall. For example, for Chebyshev grid spacing in the $y-$direction and constant grid spacing in the $x-$ and $z-$directions, the choice $\Delta = (\Delta_x \Delta_y \Delta_z)^{1/3}$ gives $\tau_{12} = \nu(\Delta)S_{12} \sim y^{2/3}$ while the choice $\Delta = (\Delta_x^2 + \Delta_y^2 + \Delta_z^2)^{1/2}$ gives $\tau_{12}$ constant at the wall.

To better predict wall regions in simulations of channel flow, Moin and Kim (1982) used

$$\nu(\Delta) = c_s f(y)\Delta^2 S, \qquad (34)$$

where the wall function $f(y)$ monotonically increases with $y$. The problem with this type of wall function is that it does not account for the strongly inhomogeneous, "spotty" nature of the wall turbulence, and therefore underpredicts $\nu(\Delta)$ at locations with increased turbulence activity.

### 3.4.1 The Dynamic Model

In a recent attempt to improve the Smagorinsky model, Germano et al. (1991) have used the relation

$$\tau_{ij}(\Delta) - \frac{\delta_{ij}\tau_{ll}(\Delta)}{3} = -\nu(\Delta)S_{ij}, \qquad (35)$$

where, again, $S_{ij}$ is the rate of strain tensor averaged over the scales $l < \Delta$ and $S = (S_{ij}S_{ij})^{1/2}$. They introduce a second filter width $\Delta_2 > \Delta$. The operator $\mathcal{L}_{ij} = \tau_{ij}(\Delta_2) - \overline{\tau_{ij}(\Delta)}$, where filtering over the scales $l < \Delta_2$ is denoted by overbar, is used to express the coefficient $c_s$ in (34) as

$$c_s = -\frac{\mathcal{L}_{ij}S_{ij}}{\Delta_2^2\bar{\bar{S}}\bar{S}_{mn}S_{mn} - \Delta^2\overline{SS_{pq}}S_{pq}}, \tag{36}$$

where $\mathcal{L}_{ij}$ is the stress tensor of the resolved scales $\Delta_2 > l > \Delta$. The name "dynamic model" comes from the fact that the kinematic identity (36) is used to calculate $c_s$ as a function of space and time, without invoking any dynamical constraints additional to those implied by assumptions 1–3 of Section 3. If these assumptions were exact the ratio (36) would be constant.

Strictly speaking, relation (36) can be used only if both $\Delta$ and $\Delta_2$ are in the inertial range of scales. Nevertheless, Germano et al. (1991) extrapolated (36) into wall regions of turbulent flows where the filter widths are much smaller than the dissipation scale $l_d$, which is of the order of the distance from the wall to the buffer region. In their simulations, Eq. (36) produced satisfactory results when

1. the filter widths were both in the inertial range of scales; as already discussed, $c_s$ is constant in this case;
2. the filter widths were both in the dissipation range of scales; then $\mathcal{L}_{ij}$ very rapidly approaches zero since $\tau_{ij}(\Delta_2) \approx \overline{\tau_{ij}(\Delta)} \approx \tau_{ij}(\Delta) \approx \langle v_i v_j \rangle$; filtering out the scales in the dissipation range does not significantly affect $\langle v_i v_j \rangle$, which is dominated by the large-scale fluctuations;
3. $\Delta_2$ was in the inertial range and $\Delta$ was in the dissipation range; then $\tau(\Delta_2) \ll \tau(\Delta)$ and $\langle \mathcal{L}_{ij} \rangle \approx \langle \tau_{ij}(\Delta) \rangle \propto y^3$ in the wall region of a plane channel flow, where $y$ is the distance from the wall and $\langle \rangle$ denotes the average over $x$ and $z$.

The additional averaging procedures, introduced in Germano et al. (1991) in order to ensure the positivity of the denominator in (36), effectively construct a wall function. However, the dynamical meaning of this model in the wall region is rather unclear. The model depends very strongly on the properties of the grid and on the ratio $\Delta_2/\Delta$. Another problem is that in the LES of plane channel flow presented in Germano et al. (1991), the second filter was introduced only in the $x-z$-plane, and not in the $y$-direction, which leaves open the question of implementation in complex geometries.

On the other hand, if both $\Delta$ and $\Delta_2$ are in the dissipation range, then (36) is no longer an identity and the function $c_s$ depends critically on viscous effects and the geometric character of the grid in the viscous sublayer. It is easy to see that in this case $c_s$ rapidly approaches zero, producing an effective wall function in the viscous regions of the flow. It is this effect that explains the success of the model (36) in describing low Reynolds number wall flows. On the other hand, at high Reynolds numbers the problem of resolving wall streaks remains, and the application of the model may become rather problematic. If only one of the filters $\Delta$ and $\Delta_2$ is in the inertial range, while the other is in the dissipation range, the model may produce an uncontrolled behavior of $c_s$, which may become unrealistically large or even negative, thereafter creating numerical instabilities. Indeed, problems with negative $c_s$ are now being observed with this model. We believe that more consideration should be given to the dynamical properties in order to construct a subgrid model working in the wall regions of high Reynolds number flows.

### 3.4.2 Dissipation-Range Modifications to the Subgrid-Scale Viscosity

Here we show how one can modify the Smagorinsky and RNG models to account for the properties of the dissipation-range spectrum. Such modifications can improve the performance of LES models in wall regions where $\Delta \ll l_d = 2\pi/\Lambda_o$.

It has been shown by Kraichnan (1959) and Orszag (1977), and confirmed by numerical experiments (Chen et al., 1993), that the energy spectrum in a turbulent flow is given by

$$E(k) = C_K \bar{\mathcal{E}}^{2/3} k^{-5/3} \left[1 + \alpha\left(\frac{k}{\Lambda_o}\right)^{14/3}\right] \exp(-\beta k/\Lambda_o), \qquad (37)$$

where $\alpha = O(10^{-1})$ and $\beta = O(10)$ are numerical coefficients. When $k \ll \Lambda_o$, the spectrum (12) is recovered. In the dissipation range, where $k > \Lambda_o$, the effective Reynolds number is small and the contribution of these scales to the effective viscosity can be taken into account perturbatively. In this case, an approximate expression for the correction to viscosity $\delta\nu$ due to elimination of scales $k > 2\pi/\Delta \gg \Lambda_o$ is given by

$$\delta\nu(\Delta) \approx \frac{2}{3} \int_{2\pi/\Delta}^{\infty} \frac{E(q)\,dq}{\nu_o q^2}. \qquad (38)$$

Integrating Eq. (38) with the energy spectrum $E(k) \sim C_K \bar{\mathcal{E}}^{2/3}$ $\times\, \alpha(\Lambda_o)^{-14/3} k^3 \exp(-\beta k/\Lambda_o)$ yields

$$\delta\nu(\Delta) \sim \frac{2\alpha C_K \bar{\mathcal{E}}^{2/3}}{3\nu_o \Lambda_o^{14/3}} \frac{2\pi\Lambda_o}{\Delta} \exp(-2\pi\beta/(\Delta\Lambda_o)) \qquad (39)$$

for $2\pi/\Delta \gg \Lambda_o$. Combining the relation (17) for $\Lambda_o$ with Eq. (39), we obtain the viscosity correction

$$\delta\nu \sim 8\nu_o \frac{l_d}{\Delta} \exp(-\frac{\beta l_d}{\Delta}). \qquad (40)$$

Matching the expression (40), valid for $k/\Lambda_o \gg 1$, with the RNG formula (19), which is valid in the range $k/\Lambda_o \to 0$, where it yields the Smagorinsky model with $c_s = 0.003$, we obtain the subgrid model

$$\nu(\Delta) = \nu_o + \nu_o\left[\frac{c_s\Delta^2}{\nu_o}S + 8\frac{l_d}{\Delta}\right]\exp(-\frac{\beta l_d}{\Delta}). \tag{41}$$

The relation (41) is to be used with the local value of $l_d$ found from the relation

$$\frac{l_d}{2\pi} = 0.2\left(\frac{\nu_o^3}{\bar{\mathcal{E}}(\Delta)}\right)^{1/4} \approx 0.2\left(\frac{\nu_o^3}{\nu(\Delta)S_{ij}S_{ij}}\right)^{1/4}. \tag{42}$$

Model (41)–(42) is conceptually similar to the original RNG model (19), but accounts for the information about the dissipation range derived from dynamical constraints. Thus it might be more accurate in wall regions than is (19), since it accounts for the small-scale structure of turbulence in a more systematic way. Note that if one rewrites (41) in the form of the Smagorinsky model, then $c_s$ is a function of $\Delta$.

## 3.5 Example of LES for Complex Flows: Turbulent Combustion

The versatility of LES methods and their ability to solve complex flows of real-world engineering interest can be illustrated with the example of turbulent combustion. Turbulence has a profound effect on the physicochemical dynamics of combustion, controlling both the energetic efficiency and waste product generation of the processes. The study of the effects of turbulent flow fields on premixed flame propagation is of great practical importance because turbulence may increase the propagation rate $(u_T)$ of the turbulent flame to well above the laminar burning velocity $(u_0)$ in the same mixture. Indeed, premixed combustion is important in design problems ranging from automobile engines to environmental protection equipment. The enhanced flame speed $u_T$ increases the heat release rate and, thus, the power available from a combustor or engine of a given size. On the other hand, increasing turbulence beyond a certain level increases the mass consumption rate very little, if any, and in some cases may lead to "quenching" of the flame. Quenching is particularly pronounced when $u_0$ is small compared with the rms velocity fluctuation of the turbulent flow $u_{rms}$.

From a fundamental point of view, the problem of turbulent combustion involves major difficulties related to the enormous number of degrees of freedom needed to resolve the turbulence, as well as the strong nonlinearity and stiffness of the equations of chemical kinetics. Turbulence and combustion processes are strongly coupled. Turbulence causes wrinkles on flame fronts and can change the front width. Combustion, on the other hand, generates secondary flows which contribute to the complexity of the velocity field. It is the complexity of the velocity field and the flame front, that, in fact, enables the use of modern statistical physics methods to derive large-scale effective equations for flame front propagation which then can be solved numerically. The possibility of the statistical description of small-scale fluctuations in turbulence enables us to derive subgrid models for LES. While the first LESs of non-reactive incompressible flows were made twenty years ago by Deardorff, the first LESs of the combustion process were made by us in 1990–1991. To advance to the point at which LES based on a subgrid model of flame propagation had become feasible, it was necessary to develop a general approach based on RNG methods.

One of the most fundamental problems of combustion theory is to determine the burning velocity of turbulent premixed flames in the regime of a thin reaction zone $b$. In this case, the flame surface can be represented as a thin wrinkled sheet propagating in the direction of the unburned material with a velocity $u_T$ which is larger than the laminar flame velocity $u_0$. In the regime of a thin flame, $b \ll l_d$, where $l_d$ is the Kolmogorov dissipation scale, the smallest excited scale of turbulence. The theory must yield the dependence of $u_T$ on the rms velocity $u_{\text{rms}}$. Experiment indicates that the turbulent flame velocity can be expressed in terms of the laminar velocity $u_0$ as

$$\frac{u_T}{u_0} = f\left(\frac{u_{\text{rms}}}{u_0}\right).$$

Here, $f(x)$ is believed to behave like $f \propto x^{\alpha}$ with $\alpha < 1$ when $x > 1$. However, theoretical work directed toward a derivation of $f(x)$ conducted during the past 50 years has led to various values of $\alpha$, ranging from 0.5 to 2. It is easy to show that in the thin-flame regime $u_T/u_0 = s_T/s_0$, where $s_T/s_0$ is the ratio of the areas of the wrinkled to the plane flame front. Thus, the concept of turbulent flame speed accounts for increased consumption of fuel due to increased surface areas.

In the thin-flame approximation, the position and shape of flame fronts are determined by the eikonal equation

$$\frac{\partial G}{\partial t} + \mathbf{v} \cdot \nabla G = u_0 |\nabla G|, \tag{43}$$

where $u_0$ is the laminar flame speed. The level surface $G(\mathbf{x}, t) = 1$ determines the location of the front at each time $t$. Equation (43) is difficult to solve numerically; it involves all the typical difficulties of a turbulence problem including the necessity to resolve small-scale wrinkles. Although (43) does not take into account processes like heat release and thermal expansion, it captures the key physical properties of combustion in the thin-flame limit and provides results which agree quite well with experimental data. The RNG description of the average behavior of fronts described by (43) was developed in Yakhot (1988). It was shown that the LES dynamical equation for premixed turbulent combustion is just Eq. (43) with $u_0$ replaced by $u_{turb}$, the turbulent flame speed, and $\mathbf{v}$ replaced by the filtered velocity $\mathbf{v}^<$ obtained from the RNG theory. In Yakhot's (1988) model the effective flame speed is given by

$$\frac{u_{turb}}{u_0} = \exp\left[-\frac{(u^{sg})^2}{u_{turb}^2}\right],$$

where $u^{sg}$ is the subgrid turbulence level. Thus, LES of premixed combustion should be done in two steps:

1. LES of the velocity field to determine the $\mathbf{v}^<$;
2. solution of the LES form of (43) (with $u_0$ replaced by $u_{turb}$ and $\mathbf{v}$ replaced by $\mathbf{v}^<$) to determine the position and shape of the flame front.

The effects of heat release can be taken into account by including the equations of heat transfer with a heat source at the propagating flame front. This has been done by Menon et al. (Chapter 14, this volume), who used the RNG-based subgrid model to perform LES studies of unstable combustion in an axisymmetric ramjet combustor. Both stable and unstable combustion were simulated with evidence of decaying and growing pressure oscillations, respectively. The observed flow features of unstable combustion, such as a propagating hooked flame and its phase relation with the pressure oscillation, agree with observations. These simulations establish the evidence that the RNG-based subgrid model contains the essential physics of combustion instability.

## 3.6 Conclusions

The LES approach bridges the most fundamental and the most applied fields of turbulence research. A theoretical physicist will examine the details of renormalization of transport coefficients. An engineer will request information on complex flows of practical interest, and require them to be obtained within practically reasonable computational time. A fluid mechanician will thoroughly analyze quantitative data on a flow pattern. As it is directly derived from the first principles, the LES approach has the capacity to address all these issues. Its potential to address them in full will depend largely on our ability to develop an adequate description of anisotropic turbulence production and dissipation phenomena in near-wall regions.

## Acknowledgments

We are grateful to Leslie Smith for useful comments. This work was supported by the AFOSR under Contract 90-124, DARPA under Contract F49620-91-C-0059, NASA-Lewis Research Center under Contract NAS3-26702, and the ONR under Contract N00014-92-C-0089.

## References

CHEN, S., DOOLEN, G., HERRING, J.R., KRAICHNAN, R., ORSZAG, S.A., AND SHE, Z.-S. (1993) Far-dissipation range of turbulence. *Phys. Rev. Lett.* **70**, 3051–3054.

DANNEVIK, W., YAKHOT, V., AND ORSZAG, S.A. (1987) Analytical theories of turbulence and the $\epsilon$-expansion. *Phys. Fluids A* **30**, 2021–2029.

DEAN, R.B. (1978) Reynolds number dependence of skin friction and other bulk variables in two-dimensional rectangular duct flow. *Trans. ASME J. Fluids Eng.* **100**, 215–223.

DEARDORFF, J.W. (1970) A numerical study of three-dimensional turbulent channel flow at large Reynolds numbers. *J. Fluid Mech.* **41**, 453–480.

ECKELMANN, H. (1974) The structure of the viscous sublayer and the adjacent wall region in a turbulent channel flow. *J. Fluid Mech.* **65**, 439–459.

GERMANO, M., PIOMELLI, U., MOIN, P. AND CABOT, W.H. (1991) A dynamic subgrid-scale eddy viscosity model. *Phys. Fluids A* **3**, 1760–1765.

HUSSAIN, A.K.M.F. AND REYNOLDS, W.C. (1975) Measurements in fully developed turbulent channel flow. *J. Fluid Eng.* **97**, 568–578.

KRAICHNAN, R.H. (1959) The structure of isotropic turbulence at very high Reynolds numbers. *J. Fluid Mech.* **5**, 497–543.

KREPLIN, H. AND ECKELMANN, M. (1979) Behavior of the three fluctuating velocity components in the wall region of a turbulent channel flow. *Phys. Fluids* **22**, 1233–1239.

KIM, J., MOIN, P., AND MOSER, R.D. (1987) Turbulence statistics in fully
    developed channel flow at low Reynolds number. *J. Fluid Mech.*
    **177**, 133–166.

MOIN, P., AND KIM, J. (1982) Numerical investigation of turbulent chan-
    nel flow. *J. Fluid Mech.* **118**, 341–377.

ORSZAG, S.A. (1970) Analytical theories of turbulence. *J. Fluid Mech.*
    **41**, 363–386.

ORSZAG, S.A. (1977) *Statistical theory of turbulence: Les Houches Sum-
    mer School in Physics.* Ed. R. Balian and J.-L. Peabe, pp. 237–
    374. Gordon and Breach.

PANDA, R., SONNAD, V., CLEMENTI, E., ORSZAG, S.A., AND YAKHOT, V.
    (1989) Turbulence in a randomly stirred fluid. *Phys. Fluids A* **1**,
    1045–1053.

PIOMELLI, U., AND ZANG, T.A. (1991) Large-eddy simulation of transi-
    tional channel flow. *Comput. Phys. Comm.* **65**, 224–230.

PROUDMAN, I., AND REID, W.H. (1954) On the decay of a normally dis-
    tributed and homogeneous turbulent velocity field. *Phil. Trans.
    Roy. Soc.* **A247**, 163–189.

SMAGORINSKY, J. (1963) General circulation experiments with the prim-
    itive equations. I. The basic experiment. *Mon. Weather Rev.* **91**,
    99–164.

YAKHOT, A., ORSZAG, S.A. AND YAKHOT, V. (1989) Renormalization
    group formulation of large-eddy simulations. *J. Sci. Comp.* **4**,
    139–158.

YAKHOT, V. AND ORSZAG, S.A. (1986a) Renormalization group analysis
    of turbulence, Part I: Basic theory. *J. Sci. Comp.* **1**, 3–51.

YAKHOT, V. AND ORSZAG, S.A. (1986b) Renormalization-group analysis
    of turbulence. *Phys. Rev. Lett.*, **57** 1722–1724.

YAKHOT, V., AND ORSZAG, S. A. (1987) Renormalization group and local
    order in strong turbulence. *Nucl. Phys. B* **2**, 417–440.

YAKHOT, V. (1988) Propagation velocity of premixed turbulent flames.
    *Comb. Sci. Technol.* **60**, 191–214.

# 4

---

# Some Contributions of Two-Point Closure to Large Eddy Simulations

JACKSON R. HERRING AND ROBERT M. KERR

## 4.1 Introduction

Any precise distinction between large-scale, deterministic large eddy simulation (LES) scales and statistically specified "subgrid scales" is lost after the predictability time, unless the distinction corresponds to a stable statistical symmetry of the problem (see, e.g., Fox and Lilly, 1972; Herring, 1978). This dilemma does not occur for two-point closures, and we describe in this chapter some simple two-point closures (principally the test field model, or TFM; Kraichnan, 1971) and note how such methods may be used to place bounds on LES computed velocity fields. The approximate form of these ideas is a "hyperviscosity" prescription, as described by Kraichnan (1976) and Chollet and Lesieur (1981).

We then present – in summary form – current information on the accuracy of statistical theories, principally the TFM and related theories, comparing theoretical predictions with recent direct numerical simulations (DNS) of Kerr. These comparisons suggest that although such theories may accurately represent inertial range spectra, their errors in tracking the spectra of an initial large-scale pulse may be substantial.

In the context of a more fundamental two-point closure scheme, the direct interaction approximation, or DIA (Kraichnan, 1959), we discuss briefly the origin of "stochastic backscatter," noting that its magnitude is uncertain and may be easily overestimated. This overestimation is attributed to the fact that the real flow's acceleration may be considerably smaller than its Gaussian estimation.

We also describe the application of these ideas to two-dimensional and quasi-geostrophic turbulence. Here, the characterization of the eddy viscosity (based on two-point closure) is simpler, and we are able to delineate the role of strain and vorticity in the composition of eddy viscosity in a more precise way.

## 4.2 An Overview of Two-Point Closure-Based LES Models

LES seems at present an inevitable tool for studying high Reynolds number flows, yet its derivation as a theoretical procedure is unsatisfactory. As an illustration of this point, consider the following avenue to LES via the statistical theory of turbulence, assuming that there exists a closure furnishing useful approximations to ensemble averages $\langle q(\mathbf{x}, t) \rangle \equiv Q(\mathbf{x}, t)$, $\langle q(\mathbf{x}, t) q(\mathbf{x}', t) \rangle \equiv Q(\mathbf{x}, \mathbf{x}', t)$, where $q$ is some turbulence quantity. Imagine that the Lagrangian DIA (Kraichnan, 1964) or the theory of Kaneda (1981) serves this role. Then we may try to derive an LES model by setting up the following initial value problem:   for $\mid \mathbf{k} \mid \leq k_{\mathrm{LES}}$ ($\mathbf{k}$ is the wave number), let the statistical distribution of $q(\mathbf{k}, 0)$ be sharp (the same for all realizations), and let $q(\mathbf{k}, 0)$, $k \geq k_{\mathrm{LES}}$ be distributed (say, Gaussianly). In this example we refer to the ranges $0 < k < k_{\mathrm{LES}}$ as large eddy, and $k_{\mathrm{LES}} < k < \infty$ as subgrid scales. If the distinction between the deterministically specified region $k \leq k_{\mathrm{LES}}$ and its statistical counterpart is preserved in time, the closure also gives an LES prescription. The problem, of course, is that the distinction grows progressively less sharp (because of unpredictability), and in time the encroachment of the statistically fuzzy region into the deterministic is such that the latter is eventually gone. The full burden of describing the flow dynamics is then placed on the fluctuating field. If the division between "sharp" and statistical corresponds to a statistical symmetry of the flow (e.g., two-dimensional rolls) and if the large eddy configuration is stable, we can then proceed. Such statistical symmetries are just those used to reduce the computational labor in solving the statistical theory in the first place. From this perspective, the above idea of the "large eddy" is simple-minded; we should, instead, talk about those large-scale structures which are observed to be dominant or long-lived. If averaging is defined in that sense, the same formal algebra that produces the DIA could be invoked to yield an LES closure and we would have a "derived" LES. Yet these structures are plastic and ephemeral. Under these conditions, it is difficult to imagine how to carry out the above program.

With these comments as background, let us see what alternatives two-point closures may offer. We assume such methods comprise closed approximations for ensemble means $\langle u_i(\mathbf{x}, t)\rangle$ and $\langle u_i(\mathbf{x}, t)u_j(\mathbf{x}', t')\rangle$. We focus on isotropic (in three dimensions) homogeneous turbulence, for which $U(k, t) \equiv \int d(\mathbf{x} - \mathbf{x}')\langle u_i(\mathbf{x}, t)u_i(\mathbf{x}', t)\rangle \exp(i\mathbf{k} \cdot (\mathbf{x} - \mathbf{x}'))$ suffices to describe completely the second-moment level. Then, for a given computational domain $(k_0 \le k \le k_{\max})$, closure consists of equations of motion for $U(k, t)$:

$$(\partial_t + 2\nu k^2)U(k, t) = T(k, \{U\}, k_0, k_{\max}). \tag{1}$$

(Note that $T(k) \equiv T(k)/(2\pi k^2)$, where $T(k)$ is the energy transfer function.) To be concrete, let us write a simplified form for $T$, in which the specific functional dependence of $T$ on $U(k, t)$ is manifest.

After some rather heavy approximations (Herring et al., 1982), TFM equations become [1]

$$T = \int_{k_0}^{k_{\max}} dp \int_{\max(k_0, |k-p|)}^{\min(k_{\max}, k+p)} U(q, t)\, [U(p, t) - U(k, t)]\, \Theta(k, p, q)\, dq, \tag{2}$$

$$\Theta(k, p, q) = \pi(pq/k)[(p^2 - q^2)(k^2 - q^2) + k^2 p^2]/[\eta(k) + \eta(p) + \eta(q)], \tag{3}$$

$$\eta(k) = \nu k^2 + \pi \int_\Delta dp\, dq\, (kpq)(1 - y^2)(1 - z^2)U(q, t)/[\eta(k) + \eta(p) + \eta(q)], \tag{4a}$$

$$(y, z) = [\sin(\mathbf{k}, \mathbf{p}), \sin(\mathbf{k}, \mathbf{q})]. \tag{4b}$$

In (4a), $\int_\Delta$ stands for the same $(dp\, dq)$ integral as in (2). The integral in (4a) may be approximated (Herring et al., 1982) as

$$\eta^2(k) = \lambda^2 \int_0^k p^2 E(p)\, dp + \lambda^2 k^2 \eta(k) \int_k^\infty dp\, E(p)/\eta(p), \tag{4c}$$

with $\lambda \sim 1$. We remark that (4c) is indicated by the TFM (Kraichnan, 1971), and also by the more recent Lagrangian renormalization approximation (LRA) of Kaneda (1981). The underlying physics of (4) is that pressure fluctuations account for the dephasing of the triple moments contributing to $T$; thus, the $(dp\, dq)$ integral in (4a) is the same as in the quasi-normal evaluation of the pressure variance (Batchelor, 1952). Equation (4c) represents a bridge between the eddy-damped quasi-normal approximation, or EDQNM (Lesieur, 1987), and the DIA

[1] In the more rigorous TFM, (3) should be replaced by the relation $\{\partial_t + [\eta(k) + \eta(p) + \eta(q)]\}\Theta(k, p, q) = B(k, p, q)$, and (4a) for $\eta(k)$ should be coupled to a compressive relaxation $\eta^c(k)$. In (1)–(4), we have simply put $\eta^c = 2\eta$, and have used an adiabatic relation, $\partial_t \Theta = 0$.

(Kraichnan, 1959); the EDQNM evaluates $\eta(k)$ with only the first term on the right hand side, whereas the DIA has only the second. So much for the zoology of two-point closures. Note that the second term of (2) is essentially an eddy viscosity term, since it is $\sim U(k)$. In reality, $k_{max} = \infty$.

Underlying intensity equations such as (1)–(4) are amplitude equations (for $\mathbf{u}(\mathbf{k}, t)$) which ensure realizability for any application. There are two types of such models: (1) random coupling models, or RCM (Kraichnan, 1961), in which equations like (1)–(4) are shown to result from a random modification of the Fourier coefficients entering the nonlinear terms (i.e., random $\pm$ modulations of $C(\mathbf{k}, \mathbf{p}, \mathbf{q})$, where $\sum_{k=p+q} C(\mathbf{k}, \mathbf{p}, \mathbf{q}) u_i(\mathbf{p}) u_j(\mathbf{q}) \equiv \{\mathbf{u} \cdot \nabla \mathbf{u}\}_{FT}$); and (2) Langevin-type models (Kraichnan, 1971; Leith and Kraichnan, 1972; Herring and Kraichnan, 1972), in which the dynamics stemming from $\mathbf{u} \cdot \nabla \mathbf{u}$ is modeled as a random force ($\equiv \mathbf{w} \cdot \nabla \mathbf{w}$, where $\mathbf{w}$ is a white-noise, incompressible Gaussian vector) plus an associated eddy damping. Here, $\{\}_{FT}$ means the Fourier transform of the bracketed quantity. In Langevin models, the variance of $\mathbf{w}$ is constrained to be identical to that of $\mathbf{u}$. We display only the Langevin model here, since it figures in the LES ideas about stochastic backscatter, to be discussed later:

$$[\partial_t + \nu k^2 + \mu(k, t)] u_i(\mathbf{k}, t) = -i[P_{ij}(\mathbf{k}) k_m$$
$$+ P_{im}(\mathbf{k}) k_j] \sum_{k=p+q} [\theta(\mathbf{k}, \mathbf{p}, \mathbf{q})]^{1/2} \sigma(t) w_j(\mathbf{p}, t) w_m(\mathbf{q}, t), \qquad (5)$$

$$\{\sigma(t)\sigma(t')\} = 2\delta(t - t'). \qquad (6)$$

Navier–Stokes is (5) with $\theta = \sigma = 1$, $\mathbf{w} = \mathbf{u}$. The physics of (5)–(6) is that the nonlinearities act partially as a random force (the right hand side of (5)) and partially as an eddy viscosity $\mu(k)$, a division of labor familiar from Fokker–Planck. It is sobering to note that the dynamics underlying the two types of closure models are profoundly different: non-Gaussian for the RCM (Chen et al., 1989) but completely Gaussian for the Langevin model.

Except for extracting various asymptotic laws, equations (1)–(4) must in general be solved numerically, a topic that has already evoked considerable effort (Kraichnan, 1964; Leith, 1971; Lesieur and Schertzer, 1978; Herring, 1984). If we are interested in large $R_\lambda (\equiv \langle u^2 \rangle^{1/2} \lambda/\nu)$ flows, we know that the computational domain $(k_0, k_{max}) \to \infty$, since $k_{max} \sim k_s \equiv (\epsilon/\nu^3)^{1/4}$. Here, $\epsilon$ is the kinetic energy dissipation, $\nu$ the viscosity, $k_s$ the Kolmogorov scale, and $\lambda$ the Taylor microscale. If the computational effort becomes large, it is natural to ask if we may use high wave number asymptotics and employ equations (1)–(4) numerically on a smaller domain $(k_0, k_{LES})$, $k_{LES} \ll k_s$. As we shall see, "high wave number asymptotics" turn out to mean simply an eddy viscosity, $\nu_{eddy}(k, t)$, whose form is implied by (2)–(4) in such a way that $E(k, t) \sim k^{-5/3}$, $k \geq k_{LES}$ (Leith, 1971). This will require that $\nu_{eddy}(k)$ increases rapidly as $k \to k_{LES}$ to prevent $E(k, t)$ from "tailing up" near $k_{LES}$, since $k_{LES} \ll k_s$.

The two-point closure version of LES starts from this requirement that the computational domain of (2)–(4), $(k_0, k_{max})$, be compressed to $(k_0, k_{LES})$, as discussed above. Thus LES is an attempt to map large $R_\lambda$ problems onto a modest $R_\lambda$ problem, an implicit theme in most LES. In this connection, as a rule of thumb, the value of $R_\lambda$ up to which a numerical simulation is valid is proportional to $(k_s - k_0)$. We could hope that using LES one could replace $k_s$ by $k_{LES}$. Formally we eliminate from (2)–(4) variables $U(k), \eta(k)$ in the domain $k \in [k_{LES}, k_{max}]$ in favor of ingredients $U(k), \eta(k)$ from $k \in [k_0, k_{LES}]$, and thereby obtain equations for $U_{LES}(k, t)$ alone. Here, the tag "LES" means simply that $k \in [k_0, k_{LES}]$. This elimination may need to include time integrals, if the flow is rapidly evolving. Notice that $U_{LES}(k, t)$ describes the variance of the flow field, and hence the connection with the actual flow as assumed in the first paragraph is lost. However, so is the associated predictability problem described there.

For the case of homogeneous and isotropic flows, the elimination of the subgrid scales in the manner described above results approximately in the following eddy viscosity and conductivity (Chollet and Lesieur, 1981; Chollet, 1983, 1985):

$$\nu(k) = [0.267 + 9.21 \exp(-3.03 k_{LES}/k)][E(k_{LES})/k_{LES}]^{1/2}, \quad (7)$$

$$\kappa(k) = 0.6\nu(k). \quad (8)$$

Equations (7)–(8) follow from an application of the EDQNM approximation (Orszag, 1977), and we refer to Chollet's papers for complete details.

## 4.3 Accuracy Limits of Two-Point Closure

Before examining how two-point closure (in the guise of TFM) may be used for LES, we must note its accuracy and certain of its fundamental drawbacks; we should not expect a resulting LES to exceed the theory from which it was derived. There are two aspects to this problem: (a) high $R_\lambda$ assessments, where the comparison data sets are drawn from wind tunnels and geophysical observations; and (b) comparisons of theory with initial data problems (perhaps DNS) whose solution we know more or less exactly. The evolution of these solutions should correspond – in some idealized way – to typical flow situations we want to simulate. Generally, item (b) implies that sufficiently low values of $R_\lambda$ are used in numerical simulations so that all scales up to the Kolmogorov scale, $k_s$, are accurately resolved.

At high Reynolds numbers (item (a)) the theory compares well with the experimental inertial range of isotropic turbulence, where such data are available. This in itself is no achievement for TFM, since it has an arbitrary constant. Agreement in the intermittency dominated dissipation range (e.g., the $R_\lambda \sim 2000$ atmospheric observations of Champagne et al., 1977) is not so good; here TFM is in reasonable agreement with the data for $k \leq k_s$, but at $k > k_s$ it progressively underestimates the experimental value.

Flows under item (b) should include rapidly evolving flows that may typically follow from (large-scale) instabilities. We would additionally want to describe a turbulent burst – at least its homogeneous analog – if such exists.

One initial spectrum $E(k,0)$ that satisfies the above criterion is that studied via DNS by Lesieur and Rogallo (1989) and compared with TFM by Herring (1990):

$$E(k, t = 0) = k^8 \exp(-k^2/16). \qquad (9)$$

Figure 1 compares the DNS of Lesieur and Rogallo with the TFM for eddy viscosity and conductivity given by (7). In fact, Lesieur and Rogallo – and Fig. 1 – use (7) with 0.267 replaced by 0.050, for $t \leq 5.683 \times 10^{-4}$. Thus, for early times, the flow is essentially inviscid. We should note that the determination of $\nu_{eddy}(k)$ is for EDQNM rather than the TFM; however, detailed comparison of these theories shows little quantitative differences (Herring et al., 1982).

We see here an especially pronounced tendency, during the early, essentially inviscid phase, for the DNS spectra to evolve with more energy concentrated at intermediate scales than TFM, whose spectra tend rather to be a power law. The tendency for the DNS data to "curl up" at high $k$ is, of course, a result of a lack of resolution in the DNS. At later times, it also reflects the fact that there is a discrepancy between theory and simulations; more "hyperviscosity" is needed in a formula like (7) to prevent reflection of energy from the high wave number cutoff $k_{LES}$. At least for the present resolution, $k_{LES} = 64$. On the other hand,

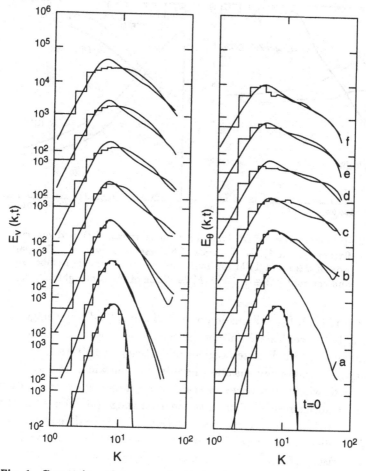

Fig. 1 Comparison of energy and scalar variance spectra for TFM (smooth curves) with the DNS of Lesieur and Rogallo (histograms) at a sequence of times. $E_v(k, t)$ at left; $E_\theta(k, t)$ at right. Initial conditions, (9). (From Herring, 1990.)

the TFM does not develop a "curled-up" region at $k_{\rm LES}$. This should not be surprising; (7) was derived so that TFM (really EDQNM) would have no such region.

The tendency for DNS to develop more large-scale energy than the statistical theory predicts may portend a serious problem for the latter. We recall that recent investigations of inviscid flows (Kerr, 1991a; Pumir and Siggia, 1990) indicate that the Euler equations do not develop a finite-time enstrophy singularity as indicated by closures, such as TFM

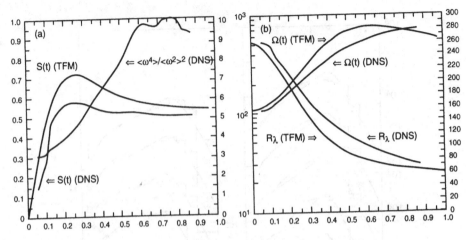

Fig. 2   (a) Comparison of DNS and TFM for skewness, $S(t)$, and vorticity flatness $f \equiv \langle \omega^4 \rangle / \langle \omega^2 \rangle^2$. Scale of flatness is at right. (b) Comparison of DNS and TFM for Taylor microscale Reynolds number $R_\lambda$ and enstrophy $\Omega \equiv \int_0^\infty k^2 E(k)\, dk$. Scale of $R_\lambda$ is at right.

(see Lesieur, 1987, for a discussion of why closures tend to be singular). This may be viewed as a failure of the statistical theory to resolve and discriminate properly between regions of vorticity and strain.

Perhaps the important question here is the following: In real flows with $\nu \to 0$, where – in terms of scale size – does the overestimation of energy transfer reside? Is it in the inertial range (as inferred in the discussion of the Lesieur–Rogallo simulation) or is it confined to a portion of the dissipation range? To discuss this point, we show in Figs. 2–4 comparisons of $256^3$ DNS for initial $R_\lambda(t = 0) = 258$, $\nu = 0.0125$ ( $R_\lambda(t = 0.9) = 60$). Figures 2a,b give information on the DNS non-Gaussian behavior, velocity derivative skewness, $S(t)$, and vorticity flatness, $f \equiv \langle \omega^4 \rangle / \langle \omega^2 \rangle^2$, while Fig. 3 compares DNS with TFM for the compensated spectrum, $E(k,t)k^{5/3}/\epsilon^{2/3}$, and Fig. 4 shows energy transfer $T(k,t)$. Spectral quantities are shown at times $t = 0.25$, 0.900. To attain such high initial $R_\lambda$, only the modes with $k \leq 3$ were excited in the initial data. This implies a considerable excitation of the lowest available modes during the course of the run. Note at the early time ($t = 0.25$) a rather serious discrepancy between TFM and DNS in the dissipation range (the rightmost portion of the positive lobe of $T(k,t)$), with more extreme divergence for larger $k$ (here $k_s = 28.3$). At this time, there is only a hint of an inertial range, but at the maximum of $C_{\text{Kol}}(k) \equiv k^{5/3}E(k,0.25)/\epsilon^{2/3}$, TFM and DNS are in reasonable ac-

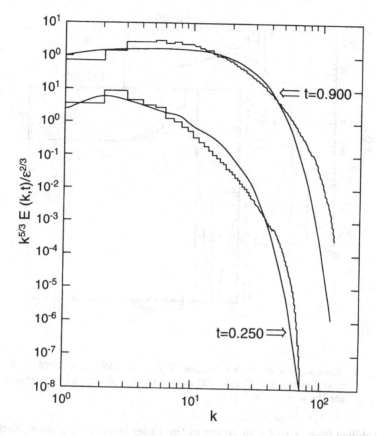

Fig. 3  Comparison of DNS with TFM for compensated spectra,
$k^{5/3}E(k,t)/\epsilon^{2/3}$, for $t = 0.25$ (bottom) and $t = 0.900$ (top). Initial spectra are such that only modes with $k \leq 3$ are excited, and $R_\lambda(0) = 258$ ($\nu = 0.025$).

cord, insofar as the numerics permit an assessment. At the later time ($t = 0.900$), the maximum of $C_{Kol}(k)$ has broadened nicely, so that in the TFM the region of constancy extends from $k \simeq 2$ to $k \simeq 10$, and the value of $C_{Kol}^{TFM} \simeq 1.72$ is near its asymptotic value of 1.76. On the other hand, $C_{Kol}^{DNS}(k) \sim 3.9$, with a less well defined plateau. At the same time, the dissipation range discrepancy has significantly decreased, except for the divergence at large $k$, beyond $k_s = 46.0$.

The combined DNS–TFM results presented here are consistent with the idea that the development of strong non-Gaussianity associated with this initial data problem leads at short times to an inertial range discrepancy of about a factor of 2, during the time before viscosity can

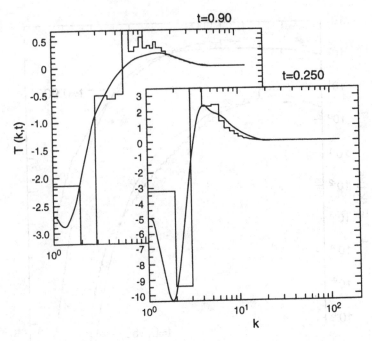

Fig. 4   Comparison of energy transfer $T(k,t)$ for DNS and TFM. $T(k, 0.900)$ at top and $T(k, 0.250)$ at bottom. Conditions of comparison are the same as in Fig. 3.

destabilize those structures innate to the Euler ($\nu = 0$) equations. Such structures imply a heaping-up of energy at large scales and a shutting-down of the energy transfer mechanism – at small scales, as seen in Fig. 4a, which compares $T(k, 0.250)$ for DNS and TFM. The idea that the large DNS value of $C_{\text{Kol}}$ is attributable to the "bump" (Chollet and Lesieur, 1981) is vitiated by the fact that the same "bump" dynamics (a dearth of intermediate-scale triad interactions near the beginning of the dissipation range) is present also in the TFM. Of course, it could be that these intermediate-range interactions are much more significant in real flows than in the statistical theory.

## 4.4 Comments on Closure Studies
### of the Eddy Prandtl Number

Two additional points may be gleaned from comparisons of this type: (1) the scalar field, in both the DNS and TFM, has during all of its decay a spectral slope shallower than $k^{-5/3}$; and (2) the decay rate for the total scalar energy approaches its asymptotic value only at long times – much longer than the final time of the DNS. (This last point is for TFM alone.) The first point, as noted by Lesieur and Rogallo (1989), has implications for the eddy Prandtl number frequently needed in various forms of LES (see, e.g., Deardorff, 1970; Moeng and Wyngaard, 1988). In TFM, as in renormalization group (RNG) theory, there emerges a value for $P_r^{\text{eddy}} = \nu_{\text{eddy}}/\kappa_{\text{eddy}}$, for which Lesieur and Rogallo employ (7)–(8).

We shall now discuss the scalar results in more detail, examining the issue of the eddy Prandtl number, $P_r^{\text{eddy}}$, both that inferred from closure and that used in applications. To discuss this point we need the scalar-energy spectral equations, companions to (1)–(4). These are

$$(\partial_t + 2\kappa k^2)U_\theta(k,t) = T_\theta(k,t), \tag{10}$$

$$T_\theta(k,t) = \int_\Delta dp\, dq\, U(q,t)[U_\theta(p,t) - U_\theta(k,t)]\Theta_\theta(k,p,q), \tag{11}$$

$$\Theta_\theta(k,p,q) = \pi(p^3 q/k)\sin^2(\mathbf{p},\mathbf{q})/\{g_\theta^2[\eta^c(k) + \eta^c(p)] + \tilde{g}_\theta^2\eta(q)\}. \tag{12}$$

Here, $U_\theta = \langle \theta^*(\mathbf{k},t)\theta(\mathbf{k},t)\rangle$, $\theta(\mathbf{k},t)$, being the scalar field, and $\eta^c(k) \simeq 2\eta(k)$ is a relaxation frequency of a compressive "test field" which, along with $g_\theta$, measures triple-moment relaxation for $\theta$. The constant $\tilde{g}_\theta$ ($\equiv 1$, for TFM) is introduced here to compare results with the EDQNM results of Larcheveque and Lesieur (1981). The significance of $(g_v, g_\theta)$ may be inferred by recalling that the theory's (stationary state) characterization of (Lagrangian) moment relaxation is given by

$$\langle\theta(-\mathbf{k},t)\theta(\mathbf{k},t')\rangle = \langle\theta(-\mathbf{k},t)\theta(\mathbf{k},t)\rangle \exp[-g_\theta^2\eta(k)\,|\,t - t'\,|]. \tag{13}$$

The choice $(g_v, g_\theta) = (1.5, 0.5)$ yields Kolmogorov and Corrsin-Oboukhov constants, 1.7 and 0.6, respectively. As noted above, the TFM implies $\tilde{g}_\theta = 1$. Using the asymptotic forms of (2) and (10) appropriate for $k \to 0$ (Kraichnan, 1976) gives

$$\nu_{\text{eddy}} = \left(\frac{2}{15}\right)\int_0^\infty \frac{dp}{\eta(p)}\{E_v(p) + \left(\frac{1}{4}\right)\frac{d}{dp}[pE_v(p)]\}, \tag{14a}$$

$$\kappa_{\text{eddy}} = \left(\frac{2}{3}\right)\int_0^\infty E_v(p)\frac{dp}{[\eta(p) + g_\theta^2\eta^c(p)]}, \tag{14b}$$

$$P_r^{\text{eddy}} = \left(\frac{1}{6}(2g_\theta^2 + \tilde{g}_\theta^2)\right). \tag{14c}$$

Then, with $(g_\theta, \tilde{g}_\theta) = (0.5, 1)$ we find $P_r^{\text{eddy}} = 0.3$, a small value compared with the oft-cited range $0.6 \le P_r^{\text{eddy}} \le 0.9$ (see, e.g., Fulachier and

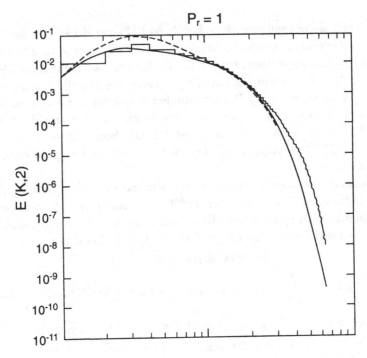

Fig. 5   Comparison of $E_\theta(k, t = 1)$ for two choices of TFM scalar parameters, as described in the text. Solid line is case (a); dashed, case (b). (After Herring and Métais, 1992.)

Dumas, 1976). However, this value, ~0.3, is much in vogue with meteorologists (Deardorff, 1970; Moeng and Wyngaard, 1988). We inquire as to why this apparent disparity is tolerable (note that the EDQNM study of Larcheveque and Lesieur (1981) uses $(g_\theta, \tilde{g}_\theta) = (1.5, 0.0, 3.62)$, for which $P_r^{eddy} = 0.60$, as in (8)).

With respect to fixing $P_r^{eddy}$, we may appeal to DNS experiments. A comparison of the two choices $(g_\theta, \tilde{g}_\theta) = (0.5, 1.0)$ (case (a), solid line) with $(g_\theta, \tilde{g}_\theta) = (0.0, 2.77)$ (case (b), dashed; the EDQNM of Larcheveque and Lesieur) is given in Fig. 5. The results of DNS are represented by a histogram. The agreement favors case (a), although calculations at higher $R_\lambda$ would be useful in order to be certain.

What, then, do we make of the $P_r^{eddy} \sim 0.3$ for case (a)? We argue that the asymptotic expansion for $(\nu_{eddy}, \kappa_{eddy})$, which yields (14c), is indifferent to large-scale variations in the distribution of fluid motions, and that this distribution is vital in applying eddy concepts to practical situations. To put the issue in perspective, consider the idealized case in which scales of $\theta$ are much larger than those of $\mathbf{u}$. Figure 6 gives spectra $E_v(k,t), E_\theta(k,t)$, for initial spectra $E_{v,\theta}(k,0) = k^4/(a_{v,\theta}+k)^{-17/3}, a_\theta = a_v/100, a_v = 0.5, R_\lambda \sim 300$ (see Herring et al., 1982, for a more complete discussion). We note that in the small $k$ region, where $E_v(k,t)$ has yet to be significant, $\theta$ decays under eddy conductivity alone. The numerics give

$$\kappa_{\text{eddy}} = 1.16\epsilon^{1/3}L_v^{4/3}, \tag{15}$$

where $L_v$ is the integral scale of the velocity field. If we combine (15) with the numerical calculation of the eddy viscosity, we find $P_r^{\text{eddy}} = 0.8$, substantially larger than that given by (14c). Notice further – in this connection – that we may write $\nu_{\text{eddy}} = a(s)u_{\text{rms}}L$, $u_{\text{rms}}$ being the rms velocity, $L$ the longitudinal integral scale of the turbulence, and $a(s)$ a constant factor determined by the shape of the large scales of $E_v(k,t)$ (Larcheveque and Lesieur, 1981),

$$E_v(k) \to k^s, \tag{16a}$$

$$a(2,3,4) = (1.58, 1.24, 1.11). \tag{16b}$$

The above applies for case (a).

Before going on to other topics, we should note that the TFM version of two-point closure departs significantly from DNS results for small $P_r$ ($P_r \leq \frac{1}{4}$) (see Herring and Kerr, 1989; Herring and Métais, 1992). Comparisons of energy transfer functions $T_v(k,t), T_\theta(k,t)$ are presented in Fig. 7 for $P_r = 8, 2, 1, 0.25, 0.0625$. Initial spectra were set to $(E_v, E_\theta) = Ak^4 \exp[-2(k/k_0)^2]$, where $A$ and $k_0$ are such that $R_\lambda(0) = 35.0$. The simulation was started from random initial data and comparisons were made at $t = 1$, which is about three times the large-scale eddy circulation time scale, $\tau = L_v u_{\text{rms}}(0)$. Similar discrepancy has also been noted by Kerr (1985), who observed that the mixed scalar skewness in DNS did not decrease with $P_r$ in accordance with the theory of Batchelor et al. (1959). The discrepancy at low $k$ in Fig. 6f may be attributed to sample errors; there are, after all, only two wave number bins, where $T(k) \leq 0$. But sampling errors at larger $k$ (where $T(k) \geq 0$) are small, and the TFM underestimation of $T(k,t)$ is significant. The extent of the DNS–TFM discrepancy may be judged by comparing DNS and TFM for the mixed scalar skewness,

$$S_{uu\theta} \equiv \frac{\langle(\partial u/\partial x)(\partial \theta/\partial x)^2\rangle}{\langle(\partial u/\partial x)^2\rangle^{1/2}\langle(\partial\theta/\partial x)^2\rangle} \sim \int_0^\infty k^2\,dk\,T(k).$$

For $P_r = \frac{1}{16}$ this is 0.286 and 0.177 for DNS and TFM, respectively.

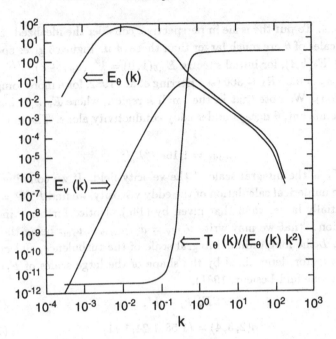

Fig. 6    $E_v(k,t)$ and $E_\theta(k,t)$, according to TFM, for large $R_\lambda \sim 500$, after several eddy turnover times into the decay. Here, the initial scale of $\theta$ is $\sim 100$ times that of $v$. This figure also gives $T_\theta(k,t)/(E_\theta(k,t)k^2)$, with scale to the right.

On the other hand, the DIA gives $S_{uu\theta} = 0.260$, a point to which we return shortly.

We believe that the reason TFM gives increasingly poor results as $P_r \to 0$ is associated with the fact that Markovian closures, such as TFM, underestimate energy (or scalar variance) transport as compared with their non-Markovian counterparts. As noted by Herring and Kerr (1982), such an underestimation increases rapidly as $R_\lambda \to 0$ (see their Fig. 8). According to that study, the TFM begins to underestimate severely the velocity derivative skewness, $S_{uuu}$, for $R_\lambda \le 5$. On the other hand, the DIA, which is not Markovianized, compares well with DNS, down to the lowest $R_\lambda \sim 0.5$ investigated.

The scalar equivalent of $R_\lambda$ is the Peclet number,

$$P_\lambda \equiv \frac{[\langle u^2\rangle\langle\theta^2\rangle/\langle(\partial\theta/\partial x)^2\rangle]^{1/2}}{P_r\nu}.$$

By analogy with the remarks of the last paragraph, we may expect the Markovianization errors to become significant for $P_\lambda \sim\!\!\leq 5$. For the

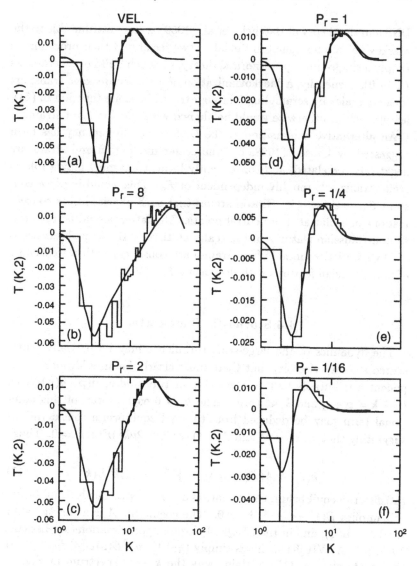

Fig. 7   Comparison of TFM values of $T_\theta(k, t = 1)$ (smooth curves) to $128^3$
DNS (histogram) of Kerr for $P_r = (8, 2, 1, 0.25, 0.0625)$, as labeled
with $T_v(k, t)$ in the upper left hand corner. (After Herring and
Métais, 1992.)

study described in Fig. 6, and for $P_r = 0.25$, $P_\lambda$ is 6.59, whereas for
$P_r = 0.0625$, $P_\lambda = 3.03$.

These observations suggest that the discrepancy observed between

DNS and the theory of Batchelor et al. (1959) is more attributable to the Markovianization implicit in the latter (we recall that their procedure is simply a single-time, quasi-normal theory, to which TFM degenerates as $P_\lambda \to 0$). Generally, a Markovianization of a theory weakens transport to small scales by scrambling (through the white-noise function in (5)), features which otherwise would be aligned with the dominant strain.

An alternative explanation of the DNS–TFM discrepancy has been suggested by Gibson (1968a,b): the scalar dissipation and strain are negatively correlated, and consequently local strain must play a role in scalar transport, roughly independent of $P_r$. This would imply a constant $S_{uu\theta}$ as $P_r \to 0$. Gibson argues that such physics imply concentrations of scalar at flow neutral points. The latter should be manifest as non-Gaussian features in the scalar statistics. But our observations of DNS statistics (mainly histograms) are consistent with a weakening of non-Gaussian features with decreasing $P_r$.

### 4.5 Stochastic Backscatter

The dynamics of the large-scale turbulence may be partially represented as eddy viscosity, but there must clearly be an additional component attributable to the interaction of two modes, $\mathbf{u}(\mathbf{p}), \mathbf{u}(\mathbf{q})$, such that $\mathbf{k} = \mathbf{p} + \mathbf{q}$ and $k \ll (p, q)$. The $k \to 0$ asymptotics of this additional term may be deduced from thermal equilibrium considerations suggesting that $E(k) \sim k^2$ and $T(k,t) = (\partial_t + 2\nu k^2)E(k,t) = 0$. Thus

$$T(k,t) \to A\{E(k), \eta(k) \dots\}k^4 - \nu_{\text{eddy}}k^2 E(k) \tag{17}$$

satisfies the equilibrium constraint, and – if trusted out of equilibrium – it implies $E(k) \to Ak^4$, $k \to 0$. The coefficient $A$ is a functional of $\mathbf{u}(\mathbf{p})$ and $\mathbf{u}(\mathbf{q})$, and in the EDQNM closure approximation (see Lesieur, 1990, p. 182, VII-6-11) this is simply $(\frac{7}{15}) \int_0^\infty p^{-2}E(p)E(p)\,dp/\eta(p)$. If one trusts closure, (17) explains why the $k \to 0$ spectrum in Fig. 1 goes like $k^4$ as $t \to \infty$; it also underlies the decay laws for total energy $E(t) \sim t^{-1.37}$, as originally proposed by Lesieur and Schertzer (1978). If $E(p), E(q)$ are both subgrid-scale, and $k$ is in the resolved scales, then we may refer to this input term as "stochastic backscatter." The incorporation of this effect into LES has been proposed and examined by Leith (1990) and Bertoglio (1977). It has also been tagged the "beating" term, since it arises from the interaction of two random modes of neighboring scales interacting to contribute to a scale much larger than either. If $k$ is not small, then we must use the more general form as given by the first term in (2). The qualitative physics of this term may be (*in part*) inferred from the Langevin amplitude model (5)–(6), a random (Gaussian) estimation of the Eulerian acceleration. However, the real dynamics are not Gaussian, and two-point closure (DIA) may be invoked to estimate the degree to which non-Gaussian modal coherency may in reality reduce the size of this term. What we have in mind here is that flows may be nonturbulent in patches (a possible form of coherent structures; Moffatt, 1990), and within such a patch the "backscatter" is nonstochastic.

An indicator of the extent to which such modal coherency reduces the backscatter term is $\langle(\partial\mathbf{u}/\partial t)^2\rangle(\mathbf{k})/\{\langle(\partial\mathbf{u}/\partial t)^2\rangle\}_G(\mathbf{k})$, where the subscript $G$ indicates an evaluation of bilinear terms entering $\partial\mathbf{u}/\partial t$ as if Gaussian. Both two-dimensional (Herring, 1980) and three-dimensional calculations (Chen et al., 1989) indicate this reduction to be by about a factor of 2, with the dominant suppression at scales near the energy containing-range. In three dimensions, the closure estimates are in fairly good agreement with the DNS.

## 4.6 Two- and Quasi-Two-Dimensional Turbulence

In three dimensions, local isotropy at small scales seems plausible by the observation that the return to isotropy proceeds at the local eddy turnover rate: $\eta(k) \sim [\int_0^k p^2\, dp\, E(p)]^{1/2}$. In an $k^{-5/3}$ range this increases with $k$, but the force driving the flow toward anisotropy does not. In two dimensions, where $E(k) \sim k^{-3}$ or steeper, this is not so; and it is more reasonable that anisotropy remains constant, or only very slowly decreases as $k \to \infty$. If these scales are a coherent part of the large scales, then we do not have the sort of universality which underlies much of LES modeling in three dimensions. However, since two dimensions is a simpler domain, we shall use it to discuss certain issues such as the need to discriminate between strain and vorticity in statistical theories, and finally to give a simple physical interpretation of (1)–(4) for anisotropic flows.

We return now to the theme of the introduction: let us suppose that a large-scale field, $\mathbf{V}(\mathbf{x})$, is stable or, at most, slowly evolving, and introduce a more rapidly evolving fluctuating field $\mathbf{v}$. We identify the large-scale field with the statistically sharp field discussed there, and – as a first step – ask where the turbulent field $\mathbf{v}$ grows, or decays. This setting is basically inhomogeneous and anisotropic, but we shall maintain the fiction that it is possible to write equations of motion for the energy spectrum and for the associated Reynolds stress spectrum of the turbulence. These will depend parametrically on position, $\mathbf{X}$. We adapt a compact notation in which $\langle u_i(-\mathbf{k}, \mathbf{X})u_j(\mathbf{k}, \mathbf{X})\rangle$ is representable in terms of $\{U_0(k, \mathbf{X}), U_2(k, \mathbf{X})\}$, where the correspondence between $\langle u_i u_j\rangle(\mathbf{k}, \mathbf{X}, t)$ is

$$\left(\frac{1}{2}\right)\langle(v_1 + iv_2)^2\rangle(\mathbf{X}, t) = \pi \int_0^\infty k\,dk\,U_2(k, \mathbf{X}, t), \qquad (18)$$

$$\left(\frac{1}{2}\right)\langle|\,v_1 + iv_2\,|^2\rangle(\mathbf{X}, t) = \pi \int_0^\infty k\,dk\,U_0(k, \mathbf{X}, t). \qquad (19)$$

Associated with $\mathbf{V}$ are strain and vorticity fields, which we write in a notation analogous to (18)–(19):

$$S(\mathbf{X}, t) \equiv -\partial_1 V_1 + \partial_2 V_2 + i(\partial_2 V_1 + \partial_1 V_2) = i(\partial_1 + i\partial_2)^2\Psi, \qquad (20)$$

$$\omega(\mathbf{X}, t) \equiv -\partial_2 V_1 + \partial_1 V_2 \equiv |\,\partial_1 + i\partial_2\,|^2\,\Psi. \qquad (21)$$

Let us consider now the fate of a turbulent eddy, with isotropic component $U_0$ and anisotropic part $U_2$, subjected to large-scale strain and vorticity, $(S, \omega)$. The strain stretches an isotropic element to an anisotropic shape, and the vorticity should simply rotate the turbulent element (locally), hence affecting only $U_2(k, \mathbf{X}, t)$. The action of strain on $U_2(k)$ should be representable as a wave number derivative, homogeneous, of degree zero (since the scale of the strain is immaterial). Moreover, for the case in which the flow is isotropic, and enstrophy is equipartitioned ($U_0 \sim k^{-2}, U_2 = 0$), we expect no anisotropy to be generated by $S$. This suggests

$$(D/Dt + 2\nu k^2)U_2(k, \mathbf{X}) = i\omega U_2 + (S/2)[1 + k\partial/\partial(2k)]U_0$$
$$- \{\rho U_2\} + CC + \cdots. \qquad (22)$$

Similarly, the equation for $U_0$ is

$$(D/Dt + 2\nu k^2)U_0(k, \mathbf{X}) = S^*[1 + k\partial/\partial(4k)]U_2 + CC\cdots. \qquad (23)$$

In (22)–(23), $D/Dt = \partial_t + \mathbf{V}\cdot\nabla$ and $CC$ means complex conjugate. The term encased in braces in (22) represents relaxation to isotropy and has been added without any justification. These equations may be derived via a Wentzel, Kramers, and Brillouin (WKB) approximation on the full, two-point Reynolds stress equation of motion (Herring, 1975). Only those terms within the "rapid distortion" approximation (mean fluctuating-field interactions) are included, with the exception of the term proportional to $\rho$.

It is interesting to note that if we assume that $DU_2/Dt$ is rapid (i.e., we simply balance $-\rho U_2(k)$ against the vorticity "tumbling" term), we may eliminate $U_2(k)$ in (23), and find

$$(D/Dt + 2\nu(k))U_0 = \mid S \mid^2 \rho/(\rho^2 + \omega^2)[1 + k\partial/\partial(4k)][1 + k\partial/\partial(2k)]U_0(k).$$
(24)

We remark that setting the right hand side of (24) equal to zero would mean that $E_0(k) = 2\pi k U_0(k) \sim k^{-3}$. Notice that regions of excess strain $S$ (over vorticity $\omega$) imply rapid transfer (to ever smaller scales), whereas regions where $\mid \omega \mid >> \mid S \mid$ are stable. This is similar to the "Weiss" criterion (Weiss, 1981) for discriminating zones of stability in two-dimensional flow. (Weiss' stability criterion is simply $I \equiv S^2 - \omega^2 < 0$). In three dimensions, $I$ is the second invariant of $S$, a point explored by Hunt et al. (1988) and by Nelkin and Tabor (1990).

If we now try to integrate (24) into the homogeneous turbulence theory, we have a fundamental problem in that for homogeneous flows, there is no way to discriminate between strain and vorticity, at least at the level of second-order moments. Thus the mechanism for discriminating between stable and active regions is lost, unless we have equations that involve more than second-order moments and associated Green's functions.

We now consider – very briefly – the physical content of theories such as TFM for anisotropic turbulence. In this case, we introduce angular harmonics such as $(U_0, U_2 \ldots)$ into (2) (which we do not record here for the two-dimensional case). The resulting equation for the anisotropy measurement, $U_2(k)$, will contain an input from the first term $(\sim U_0(p)U_2(q))$ and an eddy viscous term $(\sim U_0(q)U_2(k))$, whose general structure (in two dimensions) is

$$\frac{DU_2(k)}{Dt} = \int_0^\infty dp\, R(k,p)U_2(p) - \mu(k)U_2(k),$$
(25)

where the kernel $R(k,p)$ and $\mu(k)$ depend only on the background isotropic field. Generally, $R(k,p)$ combines a broad-band, large-scale contribution and a sharp pulse centered at $k$, which cancels much of the eddy drain term, so that the net drain is $\sim \eta(k)$. If we further make the usual diffusion approximation to the integral in (25) (see Herring, 1975, for details), we obtain for $E_2(k) \equiv \pi k U_2(k)$,

$$\frac{DE_2(k)}{Dt} = -\frac{1}{8}\partial_k E_0(k) \int_0^k \frac{p^2\, dp\, E_2(p)}{[\eta(p) + 2\eta(k)]}$$
$$- \eta_D(k)E_2(k),$$
(26)

where a numerical integration is required to determine $\eta_D(k)$. We expect an equation like (26) on heuristic grounds. Anisotropy is produced by distorting $k$-sized isotropic elements by large-scale strain, and destroyed by local interactions. It is interesting to note that a pulse of large-scale anisotropy immediately produces small-scale anisotropy, although such may disappear rapidly.

The idea that large and small scales may be more strongly correlated in two dimensions than in three must vitiate LES for this case. This lack of universality in two-dimensional flows has recently been noted by the numerical study of Maltrud and Vallis (1990). The reality of two- and quasi-two-dimensional flows may be even more exotic. The discussion following (17) suggests that active regions of excessive strain may attenuate the flow, leaving regions of excessive vorticity to survive. Numerical calculations have shown that, in fact, this does happen for both two-dimensional and quasi-geostrophic flows (Fornberg, 1977; Basdevant et al., 1981; McWilliams, 1984, 1988). The presence of such an extreme intermittency (in McWilliams' calculations the Kurtosis of vorticity is $\sim$60) is inhospitable for simple moment closures such as those discussed here. Thus the comparisons of DNS to TFM for freely decaying two-dimensional turbulence as presented in Herring and McWilliams (1984) show the latter to represent reality poorly, at late time. The TFM develops a convincing $k^{-3}$ range, whereas the DNS seems not to have a power-law range; indeed, it is not clear that its energy spectrum is in any sense universal but may reflect features of the initial data.

## 4.7 Concluding Comments

We began by noting logical difficulties inherent in attempts to split turbulence dynamics simultaneously into deterministic and statistical components. It may be that more progress would be forthcoming if we had simply sorted out the dynamics according to time scale, introducing projection operators that separate the longer from the shorter scales. We noted that two-point closure, as developed by Kraichnan (1976) and Chollet and Lesieur (1981), offered an alternative free of such problems, but the price is the decoupling of the pointwise dynamics from the statistical calculation; at best, we may only assert that the two-point version of LES yields accurate variance information for the real flow, that is, if the closure is accurate, an issue we shall discuss in the next paragraph. We have further neglected in our purview nonsimultaneous time statistics, that is, correlations of the type $\langle q(\mathbf{x}, t) q(\mathbf{x}', t') \rangle, t \neq t'$. With respect to Eulerian time differences, their inclusion would bring in large-scale sweeping, and it is difficult to see that this would be decisive. If we deal with Lagrangian time differences, the improvement may be substantial, but such a program seems difficult.

The accuracy of two-point closure for high $R_\lambda$ flows is suspect principally on two grounds. First, it is unable to follow the development of a large-scale pulse as it evolves toward its $\nu \to 0$ singularity. For an initial value problem of this kind, real flows (as judged by $256^3$ DNS) appear to evolve with at most an exponential growth in enstrophy (Kerr, 1991a), whereas closure predicts a finite-time ($\nu \to 0$) singularity. The issue is that simple (close to Gaussian) closures such as (1)–(4) are unable to follow highly non-Gaussian events, such as those involved in the vorticity strain correlation that suppresses rapid enstrophy growth and – at the same time – suppresses derivative skewness, forcing the small scales toward quasi-two-dimensional structures. We do not, however, know how this defect translates into practice; the question is how much of the flow consists of bursting phenomena most appropriately modeled by the simple large-scale Gaussian initial data.

The other issue here is the low Prandtl number regime. Here the problem is less exotic and shows up dramatically at $R_\lambda \sim 20$; TFM closure predicts a rapid decrease of scalar variance transfer with decreasing $P_r$, whereas the DNS of Kerr (1985), for example, suggests a leveling off of transfer. We have suggested that this problem is attributable to the "Markovianization" of the simpler version of statistical theory, such as TFM.

It is often remarked that LES is attempting to map a large $R_\lambda$ problem into one of modest $R_\lambda$. The two-point closure route is to introduce a hyperviscosity (7), which is different from that used in much oceanographic and meteorological literature; the main difference is the presence in (7) of $E(k_{\text{LES}})$, shutting down dissipation if no energy resides near $k_{\text{LES}}$. Yet if this is the case, and if the inertial range is multifractal, as claimed by some (Argoul et al., 1990), then LES has bought us only a modest increment unless we learn to couple an LES with the coherent structures that comprise the fractals.

Our discussion has focused primarily on homogeneous flows, with little mention of applications of two-point closure to flows with boundaries. In the homogeneous context we have well-established, simple models (such as the TFM or EDQNM) that can be compared with direct numerical simulations. Of course, such idealizations cannot be obtained in the laboratory. For bounded flows, the two-point closure exists (see, e.g.,

Kraichnan, 1964), but the formalism is so unwieldy that it would seem nearly impossible to implement without additional, probably compromising approximations. (See, however, Dannevik, 1984, who made some progress at low Rayleigh numbers.) But it may be that DNS studies – coupled with detailed comparisons with recent accurate laboratory data (Heslot et al., 1987) – would give very useful information, especially if incipient asymptotics of high $R_\lambda$ could be reached in the comparisons. For example, in the case of convection between heated and cooled plates, we now have results from DNS that are in excellent agreement with the experiments (Kerr, 1991b). The results are, in fact, at sufficiently high Rayleigh and Reynolds numbers that inertial range asymptotics begin to be manifest. This type of database could be used to develop and test models for thermal boundary layers, as well as their generalizations for shear-driven boundary layers. The convection problem is found to have shear-driven boundary layers that come from strong shears associated with the convergent surface flow that feeds fluid into thermal plumes. Such large-scale flow above the surface appears essential in explaining a nonclassical scaling regime.

Significant computing resources were necessary to achieve good agreement between the direct simulations and the experiments, although qualitative agreement is often achieved with significantly fewer computational resources. This suggests several tests for large eddy simulations. One test would use the minimum resolution needed to get qualitatively correct results, but use LES to make up the difference. A test of boundary layer parameterizations would be not to resolve the boundary layer at all, and instead replace it with a wall model. In both cases the objective would be to reproduce the heat flux and other large-scale quantities obtained in the experiments and the fully resolved direct simulations. The applicability of homogeneous models to real anisotropic physical situations could also be studied.

## Acknowledgments

We are grateful to R. H. Kraichnan for several helpful discussions. The National Center for Atmospheric Research is sponsored by the National Science Foundation.

# References

ARGOUL, F., ARNÈDO, A., GRASSEAU, G., GAGNE, Y., HOPFINGER, E.J. AND FRISCH, U. (1990) Wavelet analysis reveals the multi-fractal nature of the Richardson cascade. *Nature* **338**, 51–53.

BASDEVANT, C., LEGRAS, B., SADOURNEY, R. AND BELAND, B. (1981) A study of barotropic model flows: Intermittency, waves and predictability. *J. Atmos. Sci.* **38**, 2305–2326.

BATCHELOR, G.K. (1952) Pressure fluctuations in isotropic turbulence. *Proc. Camb. Phil. Soc.* **47**, 359–374.

BATCHELOR, G.H., HOWELLS, I.D. AND TOWNSEND, A. (1959) Small-scale variation of convected quantities like temperature in turbulent fluid. Part 2: The case of large conductivity. *J. Fluid Mech.* **5**, 134–139.

BERTOGLIO, J.P. (1977) A stochastic model for sheared turbulence. In *Macroscopic Modeling of Turbulent Flows*. Lecture Notes in Physics, Vol. 230. Ed. U. Frisch, J.B. Keller, G. Papanicolaou, and O. Pironneau, pp. 100–119. Springer-Verlag.

CHAMPAGNE, F.H., FRIEHE, C.A., LaRUE, J.C. AND WYNGAARD, J.C. (1977) Flux measurements, flux estimation techniques and fine-scale turbulence measurements in the unstable surface layer over land. *J. Atmos. Sci.* **34**, 515–530.

CHEN, H.D., HERRING, J.R., KERR, R.M. AND KRAICHNAN, R.H. (1989) Non-Gaussian statistics in isotropic turbulence. *Phys. Fluids A1*, 1844–1854.

CHOLLET, J.P. (1983) Turbulence tridimensionalle isotrope: Modelization statistique des petit échelles et simulation numérique des grandes éschelles. Thesis, Grenoble.

CHOLLET, J.P. (1985) Spectral closure to derive a subgrid scale modeling for large eddy simulations. In *Macroscopic Modeling of Turbulent Flows*. Lecture Notes in Physics, Vol. 230, Ed. U. Frisch, J.B. Keller, G.B. Papanicolaou, and O. Pironneau, pp. 161–176. Springer-Verlag.

CHOLLET, J.P. AND LESIEUR, M. (1981) Parameterization of small scales of three-dimensional isotropic turbulence utilizing spectral closure. *J. Atmos. Sci.* **38**, 2747–2757.

DANNEVIK, W.P. (1984) Two-point closure study of covariance budgets for turbulent Rayleigh Bernard convection. Ph.D. thesis, St. Louis University.

DEARDORFF, J.W. (1970) A numerical study of three-dimensional turbulent channel flow at large Reynolds numbers. *J. Fluid Mech.* **41**, 453–480.

FORNBERG, B. (1977) A numerical study of 2-D turbulence. *J. Comp. Phys.* **2**, 1–31.

FOX, D.G. AND LILLY, D.K. (1972) Numerical simulation of turbulent flow. *Rev. Geophys. Space Phys.* **10**, 51–72.

FULACHIER, L. AND DUMAS, R. (1976) Spectral analogy between temperature and velocity fluctuations in a turbulent boundary layer. *J. Fluid Mech.* **77**, 257–277.

GIBSON, C.H. (1968a) Fine structure of scalar fields mixed by turbulence. I. Zero gradient points and minimal gradient surfaces. *Phys. Fluids* **11**, 2305–2315.

GIBSON, C.H. (1968b) Fine structure of scalar fields mixed by turbulence. II. Spectral theory. *Phys. Fluids* **11**, 2316–2327.

HERRING, J.R. (1975) Theory of two-dimensional anisotropic turbulence. *J. Atmos. Sci.* **32**, 2254–2271.

HERRING, J.R. (1978) Subgrid scale modeling: An introduction and overview. In *Turbulent Shear Flow I.* Ed. B.E. Launder, F.W. Schmidt, and J.H. Whitelaw, pp. 347–351. Springer-Verlag.

HERRING, J.R. (1980) A note on Owens' mesoscale eddy simulation. *J. Phys. Oceanogr.* **10**, 804–806.

HERRING, J.R. (1984) Some contributions of two-point closure to turbulence. In *Frontiers in Fluid Mechanics.* Ed. S.H. Davis and J.L. Lumley, pp. 68–86. Springer.

HERRING, J.R. (1990) Comparison of closure to spectral-based large-eddy simulations. *Phys. Fluids A* **2**, 979–983.

HERRING, J.R. AND KRAICHNAN, R.H. (1972) Comparison of some approximations for isotropic turbulence. In *Lecture Notes in Physics,* Symposium on Statistical Models and Turbulence. Ed. M. Rosenblatt and C. Van Atta, 148–193. Springer-Verlag.

HERRING, J.R. AND KERR, R.M. (1982) A comparison of direct numerical simulation with two-point closures for isotropic turbulence convecting a passive scalar. *J. Fluid Mech.* **118**, 205–219.

HERRING, J.R., SCHERTZER, D., LESIEUR, M., NEWMAN, G.R., CHOLLET, J.P. AND LARCHEVEQUE, M. (1982) A comparative assessment of spectral closure as applied to passive scalar diffusion. *J. Fluid Mech.* **124**, 411–437.

HERRING, J.R. AND KERR, R.M. (1989) Numerical simulation of turbulence. In *Theoretical and Applied Mechanics,* Ed. P. Germain, M. Piau, and D. Caillerie, pp. 101–116. North-Holland.

HERRING, J.R. AND MÉTAIS, O. (1992) Spectral transfer and bispectra for turbulence with passive scalars. *J. Fluid Mech.* **235**, 103–121.

HESLOT, F., CASTAING, B. AND LIBCHABER, A. (1987) Transition to turbulence in helium gas. *Phys. Rev. A* **36**, 5870–5873.

HUNT, J.C.R., WRAY, A.A. AND MOIN, P. (1988) Eddies, streams, and convergence zones in turbulent flows. In *Studying Turbulence Using Numerical Simulation Database II*, Proceedings of the 1988 Summer Program, Center for Turbulence Research, Report CTR-S88, p. 1193. NASA Ames Research Center, Stanford University.

KANEDA, Y. (1981) Renormalized expansions in the theory of turbulence with the use of the Lagrangian position function. *J. Fluid Mech.* **107**, 131–145.

KERR, R.M. (1985) Higher order derivative correlations and the alignment of small-scale structures in isotropic numerical turbulence. *J. Fluid Mech.* **153**, 31–58.

KERR, R.M. (1991a) A scenario for the initiation of turbulence. Unpublished manuscript.

KERR, R.M. (1991b) Simulation of high Rayleigh number convection. Preprint.

KRAICHNAN, R.H. (1959) The structure of isotropic turbulence at very high Reynolds numbers. *J. Fluid Mech.* **5**, 497–543.

KRAICHNAN, R.H. (1961) Dynamics of nonlinear stochastic systems. *J. Math. Phys.* **2**, 124–148.

KRAICHNAN, R.H. (1964) Direct interaction for shear and thermally driven turbulence. *Phys. Fluids* **7**, 1048–1169.

KRAICHNAN, R.H. (1971) An almost-Markovian Galilean-invariant turbulence model. *J. Fluid Mech.* **47**, 513–535.

KRAICHNAN, R.H. (1976) Eddy viscosity in two- and three-dimensional turbulence. *J. Atmos. Sci.* **33**, 1521–1536.

LARCHEVEQUE, M. AND LESIEUR, M. (1981) The application of eddy-damped Markovian closure to the problem of dispersion of particle pairs. *J. Mécanique* **20**, 113–134.

LEITH, C.E. (1971) Atmospheric predictability and two-dimensional turbulence. *J. Atmos. Sci.* **28**, 145–161.

LEITH, C.E. (1990) Stochastic backscatter in a subgrid-scale model: Plane mixing layer. *Phys. Fluids A* **2**, 297–299.

LEITH, C.E. AND KRAICHNAN, R.H. (1972) Predictability of turbulent flows. *J. Atmos. Sci.* **29**, 1041–1058.

LESIEUR, M. (1987) *Turbulence in Fluids.* Martinus and Nijhoff, 286 pp.

LESIEUR, M. (1990) *Turbulence in Fluids*, 2nd Edition. Kluwer, 412 pp.

LESIEUR, M. AND SCHERTZER, D. (1978) Amortissement auto similarité d'une turbulence à grand nombre de Reynolds. *J. Mécanique* **17**, 609–646.

LESIEUR, M. AND ROGALLO, R. (1989) Large-eddy simulation of passive scalar diffusion in isotropic turbulence. *Phys. Fluids A* **1**, 718–722.

MALTRUD, M.E. AND VALLIS, G.K. (1990) Energy spectra and coherent structures in forced two-dimensional and beta-plane turbulence. *J. Fluid Mech.* **228**, 321–342.

McWILLIAMS, J.C. (1984) The emergence of isolated vortices in turbulent flows. *J. Fluid Mech.* **146**, 21–43.

MOENG, C.-H., AND WYNGAARD, J.C. (1988) Spectral analysis of large eddy simulations of the convective boundary layer. *J. Atmos. Sci.* **45**, 3573–3587.

MOFFATT, K.H. (1990) Fixed points of turbulent dynamical systems and suppression of nonlinearity. In *Whither Turbulence?* Lecture Notes in Physics, Vol. 357. Ed J.L. Lumley, pp. 250–257. Springer-Verlag.

NELKIN, M. AND TABOR, M. (1990) Time correlations and random sweeping in isotropic turbulence. *Phys. Fluids A2*, 81–83.

ORSZAG, S.A. (1977) Statistical theory of turbulence. In *Fluid Dynamics 1973, Les Houches Summer School of Theoretical Physics*. Ed. R. Balian and J.L. Peube, pp. 237–374. Gordon and Breach.

PUMIR, A. AND SIGGIA, E. (1990) Collapsing solutions to the 3-D Euler equations. *Phys. Fluids A2*, 220–241.

WEISS, J. (1981) The dynamics of enstrophy transfer in two-dimensional hydrodynamics. La Jolla Institute Report, 123 pp.

# 5

# Stochastic Backscatter Formulation for Three-Dimensional Compressible Flows

CECIL E. LEITH

## 5.1 Introduction

A subgrid-scale model with stochastic backscatter supplementing the well-known Smagorinsky eddy viscosity is formulated in the context of a three-dimensional (3D) large eddy simulation (LES) of compressible hydrodynamics. This natural extension of earlier work with a 2D LES of a shear mixing layer has been implemented on a BBN TC2000 highly parallel computer. Timing studies show that this relatively new computer architecture is well suited to such a simulation.

The nonlinearity of fluid flow leads typically to the excitation of many interacting scales of motion which, unless strongly damped by viscosity, become chaotic. The resulting turbulent motion has important transport properties which cannot be reliably predicted from statistical theories of turbulence in complicated flow configurations of practical interest. Instead one turns to numerical simulation in order to generate realizations of a flow from which average transport properties may be extracted. The Reynolds number measures the importance of nonlinearity compared with linear viscous effects. For the high Reynolds numbers that occur in many applications, the range of excited scales can far exceed the range feasible for direct numerical simulation (DNS) on present or future computers. Observations show, however, that the turbulent transport of interest is carried primarily by the largest scales of motion. This has led to the hope that LES will be adequate for practical use.

In LES, the turbulence problem is reduced to treating, as well as one can, the effect of the unresolved subgrid scales on those explicitly

computed. In the earliest models of geophysical fluid flows, which have
essentially infinite Reynolds number, it was recognized (Phillips, 1959)
that unless some artificially large viscosity was introduced, the natural
nonlinear cascade toward small scales would produce nonsensical behavior. The simplest cure is to introduce a linear viscosity large enough
that the associated Kolmogorov dissipation scale lies within the resolved
range. In practice this has been found to damp unnecessarily the large
eddies of interest, and a pragmatic approach has been to use instead a
linear hyperviscosity proportional to some high power of the Laplacian
and thus more concentrated on the barely resolvable scales. Such linear
viscosities suffer from requiring prior knowledge of the turbulent nature
of the flow and from being global in nature. They are thus not well
suited to inhomogeneous turbulence.

Over a quarter century ago, Smagorinsky (1963) introduced a nonlinear eddy viscosity tied to the grid size and to the estimated local
subgrid-scale turbulent kinetic energy production and cascade rate. It
had the decided advantage of adjusting itself to provide a local viscosity of the needed strength without prior knowledge of the turbulence.
A similar artificial viscosity had been devised by von Neumann and
Richtmyer (1950) to treat shocks in compressible flows. Although a
deeper understanding of the problem has led to the development of many
subgrid-scale viscosity prescriptions that are much more elaborate, none
has been much more successful.

Subgrid-scale viscosity is, of course, designed to account for the mean
damping effect of unresolved eddies on the larger, resolved scales of motion. But there is another effect that is qualitatively different. Subgrid-
scale eddies also induce a random forcing of the large scales through
nonlinear interaction. Although such stochastic backscatter has been
understood theoretically for some time (Kraichnan, 1976; Leslie and
Quarini, 1979) in the context of homogeneous turbulence, there has been
little experience with the consequences for LES. One immediate consequence of adding backscatter to a subgrid-scale model is that the LES
acquires a stochastic nature. An ensemble of simulations is needed to
reveal the mean and fluctuating components of the explicitly computed
flow. It is unfortunate that the large eddies are not deterministic, but
this fact is completely consistent with the chaotic nature of turbulent
flows that renders them of limited deterministic predictability (Leith
and Kraichnan, 1972) on all scales. A detailed description of a subgrid-
scale model that includes both a Smagorinsky viscosity and stochastic
backscatter is given in Section 5.2.

An extensive analysis of the theory and application of eddy viscosity and stochastic backscatter in simulations of homogeneous turbulence is provided by Chasnov (1991). He shows first that a DNS provides damping and forcing effects that are consistent with the predictions of the eddy-damped quasi-normal Markovian (EDQNM) model of homogeneous isotropic turbulence. He then uses the EDQNM model to specify the damping and stochastic forcing terms in an LES of the Kolmogorov inertial subrange. He succeeds in producing a consistent Kolmogorov spectrum, although with a value of the Kolmogorov constant Ko $\approx$ 2.1, which is somewhat higher than other estimates. The traditional approach to such an LES is to account for backscatter by diminishing the damping term, but Chasnov finds that such an approach produces no well-defined Kolmogorov spectrum and thus an ill-defined, but somewhat lower value of Ko. The explicit use of stochastic backscatter has thus been shown to lead to a more consistent LES of the Kolmogorov inertial subrange in a spectral transform model of homogeneous turbulence.

Numerical evidence of the importance of backscatter is given by Piomelli et al. (1991) with filtering experiments in DNS of turbulent channel flows. They show that at many grid points the subgrid-scale energy transfer is, in fact, into the larger, unfiltered scales of motion. As the channel wall is approached, the eddies are smaller and accurate subgrid-scale modeling becomes increasingly important. It is indeed near the wall that Piomelli et al. find backscatter effects to be the most pronounced. They have also shown the dependence of backscatter effects on the definition of the filter used to distinguish resolved and subgrid scales of motion.

Mason and Thomson (1992) have applied stochastic backscatter in conjunction with a Smagorinsky eddy viscosity for 3D LES of the turbulent planetary boundary layer of the atmosphere. The formulation in the interior of the flow is essentially the same as that given in this chapter, but modifications are introduced to treat the smaller eddies close to the surface. A long-standing problem with traditional LES for this case has been the development of unrealistically large velocity gradients near the surface. Mason and Thomson show that the inclusion of stochastic backscatter with a prescribed strength cures this particular problem without introducing other errors elsewhere.

A measure of the strength of the stochastic backscatter is the dimensionless backscatter constant $C_B$ defined such that the mean backscattered energy transfer rate is $C_B\epsilon$, and the eddy viscous damping rate is $(1 + C_B)\epsilon$, where $\epsilon$ is therefore the net energy transfer rate to small

scales. Mason and Thomson (1992) find that $C_B = 1.0$ gives the best
simulation for their problem. They also quote an unpublished evaluation
by Chasnov giving $C_B = 1.4$ based on the inertial range integrals in his
paper (Chasnov, 1991).

Another important contribution in the paper by Mason and Thomson
(1992) is a discussion of stochastic backscatter of passive scalar variance. They carry through an analogous analysis for a passive scalar
embedded in their turbulent fluid and find that a suitable value for the
corresponding scalar backscatter constant is $C_{B\theta} = 0.5$.

Stochastic backscatter was applied successfully to the plane shear
temporal mixing layer (Leith, 1990) induced by Kelvin–Helmholtz instability, but in this case the LES was 2D in spite of the use of a 3D
subgrid-scale model. Such an inconsistency is partially justified for the
shear mixing layer, which is known to generate primarily 2D large eddies.
The principal purpose of the present paper is to formulate a consistent
3D LES for compressible fluid flow.

The 3D compressible Eulerian hydrodynamic equations are given in
Section 5.3. They have been implemented with a standard two-step
Lax–Wendroff algorithm to run in parallel on up to 50 nodes presently
available on the BBN TC2000 parallel computer at the Lawrence Livermore National Laboratory. The whole domain of the calculation is
decomposed into subdomains, one for each node. Computation is carried out in parallel for the volume of each subdomain, whereas only
border information need be communicated in parallel between nodes.
The resulting surface to volume ratio benefit reduces the time spent in
communication relative to that in arithmetic to satisfactorily low levels
for calculations of reasonable size. Details of the Lax–Wendroff scheme
and its parallel implementation are provided in Section 5.4 for those
interested in the use of such relatively new computer architectures.

## 5.2 Subgrid-Scale Model

The most important requirement of a subgrid-scale model is that it
have some mechanism to simulate the mean turbulent energy cascade
from large scale to small across the resolution scale of the LES hydrodynamics model. If, as is usually assumed, the resolution length scale $\lambda$
or wave number scale $\kappa = 1/\lambda$ lies within the Kolmogorov inertial range
with energy cascade rate $\epsilon$ whose energy spectrum is given by the well-
known law $E(k) = \alpha\epsilon^{2/3}k^{-5/3}$, then there is some hope for universality

of behavior, and simple dimensional scaling arguments can be used to deduce an appropriate eddy viscosity.

The resolution length scale $\lambda$ of the model is, of course, also the length scale of the largest unresolved eddies. Let $K$ be the specific turbulent kinetic energy of the unresolved eddies given by

$$K = \int_\kappa^\infty E(k)dk \sim \epsilon^{2/3}\kappa^{-2/3} \sim \epsilon^{2/3}\lambda^{2/3}. \qquad (1)$$

We assume here, for simplicity, that the Reynolds number is infinite and thus that the inertial range extends to indefinitely high wave numbers $k$. The symbol $\sim$ means equality to within a dimensionless constant factor.

In simple eddy viscosity models, local turbulence with length scale $\lambda$ and specific kinetic energy $K$ induces an eddy viscosity with coefficient $\nu_T = C_1 K^{1/2}\lambda \sim K^{1/2}\lambda$. This is the product of an rms eddy velocity and a mixing length multiplied by an adjustable dimensionless coefficient $C_1$ that cannot be determined through this kind of dimensional scaling analysis. Similarly, from Eq. (1), the turbulent dissipation rate may be estimated by $\epsilon \sim K^{3/2}/\lambda$.

The Smagorinsky (1963) eddy viscosity is based on a local balance between shear production of subgrid-scale turbulent kinetic energy $K$ and its removal by cascade and viscous dissipation at the rate $\epsilon$. The explicitly resolved deviatoric strain rate tensor components are given by

$$S_{ij} = v_{i,j} + v_{j,i} - \tfrac{2}{3}v_{k,k}\delta_{ij}, \qquad (2)$$

where $v_i$ are the resolved velocity vector components with $x_j$ derivatives $v_{i,j}$ and $\delta_{ij}$ are the Kronecker tensor components. The summation convention applies. For any eddy viscosity coefficient $\nu_T$, the local eddy viscous stress tensor components are $\nu_T S_{ij}$, and the local shear production of subgrid-scale turbulence kinetic energy $K$ is given by the eddy viscous stress work $\nu_T S_{ik}v_{i,k} = \nu_T S^2$, where the local scalar strain rate $S$ is defined such that $S^2 = S_{ij}v_{i,j} = (S_{ij}S_{ij})/2$. The local balance condition for $K$ is therefore that $\epsilon = \nu_T S^2$, or $K^{3/2}/\lambda \sim K^{1/2}\lambda S^2$, whence $K \sim (\lambda S)^2$ and $\nu_T \sim \lambda^2 S$. The local eddy frequency is, in this case, given by the local strain rate $S$. The Smagorinsky eddy viscosity is defined as

$$\nu_T = (C_S\lambda)^2 S, \qquad (3)$$

where $C_S \approx 0.2$ is the traditional Smagorinsky constant.

The Smagorinsky eddy viscosity is generalized in two ways in order to apply it to the compressible flow model considered here. In addition to the shear source of $K$ we also include a buoyancy source which depends on the scalar product of the local acceleration and eddy mass flux. We define a quantity

$$B^2 = -\left(\frac{1}{\text{Sc}}\right)\left(\frac{\rho_{,j}}{\rho}\right)\left(\frac{p_{,j}}{\rho}\right) \tag{4}$$

if it is positive; otherwise it vanishes. Here Sc $\approx 0.7$ is a Schmidt number for eddy mass diffusion. The shear and buoyancy sources of $K$ can now be combined as $\nu_T(S^2+B^2)$. There is also a compressive eddy work term depending on the product of the divergence $D = v_{i,i}$ and the turbulent pressure $\frac{2}{3}K$. The generalized balance equation for $K$ becomes

$$\nu_T(S^2 + B^2) - \tfrac{2}{3}KD = \epsilon, \tag{5}$$

the generalized eddy frequency becomes

$$\mathbf{S} = (S^2 + B^2 + C_D{}^2D^2)^{1/2} - C_D D, \tag{6}$$

and the generalized eddy viscosity is

$$\nu_T = (C_S\lambda)^2\mathbf{S}. \tag{7}$$

By choosing the dimensionless coefficient $C_D \approx 10$ one finds that in the limit of the strong compression found in a shock the eddy viscosity becomes the artificial viscosity introduced by von Neumann and Richtmyer (1950). Note that for $D > 0$ the compression terms in Eq. (6) tend to cancel. Test calculations indicate, however, that such a large value of $C_D$ leads to numerical instability in the Lax–Wendroff scheme described in Section 4. For current simulations of buoyancy-driven flows at low Mach number, $C_D$ has, therefore, been set to zero, and only the buoyancy generalization is being used.

We next consider the stochastic backscatter, which, for isotropic homogeneous turbulence in three dimensions, is known (Kraichnan, 1976; Chasnov, 1991) to have a $k^4$ spectrum to lowest order in wave number $k$. This means that backscatter is concentrated on scales of motion that are only a little larger than the resolvable limit. In a 3D fluid dynamics code, the lowest order effect of stochastic backscatter may be achieved by introducing on the computational grid an isotropic space- and time-white random acceleration vector potential from which is derived a nondivergent random acceleration term to be added to the momentum equation.

The mean of the random acceleration potential is zero. Its variance is determined by the following dimensional scaling argument. The space derivative of a vector potential $\phi_i[L^2T^{-2}]$ is an acceleration $a_j[LT^{-2}]$ with the dimensions shown. The space–time covariance of $\phi_i$ is given by

$$\Phi_{ik}(\mathbf{x}, t; \mathbf{x}', t') = < \phi_i(\mathbf{x}, t)\phi_k(\mathbf{x}', t') > [L^4T^{-4}] \tag{8}$$
$$= \Psi(\mathbf{x}, t)\delta(\mathbf{x} - \mathbf{x}')\delta(t - t')\delta_{ik}.$$

Equation (8) implicitly defines a variance function $\Psi(\mathbf{x}, t)$ for the assumed isotropic space- and time-white random process. It may be written explicitly as

$$\Psi(\mathbf{x},t) = \tfrac{1}{3} \int dt' \int d\mathbf{x}' \; \Phi_{kk}(\mathbf{x},t;\mathbf{x}',t')[L^7 T^{-3}]. \tag{9}$$

As with eddy viscosity, one now constructs a dimensionally correct expression for $\Psi(\mathbf{x},t)$ in terms of the local eddy frequency $S[T^{-1}]$ and the resolution length scale $\lambda[L]$. To within a dimensionless coefficient, this must then be $\Psi(\mathbf{x},t) \sim S^3 \lambda^7$. In finite difference approximation, $\phi_k$ can be space- and time-white only on resolved scales. In fact, the integral of Eq. (9) becomes, in finite approximation,

$$\Psi(\mathbf{x},t) = \tfrac{1}{3} < \phi_k \phi_k > \lambda^3 \delta t. \tag{10}$$

The dimensionally proper scaling is therefore achieved if the random acceleration potential for each component and at each grid point and time step are chosen independently as

$$\phi_k = C_b (S\delta t)^{3/2} (\lambda/\delta t)^2 g_k [L^2 T^{-2}], \tag{11}$$

where each component of $g_k$ is a unit Gaussian random number, i.e., drawn from a population with zero mean and unit variance. All dimensionless coefficients in this derivation are lumped into the single one $C_b$, which is adjustable but should be of order 1.

The inclusion of shock effects in the eddy frequency $S$ used in Eq. (11) may lead to too much random backscatter noise being induced by shock passage. Numerical experiments are needed to devise, if necessary, an appropriate modification of the scheme.

The procedure just described is based on the implicit assumption that the resolution scale $\lambda$ is the same as the grid interval, say $\delta x$. But random disturbances on the scale $\delta x$ are so poorly treated by finite-difference schemes that it may be better to run a smoothing average filter over the field of $\phi_k$'s chosen above so that the filtered field has spatial correlation extending over about $2\delta x$. We may still take $\lambda = \delta x$ and absorb the filter effect into the constant $C_b \approx 0.4$.

## 5.3 Compressible Hydrodynamic Equations

The equations of motion for 3D compressible hydrodynamics are formulated in conservation form as

$$\partial \mathbf{U}/\partial t + \mathbf{F}_{i,i} = \mathbf{K} \tag{12}$$

in terms of the 5-vector of predicted variables

$$\mathbf{U} = \begin{pmatrix} \rho \\ \rho v_1 \\ \rho v_2 \\ \rho v_3 \\ \rho e \end{pmatrix}, \tag{13}$$

three components of the 5-vector of fluxes

$$\mathbf{F}_i = \begin{pmatrix} \rho v_i \\ \rho v_1 v_i - \tau_{1i} + \epsilon_{1ij}\rho\phi_j \\ \rho v_2 v_i - \tau_{2i} + \epsilon_{2ij}\rho\phi_j \\ \rho v_3 v_i - \tau_{3i} + \epsilon_{3ij}\rho\phi_j \\ \rho e v_i - \tau_{ij} v_j + \sigma_i \end{pmatrix}, \tag{14}$$

and the 5-vector of sources

$$\mathbf{K} = \begin{pmatrix} 0 \\ \rho f_1 \\ \rho f_2 \\ \rho f_3 \\ \rho f_i v_i + \rho q \end{pmatrix}. \tag{15}$$

The variables that appear in these equations are density $\rho$, velocity $v_i$, total specific energy $e$, stress tensor $\tau_{ij}$, stochastic backscatter potential $\phi_k$ of Eq. (11), diffused energy flux $\sigma_i$, external specific body force $f_k$, and external specific energy source $q$. The total specific energy is made up of internal specific energy $i$ and kinetic energy so that $e = i + v_i v_i/2$. The summation convention applies, and $\epsilon_{ijk}$ is the standard alternating tensor. The stress tensor is made up of an isotropic part involving the gas pressure $p$ assumed to satisfy a $\gamma$-law equation of state, $p = (\gamma-1)\rho i$, and an eddy viscous part so that

$$\tau_{ij} = -p\delta_{ij} + \rho\nu_T S_{ij}, \tag{16}$$

where $\nu_T$ is the generalized Smagorinsky eddy viscosity coefficient given by Eq. (7). The eddy diffusion coefficient $\kappa_T$ for internal energy is assumed to be proportional to the eddy viscosity coefficient; thus $\kappa_T = \nu_T/\mathrm{Pr}$, where Pr is the dimensionless eddy Prandtl number. The eddy flux of internal energy is given by

$$\sigma_k = -\kappa_T i_{,k}. \tag{17}$$

Note that in this formulation stochastic backscatter is introduced in Eq. (14) as a random stress. For constant density this reduces to the nondivergent acceleration discussed in Section 5.2 and in an earlier note

(Leith, 1990). The present formulation ensures that stochastic backscatter conserves momentum exactly rather than only statistically.

## 5.4 Numerical Scheme

The finite-difference model for the integration of the equations of motion is based on the Lax–Wendroff predictor–corrector scheme (Richtmyer and Morton, 1967). Let $\mathbf{Uo}$ be the old values of $\mathbf{U}$ at the beginning of the time step, $\mathbf{Up}$ the predicted values appropriate for a half time step, and $\mathbf{U}$ the new values at the end of the step. The three-dimensional domain of the model is divided by a regular Cartesian mesh into a three-dimensional array of cubical cells. The fields $\mathbf{U}$ and the fluxes $\mathbf{F}_i$ based on them by Eq. (14) are defined as cell-centered quantities. Spatial differences are obtained by differencing neighbor values to east and west for $x_1$, to north and south for $x_2$, and up and down for the $x_3$-direction. In obvious notation the predictor step becomes

$$\mathbf{Up} = (\mathbf{Uo}^e + \mathbf{Uo}^w + \mathbf{Uo}^n + \mathbf{Uo}^s + \mathbf{Uo}^u + \mathbf{Uo}^d)/6$$
$$- (\delta t/4\delta x_1)[\mathbf{F}_1(\mathbf{Uo}^e) - \mathbf{F}_1(\mathbf{Uo}^w)]$$
$$- (\delta t/4\delta x_2)[\mathbf{F}_2(\mathbf{Uo}^n) - \mathbf{F}_2(\mathbf{Uo}^s)]$$
$$- (\delta t/4\delta x_3)[\mathbf{F}_3(\mathbf{Uo}^u) - \mathbf{F}_3(\mathbf{Uo}^d)]$$
$$+ (\delta t/2)\mathbf{K}(\mathbf{Uo}), \tag{18}$$

and the corrector step

$$\mathbf{U} = \mathbf{Uo} - (\delta t/2\delta x_1)[\mathbf{F}_1(\mathbf{Up}^e) - \mathbf{F}_1(\mathbf{Up}^w)]$$
$$- (\delta t/2\delta x_2)[\mathbf{F}_2(\mathbf{Up}^n) - \mathbf{F}_2(\mathbf{Up}^s)]$$
$$- (\delta t/2\delta x_3)[\mathbf{F}_3(\mathbf{Up}^u) - \mathbf{F}_3(\mathbf{Up}^d)]$$
$$+ (\delta t)\mathbf{K}(\mathbf{Up}). \tag{19}$$

At the end of the time step, an update $\mathbf{Uo} = \mathbf{U}$ is needed to prepare for the next time step.

As a simple, specific example of the parallel implementation of this algorithm, consider the numerical simulation of compressible hydrodynamics in a periodic cube subdivided into $48 \times 48 \times 48 = 110,592$ cubical cells. Make a 2D decomposition into $3 \times 3 = 9$ columns, each consisting of $16 \times 16 \times 48 = 12,288$ cells, and assign each column to one processor node. Owing to the compact nature of the finite-difference stencil, each node needs information only from a border layer of cells in each of four neighboring nodes. The cell arrays on each node are expanded to $18 \times 18 \times 50 = 16,200$ in order to include such outer border cells. The

*Cecil E. Leith*

data for such outer border cells are obtained by an exchange between inner and outer border cells on adjacent nodes. Such data exchanges require internodal communication that is independent of the details of the numerical algorithm as long as it involves an explicit forward time step calculation based at most on a $3 \times 3 \times 3 = 27$ cell stencil.

The computational time step cycle reduces then to a sequence of arithmetic evaluation sweeps interleaved with communication sweeps all done in parallel. The nature and timing of the sequence in an implementation of this simple example on the BBN TC2000 are as follows:

| Time cycle | sec |
|---|---|
| Gaussian | |
|     Generate Gaussian ($18 \times 18 \times 50 \times 3 = 48,600$) | 3.8 |
|     Exchange borders ($4 \times 18 \times 50 \times 3 = 10,800$) | 0.1 |
| Predictor step | |
|     Compute fluxes ($16 \times 16 \times 48 \times 5 \times 4 = 245,760$) | 1.6 |
|     Exchange borders ($4 \times 18 \times 50 \times 5 \times 4 = 72,000$) | 0.6 |
|     Advance hydro ($16 \times 16 \times 48 \times 5 = 61,440$) | 1.1 |
|     Exchange borders ($4 \times 18 \times 50 \times 5 = 18,000$) | 0.2 |
| Corrector step | |
|     Compute fluxes ($245,760$) | 1.6 |
|     Exchange borders ($72,000$) | 0.6 |
|     Advance hydro ($61,440$) | 0.8 |
|     Exchange borders ($18,000$) | 0.2 |
| Update | 0.1 |
| Total | 10.7 |
|     Total exchange | 1.7 |

In parentheses are shown either the number of 64-bit results computed or the number of 64-bit words transmitted. For this calculation the computational speed per node is about 1 Mflop per second and the communication time is about 9 $\mu$s per word.

In the time cycle the first step produces a field of random Gaussian $g_k$ for which border values are then exchanged with neighbors in order to provide a consistent basis for the generation of the backscatter potentials $\phi_k$ as part of the flux calculations. The same $g_k$ field is used for both the predictor and corrector step.

The extra price paid for the use of parallel computing is communication, which in this example takes about 15% of the total time. For the domain-decomposition message-passing paradigm used here the scaling properties are quite simple. If, for example, we double each dimension of the cube to $96 \times 96 \times 96 = 884,736$ cells, and at the same time increase the node array to $6 \times 6 = 36$, so that there are now $16 \times 16 \times 96$ cells on each node, both arithmetic time and communication time are doubled, and the ratio remains the same. If, on the other hand, for the larger cube we still use a $3 \times 3 = 9$ array of nodes, then the arithmetic time is increased eightfold whereas that for communication only fourfold. This reflects the surface to volume ratio benefit enjoyed by communication relative to arithmetic.

### 5.5 Outlook

Although the basic LES model has been tested against simple known solutions such as acoustic standing waves, it has not yet been applied to the study of turbulent mixing layers. Before a large investment in computing time is made for such studies, better visualization tools are needed, and this is where the present effort is focused. Early studies are expected to be of a 3D version of the shear mixing layer calculations made earlier in 2D and of the buoyancy-driven mixing layer induced by Rayleigh–Taylor instability.

## Acknowledgments
The staff of the Lawrence Livermore National Laboratory Massively Parallel Computing Initiative provided outstanding support for this effort. Eugene Brooks, as leader of the Initiative, and Tammy Welcome, as developer of the Livermore Message Passing System, were particularly helpful. Paul Amala developed a single-node version of the Lax–Wendroff scheme that was the starting point of the parallel version used here. This work was performed under the auspices of the U.S. Department of Energy by the Lawrence Livermore National Laboratory under Contract No. W-7405-ENG-48.

## References
CHASNOV, J.R. (1991) Simulation of the Kolmogorov inertial subrange using an improved subgrid model. *Phys. Fluids A* **3**, 188–200.

KRAICHNAN, R.H. (1976) Eddy viscosity in two and three dimensions. *J. Atmos. Sci.* **33**, 1521–1536.

LEITH, C.E. (1990) Stochastic backscatter in a subgrid-scale model: Plane shear mixing layer. *Phys. Fluids A* **2**, 297–299.

LEITH, C.E. AND KRAICHNAN, R.H. (1972) Predictability of turbulent flows. *J. Atmos. Sci.* **29**, 1041–1058.

LESLIE, D.C. AND QUARINI, G.L. (1979) The application of turbulence theory to the formulation of subgrid modelling procedures. *J. Fluid Mech.* **91**, 65–91.

MASON, P.J. AND THOMSON, D.J. (1992) Stochastic backscatter in large-eddy simulations of boundary layers. *J. Fluid Mech.* **242**, 51–78.

PHILLIPS, N.A. (1959) An example of nonlinear computational instability. In *The Atmosphere and the Sea in Motion,* Rossby Memorial Volume, pp. 501–504, Rockefeller Institute Press.

PIOMELLI, U., CABOT, W.H., MOIN, P., AND LEE, S. (1991) Subgrid-scale backscatter in turbulent and transitional flows. *Phys. Fluids A* **3**, 1766–1771.

RICHTMYER, R.D. AND MORTON, K.W. (1967) *Difference Methods for Initial-Value Problems,* Wiley-Interscience.

SMAGORINSKY, J. (1963) General circulation experiments with the primitive equations: The basic experiment. *Mon. Wea. Rev.* **91**, 99–165.

VON NEUMANN, J. AND RICHTMYER, R.D. (1950) A method for the numerical calculation of hydrodynamic shocks. *J. Appl. Phys.* **21**, 232–237.

# PART TWO

## LARGE EDDY SIMULATION IN ENGINEERING

# 6

---

# Applications of Large Eddy Simulations in Engineering: An Overview

### UGO PIOMELLI

## 6.1 Introduction

Over twenty years have passed since the first large eddy simulation (LES) results by Deardorff (1970) were published. During this period, this technique has matured considerably: the underlying theory has been advanced, new models have been developed and tested, more efficient numerical schemes have been used. The progress in computer power and memory has made possible the application of LES to a variety of flows, compressible and incompressible, including heat transfer, stratification, passive scalars and chemical reactions.

Despite the notable advancements made since the 1970s, LES cannot yet be considered an engineering tool in the sense that the vast majority of problems tackled with this technique are still "building-block" flows: problems that isolate one or two physical phenomena of engineering interest in simplified geometries. To date, very rarely has LES been applied to actual engineering configurations, such as flows in complex geometries and at high Reynolds numbers.

This chapter reviews the development of LES in order to put the present state of this technique in historical perspective. Since it is aimed more at giving an overview of the current state of LES than at discussing in detail all past contributions to the field, the reader is referred to the review articles by Rogallo and Moin (1984), Yoshizawa (1987) and Reynolds (1990) and the book by Lesieur (1990) for more in-depth analyses of the literature.

## 6.2 Past Applications

In large eddy simulations the large energy-carrying scales are directly computed, and only the effects of the small subgrid ones are modeled. The large-scale quantities (indicated by an overbar) are defined by the filtering operation,

$$\overline{f}(\mathbf{x}) = \int f(\mathbf{x}')G(\mathbf{x},\mathbf{x}')\,d\mathbf{x}', \tag{1}$$

in which $G$ is the filter function and the integral is extended over the entire domain. Filter functions commonly used include the Gaussian, the sharp Fourier cutoff and the top hat in real space (Leonard 1974). Applying the filtering operation to the appropriate set of governing equations, one obtains the filtered set of governing equations. For incompressible, isothermal flows these are the filtered continuity and Navier–Stokes equations, given, in dimensionless form, by

$$\frac{\partial \overline{u}_i}{\partial x_i} = 0, \tag{2}$$

$$\frac{\partial \overline{u}_i}{\partial t} + \frac{\partial}{\partial x_j}(\overline{u}_i\overline{u}_j) = -\frac{\partial \overline{p}}{\partial x_i} - \frac{\partial \tau_{ij}}{\partial x_j} + \frac{1}{Re}\frac{\partial^2 \overline{u}_i}{\partial x_j \partial x_j}. \tag{3}$$

Equations (2)–(3) govern the evolution of the large scales. The effects of the small scales appear in the subgrid-scale (SGS) stresses,

$$\tau_{ij} \equiv \overline{u_i u_j} - \overline{u}_i\overline{u}_j, \tag{4}$$

that must be modeled. The SGS stresses are often decomposed into three parts (Leonard 1974): the resolvable part, also known as "Leonard stresses," $L_{ij} \equiv \overline{\overline{u}_i\overline{u}_j} - \overline{u}_i\overline{u}_j$, the cross terms, $C_{ij} \equiv \overline{\overline{u}_i u'_j} + \overline{\overline{u}_j u'_i}$, and the SGS Reynolds stresses, $R_{ij} \equiv \overline{u'_i u'_j}$. Most of the existing models are of the eddy viscosity type: they assume proportionality between the anisotropic part of the SGS stress tensor, $\tau_{ij}^a = \tau_{ij} - \delta_{ij}\tau_{kk}/3$, and the large-scale strain rate tensor, $\overline{S}_{ij}$:

$$\tau_{ij}^a = -2\nu_T \overline{S}_{ij} = -\nu_T \left( \frac{\partial \overline{u}_i}{\partial x_j} + \frac{\partial \overline{u}_j}{\partial x_i} \right). \tag{5}$$

Dimensional arguments suggest that the eddy viscosity, $\nu_T$, should be given by the product of a velocity scale, $q$, and a length scale, $\ell$; $\ell$ is usually related to the filter width $\Delta$; various models differ in their prescription for $q$.

Perhaps the principal effect of the SGS model on the resolved scales can be best explained by considering the transport equation for the resolved kinetic energy, $\overline{q}^2 = \overline{u}_i\overline{u}_i$:

$$\frac{\partial \overline{q}^2}{\partial t} + \frac{\partial}{\partial x_j} \left( \overline{q}^2 \overline{u}_j \right) = \frac{\partial}{\partial x_j} \left( -2\overline{p}\,\overline{u}_j - 2\overline{u}_i \tau_{ij} + \frac{1}{Re} \frac{\partial \overline{q}^2}{\partial x_j} \right)$$
$$- \frac{2}{Re} \frac{\partial \overline{u}_i}{\partial x_j} \frac{\partial \overline{u}_i}{\partial x_j} + 2\tau_{ij} \overline{S}_{ij}. \qquad (6)$$

The SGS dissipation, $\epsilon_{SGS} \equiv \tau_{ij} \overline{S}_{ij}$, represents the energy transfer between resolved and subgrid scales. If it is negative, the subgrid scales remove energy from the resolved ones (forward scatter); if it is positive, they inject energy into the resolved scales (backscatter). In three-dimensional engineering turbulent flows, energy cascades from the large to the small scales (on the average). Thus, the main function of the SGS model is to drain the appropriate amount of energy from the resolved scales. In eddy viscosity models, $\epsilon_{SGS} = -2\nu_T \overline{S}_{ij} \overline{S}_{ij} = -\nu_T |\overline{S}|^2$; as long as $\nu_T$ is non-negative, $\epsilon_{SGS} \leq 0$ and the model is said to be "dissipative."

Similar considerations apply to compressible flows; the difference between incompressible and compressible flow formulations is that, instead of spatial filtering, it is customary to introduce Favre filtering, $\tilde{f} = \overline{\rho f}/\overline{\rho}$, in the latter. The Favre-filtered SGS stresses and heat flux are then

$$\tau_{ij} = \overline{\rho} \left( \widetilde{u_i u_j} - \tilde{u}_i \tilde{u}_j \right), \qquad (7)$$
$$Q_k = c_p \overline{\rho} \left( \widetilde{u_k T} - \tilde{u}_k \tilde{T} \right), \qquad (8)$$

where $T$ is the temperature and $c_p$ is the specific heat at constant pressure.

### 6.2.1 Smagorinsky Model

LES was first applied by Deardorff (1970), who carried out a simulation of the turbulent flow in a channel at infinite Reynolds number. Approximate boundary conditions were used to ensure that the logarithmic law-of-the-wall was obeyed; the dynamics of the wall layer, in which most of the turbulence is produced, were not computed. The mean velocity and Reynolds stress profiles were within 30–50% of the experimental values.

To parameterize the effect of the small subgrid scales on the large scales, Deardorff used the eddy viscosity model introduced by Smagorinsky (1963) and further developed by Lilly (1966). This model assumes $\ell = C_S \Delta$ (where $C_S$ is the Smagorinsky constant and $\Delta$ is the filter width); the velocity scale is obtained by assuming that the small scales are in equilibrium, and that energy production and dissipation are in balance. This yields

$$\nu_T = (C_S \Delta)^2 |\overline{S}|. \qquad (9)$$

Lilly (1966) determined that, for homogeneous isotropic turbulence with cutoff in the inertial subrange, $C_S \simeq 0.17$. In the presence of mean shear, however, Deardorff (1970) found this value to cause excessive damping of large-scale fluctuations, and in his simulation he used $C_S = 0.094$. A historical overview of the Smagorinsky model and of its relation to other theories can be found in Chapter 1 in this volume.

In the wake of the interest raised by these calculations, considerable effort was spent in developing theoretical foundations for large eddy simulations. The Stanford–Ames group applied LES to a variety of homogeneous unbounded flows (Kwak, Reynolds and Ferziger 1975; Shaanan, Ferziger and Reynolds 1975; Clark, Ferziger and Reynolds 1979), developing at the same time the *a priori* test in which the results of a direct numerical simulation (DNS) are filtered to yield *exact* large-scale and SGS fields that can be used to study the accuracy of a model or the physical behavior of cross terms, Leonard stresses and SGS Reynolds stresses. As a result of the *a priori* investigations, Clark et al. (1979) found that eddy viscosity models of the Smagorinsky type predict the global energy transfer from large to small scales with acceptable accuracy, but fail to predict the local stresses. McMillan, Ferziger and Rogallo (1980) found that the value of the Smagorinsky constant, $C_S$, must be lowered in the presence of strain or shear. Furthermore, in these cases the correlation between modeled and exact stresses and dissipation becomes very poor, of the order of 10%.

Moin, Reynolds and Ferziger (1978) and Moin and Kim (1982) performed simulations of turbulent channel flows in which the wall layer was resolved. To force the SGS stresses to vanish at the solid boundary and to account for the decreased size of the small-scale structures due to the presence of the wall, the length scale, $\ell$, in the Smagorinsky model (9) had to be modified. Near the wall, Moin et al. (1978) matched $C_S \Delta$ to a mixing length, $\ell = \kappa y_w$ (where $y_w$ is the distance from the nearest wall), while Moin and Kim (1982) used damping functions of the Van Driest (1956) type. Moin and Kim (1982) also used an SGS stress model similar to Schumann's (1975) (see below) that divided the SGS stresses into a locally isotropic part and an anisotropic part. The mean velocity, Reynolds stresses, turbulent intensities and higher order statistics predicted by Moin and Kim (1982) agreed well with experimental results, although the streak spacing was too large.

Mason and Callen (1986) studied the effect of numerical resolution on the optimal value of the Smagorinsky constant, $C_S$, by performing various simulations of turbulent channel flow. They concluded that, with $C_S = 0.2$, the LES gives accurate results as long as the mesh resolution is adequate. In their simulation, however, the wall layer, which is the region where the largest shear occurs, was bypassed.

Horiuti (1987) analyzed the numerical method used by Moin et al. (1978) and Moin and Kim (1982) and showed that the use of the rotational form of the Navier–Stokes equations coupled with finite differences in the normal direction produces large truncation errors near the wall. He suggested using the conservative skew-symmetric scheme by Arakawa (1966) instead; Zang (1991) found that this scheme also reduces aliasing errors. Horiuti's results compared better with experiments than did those of Moin and Kim (1982).

The Smagorinsky model has also been applied to studies of transitional flows. An *a priori* study of the SGS stresses by Piomelli, Zang, Speziale and Hussaini (1990a) showed, however, that, during transition, substantial regions of the flow exhibit a reversed energy cascade (that is, energy flowing from small to large scales). The Smagorinsky model cannot predict this phenomenon (in fact, it predicts non-zero SGS stresses even in laminar flows) and damps the growth of the perturbations excessively. Piomelli et al. (1990a) and Piomelli and Zang (1991) overcame this shortcoming by introducing an additional empiricism in the form of an intermittency function that modified the Smagorinsky constant by effectively setting it to zero during the linear and early nonlinear stages of transition.

Finally, Yoshizawa (1986) used the direct interaction approximation by Kraichnan (1964) to derive a Smagorinsky-type SGS model valid in mildly compressible flows. Speziale, Erlebacher, Zang and Hussaini (1988), however, tested this model *a priori* and found that it correlates poorly with the exact stresses even in homogeneous isotropic turbulence.

*A priori* tests of the Smagorinsky model (Clark et al. 1979; McMillan et al. 1980) have shown that the modeled stresses and SGS dissipation correlate rather poorly with the exact values. The fact that this model has been fairly successful in the prediction of turbulent flows may be due to its ability to predict the global energy transfer adequately even if the local transfer is incorrect. For nonequilibrium or transitional flows, however, its shortcomings become more evident, and in more complex geometries, involving separation, reattachment, strongly three-dimensional effects and so on, its performance might become inadequate. Furthermore, the fact that various ad hoc adjustments of the length scales were required indicates that the model cannot represent turbulent fields in sheared flows, in flows near solid walls or in transitional regimes with a single universal constant. Hence, an effort has been made, particularly during the past five years, to develop more accurate models capable of accounting for more complex phenomena.

### 6.2.2 One-Equation Models

Schumann (1975) performed an LES calculation of the flow in a plane channel and in an annulus using a model that divides the SGS stresses into a locally isotropic and an anisotropic part. A transport equation for the SGS energy, $q_{SGS}^2 = \tau_{kk}$, was solved to provide the velocity scale for the eddy viscosity. Approximate boundary conditions were used.

Grötzbach (1979, 1987) used an approach similar to that of Schumann (1975) to calculate turbulent flows in the presence of heat transfer. He used a one-equation model and a finite-volume approach similar to those employed by Schumann (1975). He also developed improved approximate boundary conditions to bypass the wall layer.

Horiuti and Yoshizawa (1985) also used a transport equation for the SGS energy in the LES of turbulent channel flow. They found that the one-equation model gave more accurate results when very coarse grids were used; on finer grids, however, there was little difference between the one-equation and the Smagorinsky model results.

One-equation models have several disadvantages: first, the expense involved in solving an additional equation does not seem to be justified by improvements in the accuracy. Furthermore, the "dissipation of SGS energy" is usually taken to be proportional to $q_{SGS}^3/\Delta$. An unpublished *a priori* study by the author indicates that this assumption is extremely poor in channel flows. Finally, in laminar flows the right-hand side of the transport equation vanishes identically; this implies that, if the initial condition of a calculation consists of a large-scale perturbation (which does not contribute to the SGS energy) superimposed on a laminar profile, the subgrid-scale energy will remain equal to zero throughout the calculation.

### 6.2.3 Two-Point Closures

Two-point closures have been an alternative way to derive SGS stress models. Using the test field model (TFM), Kraichnan (1976) wrote the energy transfer spectrum in terms of an "effective eddy viscosity" $\widehat{\nu}_T(k)$ (where $k$ is the wave number), defined in Fourier space and evaluated by assuming a Kolmogorov spectrum; $\widehat{\nu}_T$ was found to exhibit a cusplike behavior near the cutoff wave-number, $K_c$, which corresponds to the local energy transfer. Chollet and Lesieur (1981) used the eddy-damped, quasi-normal Markovian (EDQNM) theory to obtain the same results. The Chollet–Lesieur (1981) model uses $[K_c E(K_c)]^{1/2}$ (where $K_c$ is the cutoff wave number and $E(k)$ is the kinetic energy spectrum) as velocity scale, and $K_c^{-1}$ as length scale for the eddy viscosity, which, in wave space, is given by

$$\widehat{\nu}_T(k) = \widehat{\nu}_T^+(k/K_c)\,[E(K_c)/K_c]^{1/2}, \tag{10}$$

where $\widehat{\nu}_T^+(k/K_c)$ can be approximated by (Chollet 1985)

$$\widehat{\nu}_T^+(x) = 0.267 + 9.21\exp(-3.03/x). \tag{11}$$

Chollet and Lesieur (1981) simulated homogeneous isotropic turbulence and recovered the Kolmogorov spectrum. The model has subsequently been extended to anisotropic flows. Dang (1985) used it to simulate turbulence subjected to two successive plane strains, while Aupoix (1986) simulated the same experiment using the simplified EDQNM model of Cambon, Jeandel and Mathieu (1981).

Among the advantages of the Chollet–Lesieur model is that it produces zero eddy viscosity as long as there is no energy near the cutoff. This property has been exploited by Dang and Deschamps (1987), who computed laminar–turbulent transition in a plane channel. Its main disadvantage is the fact that it is defined in wave space, which hampers its extension to finite-difference schemes and to complex geometries. To overcome this shortcoming, O. Métais and M. Lesieur (personal communication) derived the "structure function model." Assuming a cutoff wave number in the inertial region of a Kolmogorov spectrum, they expressed the energy spectrum at the cutoff, $E(K_c)$, in terms of the large-scale second-order velocity structure function,

$$\overline{F_2}(r) = <[\overline{u}_i(x+r) - \overline{u}_i(x)]\,[\overline{u}_i(x+r) - \overline{u}_i(x)]>, \tag{12}$$

where $<\,>$ is an appropriate spatial average, and obtained

$$\nu_T = 0.063\Delta\,[\overline{F_2}(\Delta)]^{1/2}. \tag{13}$$

This model has been used for large eddy simulations of homogeneous isotropic turbulence and performed better than (10). It has also been applied to a transitional compressible boundary layer over a flat plate by Normand and Lesieur (1990). Their results were qualitatively correct; the amplification rates, however, were overestimated by a factor of 2, perhaps due to insufficient resolution in the wall-normal direction. Chapter 9 of this volume describes other applications of the structure function model. For a comparison of the predictions of TFM and related theories with DNS results, see Chapter 4.

### 6.2.4 Scale Similarity Models

Bardina, Ferziger and Reynolds (1980) introduced the mixed model, which is based on the assumption that the main interaction between resolved and SGS eddies takes place between the smallest resolved eddies and the largest SGS ones. The SGS velocity is approximated by the difference between the filtered and twice-filtered velocities, $\bar{u}_i - \bar{\bar{u}}_i$. This gives a model of the form

$$\tau_{ij}^a = C_B \left[ (\bar{u}_i \bar{u}_j - \bar{\bar{u}}_i \bar{\bar{u}}_j) - \delta_{ij} (\bar{u}_k \bar{u}_k - \bar{\bar{u}}_k \bar{\bar{u}}_k)/3 \right] - 2\nu_T \bar{S}_{ij}, \qquad (14)$$

in which a Smagorinsky-like term (with $\nu_T$ given by (9)) is introduced to account for the proper energy dissipation; the model is Galilean-invariant if $C_B = 1$ (Speziale 1985). Bardina et al. (1980) found that the mixed model gives good correlation between modeled and exact stresses even in flows with nonzero mean shear. Bardina, Ferziger and Reynolds (1983) subsequently used the model for simulations of homogeneous turbulence in a rotating coordinate frame and of sheared turbulence.

Piomelli, Moin and Ferziger (1988) examined the relationship between the filter and the SGS model. They found that the mixed model gives more accurate results than the Smagorinsky model when a Gaussian filter is used, while the Smagorinsky model is fairly accurate when coupled with a sharp Fourier cutoff filter.

Speziale et al. (1988) proposed a compressible extension of the mixed model (14), which correlates better with the exact stresses than Yoshizawa's (1986) model. Since, however, in their initial conditions both the velocity divergence and its time derivative were zero, the velocity fields used in their tests were essentially incompressible.

Extensive *a priori* and *a posteriori* tests of the compressible mixed model for homogeneous isotropic turbulence with strong compressibility effects have been performed by Erlebacher, Hussaini, Speziale and Zang (1990) and by Zang, Dahlburg and Dahlburg (1991). The compressible mixed model was found to produce correct turbulence statistics for a variety of cases which included strong compressibility effects. The model has also been used by Ragab in his simulation of a two-dimensional mixing layer (see Chapter 13, in this volume).

A drawback of this model, however, is the fact that its formulation requires the use of the Gaussian filter. Its extension to finite-difference schemes seems, therefore, difficult.

### 6.2.5 Renormalization Group Theory

Recently, much work has concentrated on the application of the renormalization group (RNG) theory of Yakhot and Orszag (1986) to large eddy simulations. The RNG SGS stress model is obtained by performing a recursive elimination of infinitesimal bands of small scales, which is repeated until a fixed point is reached at which the model does not change any longer. This procedure results in a total viscosity, $\nu_{tot} = \nu + \nu_T$, given by (see Yakhot, Orszag, Yakhot and Israeli 1989)

$$\nu_{tot} = \nu \left[1 + H \left(\gamma_{RNG}\gamma_{iso}\frac{\bar{\epsilon}K^{-4}}{\nu^3} - C\right)\right]^{1/3}, \qquad (15)$$

where $H(x)$ is the ramp function ($H(x) = x$ for $x > 0$, and $H(x) = 0$ otherwise), $\gamma_{RNG} = 0.12$ and $C \simeq 100$ are constants, and $\bar{\epsilon}$ is the total dissipation, approximated by $\bar{\epsilon} = 2\nu_{tot}\overline{S}_{ij}\overline{S}_{ij}$; $K$ is an inverse length scale also obtained by using the RNG theory to eliminate all scales smaller than the cutoff in a manner that accounts properly for grid anisotropy. The ad hoc factor $\gamma_{iso} = 3v^2/(u^2 + v^2 + w^2)$ ($v$ being the velocity component normal to the wall) is introduced to account for the anisotropy of the flow in the near-wall region. The RNG eddy viscosity goes to zero near the wall (although with an incorrect asymptotic behavior) or in a laminar flow without requiring damping function or intermittency factors, and is equal to the Smagorinsky eddy viscosity in regions of high turbulence activity.

Yakhot et al. (1989) applied this model to LES of turbulent channel flow, obtaining results in good agreement with DNS and experimental data. Piomelli, Zang, Speziale and Lund (1990b) used it to study forced transition in a flat-plate boundary layer. They used a standard form of the length scale (including Van Driest damping), rather than the one employed by Yakhot et al. (1989), and found that, although transition could be predicted accurately, the results were very much grid-dependent. Applications of the RNG theory to LES of the flow over a backward-facing step and of compressible and reacting flows are described in this volume by Karniadakis et al. (Chapter 8) and Orszag, Staroselsky and Yakhot (Chapter 3).

### 6.2.6 Dynamic Eddy Viscosity Model

Another SGS model that has been used in simulations of transitional flows is the dynamic eddy viscosity model of Germano, Piomelli, Moin and Cabot (1991). In this model the eddy viscosity coefficient is computed dynamically as the calculation progresses rather than imposed *a priori*, and depends on the energy content of the smallest resolved scale. This model makes use of two filters: the *grid filter*, $\overline{G}$, with filter width $\overline{\Delta}$, and the *test filter*, $\widehat{G}$, with filter width $\widehat{\overline{\Delta}} > \overline{\Delta}$. The resolved turbulent stresses, $\mathcal{L}_{ij} \equiv \widehat{\overline{u}_i \overline{u}_j} - \widehat{\overline{u}}_i \widehat{\overline{u}}_j$, which represent the contribution of the smallest resolved scales to the Reynolds stresses, can be computed from the large-scale velocity; they are related to the SGS stresses, $\tau_{ij}$, by the identity

$$\mathcal{L}_{ij} \equiv T_{ij} - \widehat{\tau}_{ij}, \tag{16}$$

where $T_{ij} \equiv \widehat{\overline{u_i u_j}} - \widehat{\overline{u}}_i \widehat{\overline{u}}_j$ are the SGS stresses associated with the coarser filter, $\widehat{G}$, obtained by applying $\widehat{G}$ to the filtered Navier–Stokes equations (3).

Substituting the same model for both $\tau_{ij}$ and $T_{ij}$ (for example, the Smagorinsky model (9) with $C_S^2$ replaced by an unknown coefficient $C$) into (16), one obtains

$$\mathcal{L}_{ij}^a = -2C M_{ij}, \tag{17}$$

where $\mathcal{L}_{ij}^a = \mathcal{L}_{ij} - \delta_{ij} \mathcal{L}_{kk}/3$ is the anisotropic part of $\mathcal{L}_{ij}$, and

$$M_{ij} = \widehat{\overline{\Delta}}^2 |\widehat{\overline{S}}| \widehat{\overline{S}}_{ij} - \overline{\Delta}^2 \widehat{|\overline{S}| \overline{S}_{ij}}. \tag{18}$$

To determine the scalar coefficient $C$ from the five independent equations represented by (17), Germano et al. (1991) contracted it with $\overline{S}_{ij}$ to yield

$$C = -\frac{1}{2} \frac{\mathcal{L}_{kl} \overline{S}_{kl}}{M_{mn} \overline{S}_{mn}}; \tag{19}$$

furthermore, to overcome numerical problems caused by the vanishing of the denominator of (19), they recommended taking an appropriate average of (19); the value of $C$ was then substituted back into the model to yield

$$\tau_{ij}^a = \frac{< \mathcal{L}_{kl} \overline{S}_{kl} >}{< M_{mn} \overline{S}_{mn} >} \overline{\Delta}^2 |\overline{S}| \overline{S}_{ij}, \tag{20}$$

where $< >$ indicates plane averaging. For flows that are not homogeneous in a plane, Germano et al. (1991) recommended localized space or time averaging to replace the plane averaging in (20).

Lilly (1992) suggested an alternative contraction that minimizes the difference between the modeled and exact stresses in the least-squares sense; this approach gives

$$\tau_{ij}^a = \frac{<\mathcal{L}_{kl}M_{kl}>}{<M_{mn}M_{mn}>}\overline{\Delta}^2|\overline{S}|\overline{S}_{ij}, \qquad (21)$$

which has the advantage of the denominator of (21) being positive definite and vanishing only when all the components of $M_{ij}$ vanish (in which case the numerator should also vanish).

The SGS stresses given by (20) or (21) vanish in laminar flows and at solid boundaries, and have the correct asymptotic behavior in the near-wall region without damping functions or intermittency factors; the model can also be formulated in real space or adapted to finite-difference calculations. Furthermore, the model is capable of providing backscatter. It has been used in LES of transitional and fully-developed turbulent channel flows, in which it gave more accurate results than either the Smagorinsky or the RNG model in the formulation of Piomelli et al. (1990b). This model has also been used for calculation of detuned transition in a plane channel (Garg and Piomelli 1991), and has yielded results in good agreement with secondary instability theory and experiments. Additionally, it has been extended to compressible flows by Moin, Squires, Cabot and Lee (1991), who tested the compressible extension both *a priori* and *a posteriori* in the LES of isotropic turbulence. They examined two cases: an almost incompressible one and one with high turbulent Mach number. The LES results compared well with the DNS in both cases.

## 6.3 Present State

In the hierarchy of methods for the solution of fluid flow problems, LES occupies an intermediate position between DNS and solution of the Reynolds-averaged Navier–Stokes equations (RANS). It will be successful as an engineering tool if its advantages over other techniques are exploited rather than if it is used to replace either DNS or RANS modeling.

The principal advantage of LES over DNS is the fact that it allows one to compute flows at Reynolds numbers much higher than those feasible in DNS, or at the same Reynolds numbers but at a considerably smaller expense. One should not expect to be able to extract from LES the same information as can be extracted from DNS, since modeling the small scales affects high-order statistics more than the lower-order ones. Thus, LES is expected to be more reliable for first and second moments, and to reproduce qualitatively the basic structures of the flows (existence of shear layers, vortical structures and so on). Flatness, skewness, vorticity and other derivative statistics should be expected to be correct only qualitatively.

Large eddy simulation is considerably more expensive than RANS techniques for flows that are one- or two-dimensional in the mean and steady. For this reason, it should be applied to problems in which its cost is comparable to that of the solution of the RANS equations or to problems in which lower-level turbulence models fail. Such problems include unsteady or three-dimensional boundary layers, vortex–boundary layer interactions, separated flows and flows involving geometries with sharp corners (in square ducts, for example). Although in the near future LES will still be limited to fairly simple geometries, significantly more complex flows should be studied than those examined so far. Large eddy simulation of these flows can also be used to provide data for the development of more accurate lower-level models (especially pressure statistics, which are difficult to measure experimentally).

So far, however, LES has been more a postdictive than a predictive tool. Useful physical insight and guidance for modeling have been obtained from the results of various LES calculations (Bardina et al. 1983; Moin and Kim 1985; Kim and Moin 1986), but, by and large, LES has been applied only to "building-block" flows: fairly low Reynolds number flows in simple geometries, for which experimental data are available and which isolate a physical feature of interest (strained or sheared turbulence, channel flows, boundary layer flows).

One important reason for this lack of practical applications is certainly the large amount of effort that was concentrated on the direct simulation of turbulence at the expense of LES. The past four or five years, however, have seen a renewed interest in the application of LES to engineering flows, for which DNS will be unfeasible for many years. It is hoped that this interest will result in greater application of LES.

At present, LES research in engineering continues in many areas, with a particular focus on the following: the continued effort to devise more accurate models; the study of transitional flows; and, finally, the application of LES to compressible flows.

The three models that are receiving the most attention are the RNG-based model (15), the structure function model (13) and the dynamic model (21). These models are being tested and evaluated in a variety of flows, incompressible and compressible, and their application in more complex geometries is also under way.

One issue that has recently become a subject of attention is backscatter modeling. Most SGS models are absolutely dissipative; the actual SGS stresses, however, may transfer energy to the large scales (backscatter) at a given location, and while, on average, energy is transferred from large to small scales, reversed energy flow from small to large scales may also occur intermittently. Backscatter can also be significant in quasi-two-dimensional geophysical flows, in which an inverse energy cascade may exist. Piomelli et al. (1990a) showed that during the early nonlinear stages of transition energy is transferred from small to large scales even in the mean; failure to account for this phenomenon can cause inaccurate prediction of the growth of the perturbations. The Smagorinsky model predicted decay of the perturbations even in instances in which the flow should have been unstable. Most of the existing models are absolutely dissipative; only few of them, such as the mixed model and the dynamic model, are capable of predicting backscatter. The RNG model also has such a capability; for instance, a backscatter force has been derived by L. Smith (personal communication) using the RNG formalism, but it has not been used in numerical simulations so far. Leith (1990), Chasnov (1991) and P. Mason and D. Thomson (personal communication) have used models that include a stochastic backscatter force, which have given improved results over purely dissipative models. Piomelli, Cabot, Moin and Lee (1991) used direct numerical simulations of turbulent channel flow and compressible isotropic turbulence to study backscatter *a priori*. They found that the backscatter and forward-scatter contributions to the SGS dissipation are comparable, and each is often much larger than the total SGS dissipation. Backscatter, however, depends on the type of filter used, being largest for the sharp Fourier cutoff, smallest for the Gaussian and intermediate for the top hat in real space. This finding indicates that perhaps backscatter modeling may be less critical in calculations that use finite-difference schemes than in those that use spectral methods.

The application of LES to transitional flows is relatively recent, perhaps because most of the transition process can be captured reasonably well by coarse direct simulations, and it was felt that there was no need for SGS modeling. Piomelli and Zang (1991), however, showed that the presence of an SGS stress model significantly improves the numerical results at the late stages of transition, where peak heat transfer and wall stress occur. Furthermore, the direct simulation of spatially developing flows is extremely expensive, given the extent of the transitional region (especially in compressible flows), and LES may be the only tool capable of providing velocity and temperature fields for these problems. Similar considerations apply to the case of natural transition, especially when started from low-amplitude disturbances. In the simulation of detuned transition by Garg and Piomelli (1991), for example, the computational domain included nine Tollmien–Schlichting wavelengths. A direct simulation of this problem would have required almost two thousand grid points in the streamwise direction, which is obviously unfeasible. A review of applications of LES to transitional problems can be found in Chapter 11 in this volume.

Although the first applications of LES to compressible and reacting flows were relatively recent, much progress has already been made in this area. All of the most popular models have been extended to compressible flows, and a number of test cases have been examined (see Chapters 3, 5, 9, 12 and 13, in this volume). While many of the concerns that apply to incompressible applications (model accuracy, backscatter and so on) extend also to compressible flows, additional difficulties are due to the fact that the equations are more complex and shock wave interactions and eddy shocklets may occur; furthermore, finite differences introduce artificial dissipation, the effect of which must be studied and compared withe that of the SGS dissipation. Aliasing errors can also be more significant, owing to the presence of triple products in the equations of motion. The simulation and modeling of flows including chemical reactions is still in its infancy. See the work of Madnia and Givi (Chapter 15), Menon et al. (Chapter 14) and Orszag, Staroselsky and Yakhot (Chapter 3) for applications of LES to combusting flows.

Although large eddy simulations in engineering have not enjoyed a rapid development similar to that of direct simulations, the renewed activity in LES justifies an optimistic view of the future of this technique. One would hope that 10 years from today LES of engineering problems will be routinely performed on a desktop workstation. To achieve this end a continued effort is required by the research community, involving increased interactions among its members. Experiments, direct simulations and numerical analysis are all necessary for further progress. The interaction between geophysical scientists and engineers, which has not been particularly noteworthy so far, must be enhanced to facilitate the development of LES into a practical tool useful in both fields.

## Acknowledgments

The author received support from the Office of Naval Research under Grant N00014-89-J-1531 and by the NASA Langley Research Center under Grant NAG-1-1089.

## References

ARAKAWA, A. (1966) Computational design for long-term numerical integration of the equations of fluid motion: Two-dimensional incompressible flow. Part I. *J. Comput. Phys.* 1, pp. 119–143.

AUPOIX, B. (1985) Subgrid scale models for homogeneous anisotropic turbulence. In *Finite Approximations in Fluid Mechanics.* Ed. E.H. Hirschel, pp. 37–66 (Vieweg).

BARDINA, J., FERZIGER, J.H., AND REYNOLDS, W.C. (1980) Improved subgrid scale models for large eddy simulation. AIAA Paper 80–1357.

BARDINA, J., FERZIGER, J.H., AND REYNOLDS, W.C. (1983) Improved turbulence models based on large eddy simulation of homogeneous, incompressible turbulent flows. Report TF-19, Stanford University, Dept. of Mechanical Engineering.

CAMBON, C., JEANDEL, D., AND MATHIEU, J. (1981) Spectral modelling of homogeneous non-isotropic turbulence. *J. Fluid Mech.* 104, pp. 247–262.

CHASNOV, J.R. (1991) Simulation of the Kolmogorov inertial subrange using an improved subgrid model. *Phys. Fluids A* 3, pp. 188–200.

CHOLLET, J.-P. (1985) Two-point closure used for a subgrid scale model in large eddy simulation. In *Turbulent Shear Flows 4.* Ed. L.J.S. Bradbury, F. Durst, B.E. Launder, F.W. Schmidt, and J.H. Whitelaw, pp. 62–72 (Springer-Verlag).

CHOLLET, J.-P. AND LESIEUR, M. (1981) Parameterization of small scales of three-dimensional isotropic turbulence utilizing spectral closures. *J. Atmos. Sci.* 38, pp. 2747–2757.

CLARK, R.A., FERZIGER, J.H., AND REYNOLDS, W.C. (1979) Evaluation of subgrid-scale models using an accurately simulated turbulent flow. *J. Fluid Mech.* 91, pp. 1–16.

DANG, K.T. (1985) Evaluation of simple subgrid-scale models for the numerical simulation of homogeneous turbulence. *AIAA J.* **23**, pp. 221–227.

DANG, K.T. AND DESCHAMPS, V. (1987) Numerical simulation of transitional channel flow. In *Numerical Methods in Laminar and Turbulent Flows.* Ed. C. Taylor, W.G. Habashi, and M.M. Hafez, pp. 423–434 (Pine Ridge).

DEARDORFF, J.W. (1970) A numerical study of three-dimensional turbulent channel flow at large Reynolds numbers. *J. Fluid Mech.* **41**, pp. 453–480.

ERLEBACHER, G., HUSSAINI, M.Y., SPEZIALE, C.G., C.G., AND ZANG, T.A. (1990) Toward the large-eddy simulation of compressible turbulent flows. ICASE Report 90-76.

GARG, R. AND PIOMELLI, U. (1991) Large-eddy simulation of detuned transition. Presented at the 4th Int. Symp. on Computational Fluid Dynamics, Davis, CA.

GERMANO, M., PIOMELLI, U., MOIN, P., AND CABOT, W.H. (1991) A dynamic subgrid-scale eddy viscosity model. *Phys. Fluids A* **3**, pp. 1760–1765.

GRÖTZBACH, G. (1979) Numerical investigation of radial mixing capabilities in strongly buoyancy-influenced vertical, turbulent channel flows. *Nucl. Eng. Design* **54**, pp. 49–66.

GRÖTZBACH, G. (1987) Direct numerical and large eddy simulation of turbulent channel flows. In *Encyclopedia of Fluid Mechanics*, Vol. 6. Ed. N.P. Cheremisinoff, pp. 1337–1391 (Gulf Publishing).

HORIUTI, K. (1987) Comparison of conservative and rotational forms in large eddy simulation of turbulent channel flow. *J. Comput. Phys.* **71**, pp. 343–370.

HORIUTI, K. AND YOSHIZAWA, A. (1985) Large eddy simulation of turbulent channel flow by 1-equation model. In *Finite Approximations in Fluid Mechanics.* Ed. E.H. Hirschel, pp. 119–134 (Vieweg).

KIM, J. AND MOIN, P. (1986) The structure of the vorticity field in turbulent channel flow. Part 2. Study of ensemble-averaged fields. *J. Fluid Mech.* **162**, pp. 339–363.

KRAICHNAN, R.H. (1964) Direct interaction approximation for shear and thermally driven turbulence. *Phys. Fluids* **7**, pp. 1048–1062.

KRAICHNAN, R.H. (1976) Eddy viscosity in two and three dimensions. *J. Atmos. Sci.* **33**, pp. 1521–1536.

KWAK, D., REYNOLDS, W.C., AND FERZIGER, J.H. (1975) Three-dimensional time-dependent computation of turbulent flows. Report TF-5, Stanford University, Dept. of Mechanical Engineering.

LEITH, C.E. (1990) Stochastic backscatter in a subgrid-scale model: plane shear mixing layer. *Phys. Fluids A* 2, pp. 297–299.

LEONARD, A. (1974) Energy cascade in large-eddy simulations of turbulent fluid flows. *Adv. Geophys.* 18A, pp. 237–248.

LESIEUR, M. (1990) *Turbulence in Fluids.* (Kluwer).

LILLY, D.K. (1966) On the application of the eddy viscosity concept in the inertial sub-range of turbulence. NCAR Manuscript 123.

LILLY, D.K. (1991) A proposed modification of the Germano subgrid-scale closure method. *Phys. Fluids A* 4, pp. 633–635.

MASON, P.J. AND CALLEN, N.S. (1986) On the magnitude of the subgrid-scale eddy coefficient in large-eddy simulation of turbulent channel flow. *J. Fluid Mech.* 162, pp. 439–462.

MCMILLAN, O.J., FERZIGER, J.H., AND ROGALLO, R.S. (1980) Tests of new subgrid-scale models in strained turbulence. AIAA Paper 80-1339.

MOIN, P., REYNOLDS, W.C., AND FERZIGER, J.H. (1978) Large eddy simulation of incompressible turbulent channel flow. Report TF-12, Stanford University, Dept. of Mechanical Engineering.

MOIN, P. AND KIM, J. (1982) Numerical investigation of turbulent channel flow. *J. Fluid Mech.* 118, pp. 341–377.

MOIN, P. AND KIM, J. (1985) The structure of the vorticity field in turbulent channel flow. Part 1. Analysis of instantaneous fields and statistical correlations. *J. Fluid Mech.* 155, pp. 441–464.

MOIN, P., SQUIRES, K., CABOT, W.H., AND LEE, S. (1991) A dynamic subgrid-scale model for compressible turbulence and scalar transport. *Phys. Fluids A* 3, pp. 2746–2757.

NORMAND, X. AND LESIEUR, M. (1990) Numerical experiments on transition in the compressible boundary layer over an insulated flat plate. *Theor. Comput. Fluid Dynamics* 3, pp. 231-252.

PIOMELLI, U., MOIN, P., AND FERZIGER, J.H. (1988) Model consistency in large eddy simulation of turbulent channel flows. *Phys. Fluids* 31, pp. 1884–1891.

PIOMELLI, U., ZANG, T.A., SPEZIALE, C.G., AND HUSSAINI, M.Y. (1990a) On the large eddy simulation of transitional wall-bounded flows. *Phys. Fluids A* 2, pp. 257–265.

PIOMELLI, U., ZANG, T.A., SPEZIALE, C.G., AND LUND, T.S. (1990b) Application of renormalization group theory to the large-eddy simulation of transitional boundary layers. In *Instability and Tran-*

*sition.* Ed. M.Y. Hussaini and R.G. Voigt, Vol. 2, pp. 480–496 (Springer-Verlag).

PIOMELLI, U., CABOT, W.H., MOIN, P., AND LEE, S. (1991) Subgrid-scale backscatter in turbulent and transitional flows. *Phys. Fluids A* 3, pp. 1766–1771.

PIOMELLI, U. AND ZANG, T.A. (1991) Large-eddy simulation of transitional channel flow. *Comput. Phys. Comm.* 65, pp. 224–230.

REYNOLDS, W.C. (1990) The potential and limitations of direct and large eddy simulations. In *Whither turbulence? Turbulence at the crossroads.* Ed. J.L. Lumley, pp. 313–342 (Springer-Verlag).

ROGALLO, R.S. AND MOIN, P. (1984) Numerical simulation of turbulent flows. *Ann. Rev. Fluid Mech.* 16, pp. 99–137.

SCHUMANN, U. (1975) Subgrid scale model for finite difference simulation of turbulent flows in plane channels and annuli. *J. Comput. Phys.* 18, pp. 376–404.

SHAANAN, S., FERZIGER, J.H., AND REYNOLDS, W.C. (1975) Numerical simulation of turbulence in the presence of shear. Report TF-6, Stanford University, Dept. of Mechanical Engineering.

SMAGORINSKY, J. (1963) General circulation experiments with the primitive equations. I. The basic experiment. *Month. Wea. Rev.* 91, pp. 99–164.

SPEZIALE, C.G. (1985) Galilean invariance of subgrid scale stress models in the large eddy simulation of turbulence. *J. Fluid Mech.* 156, pp. 55–62.

SPEZIALE, C.G., ERLEBACHER, G., ZANG, T.A., AND HUSSAINI, M.Y. (1988) The subgrid-scale modeling of compressible turbulence. *Phys. Fluids A* 31, pp. 940–942.

VAN DRIEST, E.R. (1956) On the turbulent flow near a wall. *J. Aerospace Sci.* 23, pp. 1007–1011.

YAKHOT, V. AND ORSZAG, S.A. (1986) Renormalization group analysis of turbulence. I. Basic theory. *J. Sci. Computing* 1, pp. 3–51.

YAKHOT, A., ORSZAG, S.A., YAKHOT, V., AND ISRAELI, M. (1989) Renormalization group formulation of large-eddy simulation. *J. Sci. Computing* 4, pp. 139–158.

YOSHIZAWA, A. (1986) Statistical theory for compressible turbulent shear flows, with the application to subgrid modeling. *Phys. Fluids* 29, pp. 2152–2164.

YOSHIZAWA, A. (1987) Large eddy-simulation of turbulent flows. In *Encyclopedia of Fluid Mechanics*, Vol. 6. Ed. N.P. Cheremisinoff, pp. 1277–1297 (Gulf Publishing).

ZANG, T.A. (1991) On the rotational and skew-symmetric forms for incompressible flow simulations *Appl. Num. Math.*, 7, pp. 27–40.

ZANG, T.A., DAHLBURG, R.B., AND DAHLBURG, J.P. (1991) Direct and large-eddy simulation of three-dimensional compressible Navier–Stokes turbulence. NRL Memorandum Report 6799.

# INCOMPRESSIBLE FLOWS

# 7

## Large Eddy Simulation of Scalar Transport with the Dynamic Subgrid-Scale Model

WILLIAM CABOT AND PARVIZ MOIN

### 7.1 Introduction

A new subgrid-scale (SGS) model that dynamically adjusts itself to flow conditions has recently been generalized for scalar transport problems and applied to large eddy simulations in channel flow. The dynamic SGS concept developed by Germano, Piomelli, Moin, and Cabot (1991) utilizes the high-wave-number information contained in a large eddy simulation (LES) to extrapolate the effect of the unresolved scales on the resolved scales, in principle at each point in space and at each instant in the time integration. When applied, for example, to the Smagorinsky (1963) eddy viscosity model, the constant of proportionality that appears in the model becomes a dynamically determined coefficient, which adjusts itself to flow conditions without any ad hoc prescriptions or modifications (such as damping functions near walls). This model, originally developed for incompressible flows, has been extended to compressible flows and scalar transport problems by Moin, Squires, Cabot, and Lee (1991).

In this chapter we present a brief formulation of the dynamic SGS model for incompressible scalar transport using an eddy viscosity/diffusivity formulation, and we discuss some of the alternative formulations possible in the model. We then report on results from the model's use in *a priori* tests on direct numerical simulation (DNS) data and on results from LES of flows with passive scalars. The results of these large eddy simulations are compared with those from DNS, and the performance of different formulations of the dynamic model is discussed.

## 7.2 The Dynamic SGS Model
### 7.2.1 Mathematical Formulation

The filtered governing equations for the LES of a passive scalar $\theta$ in an incompressible flow are

$$\frac{\partial \overline{u}_i}{\partial t} + \frac{\partial}{\partial x_j}(\overline{u}_i \overline{u}_j) = -\frac{\partial \overline{p}}{\partial x_i} + \frac{\partial}{\partial x_j}\left(2\nu \overline{S}_{ij} - \tau_{ij}\right), \qquad (1)$$

$$\frac{\partial \overline{u}_i}{\partial x_i} = \overline{S}_{ii} = 0, \qquad (2)$$

$$\frac{\partial \overline{\theta}}{\partial t} + \frac{\partial}{\partial x_k}(\overline{\theta}\,\overline{u}_k) = \frac{\partial}{\partial x_k}\left(\alpha \frac{\partial \overline{\theta}}{\partial x_k} - q_k\right) + \overline{Q}, \qquad (3)$$

where the overbar denotes the filtering operation. In these equations, $u_i$ are the velocity components, $Q$ is the scalar source term, and $\nu$ and $\alpha$ are the molecular viscosity and diffusivity. The filtered strain rate tensor is defined by

$$\overline{S}_{ij} \equiv \frac{1}{2}\left(\frac{\partial \overline{u}_i}{\partial x_j} + \frac{\partial \overline{u}_j}{\partial x_i}\right). \qquad (4)$$

All of the terms in equations (1)–(4) are resolved in the LES except the residual stress tensor

$$\tau_{ij} = \overline{u_i u_j} - \overline{u}_i \overline{u}_j \qquad (5)$$

and the residual scalar flux

$$q_k = \overline{\theta u_k} - \overline{\theta}\,\overline{u}_k, \qquad (6)$$

which must be modeled. We will consider one of the simpler models that is commonly invoked, the Smagorinsky eddy viscosity model,

$$\tau_{ij} - \tfrac{1}{3}\delta_{ij}\tau_{kk} = -2\nu_t \overline{S}_{ij}, \quad \nu_t = C\Delta^2|\overline{S}|, \qquad (7)$$

and an analogous eddy diffusivity model (cf. Eidson 1985; Erlebacher, Hussaini, Speziale, and Zang 1990)

$$q_k = -\alpha_t \frac{\partial \overline{\theta}}{\partial x_k}, \quad \alpha_t = C_\theta \Delta^2 |\overline{S}|, \qquad (8)$$

where $|\overline{S}| \equiv |2\overline{S}_{ij}\overline{S}_{ij}|^{1/2}$, and where $C_\theta$ is usually expressed in terms of an SGS turbulent Prandtl number as $C/Pr_t$. $\Delta$ is the filter width, which is generally related to the mesh size of the resolved field in the LES; the dependence of the dynamic SGS model and LES results on the exact definition of $\Delta$ will be discussed below.

Whereas $C$ and $Pr_t$ were previously assigned as flow-dependent constants in an LES, the dynamic SGS model uses flow characteristics at small *resolved* scales to extrapolate their (spatially and temporally varying) values at subgrid scales, assuming scale similarity. If the resolved

LES fields are filtered again at a larger scale by a "test" or "high-pass" filter (denoted by a circumflex), their residual stress and scalar flux are

$$T_{ij} = \widehat{\overline{u_i u_j}} - \widehat{\overline{u}}_i \widehat{\overline{u}}_j \, , \tag{9}$$

$$Q_k = \widehat{\overline{\theta u_k}} - \widehat{\overline{\theta}} \, \widehat{\overline{u}}_k \, . \tag{10}$$

In the dynamic SGS model, these are modeled in the same manner as the residual terms in (7) and (8),

$$T_{ij} - \tfrac{1}{3}\delta_{ij} T_{kk} = -2\widehat{\nu}_t \widehat{\overline{S}}_{ij} \, , \quad \widehat{\nu}_t = C\widehat{\Delta}^2 |\widehat{\overline{S}}| \, , \tag{11}$$

$$Q_k = -\widehat{\alpha}_t \frac{\partial \widehat{\overline{\theta}}}{\partial x_k} \, , \quad \widehat{\alpha}_t = C_\theta \widehat{\Delta}^2 |\widehat{\overline{S}}| \, , \tag{12}$$

where $|\widehat{\overline{S}}| \equiv |2\widehat{\overline{S}}_{ij}\widehat{\overline{S}}_{ij}|^{1/2}$ and $\widehat{\Delta}$ is the test filter width. (Note that the coefficients $C$ and $C_\theta$ have been assumed to be independent of the filtering process; this is strictly true only if they are spatially uniform.) Both $\tau_{ij}$ and $T_{ij}$, and both $q_k$ and $Q_k$, contain unresolved terms, but the differences

$$\mathcal{L}_{ij} = T_{ij} - \widehat{\tau}_{ij} = \widehat{\overline{u_i}\overline{u}_j} - \widehat{\overline{u}}_i \widehat{\overline{u}}_j \, , \tag{13}$$

$$\mathcal{F}_k = Q_k - \widehat{q}_k = \widehat{\overline{\theta}\,\overline{u}_k} - \widehat{\overline{\theta}}\,\widehat{\overline{u}}_k \tag{14}$$

are computable from the resolved field. By substituting the eddy viscosity/diffusivity models for terms in (13) and (14), one finds that the coefficients $C$ and $C_\theta$ are determined by the relations

$$\mathcal{L}_{ij} - \tfrac{1}{3}\delta_{ij}\mathcal{L}_{kk} = -C\Delta^2 \mathcal{M}_{ij} \, , \quad \mathcal{M}_{ij} = 2\left[ (\widehat{\Delta}/\Delta)^2 |\widehat{\overline{S}}|\widehat{\overline{S}}_{ij} - \widehat{|\overline{S}|\overline{S}_{ij}} \right] \, , \tag{15}$$

$$\mathcal{F}_k = -C_\theta \Delta^2 \mathcal{H}_k \, , \quad \mathcal{H}_k = (\widehat{\Delta}/\Delta)^2 |\widehat{\overline{S}}| \frac{\partial \widehat{\overline{\theta}}}{\partial x_k} - \widehat{|\overline{S}| \frac{\partial \overline{\theta}}{\partial x_k}} \, . \tag{16}$$

Note that the tensor $\mathcal{M}_{ij}$, like $\overline{S}_{ij}$ and $\widehat{\overline{S}}_{ij}$, is traceless.

In this particular model, the set of the tensor terms in (15) and the set of the vector terms in (16) are each modeled by one coefficient. In order to uniquely determine $C$, for instance, Germano et al. (1991) proposed contracting (15) with $\overline{S}_{ij}$, which gives

$$C\Delta^2 \mathcal{M}_{k\ell}\overline{S}_{k\ell} = -\mathcal{L}_{mn}\overline{S}_{mn} \, , \tag{17}$$

and an analogous contraction of (16) with $\partial\overline{\theta}/\partial x_k$ determines $C_\theta$ (Moin et al. 1991):

$$C_\theta \Delta^2 \mathcal{H}_j \frac{\partial \overline{\theta}}{\partial x_j} = -\mathcal{F}_i \frac{\partial \overline{\theta}}{\partial x_i} \, . \tag{18}$$

Recently, Lilly (1992) suggested applying least squares analyses on the components of (15) and (16), which is equivalent to contracting (15) and (16) with $\mathcal{M}_{ij}$ and $\mathcal{H}_k$, respectively, giving

$$C\Delta^2 \mathcal{M}_{k\ell}\mathcal{M}_{k\ell} = -\mathcal{L}_{mn}\mathcal{M}_{mn}, \tag{19}$$

$$C_\theta \Delta^2 \mathcal{H}_j \mathcal{H}_j = -\mathcal{F}_i \mathcal{H}_i. \tag{20}$$

We will describe the difference in LES results between both contractions below. Note that these coefficients do not require ad hoc wall functions to ensure their proper behavior near solid boundaries, and that they vanish – as they should – in the laminar regime.

### 7.2.2 Local versus Global Application of Model Coefficients

The equations for the model coefficients, which depend on time and space, are evaluated in an LES at each time step from the flow field at the previous time step. This allows the coefficients to adjust to conditions in the flow. While equations (17)–(20) give coefficients that are in principle fully spatially dependent, Germano et al. (1991) found that $\mathcal{M}_{k\ell}\overline{S}_{k\ell}$ in (17) frequently traversed zero, resulting in an ill-conditioned function for $C$. They found instead that assuming $C$ to be a function only of time and the inhomogeneous direction(s) (viz., the direction normal to the walls in channel flows) and averaging (17) over the homogeneous spatial directions gave a well-conditioned function for $C$. This spatial averaging, denoted by angle brackets, gives global expressions for the coefficients from (17) and (18) as

$$C\Delta^2 = -\langle \mathcal{L}_{mn}\overline{S}_{mn}\rangle/\langle \mathcal{M}_{k\ell}\overline{S}_{k\ell}\rangle, \tag{21}$$

$$C_\theta \Delta^2 = -\left\langle \mathcal{F}_i \frac{\partial \overline{\theta}}{\partial x_i}\right\rangle \bigg/ \left\langle \mathcal{H}_j \frac{\partial \overline{\theta}}{\partial x_j}\right\rangle, \tag{22}$$

or from (19) and (20) as

$$C\Delta^2 = -\langle \mathcal{L}_{mn}\mathcal{M}_{mn}\rangle/\langle \mathcal{M}_{k\ell}\mathcal{M}_{k\ell}\rangle, \tag{23}$$

$$C_\theta \Delta^2 = -\langle \mathcal{F}_i \mathcal{H}_i\rangle/\langle \mathcal{H}_j \mathcal{H}_j\rangle. \tag{24}$$

For flows in which one is interested in the statistical steady state, a temporal average could also in principle replace the spatial average. As previously noted, taking the coefficients in equations (17)–(20) as fully spatially dependent functions is not self-consistent with their derivation; it is not known, however, the extent to which this inconsistency affects the ill-conditioning of the local coefficients.

The contraction proposed in equations (19) and (20) has the obvious advantage of having the terms $\mathcal{M}_{k\ell}\mathcal{M}_{k\ell}$ and $H_j H_j$ be positive definite everywhere in space and holds the possibility of being able to use locally determined, well-conditioned model coefficients; this would be a boon for performing the LES of more complex geometries that lack homogeneous directions. However, it is found, e.g., in channel flow that $\mathcal{M}_{k\ell}\mathcal{M}_{k\ell}$ in (19) becomes frequently small enough (compared with $\mathcal{L}_{mn}\mathcal{M}_{mn}$) to make locally determined values of $C$ still ill-behaved. The corresponding values of $\nu_t = C\Delta^2|\overline{S}|$ have rms levels 5 to 10 times the mean values, which is actually quite similar to *a priori* statistics of the residual stress from DNS (Piomelli, Cabot, Moin, and Lee 1991); but this means that almost 50% of spatial points have negative SGS eddy viscosity. In addition, $\nu_t$ features large spikes 10 to 20 times greater than rms levels. This behavior is probably unphysical, perhaps arising from a misalignment of residual stress and residual strain tensors. An alternative local formulation of the dynamic model that addresses this misalignment involves modeling the residual stresses in (7) and (11) with a tensor model coefficient $C_{ij}$ such that

$$\tau_{ij} = -(C_{ik}\overline{S}_{kj} + C_{jk}\overline{S}_{ki})\Delta^2|\overline{S}|, \tag{25}$$

$$T_{ij} = -(C_{ik}\widehat{\overline{S}}_{kj} + C_{jk}\widehat{\overline{S}}_{ki})\widehat{\Delta}^2|\widehat{\overline{S}}|, \tag{26}$$

which gives, instead of (15),

$$L_{ij} = -(C_{ik}\mathcal{M}_{kj} + C_{jk}\mathcal{M}_{ki})\Delta^2. \tag{27}$$

Since $L_{ij}$ is symmetric, this equation yields six unique coefficients. *A priori* tests using this model to determine local coefficients also seem to indicate ill-conditioned behavior, however, because of frequent sign changes in the determinant of $\mathcal{M}$. Further testing of this model with LES is needed to establish whether ill-conditioned behavior occurs in self-consistent calculations.

The sign changes in $\nu_t$ that arise from locally determined SGS model coefficients have also been found to present stability problems for our numerical integration schemes using finite spatial differencing and semi-implicit time advancement. De-aliasing and local averaging (or coarse filtering) of the terms in (19) reduce the spikiness (and rms levels) by a factor of 2 or 3, but are not found to improve the local behavior or numerical stability substantially. Further work is needed to clarify to what extent locally defined SGS coefficients with positive as well as negative signs are numerically feasible. The results discussed in the rest of this chapter have been obtained from equations (21)–(24) with spatial averaging in the homogeneous directions.

### 7.2.3 Definition of Filter Widths

For the results discussed here, we have used a sharp spectral cutoff filter in homogeneous directions which truncates a Fourier series at wave number $k_i = \pi/\Delta_i$, where $\Delta_i$ is the filter width in the $i$th direction. There is some ambiguity in defining the effective filter width $\Delta$ on anisotropic grids and when explicit filtering is not performed in all directions. $\Delta$ is often defined as the geometric mean of the unidirectional filter widths (Deardorff 1970), viz.,

$$\Delta^3 = \Delta_1 \Delta_2 \Delta_3 \,, \tag{28}$$

although other alternatives are possible, e.g.,

$$\Delta^2 = \Delta_1^2 + \Delta_2^2 + \Delta_3^2 \tag{29}$$

(cf. Germano et al. 1991). In particular, Lilly (1989) has made a more rigorous determination of $\Delta$ for homogeneous flow with anisotropic grids, showing that expression (28) is satisfactory for moderate anisotropies.

In general, the dynamic SGS model is not expected to be very sensitive to exact definitions of $\Delta$, but rather to the ratio of filter widths $\widehat{\Delta}/\Delta$, which is all that enters into the SGS eddy viscosity and diffusivity through $\mathcal{M}_{ij}$ and $\mathcal{H}_k$. If all unidirectional filter width ratios $\widehat{\Delta}_i/\Delta_i$ are the same, then $\widehat{\Delta}/\Delta = \widehat{\Delta}_i/\Delta_i$ whether $\Delta$ is defined by (28) or (29). For the LES of channel flow, however, we do not explicitly filter in the direction normal to the walls ($x_2$); therefore, $\Delta_2 = \widehat{\Delta}_2$ and equation (28) gives $\widehat{\Delta}/\Delta = (\widehat{\Delta}_1\widehat{\Delta}_3/\Delta_1\Delta_3)^{1/3}$. If one were to determine the effective filter widths from the geometric means of only the filtered directions, then $\widehat{\Delta}/\Delta = (\widehat{\Delta}_1\widehat{\Delta}_3/\Delta_1\Delta_3)^{1/2}$, which is larger than the previous example and, through correspondingly larger magnitudes of $\mathcal{M}_{ij}$ and $\mathcal{H}_k$, results in lower values of eddy viscosity and diffusivity for a given flow field. For all practical purposes, using definition (29) for channel flow gives $\widehat{\Delta}/\Delta = (\widehat{\Delta}_1^2 + \widehat{\Delta}_3^2)^{1/2}/(\Delta_1^2 + \Delta_3^2)^{1/2}$ whether the inhomogeneous direction is included in the definition or not, because the mesh in the normal direction is much finer than the horizontal ones in practice, making $\Delta_2$ and $\widehat{\Delta}_2$ negligible in (29).

## 7.3 Applications

Germano et al. (1991) computed residual stresses from DNS flow fields and compared them with dynamic SGS model predictions for the same fields; these *a priori* tests indicated that a choice of $\widehat{\Delta}_i/\Delta_i = 2$ yielded the best results. They also found that the results from the LES of channel flow with $Re_\tau$ (Reynolds number based on wall stress and channel half-width) of about 400 showed little sensitivity to the size of the effective filter width ratio (in a range of 1.6 to 4), apparently due to an adjustment of the small scales to different stress levels. However, low Reynolds number LES results for channel flow (discussed below) indicate a somewhat greater sensitivity to the choice of $\widehat{\Delta}/\Delta$ and other model variants.

### 7.3.1 Results from a Priori Tests

The dynamic SGS model coefficients in a homogeneous shear flow have been estimated by Moin et al. (1991) using the DNS data of Rogers, Moin, and Reynolds (1986) for passive scalars with Prandtl numbers ($Pr = \alpha/\nu$) of 0.2, 0.7, and 2.0, and with three different (mutually orthogonal) mean gradients. The grid filter was chosen to remove the last octave of wave numbers in all three directions, and the test filter removed the next octave, giving $\widehat{\Delta}/\Delta = 2$ with either definition (28) or (29). Equations (21) and (22) were used with spatial averages performed over the entire domain, so that the resulting coefficients depended solely on time. The coefficients show significant evolution with time as the spectral distribution of energy changes. The SGS turbulent Prandtl number, $Pr_t = C/C_\theta$, tends to reach relatively constant values at large times, as seen in Figure 1 for several different cases. The hierarchy of values of $Pr_t$ for different orientations of the mean scalar gradient are the same as for the full-field turbulent Prandtl number,

$$Pr_T = \langle uv \rangle \frac{\partial \Theta}{\partial x_i} \bigg/ \langle \theta u_i \rangle \frac{\partial U}{\partial y} , \qquad (30)$$

where $\Theta$ and $U$ are the mean scalar and mean streamwise velocity and where angle brackets denote global averaging. The actual values of $Pr_t$ for this choice of grid and test filters are lower by a factor of about 2 from $Pr_T$, but larger values are obtained for coarser filters. Note that we do not expect the values of the SGS and full-field turbulent Prandtl numbers to be the same, since they are derived from different spatial scales. The values of $Pr_t = 0.4$–$0.6$ for a scalar gradient in the normal direction ($i = 2$ in (30)) are consistent with the optimal value of about 0.5 found by Erlebacher et al. (1990) and Eidson (1985).

Actual and predicted residual stresses and scalar (temperature) fluxes were also computed from DNS databases of passive scalars with $Pr = 0.1, 0.71$, and 2.0 in channel flow with isothermal walls, one hot and one cold (Case II in Kim and Moin 1989). The SGS turbulent Prandtl num-

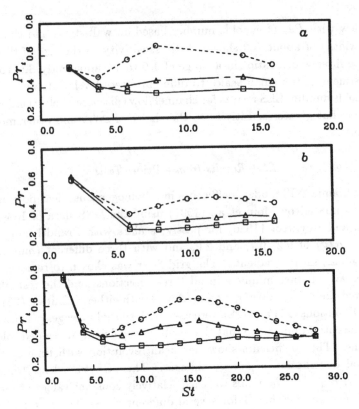

Fig. 1   Time development of SGS turbulent Prandtl numbers from *a priori*
tests of homogeneous shear flow data from DNS (Rogers et al. 1986)
for passive scalars with molecular Prandtl numbers $Pr$ of (a) 0.2,
(b) 0.7, and (c) 2.0 with time made dimensionless with the shear
rate $S$; scalar gradients are imposed in □ streamwise, o normal, and
△ spanwise directions.

bers for these cases were computed from equations (21) and (22) with
plane averaging in the horizontal directions; they are shown in Figure 2a
as functions of distance from a wall. These results were insensitive to
the definition of the filter width. For all $Pr$, $Pr_t \approx 0.5$ near midchannel,
which is similar to the values found in homogeneous shear flow with a
normal passive scalar gradient and which is about half of that found for
$Pr_T$ near midchannel (see Figure 2b). At or near the wall, $Pr_t$ rises to
values of about 1.0 for $Pr = 0.71$ and 2.0, and peaks at 1.8 for $Pr = 0.1$.
A similar behavior between $Pr_t$ and $Pr_T$ is seen for $Pr = 0.71$ and
2.0, although values for $Pr_T$ rise more gradually approaching the wall.
$Pr_t$ for $Pr = 0.1$ is larger than the values for the higher Prandtl num-

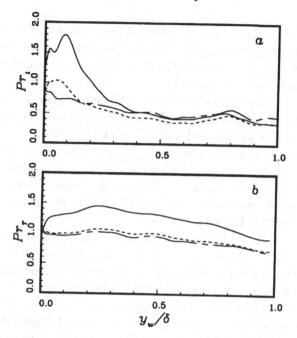

Fig. 2 Turbulent Prandtl numbers for passive scalars in channel flow from DNS data (Kim and Moin 1989) as functions of distance from the wall $y_w$ in units of channel half-width $\delta$: (a) SGS turbulent Prandtl numbers from *a priori* tests and (b) full-field turbulent Prandtl numbers. ——— $Pr = 0.1$, ---- $Pr = 0.71$, and —·— $Pr = 2.0$.

bers only near the wall, but $Pr_T$ for $Pr = 0.1$ is larger than the others throughout the channel.

We thus have some indication that the dynamic SGS model for scalar transport will provide reasonable values of eddy diffusivity under a number of flow conditions. The success of this model, though, must ultimately be gauged by its performance in large eddy simulations.

### 7.3.2 Results from LES

The dynamic SGS model for scalar transport was recently implemented in an LES channel code (described in detail by Piomelli, Ferziger, and Moin 1987). The code is spectral in the homogeneous horizontal directions and uses central finite differencing in the inhomogeneous normal direction. The mesh is stretched in the direction normal to the walls in order to resolve the near-wall region. Sharp spectral cutoff filtering is performed in the homogeneous horizontal directions. The model coefficients are obtained with plane averaging in the horizontal directions from either equations (21)–(22) (referred to hereafter as the strain rate contraction) or equations (23)–(24) (the least squares contraction); the model coefficients are therefore functions of time and normal direction. For the most part, low Reynolds number flows have been simulated to date for comparison with available DNS data. They are computed with friction Reynolds numbers ($Re_\tau = u_\tau \delta / \nu$, where $u_\tau$ is the friction velocity and $\delta$ is the channel half-width) of 150–180 on $32 \times 63 \times 64$ meshes in the streamwise, normal, and spanwise directions. Test filtering is performed by removing the highest octave of the resolved horizontal spectral field (i.e., $\widehat{\Delta}_1/\Delta_1 = \widehat{\Delta}_3/\Delta_3 = 2$). Simulations with much higher Reynolds numbers ($Re_\tau = 1200$–$1400$) are also presently being performed on $32 \times 125 \times 128$ meshes.

The LES of channel flow with passive (temperature) scalars (with $Pr = 0.1, 0.71$, and $2.0$; $Re_\tau \approx 165$) with isothermal walls, one hot and one cold, was performed using the strain rate contraction. Values of $\widehat{\Delta}/\Delta = 2^{2/3}$, based on equation (28), were used in the SGS model. Figure 3a shows that the mean streamwise velocity is about 10% higher in the log layer than the DNS results of Kim and Moin (1989) with $Re_\tau \approx 180$. In Figure 3b the mean scalar profiles (relative to the value at the wall) from the LES show fair agreement for the low-$Pr$ case, but poor agreement with the higher-$Pr$ cases, exceeding the DNS results by 10–15% in the log layer. (Quantities shown in Figure 3 and the following figures are expressed in wall units denoted by +, i.e., nondimensionalized by $u_\tau$, $\delta$, $\nu$, and the scalar scale based on the wall flux, $\alpha u_\tau^{-1} |\partial \theta / \partial y|_w$.) Figure 4 shows the streamwise and normal scalar fluxes, $\langle \theta' u' \rangle$ and $\langle \theta' v' \rangle$, for $Pr = 0.71$; compared with the DNS results, the former is substantially larger in the region corresponding to the log layer, and the latter is too small throughout (indicating that the mean scalar gradient is generally too large). Streamwise correlation lengths in the LES are also substantially larger than in the DNS, which, with the other results, suggests that this variant of the dynamic SGS model is too dissipative, at least for low Reynolds number simulations. Another measure of the mean passive scalar is the Nusselt number $Nu$, defined by Kays and Crawford (1980) as

$$Nu = 4\delta \Theta_m^{-1} \left| \frac{\partial \theta}{\partial y} \right|_w , \qquad (31)$$

where $\Theta_m$ is the mass-averaged scalar and the gradient of the scalar is evaluated at the wall. The Nusselt numbers for $Pr = 0.1, 0.71$, and $2.0$ from the DNS results are 7.2, 23.8, and 43.8, respectively, as compared

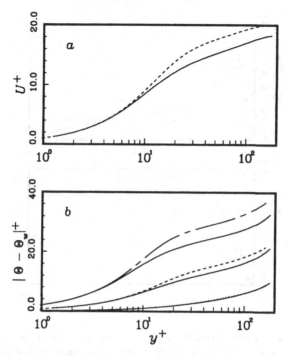

Fig. 3 Mean profiles (in wall units) from the LES of passive (temperature) scalars in channel flow with isothermal walls, one hot and one cold, using the SGS model with the strain rate contraction and $\widehat{\Delta}/\Delta = 2^{2/3}$: (a) streamwise velocity and (b) scalars (relative to the wall) for $\cdots\cdots\ Pr = 0.1$, $----\ Pr = 0.71$, and $---\ Pr = 2.0$; —— corresponding DNS results of Kim and Moin (1989).

with 7.4, 21.0, and 35.2, respectively, from the LES, showing larger discrepancies at progressively larger $Pr$.

Simulations were also performed for a passive scalar with $Pr = 0.71$, $Re_\tau \approx 140$–160, and a constant streamwise scalar gradient, making the scalar source term $Q$ in equation (3) proportional to the streamwise velocity component; this case closely resembles that with uniform $Q$ (Case I of Kim and Moin 1989). The results are compared with a recent DNS with constant streamwise scalar gradient by Kasagi, Tomita, and Kuroda (1992) and with DNS results of Kim and Moin (1989). Simulations were performed with the strain rate contraction and $\widehat{\Delta}/\Delta = 2^{2/3}$ and 2, and with the least squares contraction and $\widehat{\Delta}/\Delta = 2$. This series of model variants leads to progressively lower eddy viscosities and diffusivities, which vary by an overall factor of about 2. The results shown in Fig-

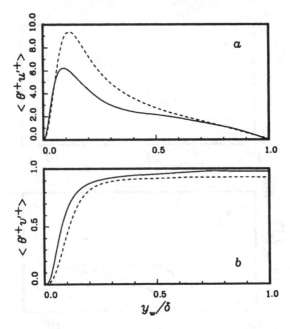

Fig. 4  Passive scalar fluxes (in wall units) for $Pr = 0.71$ as functions of
distance from the wall for the same calculations as in Figure 3: (a)
streamwise and (b) normal components; – – – – LES and ——— DNS
of Kim and Moin (1989).

ures 5–7 illustrate the effect this has on some mean and second-order
statistical quantities.

The mean streamwise velocity and scalar profiles in Figure 5 obtained
by the LES using the least squares contraction and $\widehat{\Delta}/\Delta = 2$ (the least
dissipative model variant) show much better agreement with the DNS
data than do the results for the other model variants, apparently because of the reduced SGS eddy viscosity and diffusivity. Note that the
change in SGS stresses by a factor of 2 results in changes of only about
10% in the mean streamwise velocity and mean scalar profiles. The
Nusselt numbers from the LES using the strain rate contraction and
$\widehat{\Delta}/\Delta = 2^{2/3}$ and 2 were found to be about 25% and 15% lower than
that from DNS, while $Nu$ from the LES using the least squares contraction with $\widehat{\Delta}/\Delta = 2$ was only about 3% lower than that from DNS.
The improved agreement in the scalar fluxes (including SGS contributions) between the LES for the less dissipative model variants and DNS
results is also evident in Figure 6. The agreement is quite good for
the streamwise component $\langle \theta' u' \rangle$ using the model variant with the least

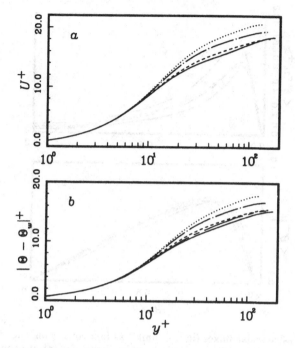

Fig. 5 Mean profiles (in wall units) of (a) streamwise velocity and (b) passive scalar from the LES of a passive scalar with $Pr = 0.71$ in channel flow with a constant streamwise scalar gradient using different variants of the SGS model: the strain rate contraction with $\widehat{\Delta}/\Delta = 2^{2/3}$ ( ········ ) and 2 ( —·— ), and the least squares contraction with $\widehat{\Delta}/\Delta = 2$ ( ---- ). The DNS results of —–– Kasagi et al. (1992) with a constant streamwise scalar gradient and ——— Kim and Moin (1989) with a constant scalar source term are shown for comparison.

squares contraction and $\widehat{\Delta}/\Delta = 2$, exceeding the DNS data by only a small amount. The improvement for the normal component $\langle\theta'v'\rangle$ from the LES with less SGS model dissipation is less pronounced, but still significant, with the position of its peak values coinciding better with those from the DNS data. The scalar intensities $\theta_{rms}$ from LES (which do not include the unknown SGS component) and from DNS are compared in Figure 7. The LES results with the strain rate contraction are significantly larger than the DNS results, but the LES with the least squares contraction and $\widehat{\Delta}/\Delta = 2$ again shows good agreement, with the LES scalar intensity only slightly exceeding the DNS results. Even though the eddy viscosity and diffusivity are reduced significantly with the least squares contraction, the LES spectra show that the dissipation

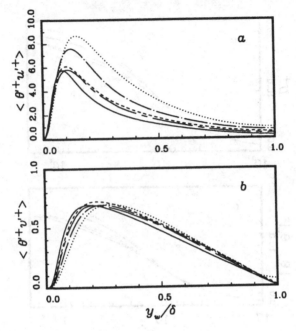

Fig. 6 Passive scalar fluxes (in wall units) as functions of distance from the wall for the same calculations as in Figure 5: (a) streamwise and (b) normal components. ⋯⋯⋯ , —·— , ----  LES; —--— DNS of Kasagi et al. (1992); and ——— DNS of Kim and Moin (1989). The failure of the dotted curve to vanish at $y_w/\delta = 1$ is due to a poor statistical sample.

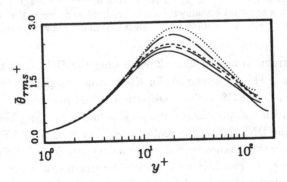

Fig. 7 Passive scalar intensities (in wall units) for the same calculations as in Figures 5 and 6: ⋯⋯⋯ , —·— , ---- LES; —--— DNS of Kasagi et al. (1992); and ——— DNS of Kim and Moin (1989). The LES results represent only the resolved field statistics here.

level is sufficient to remove small scales without any sign of excessive accumulation of energy there.

The sensitivity of the LES results to the model variants is also being explored at much higher Reynolds numbers with the LES of passive (temperature) scalars with isothermal walls, one hot and one cold. Preliminary results from simulations with $Re_\tau \approx 1200\text{--}1400$ and $Pr = 0.71$ using the SGS models with the strain rate contraction and $\widehat{\Delta}/\Delta = 2^{2/3}$ (the most dissipative variant) and with the least squares contraction and $\widehat{\Delta}/\Delta = 2$ (the least dissipative variant) indicate that the results are affected at these high Reynolds numbers in a similar manner and to about the same degree as those at low Reynolds numbers. The Nusselt number from equation (31) is about 122 in the LES using the SGS model with the strain rate contraction and $\widehat{\Delta}/\Delta = 2^{2/3}$, and it is about 176 using the least squares contraction and $\widehat{\Delta}/\Delta = 2$. The value predicted by the semiempirical model of Kays and Crawford (1980) for the same mass flux Reynolds number

$$Re_m = 2\nu^{-1} \int_{-\delta}^{\delta} U \, dy \tag{32}$$

of $1.18 \times 10^5$ is an intermediate value of $Nu = 153$. In Figure 8 we show the mean streamwise velocity and scalar profiles for the two model variants, in which the differences in the log-layer levels is seen to be comparable to those in Figure 5. The change by a factor of 2 in the SGS stresses for the two models considered results in changes of about 20% and 40% in the levels of the mean streamwise velocity and the mean scalar. The profile for $U^+$ using the least squares contraction and $\widehat{\Delta}/\Delta = 2$ is seen to agree better with the predicted logarithmic profile, $\kappa^{-1} \ln y^+ + 5.0$ with $\kappa = 0.4$; its deviation from the logarithmic profile at lower $y^+$ probably arises from a small deviation from equilibrium evident in the LES statistics and from small sample size. Thus, we have an indication that the sensitivity of the LES results to different SGS models and parameters is not strongly dependent on Reynolds number, and that the best LES results are obtained using the SGS model with the least squares contraction in equations (23) and (24) over a wide range of Reynolds numbers.

## 7.4 Conclusions

The dynamic SGS model of Moin et al. (1991) has been applied to an eddy diffusivity model for scalar transport, and it has been shown to

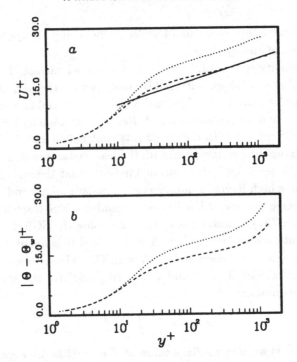

Fig. 8　Mean profiles (in wall units) of (a) streamwise velocity and (b) passive scalar (temperature) with $Pr = 0.71$ from a large Reynolds number LES of channel flow with isothermal walls, one hot and one cold, using different variants of the SGS model: ⋯⋯ the strain rate contraction with $\widehat{\Delta}/\Delta = 2^{2/3}$ and ---- the least squares contraction with $\widehat{\Delta}/\Delta = 2$. For comparison, the logarithmic profile $\kappa^{-1} \ln y^+ + 5.0$ with $\kappa = 0.4$ ( —— ) is shown in (a).

perform successfully in the LES of channel flow without the need to prescribe model constants and invoke ad hoc damping functions at the walls. Certain variants that are possible in the model, such as different prescriptions for the effective filter width and different contractions to extract scalar model coefficients, have been tested. The LES results suggest that the results are somewhat sensitive to such variants through the different levels of eddy viscosity and diffusivity that they produce. In particular, the least squares contraction in equations (23) and (24) was found to yield the least SGS dissipation of the variants considered and to perform the best with regard to matching DNS data in low Reynolds number simulations.

## Acknowledgments

We are grateful to Dr. Paul Durbin and Dr. Kyle Squires for useful discussions during the course of this work.

## References

DEARDORFF, J.W. (1970) A numerical study of three-dimensional turbulent channel flow at large Reynolds numbers. *J. Fluid Mech.* **41**, 453–480.

EIDSON, T.M. (1985) Numerical simulation of turbulent Rayleigh–Bénard convection using subgrid scale modeling. *J. Fluid Mech.* **158**, 245–268.

ERLEBACHER, G., HUSSAINI, M.Y., SPEZIALE, C.G. AND ZANG, T.A. (1990) Toward the large-eddy simulation of compressible turbulent flows. ICASE Rep. 90-76, NASA/Langley Research Center.

GERMANO, M., PIOMELLI, U., MOIN, P. AND CABOT, W.H. (1991) A dynamic subgrid-scale eddy viscosity model. *Phys. Fluids A* **3**, 1760–1765.

KASAGI, N., TOMITA, Y. AND KURODA, A. (1992) Direct numerical simulation of passive scalar field in a turbulent channel flow. *ASME J. Heat Trans.* **114**, 598–606.

KAYS, W.M. AND CRAWFORD, M.E. (1980) *Convective Heat and Mass Transfer*, 2nd ed., ch. 13. McGraw-Hill.

KIM, J. AND MOIN, P. (1989) Transport of a passive scalar in a turbulent channel flow. In *Turbulent Shear Flows, Vol. 6*. Ed. J.-C. Andre et al., pp. 86–96. Springer-Verlag.

LILLY, D. (1989) The length scale for sub-grid-scale parameterization with anisotropic resolution. In *CTR Annual Research Briefs – 1988*. Ed. P. Moin, W.C. Reynolds, and J. Kim, pp. 3–9. NASA/Ames Research Center and Stanford University.

LILLY, D.K. (1992) A proposed modification of the Germano subgrid-scale closure method. *Phys. Fluids A* **4**, 633–635.

MOIN, P., SQUIRES, K., CABOT, W. AND LEE, S. (1991) A dynamic subgrid-scale model for compressible turbulence and scalar transport. *Phys. Fluids A* **3**, 2746–2757.

PIOMELLI, U., FERZIGER, J.H. AND MOIN, P. (1987) Models for large eddy simulations of turbulent channel flows including transpiration. Rep. TF-32, Stanford University, Dept. of Mechanical Engineering.

PIOMELLI, U., CABOT, W.H., MOIN, P. AND LEE, S. (1991) Subgrid-scale backscatter in transitional and turbulent flows. *Phys. Fluids A* **3**, 1766–1771.

ROGERS, M., MOIN, P. AND REYNOLDS, W.C. (1986) The structure and modeling of the hydrodynamic and passive scalar fields in homogeneous turbulent shear flow. Rep. TF-25, Stanford University, Dept. of Mechanical Engineering.

SMAGORINSKY, J. (1963) General circulation experiments with the primitive equations. I. The basic experiment. *Mon. Wea. Rev.* **91**, 99–164.

# 8

# Renormalization Group Theory Simulation of Transitional and Turbulent Flow over a Backward-Facing Step

GEORGE E. KARNIADAKIS, STEVEN A. ORSZAG,

AND VICTOR YAKHOT

## 8.1 Introduction

Renormalization group (RNG) theory has found its way into fluid mechanics relatively recently, particularly in developing turbulence theory (Yakhot and Orszag, 1986a); however, it has already provided a series of interesting theoretical and numerical results. A milestone for RNG theory in fluid mechanics was the postulation of the so-called *correspondence principle*, stating that in the inertial range the behavior of the small-scale Navier–Stokes turbulence is statistically equivalent to the modeled Navier–Stokes equation with the addition of a random noise term. This statement has recently been derived based on a new formulation of the $\epsilon$-expansion by Migdal et al. (see Chapter 3, this volume). The correspondence principle makes it possible to use all the formalism of classical RNG theory. Recently, RNG methods have been developed by Yakhot and Orszag (1986a,b; 1987) to analyze a variety of turbulent flow problems. For homogeneous turbulent flows, such important quantities as the Kolmogorov constant, Batchelor constant, turbulent Prandtl number, rate of decay to isotropy, and skewness factor have been obtained directly from this theory in good agreement with available data.

RNG methods involve systematic approximations to the full Navier–Stokes equations that are obtained by using perturbation theory to eliminate or decimate infinitesimal bands of small-scale modes, iterating the perturbation procedure to eliminate finite bands of modes by constructing recursion relations for the renormalized transport coefficients and evaluating the parameters at a fixed point in the lowest order of a dimensional expansion around a certain critical dimension. The decimation procedure, when applied successively to the entire wave number spectrum leads to the RNG equivalent of full closure of the Reynolds-averaged Navier–Stokes equations. The resulting RNG transport coefficients are differential in character as opposed to ad hoc algebraic coefficients of conventional closure methods. All constants and functions appearing in the RNG closures are fully determined by the RNG analysis. Briefly, the RNG method provides an analytical method to eliminate small scales from the Navier–Stokes equations, thus leading to a dynamically consistent description of the large scales. The formal process of successive elimination of small scales together with rescaling of the resulting equations results in a calculus for the derivation of transport approximations in turbulent flows.

An obvious advantage of RNG models over other standard models is that they also apply in the low Reynolds number limit and thus can potentially be used to describe the late stages of transition and early turbulence. This result has been reconfirmed in the recent simulations of Yakhot et al. (1989) and Piomelli et al. (1989), who studied turbulent channel flow and boundary layers. Despite the fact that the initial conditions employed corresponded to laminar flow, transition to turbulence was faithfully simulated as in standard direct simulations. The RNG subgrid model provides an expression for the eddy viscosity with correct limits both in both the high and low Reynolds number range, namely the molecular viscosity and the Smagorinsky equation respectively. The subgrid-scale model has not yet been applied to complex geometry flows where a broader range of scales typically exists.

In this work we consider the flow over a backward-facing step and employ RNG/large eddy computations to simulate transitional and early turbulent states. A parallel study by Kaiktsis et al. (1991) investigates early transitional states of the same flow using direct simulations. Preliminary data obtained from spectral element/large eddy simulations suggest that the RNG model can be applied in the transitional and early turbulent regime, where other standard turbulence models typically fail.

This chapter is organized as follows: in Section 8.2, we present a review of the RNG formalism for homogeneous turbulence and its extension to finite systems. In Section 8.3, we describe the discretization procedure in solving the renormalized equation of motions; and finally, in Section 8.4, we present some preliminary results based on spectral element/large eddy simulations.

## 8.2 RNG Theory

In this section we develop a subgrid-scale model for inhomogeneous turbulence following the derivation presented first by Yakhot and Orszag (1986a). To this end, we first review the infrared limit of RNG theory for homogeneous turbulence. In particular, we consider the following system of equations describing the motion of an incompressible fluid stirred by an external force **f**,

$$\frac{\partial \mathbf{v}}{\partial t} + \mathbf{v} \cdot \nabla \mathbf{v} = -\frac{\nabla p}{\rho} + \nu_0 \nabla^2 \mathbf{v} + \mathbf{f}, \qquad (1a)$$

$$\nabla \cdot \mathbf{v} = 0, \qquad (1b)$$

where $p$ is the pressure, $\nu_0$ is the molecular viscosity, and $\rho$ is the density. The force has an energy spectrum given by

$$< f_j(\mathbf{k}, \omega) f_j(\mathbf{k}', \omega') > =$$
$$2(2\pi)^4 D_0 k^{-3} \left[ \delta_{ij} - \frac{k_i k_j}{k^2} \right] \delta(\mathbf{k} + \mathbf{k}') \delta(\omega + \omega'). \qquad (1c)$$

The idea of the infrared RNG method is to eliminate modes from the wave number strip defined by $\Lambda e^{-r} < k < \Lambda$, near the ultraviolet cutoff $\Lambda$. The system resulting from the elimination of these modes involves modified interaction coefficients and new nonlinearities, as well as modified viscosities and forces. The resulting equations are also rescaled and recast in a form similar to the original system (1). The first step of the RNG procedure is to Fourier transform equations (1) and solve for the transformed velocity field as follows,

$$v_l(\hat{k}) = G^0 f_l(\hat{k}) - \frac{i\lambda_0}{2} G^0 P_{lmn}(\hat{k}) \int v_m(\hat{q}) v_n(\hat{k} - \hat{q}) \, d\hat{q}, \qquad (2a)$$

where $\hat{k} = (\mathbf{k}, \omega)$ and the harmonic propagator $G^0 = (-i\omega + \nu_0 k^2)^{-1}$. The rest of the parameters are defined as

$$P_{lmn}(\mathbf{k}) = k_m P_{ln}(k) + k_n P_{lm}(k); \quad P_{lm}(k) = \delta_{lm} - \frac{k_l k_m}{k^2}. \qquad (2b)$$

Here, $\lambda_0$ is a formal parameter introduced for the iterative procedure.

Next, the transformed velocity field **v** is split into two components: $v^<(\hat{k})$ with $0 < k < \Lambda e^r$ and $v^>(\hat{k})$ with $\Lambda e^{-r} < k < \Lambda$; the former is the *resolvable* part, while the latter is to be eliminated by repeated substitution. This process of elimination generates an infinite expansion for $v^<$ in powers of $\lambda_0$. If we keep only terms up to second order in $\lambda_0$, then a correction to the bare viscosity $\nu_0$ is generated; after the integration over the frequency space $-\infty$ to $\infty$ and the wave space $\Lambda e^{-r}$

to $\Lambda$ it takes the form

$$\Delta\nu = A_d \frac{\lambda_0^2 D_0}{\nu_0^2 \Lambda^4} \frac{e^{4r} - 1}{4}, \qquad (3)$$

and $A_d$ is a constant computed exactly in terms only of the space dimensions $d$ (Yakhot and Orszag, 1986a). Thus, the viscosity resulting from elimination of the modes $v^>$ is

$$\nu = \nu_0 \left( 1 + A_d \bar{\lambda}_0^2 \frac{e^{4r} - 1}{4} \right), \qquad (4a)$$

where the dimensionless coupling constant $\bar{\lambda}_0$ is

$$\bar{\lambda}_0 = \lambda_0 \frac{D_0^{1/2}}{\nu_0^{3/2} \Lambda^2}. \qquad (4b)$$

These results are valid in the limit $r \to 0$. It is possible, however, to eliminate a finite band of modes by iterating the above procedure of eliminating an infinitesimally narrow band of modes; this will result in wave-dependent viscosity $\nu = \nu(r)$ and coupling constant $\bar{\lambda} = \bar{\lambda}(r)$. These functions can then be determined by taking the limit $r \to 0$ in equation (3) in order to obtain the differential equation

$$\frac{d\nu}{dr} = A_d \nu(r) \bar{\lambda}^2(r) \quad \text{and} \quad \bar{\lambda}^2(r) = \frac{\lambda_0^2 D_0}{\nu^3(r) \Lambda^4} e^{4r}. \qquad (5a)$$

The solution to this equation is

$$\nu(r) = \nu_0 [1 + \tfrac{3}{4} A_d \bar{\lambda}_0^2 (e^{4r} - 1)]^{1/3}, \qquad (5b)$$

$$\bar{\lambda}(r) = \bar{\lambda}_0 e^{2r} [1 + \tfrac{3}{4} A_d \bar{\lambda}_0^2 (e^{4r} - 1)]^{1/3}. \qquad (5c)$$

This equation gives a $k$-dependent viscosity in the infrared limit ($r \to \infty$) which leads to an energy spectrum similar to the Kolmogorov $k^{-5/3}$ spectrum for the inertial range. An exact comparison, in fact, with the Kolmogorov spectrum provides the relation between the amplitude of the stirring force $D_0$ and the mean dissipation rate ($\bar{\epsilon} \propto D_0$), and thus equation (5b) can be rewritten as

$$\nu(r) = \nu_0 \left( 1 + a \frac{\bar{\epsilon}}{\nu_0^3 \Lambda^4} (e^{4r} - 1) \right)^{1/3}, \qquad (6)$$

where $a = 0.120$ as computed in terms of $A_d$.

It can be shown that a more realistic model whose stirring force satisfies

$$< f_j(\mathbf{k}, \omega) f_j(\mathbf{k}', \omega') > \propto \bar{\epsilon} k^{-d} \left[ \delta_{ij} - \frac{k_i k_j}{k^2} \right] \delta(\mathbf{k} + \mathbf{k}') \delta(\omega + \omega'), \qquad (7)$$

$$\forall k \geq k_c,$$

$\forall k_c > L^{-1}$ (where $L$ is an integral scale of the flow), also gives a Kolmogorov spectrum in the limit $k \to \infty$. This is true since all contributions originating from the strip $L^{-1} < k < k_c$ are negligible in the limit $k \to \infty$. This result is the key idea in extending the above theory to finite systems and inhomogeneous turbulence. In particular, we assume that a turbulent fluid in a finite system in which the flow is locally homogeneous exhibits the Kolmogorov behavior in the intermediate range $k_c < k < k_d$, where $k_d$ denotes the dissipation cutoff. In addition, it can be shown directly from equations (1) and the definition of $\bar{\epsilon}$ that $\bar{\epsilon} \approx \epsilon^<$, which in turn can be expressed in terms of the resolvable field and the (enhanced) viscosity, i.e.,

$$\epsilon^<(\mathbf{x},t) = \nu(r) \left( \frac{\partial v^<_i}{\partial x_j} + \frac{\partial v^<_j}{\partial x_i} \right)^2. \tag{8}$$

In order to construct a subgrid-scale model based on the above formulation we identify the mesh size $\tilde{\Delta}$ as the smallest unrenormalized scale ($\tilde{\Delta} = \Lambda^{-1} e^r$) and employ a Gaussian filter of width $\Delta = 2\tilde{\Delta}$ to obtain

$$\nu = \nu_0 \left[ 1 + H \left( \frac{a\bar{\epsilon}}{(2\pi)^4 \nu_0^3} \Delta^4 - C \right) \right]^{1/3}, \tag{9}$$

where $H(x)$ is the Heaviside function, and $C \approx 100$. An alternative equation for evaluating $\Delta$ in the case of nonuniform meshes is the following:

$$\Delta^4 = \int^{\Lambda_1} \int^{\Lambda_2} \int^{\Lambda_3} \frac{dk\, dp\, dq}{(k^2 + p^2 + q^2)^{7/2}}. \tag{10}$$

Here, $\Lambda_i = \pi/\Delta_i$ (for $i = 1, 2, 3$) and $\Delta_i = 2\tilde{\Delta}$ equals twice the computational mesh in each direction respectively. One can readily evaluate the above integral by breaking it up into three asymptotic integrals and a finite triple integral that can be computed numerically. In cases of simple geometries (i.e., plane channels) the integral can be evaluated through algebraic relations (see Yakhot et al., 1989).

The assumption of local homogeneity implies that the eliminated scales are much smaller than the distance $y$ from the nearest wall, for only such scales can be isotropic. Thus, $\Delta$ must decrease as the distance to the nearest wall decreases (this is, of course, automatically true for all spectral-type discretizations). It follows then from (9) that $\nu \to \nu_0$ as $y \to 0$. On the other hand, in regions far from the wall, equation (9) reduces to

$$\nu = C_s \Delta^2 \left| \frac{\partial v^<_i}{\partial x_j} + \frac{\partial v^<_j}{\partial x_i} \right|, \tag{11}$$

where $C_s = 0.005$. This last equation is identical to the classical Smagorinsky eddy viscosity model (Smagorinsky, 1963).

The corresponding renormalized equation of motion (incorporating the full stress tensor) for $v^<$ is

$$\frac{\partial v^<}{\partial t} + v^< \cdot \nabla v^< = -\frac{\nabla p}{\rho} + \nabla \cdot \nu[\nabla v^< + \nabla^T v^<] + f^<, \qquad (12)$$

where $f^<$ refers to the renormalized force of the original system in equation (1). It can be shown that $f^< \propto k^2$ and thus is negligible (compared with the bare force, which scales as $k^{-3}$) everywhere but in the buffer region. In the next section we discuss the discretization of this equation subjected to the incompressibility constraint.

## 8.3 Spectral Element Methodology

To simplify the notation in the following we omit the superscripts, and rewrite the equation of motion as

$$\frac{\partial v}{\partial t} = -\frac{\nabla p}{\rho} + \bar{\nu}\nabla^2 v + N(v), \qquad (13a)$$

where the last term includes all nonlinear and forcing terms as well as the viscous terms, i.e.,

$$N(v) = -\tfrac{1}{2}[v \cdot \nabla v + \nabla \cdot (v \cdot v)] + f + \nabla \cdot (\nu - \bar{\nu})[\nabla v + \nabla^T v], \quad (13b)$$

where $\nu$ denotes the total viscosity given by equation (9), while $\bar{\nu}$ is an artificial (constant in space) viscosity introduced for stability reasons. The convective terms are written in skew-symmetric form for aliasing control purposes (Horiuti, 1987). Numerical solution of the above system of equations will be obtained in a three-dimensional computational domain $\Omega$. Before we proceed to the spatial discretization of the equations we first review briefly the time-stepping algorithm.

### 8.3.1 Semidiscrete Formulation

The separation of terms in equation (13) leads naturally to a splitting scheme of mixed explicit/implicit type. In particular, the recently developed high-order splitting algorithm based on mixed, stiffly stable schemes is used (Karniadakis et al., 1991). Considering first the nonlinear terms $N(v)$ we obtain,

$$\frac{\hat{v} - \sum_{q=0}^{J-1} \alpha_q v^{n-q}}{\Delta t} = \sum_{q=0}^{J-1} \beta_q[-N(v^{n-q})], \qquad (14a)$$

where $\alpha_q, \beta_q$ are implicit/explicit weight coefficients for the stiffly stable scheme of order $J$ (see Karniadakis et al., 1991). The next substep incorporates the pressure equation and enforces the incompressibility constraint as follows:

$$\frac{\hat{\hat{\mathbf{v}}} - \hat{\mathbf{v}}}{\Delta t} = -\frac{\nabla p^{n+1}}{\rho}, \tag{14b}$$

$$\nabla \cdot \hat{\hat{\mathbf{v}}} = 0. \tag{14c}$$

Finally, the last substep includes the viscous corrections and the imposition of the boundary conditions, i.e.,

$$\frac{\gamma_0 \mathbf{v}^{n+1} - \hat{\hat{\mathbf{v}}}}{\Delta t} = \bar{\nu} \nabla^2 \mathbf{v}^{n+1}, \tag{14d}$$

where $\gamma_0$ is a weight coefficient of the backward differentiation scheme employed.

The above time treatment of the system of equations (13) results in a very efficient calculation procedure as it decouples the pressure and velocity equations as in (14bc) and (14d) respectively. As regards the time accuracy of this splitting scheme a key element in this approach is the specific treatment of the pressure equation, which can be recast in the form

$$\nabla^2 p^{n+1} = \nabla \cdot \left( \frac{\hat{\mathbf{v}}}{\Delta t} \right), \tag{15a}$$

along with the consistent high-order pressure boundary condition (see Karniadakis et al., 1991)

$$\frac{\partial p^{n+1}}{\partial n} = \mathbf{n} \cdot \left[ -\sum_{q=0}^{J-1} \beta_q \mathbf{N}(\mathbf{v}^{n-q}) - Re^{-1} \sum_{q=0}^{J-1} \beta_q [\nabla \times (\nabla \times \mathbf{v}^{n-q})] \right], \tag{15b}$$

where $\mathbf{n}$ denotes the unit normal to the boundary $\Gamma$. Equation (15) therefore is a Poisson equation with constant coefficients, which can be rewritten in the standard form

$$\nabla^2 \phi = g(\mathbf{x}), \tag{16}$$

where we have defined $\phi = p^{n+1}$, and $g(\mathbf{x}) = \nabla \cdot (\hat{\mathbf{v}}/\Delta t)$. In the following section we refer to this equation in order to discuss the spatial discretization of equations (14) in three dimensions using the spectral element method.

### 8.3.2 Spatial Discretization

The spatial discretization of (14) is obtained using the spectral element methodology Patera (1984), Karniadakis et al. (1985), and Karniadakis (1989). In the standard spectral element discretization the computational domain $\Omega$ is broken up into general brick elements (hexahedra) in three dimensions, which are mapped isoparametrically to canonical cubes. The accuracy of interpolating the geometry therefore is of the same order as the accuracy of interpolating the field unknowns. Geometry, unknowns, and data are then expressed as tensorial products in terms of Legendre–Lagrangian interpolants. The final system of equations to be solved is obtained via a Galerkin variational statement. In particular, the computational domain $\Omega$ is covered with $K = 48$ spectral elements of resolution $N_1 = N_2 = N_3 = 12$ in each direction.

To illustrate the spectral element methodology in more detail let us first consider the model equation (16), which represents the elliptic contributions of the governing equations. If we define $H_0^1$, the standard Sobolev space that contains functions which satisfy homogeneous boundary conditions, and introduce test functions $\psi \in H_0^1$, we can then write the equivalent variational statement of (16) as

$$\int_\Omega \frac{\partial \psi}{\partial x_j} \frac{\partial \phi}{\partial x_j} \, ds = - \int_\Omega \psi g \, ds. \tag{17}$$

The conforming spectral element discretization corresponds to numerical quadrature of the variational form (17) restricted to the space $X_h \subset H_0^1$. The discrete space $X_h$ is defined in terms of the spectral element discretization parameters $(K, N_1, N_2, N_3)$, where $K$ is the number of spectral elements, and $N_1 - 1, N_2 - 1, N_3 - 1$ are the degrees of piecewise high-order (Legendre) polynomials in the three directions respectively that fill the space $X_h$. By selecting appropriate Gauss–Lobatto points $\xi_{pql}^k$ and corresponding weights $\rho_{pql} = \rho_p \rho_q \rho_l$, one can replace equation (17) by

$$\sum_{k=1}^{K} \sum_{p=0}^{N_1} \sum_{q=0}^{N_2} \sum_{l=0}^{N_3} \rho_{pql} J_{pql}^k \left[ \frac{\partial \psi}{\partial x_j} \frac{\partial \phi}{\partial x_j} \right]_{\xi_{pql}^k} =$$

$$- \sum_{k=1}^{K} \sum_{p=0}^{N_1} \sum_{q=0}^{N_2} \sum_{l=0}^{N_3} \rho_{pql} J_{pql}^k [\psi g]_{\xi_{pql}^k}. \tag{18}$$

Here $J_{pql}^k$ is the Jacobian of the transformation from global to local coordinates $(x, y, z) \Rightarrow (r, s, t)$ for the three-dimensional element $k$. The Jacobian is easily calculated from the partial derivatives of the geometry transformation at the nodal point $(pql)$ via collocation as follows:

$$J = z_t(x_r y_s - x_s y_r) + z_s(x_t y_r - x_r y_t) + z_r(x_s y_t - x_t y_s). \qquad (19\text{a})$$

The partial derivatives are calculated from standard Lagrangian interpolations as for example

$$(x_r)_{pql} = D_{pm} x_{mql}, \qquad (19\text{b})$$

where $x_{mql}$ is the $x$ global coordinate of node $(pql)$; here the derivative operator is defined as $D_{ij} = dh_j/dz(\xi_i)$, and $h_j$ is the Lagrangian interpolant.

The next step in implementing (18) is the selection of a basis which reflects the structure of the piecewise smooth space $X_h$. We choose an interpolant basis with components defined in terms of Legendre–Lagrangian interpolants, $h_i(r_j) = \delta_{ij}$. Here, $r_j$ represents local coordinates and $\delta_{ij}$ is the Kronecker delta symbol. It was shown by Patera (1984) and Ronquist (1988) that such a spectral element implementation converges spectrally fast to the exact solution for a fixed number of elements $K$ and $N_{1,2,3} \to \infty$, for smooth data and solution, even in nonrectilinear geometries. Having selected the basis we can proceed to write the local to the element spectral element approximations for test functions, data, and geometry and obtain the system matrix; details are presented in Karniadakis (1989).

The natural choice of solution algorithm for equation (18) is an iterative procedure; in this case the large matrices (i.e., $P^k_{ijmnpq}$) need not be stored, but instead computed during the time stepping of the matrix-vector products (i.e., $P^k_{ijpmnq}\phi_{mnq}$). Two different iterative techniques have been implemented in this context: conjugate gradient techniques and multigrid methods (Ronquist, 1988). A difference between the formulation proposed here and that of Ronquist (1988) is that the high-order splitting scheme adopted in this work results in separate, elliptic equations for the pressure and velocity that can be very efficiently and robustly solved using those iterative techniques without the need of case-dependent preconditioners or other convergence acceleration techniques.

## 8.4 Results

Here we report the results of using the RNG-based large eddy simulation model to predict transition in complex geometry flows. Earlier work by Yakhot et al. (1989) and by Piomelli et al. (1989) in channel flows and boundary layer flows respectively suggests that transitional states can

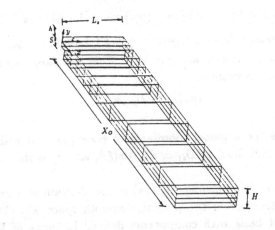

(a)

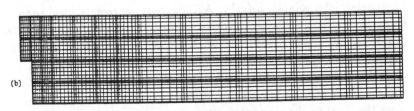

(b)

Fig. 1   (a) Geometry definition and computational domain; $h = 1$, $S = 0.94231$, $X_0 = 35$, $L_z = 2\pi$. (b) Spectral element discretization of an $x - y$ plane.

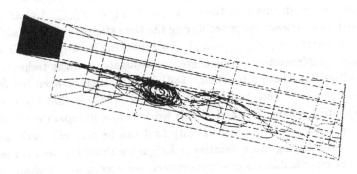

Fig. 2   Instanteneous streamlines at $Re = 2,222$.

potentially be simulated by this LES model. The problem we consider here is flow over a backward-facing step of the same geometry (expansion ratio approximately 2) as the one studied by Armaly et al. (1983); numerical simulations for the laminar regime have been recently reported

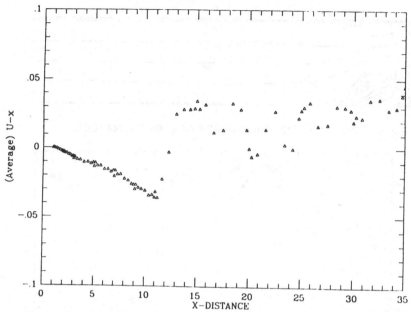

Fig. 3  Streamwise velocity profile at a distance $\Delta y = 0.013$ unit from the lower wall ($Re = 1,000$; direct simulation).

by Kaiktsis et al. (1991). The size of the separation zone is a unique function of the Reynolds number $Re$ for fixed inflow conditions, and thus this flow is a prototype for transition in massively separated wall-bounded flows. The Reynolds number is defined as $Re = \frac{2}{3}U_{max}(2h)/\nu_0$, where $h = 1$ is the height of the inlet channel, and $U_{max}$ is the maximum velocity at the inlet. The computational domain is shown in Fig. 1; the domain is $L_x = 35$ units long and the spanwise length is $L_z = 2\pi$, corresponding to a wave number $\beta = 1$. All dimensions are multiples of the inlet channel height. The inflow-prescribed velocity profile is a blunt profile of the form $U_i(y) = 1 - (2y)^4$; periodic boundary conditions are imposed in span, while the recently developed Neumann/sponge-type conditions are imposed at outflow (Tomboulides et al., 1991).

In all the results reported here, the RNG subgrid model was active throughout the computation; the production of nonzero eddy viscosity ($\nu > \nu_0$) is then an indication of transition. The first run is at Reynolds number $Re = 2,222$; instantaneous streamlines are plotted in Fig. 2 showing the existence of multiple eddies inside the separation region. In this case the subgrid model does not produce any eddy viscosity at resolution ($K = 48; N_1 = N_2 = N_3 = 12$). The flow is therefore suffi-

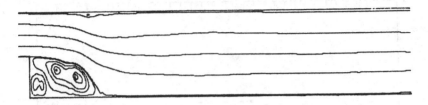

Fig. 4   Streamlines of the "equivalent" field at $Re = 4{,}444$ (LES).

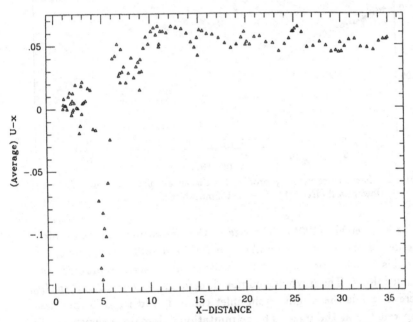

Fig. 5   Streamwise velocity profile at a distance $\Delta y = 0.013$ unit from the
lower wall $(Re = 8{,}888;$ LES).

ciently resolved using direct simulation. If we average the flow over time
and along the homogeneous spanwise direction we obtain an *equivalent*
two-dimensional steady field. At $Re = 2{,}222$ the time-averaged flow has
disappeared and only a smaller eddy is present at the step corner. This
feature seems to be a high Reynolds number effect and persists to the
fully turbulent regime. At Reynolds number $Re = 1{,}000$, which marks
the beginning of the transition process, the flow does not reverse sign
inside the separation zone. This is seen in Fig. 3, where we plot the
streamwise profile very close to the lower wall (0.013 nondimensional
unit in the vertical) from the step location to the outflow.

Starting the simulation from the velocity field generated at $Re = 2,222$ and increasing the Reynolds number to $Re = 4,444$, a nonzero eddy viscosity distribution is established at the high-strain regions of the flow, i.e., the shear layer emanating from the step corner and near the walls. The *equivalent* two-dimensional field (as defined previously) is plotted in Fig. 4; the multieddy structure is also apparent at this Reynolds number. Increasing the Reynolds number further to $Re = 8,888$ (inside the fully turbulent regime) the eddy viscosity distribution (computed by the RNG formula) becomes much broader. The corner eddy resolved by the direct simulation at $Re = 2,222$ and the LES at $Re = 4,444$ are also captured by the large eddy simulation for the fully turbulent regime (in agreement with the experimental results of Tani et al., 1967). This is shown in Fig. 5, where we plot the streamwise velocity profile very close to the lower wall (at the same location as in Fig. 3) at Reynolds number $Re = 8,888$. The positive values very close to the step corner indicate the existence of the eddy, while the other zero gives the value of the reattachment length. Similar plots of the mean streamwise velocity are shown in Fig. 6ab at Reynolds number $Re = 8,888$ inside the separation zone and in the recovery zone, respectively.

In Fig. 7 we plot the top and bottom wall static pressure profiles at $Re = 8,888$; in particular we plot the pressure coefficient $C_p = (p-p_r)/0.5\rho U_{max}^2$. Here $p_r$ is a reference pressure value taken as the top wall pressure value at the inlet. The top wall static pressure profile starts upstream of the step and increases monotonically until a location half a separation length downstream of the reattachment point; after that it reaches an asymptotic value at 0.2. The bottom wall pressure has a minimum inside the separation zone; it then rises rapidly and overshoots the top wall pressure just upstream of the reattachment point; downstream it asymptotes the top wall pressure as expected. These pressure profiles are in agreement with the results reported by Vogel and Eaton (1984). There is an important difference, however, just downstream of the reattachment point, where there is an overshoot in the top wall pressure and an unexpected oscillation in the bottom wall pressure not seen in the experiment of Vogel and Eaton (1984). It is not very clear at this point what the reasons are for such a discrepancy: numerical artifacts (e.g., inflow/outflow conditions, marginal spatial resolution, insufficient time averaging) as well as Reynolds number dependence collectively contribute to this. The experimental data correspond to $Re = 28,000$, where the flow is fully turbulent; for $Re > 20,000$, it appears that all statistical quantities depend only on the step expansion ratio and are

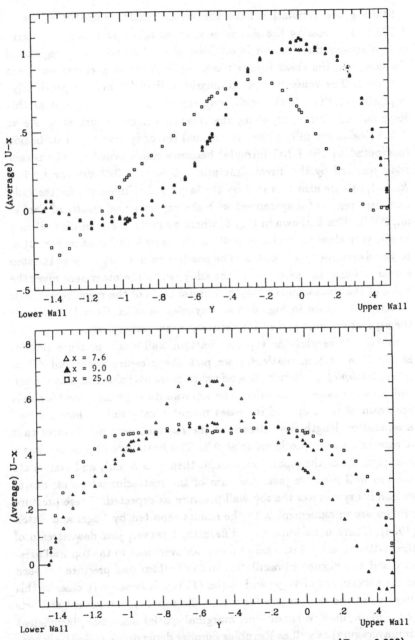

Fig. 6   (a) Mean velocity profiles inside the separation zone ($Re = 8,888$).
         (b) Mean velocity profiles in the recovery zone ($Re = 8,888$).

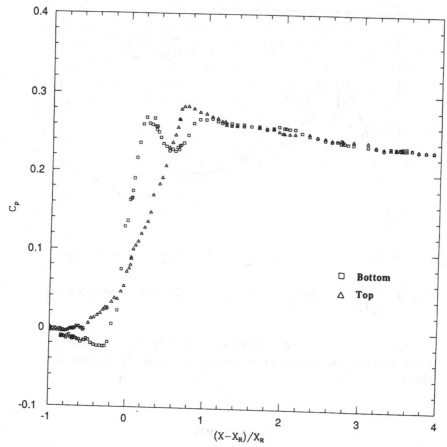

Fig. 7   Static pressure profiles along the top and bottom wall at $Re = 8,888$.

independent of the Reynolds number. Our simulation here, however, is at $Re = 8,888$, corresponding to an early turbulent flow state.

A more detailed description of the flow structure in the separation region is shown in Plate 1, where we plot color contours of the *instantaneous* streamwise velocity at two different planes along the span, at one horizontal plane close to the bottom wall and at a plane perpendicular to the flow direction, at $Re = 8,888$. The white regions are zones of separated flow. The main features of the flow are the strong spanwise variation as well as the very irregular boundary of the separation region in the lower wall as compared with laminar flows (Kaiktsis et al., 1991). In addition, a secondary separation zone (characteristic of laminar and transitional flows) attached to the upper wall appears which is responsible for the pressure distribution shown in Fig. 7. For a fully

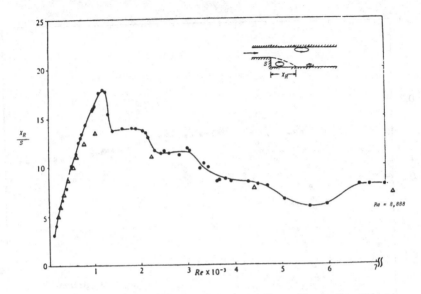

Fig. 8   Normalized separation length $X_R$ versus Reynolds number; $S$ is step height.

turbulent flow this secondary separation zone is significantly smaller or even nonexistent, as shown in the visualizations reported by Tani et al. (1967).

## 8.5 Discussion

The results presented in this chapter are preliminary and provide only a qualitative assessment of the resolution capabilities of the RNG sub-grid model in describing inhomogeneous turbulence and, in particular, separated flows. However, to the best of our knowledge, this is the first time that the *transitional* Reynolds number regime has been simulated for flow over a backward-facing step. In recent work (Morinishi and Kobayashi, 1990; Silveira-Neto et al., 1991), large eddy simulations were performed only in the *fully turbulent* regime, where the size of the separation zone is independent of the Reynolds number. Our objective in the current work was to simulate the flow in a regime where the flow state strongly depends on the Reynolds number. In particular, the separation length $X_R$ is a strong function of the Reynolds number in the laminar and transitional regime, as shown in Fig. 8, where we plot the experimental results of Armaly et al. (1983). We also include in the figure results of our computations (triangles) at the three transitional Reynolds numbers we have simulated as well as in the laminar regime. The laminar flow simulations were reported in Kaiktsis et al. (1991) using spectral element simulations and an identical computational domain at coarser resolution.

The underestimation of the separation length at Reynolds number around 1, 000 is associated with inflow conditions, as suggested by Kaiktsis et al. (1991). At higher Reynolds number our data follow the decreasing trend of $X_R$ with Reynolds number; however, they fall below the experimental curve and they underestimate the separation length by approximately 8% at $Re = 8,888$ from its steady value in fully turbulent states. One reason may be the inflow conditions imposed here corresponding to a quadratic, fully developed, time-independent velocity profile; indeed, preliminary studies with velocity profiles similar to the ones used in Kaiktsis et al. (1991) for the laminar flow (Fig. 23 in that reference) give higher values in the separation length $X_R$. Another reason may be the marginal resolution employed in the current simulations; although the minimum spacing in the direction perpendicular to the wall is 2.5 wall units in the recovery zone downstream of the separation zone, the spanwise direction was underresolved with 12 and 16 modes. In the computations reported in Morinishi and Kobayashi (1990), the coarse mesh simulation substantially overpredicted the separation length; this behavior, however, may not be universal, as both the numerical methodology and the model used in that study differ fundamentally from ours. In addition to the numerical discretization inaccuracies, the RNG subgrid model behaves differently with different discretizations, especially with the highly irregular spectral element mesh employed (Fig. 1).

In summary, we find fair agreement with the experimental results of Armaly et al. (1983); however, the important effects of mesh irregularity, inflow boundary conditions, and sufficient numerical resolution have not been fully quantified in the current work and will need to be addressed in detail in the future.

## Acknowledgments

Financial support for the current work was provided by grants from NSF (CTS-8906911 and CTS-8906432), by AFOSR Grant AFOSR-90-0261, and DARPA Contract N00014-86-K-0759. Most of the computations were performed on the Cray-Y/MP at Pittsburgh Supercomputing

Center and at the Numerical Aerodynamic Simulation Facility at NASA Ames.

# References

ARMALY B.F., DURST F., PEREIRA, J.C.F. AND SCHONUNG, B. (1983) Experimental and theoretical investigation of backward-facing step flow. *J. Fluid Mech.* 127, 473.

HORIUTI, K. (1987) Comparison of conservative and rotational forms in large eddy simulations of turbulent channel flow. *J. Comput. Phys.* 71, 343.

KAIKTSIS, L., KARNIADAKIS, G.E. AND ORSZAG S.A. (1991) Onset of three-dimensionality, equilibria, and early transition in flow over a backward-facing step. *J. Fluid Mech.* 231, 501.

KARNIADAKIS, G.E. (1989) Spectral element simulations of laminar and turbulent flows in complex geometries. *Appl. Num. Math.*, 6 85.

KARNIADAKIS, G.E., BULLISTER, E.T. AND PATERA, A.T. (1985) A spectral element method for solution of two- and three-dimensional time dependent Navier-Stokes equations. In *Finite Element Methods for Nonlinear Problems*, Ed. Springer-Verlag, 803.

KARNIADAKIS, G.E., ISRAELI, M. AND ORSZAG S.A. (1991) High-order splitting methods for the incompressible Navier-Stokes equations. *J. Comput. Phys.*, 97, 414.

MORINISHI, Y. AND KOBAYASHI, T. (1990) Large-eddy simulation of backward-facing step. In *Engineering Turbulence Modelling and Experiment*, Ed. Elsevier, p. 279.

PATERA, A.T. (1984) A spectral element method for fluid dynamics: Laminar flow in a channel expansion. *J. Comput. Phys.* 54, 468.

PIOMELLI, U., ZANG, T.A., SPEZIALE, C.G. AND LUND T.S. (1989) Application of renormalization group theory to the large-eddy simulation of transitional boundary layers. Presented at ICASE/LaRC Workshop on Instability and Transition, May 15-June 9.

RONQUIST, E.M. (1988) Optimal spectral element methods for the unsteady three-dimensional incompressible Navier-Stokes equations. Ph.D. thesis, Massachusetts Institute of Technology.

SILVEIRA-NETO, A., GRAND, D., MÉTAIS, O. AND LESIEUR, M. (1991) Large-eddy simulation of the turbulent flow in the downstream region of a backward-facing step. *Phys. Rev. Lett.* 66, 2320.

SMAGORINSKY, J. (1963) General circulation experiments with the primitive equations: I. The basic experiment. *Mon. Wea. Rev.* **91**, 99.

TANI, I., INCHI, M. AND KOMODA, H. (1967) Experimental investigation of flow separation associated with a step or a groove. Technical Report 364, Aerospace Research Institute, Tokyo University.

TOMBOULIDES, A.G., ISRAELI, M. AND KARNIADAKIS, G.E. (1991) Outflow boundary conditions for viscous incompressible flows. Presented at Minisymposium on Outflow Boundary Conditions, Stanford, CA.

YAKHOT, A., ORSZAG, S.A. AND YAKHOT, V. (1989) Renormalization group formulation of large-eddy simulations. *J. Sci. Comp.* **4**, 139.

YAKHOT, V. AND ORSZAG, S.A. (1986a) Renormalization group analysis of turbulence. I. Basic theory. *J. Sc. Comp.* **1**, 3.

YAKHOT, V. AND ORSZAG, S.A. (1986b) Renormalization group analysis of turbulence. *Phys. Rev. Lett.*, **57** 1722.

YAKHOT, V. AND ORSZAG, S.A. (1987) Relation between Kolmogorov and Batchelor constants. *Phys. Fluids* **30**, 3.

VOGEL, J.C. AND EATON, J.K. (1984) Heat transfer and fluid mechanics measurements in the turbulent reattaching flow behind a backward-facing step. Report MD-44, Stanford University.

# 9

## Spectral Large Eddy Simulation of Turbulent Shear Flows

MARCEL LESIEUR, OLIVIER MÉTAIS, XAVIER NORMAND,

AND ARISTEU SILVEIRA-NETO

### 9.1 Introduction

We use the concept of spectral eddy viscosity derived from the two-point closure to simulate the large-scale structure of the following flows: isotropic three-dimensional turbulence, stably stratified turbulence, and unsteady mixing layer.

Then, we propose a new subgrid-scale model, the *structure function model*, in which eddy viscosity is calculated with the aid of a kinetic energy spectrum evaluated *locally* in physical space. This model allows us to take into account intermittency and inhomogeneity of turbulence. The model has been applied to isotropic incompressible turbulence, incompressible flow behind a backward-facing step, and a high-supersonic boundary layer (Mach 5) spatially developing above a flat plate. For isotropic decaying turbulence, the model leads to a very good Kolmogorov $k^{-5/3}$ inertial range. In the case of the step, intense longitudinal *hairpin vortices* are found, and the predicted flow statistics are in good agreement with the experiments. For the compressible boundary layer, forced upstream by a two-dimensional wave perturbed by a small three-dimensional random white-noise perturbation, a *staggered mode* develops, leading to the breakdown of large-scale, $\Lambda$-shaped vortices.

### 9.2 EDQNM Spectral Eddy Coefficients

This work is the outcome of a theoretical effort initiated in Grenoble about 15 years ago, and aimed at deriving efficient subgrid-scale models from statistical two-point closures of turbulence, such as the eddy-damped quasi-normal Markovian (EDQNM) theory (see Orszag, 1977, and Lesieur, 1990, for a review). In Fourier space, the kinetic energy and passive temperature spectra, $\bar{E}(k,t)$ and $\bar{E}_T(k,t)$, respectively, satisfy the following two equations in the range of resolved modes $k < k_C$ ($k_C$ is a cutoff wave number):

$$\left(\frac{\partial}{\partial t} + 2\nu k^2\right)\bar{E}(k,t) = T_{<k_C}(k,t) + T_{sg}(k,t), \qquad (1)$$

$$\left(\frac{\partial}{\partial t} + 2\kappa k^2\right)\bar{E}_T(k,t) = T^T_{<k_C}(k,t) + T^T_{sg}(k,t), \qquad (2)$$

where the nonlinear transfers $T_{<k_C}$ refer to the wave number triads $[k,p,q]$ in which $p$ and $q$ are smaller than $k_C$, while at least one of the wave numbers $p$ and $q$ is larger than $k_C$ for the subgrid-scale transfers $T_{sg}$. For $k << k_C$, expansions of the isotropic EDQNM transfers with respect to the small parameter $k/k_C$ lead to (Kraichnan, 1966, 1968, 1976; Lesieur, 1990)

$$T_{sg}(k,t) = -\frac{2}{15}k^2\bar{E}(k,t)\int_{k_C}^{\infty}\theta_{0pp}\left[5E(p,t) + p\frac{\partial E(p,t)}{\partial p}\right]dp, \qquad (3)$$

$$T^T_{sg}(k,t) = -\frac{4}{3}k^2\bar{E}_T(k,t)\int_{k_C}^{\infty}\theta^T_{0pp}E(p,t)\,dp, \qquad (4)$$

where $\theta_{kpq}$ is the triple-correlation relaxation time of the EDQNM theory. This allows us to define two asymptotic eddy coefficients, the eddy viscosity and the eddy diffusivity, as functions of $k_C$ and the kinetic energy spectrum only. If it is assumed that $k_C$ lies in the Kolmogorov inertial range, then the asymptotic eddy viscosity turns out to be

$$\nu_t^{\infty} = 0.442C_K^{-3/2}\left[\frac{E(k_C)}{k_C}\right]^{1/2}. \qquad (5)$$

Here, $C_K$ is the Kolmogorov constant, which appears to be the only adjustable parameter of the EDQNM theory as far as the velocity field is concerned. For $C_K = 1.4$, the asymptotic eddy viscosity, renormalized by $[E(k_C)/k_C]^{1/2}$, is equal to 0.267. The asymptotic eddy diffusivity scales in a similar manner, but depends upon two extra adjustable constants; from the study of Herring et al. (1982), it appears that the more satisfactory choice for these constants corresponds to a turbulent Prandtl number equal to 0.6.

For $k$ close to $k_C$, the EDQNM eddy viscosity,

$$\nu_t(k|k_C) = -T_{<k_C}(k,t)/2k^2\bar{E}(k,t),$$

along with Eq. (1), presents a *cusp-like behavior*, as shown by Kraichnan (1976). The same behavior was found by Chollet (1983) for the passive scalar. For the latter, with the choice of adjustable constants leading to the asymptotic Prandtl number of 0.6, the turbulent Prandtl number remains approximately equal to 0.6 over the entire range $[0, k_C]$.

The first subgrid-scale calculations using the spectral eddy viscosity derived from (5) (with a cusp for $k$ close to $k_C$) was applied to the EDQNM kinetic energy spectrum itself by Chollet and Lesieur (1981). In their simulations of decaying turbulence with energy initially localized in low wave numbers, they were able to model, using Eq. (1), the transition to a Kolmogorov cascade across $k_C$ and the subsequent self-similar decay. The agreement with a direct EDQNM spectral calculation at very high Reynolds number was excellent for all scales of motion.

## 9.3 LES Using Spectral Eddy Coefficients

In the same article, Chollet and Lesieur (1981) used their spectral eddy viscosity for numerical solution of Navier–Stokes equation employing pseudospectral methods (Orszag and Patterson, 1972)

$$\frac{\partial}{\partial t}\hat{\mathbf{u}}(\mathbf{k},t) = \Pi(\mathbf{k}) \circ F[F^{-1}(\hat{\mathbf{u}}) \times F^{-1}(\hat{\omega})] - [\nu + \nu_t(k|k_c)]\, k^2 \hat{\mathbf{u}}(i\mathbf{k},t)\ , \quad (6)$$

where $\hat{\omega} = i\,\mathbf{k} \times \hat{\mathbf{u}}$ is the vorticity in Fourier space, and velocity field satisfies the solenoidal condition $\mathbf{k} \cdot \hat{\mathbf{u}}(\mathbf{k},t) = 0$. In Eq. (6), which is solved for $k < k_C$, the operator $F$ is the fast-Fourier transform, and $\Pi$ is the projection upon the plane perpendicular to $\mathbf{k}$. The calculation of Chollet and Lesieur (1981) involved $32^3$ Fourier modes, and showed a tendency to a $k^{-5/3}$ spectrum at the cutoff. The passive scalar equation

$$\frac{\partial}{\partial t}\hat{T}(\mathbf{k},t) = -i\mathbf{k} \cdot F[F^{-1}(\hat{T})F^{-1}(\hat{\mathbf{u}})] - [\kappa + \kappa_t(k|k_c)]\, k^2\hat{T}(\mathbf{k},t) \quad (7)$$

was first solved by Chollet (1983).

More recently, high-resolution simulations ($128^3$ modes) employing Eqs. (6) and (7) were performed by Lesieur and Rogallo (1989) and Lesieur et al. (1989). These calculations showed a self-similar decay of the kinetic energy spectrum, whereas the spectrum was closer to $k^{-2}$ than to $k^{-5/3}$ at the cutoff wave number $k_C$. The behavior of the spectral eddy viscosity, corresponding to a direct evaluation of the kinetic energy transfer across an artificial cutoff wave number $k'_C = k_C/2$, is in good agreement with the EDQNM predictions; particularly, a plateau for $k < k'_C/3$ and a rising cusp in the neighborhood of $k_C$ should be noted.

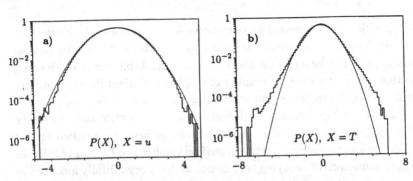

Fig. 1    Velocity (a) and passive temperature (b) PDF in a DNS of decaying isotropic turbulence at a resolution of $128^3$ Fourier modes. The functions are normalized so that their variance is equal to one. They are compared with the corresponding Gaussian distribution.

These calculations show an anomalous $k^{-1}$ behavior of the passive temperature spectrum in the energy-containing range, which may be caused by the deformation of the large-scale velocity field, in analogy with the inertial-convective range in the enstrophy cascade of two-dimensional turbulence. Indeed, it has become clear that three-dimensional isotropic turbulence possesses intense, large-scale, distorted, Kelvin-Helmholtz-type vortices (see Vincent and Ménéguzzi, 1991, and Métais and Lesieur, 1992). These vortices, marked by strong pressure lows (Métais and Lesieur, 1992), might result from the instability and roll-up of randomly oriented vortex sheets, as proposed by Moffatt (1991). Hence, the stretching of passive temperature fluctuations between these large vortices could cause the $k^{-1}$ spectrum. The eddy diffusivity, evaluated directly across $k'_C$ similarly to eddy viscosity, is very different from the EDQNM predictions: it displays a logarithmic decay with $k$ at low wave numbers (instead of the plateau), so that the spectral turbulent Prandtl number increases with $k$, from 0.3 to 0.8. This large-scale intermittency of the scalar is attested by a temperature probability distribution function (PDF) exponential in the wings, while the velocity is close to Gaussian (Métais and Lesieur, 1992). This is found both in large-eddy simulations and in direct numerical simulations, and is shown in Figure 1. The corresponding pressure PDF is exponential in the lows and Gaussian in the highs (Métais and Lesieur, 1992), which could indicate the strong correlation found in the same work between coherent vortices and depressions.

The same spectral eddy coefficients were used by Métais and Lesieur (1989, 1992) in the problem of decay of initially isotropic turbulence, submitted to the effect of mean stable stratification (within the Boussinesq approximation). Starting with high Froude number (initially, turbulence is not much affected by stratification), it was found that the large-scale intermittent character of the temperature, displayed above in the isotropic case, disappears during the initial stage of decay; the temperature pdf becomes Gaussian, the temperature spectrum loses its anomalous $k^{-1}$ range, and the spectral turbulent Prandtl number is approximately constant and equal to 0.6 over the wave-number span. It is of interest to mention the work of Batchelor et al. (1992) on homogeneous buoyancy-driven turbulence, which uses the same spectral eddy viscosity.

Finally, we stress that the spectral eddy viscosity has given good results for a three-dimensional LES of a temporal mixing layer with two fundamental vortices (Comte et al., 1989); such LES allows one to obtain the intense longitudinal vortices stretched between the primary billows, as found experimentally by Breidenthal (1981) and Bernal and Roshko (1986). The kinetic energy spectrum of the LES is very close to $k^{-5/3}$ in the small scales, as in the experiments at high Reynolds numbers. Finally, the predicted variances of the various velocity fluctuations are very close to the laboratory experiments, which is not so for direct numerical simulations.

## 9.4 The Structure Function Model

The above model of spectral eddy viscosity (including the cusp) is difficult to implement in calculations performed directly in physical space (by finite-difference methods, for instance), since the cusp cannot be approximated by an iterated Laplacian, as was shown by Chollet (1983). We therefore replace it by a constant chosen by identifying the eddy viscous flux, $2\nu_t \int_0^{k_c} p^2 E(p)dp$, with the kinetic energy flux $\epsilon$ in the Kolmogorov cascade. This gives

$$\nu_t^\infty = \frac{2}{3} C_K^{-3/2} \left[ \frac{E(k_C)}{k_C} \right]^{1/2},$$

(8)

to be compared with Eq. (5). For $C_K = 1.4$, the eddy viscosity plateau is now 0.402. However, the eddy viscosity scales with $[E(k_C)/k_C]^{1/2}$, which averages all the velocity fluctuations over the computational domain. In reality, it is well known that turbulence is highly intermittent, with spatially localized vortical structures embedded into a sea of quasi-irrotational fluid. In a calculation carried out in physical space, there is no need for any eddy dissipation in these calm areas, whereas the eddy viscosity in the active region must take into account the intensity of local turbulence activity. Accordingly, Métais and Lesieur (1992) have proposed taking an eddy viscosity that varies in physical space according to

$$\nu_t(\mathbf{x}|\Delta x) = 0.4 \sqrt{\frac{E_\mathbf{x}(k_C)}{k_C}}, \qquad (9)$$

with $k_C = \pi/\Delta x$, where $\Delta x$ is the computational grid mesh in physical space, and $E_\mathbf{x}(k_C)$ is a local kinetic energy spectrum at $\mathbf{x}$. This spectrum may be expressed in the following manner: first, one considers the local second-order velocity structure function

$$F_2(\mathbf{x}, \Delta x, t) = < \|\mathbf{u}(\mathbf{x}, t) - \mathbf{u}(\mathbf{x} + \mathbf{r}, t)\|^2 >_{\|\mathbf{r}\|=\Delta x}, \qquad (10)$$

where $< \, . \, >$ denotes a proper spatial average over all points $\mathbf{x} + \mathbf{r}$ separated from $\mathbf{x}$ by a distance $\Delta x$. Then, the local kinetic energy spectrum is related to the local structure function assuming that they both correspond to isotropic turbulence with Kolmogorov cascade (see Batchelor, 1953, and Orszag, 1977),

$$E(k) = C_K \epsilon^{2/3} k^{-5/3}, \quad F_2(r, t) = 4.82 \, C_K (\epsilon r)^{2/3}, \qquad (11)$$

and hence

$$E\left(\frac{\pi}{\Delta x}\right) = \frac{1}{4.82\pi^{5/3}} \, \Delta x \, F_2(\Delta x, t). \qquad (12)$$

Substituting Eq. (12) into Eq. (9) yields

$$\nu_t(\mathbf{x}|\Delta x) = 0.0396 \, \Delta x \, \sqrt{F_2(\mathbf{x}, \Delta x, t)}. \qquad (13)$$

Finally, this structure function has to be corrected to account for the fact that, in the LES, only the structure function of the filtered field is calculated. A correction (still assuming that turbulence at the subgrid scales follows a Kolmogorov cascade) has been introduced by Métais and Lesieur (1992), and the final expression for the eddy viscosity is

$$\nu_t(\mathbf{x}|\Delta x) = 0.063 \, \Delta x \, \sqrt{\bar{F}_2(\mathbf{x}, \Delta x, t)}, \qquad (14)$$

where $\bar{F}_2(\mathbf{x}, \Delta x, t)$ stands for the second-order structure function of the filtered velocity field, that is, the instantaneous field calculated within the LES. An analogous eddy diffusivity $\kappa_t$ may be obtained (see Métais and Lesieur, 1992) taking (quite arbitrarily) turbulent Prandtl number equal to 0.6. In fact, this Prandtl number must be equal to $C_{CO}/C_K$, where $C_{CO}$ is the Corrsin–Oboukhov constant in the temperature $k^{-5/3}$ inertial-convective range. This choice leads to $C_{CO} = 0.84$, slightly higher than the reported experimental values of about 0.7. Finally, the

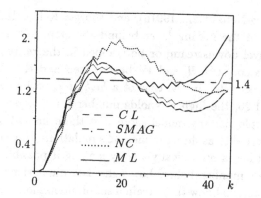

Fig. 2   Compensated kinetic energy spectra obtained in LES of decaying isotropic turbulence using respectively the Chollet–Lesieur model (C.L.), the Smagorinsky model (SMAG), the no-cusp model (NC), and the structure function model (M.L.).

molecular dynamic viscosity $\mu$ and diffusivity $\kappa$ are replaced by $\mu + \rho\nu_t$ and $\kappa + \kappa_t$ in the momentum and energy equations, respectively.

Figure 2 compares the compensated kinetic energy spectra $\epsilon^{-2/3}\,k^{5/3}$ in four large-eddy simulations of decaying isotropic turbulence using the Chollet–Lesieur (1981) model (spectral eddy viscosity with a cusp), the Smagorinsky (1963) model

$$\nu_t(\mathbf{x}, \Delta x, t) = (0.23\,\Delta x)^2 \left[\frac{1}{2}\left(\frac{\partial \bar{u}_i}{\partial x_j} + \frac{\partial \bar{u}_j}{\partial x_i}\right)\left(\frac{\partial \bar{u}_i}{\partial x_j} + \frac{\partial \bar{u}_j}{\partial x_i}\right)\right]^{1/2}, \quad (15)$$

the *no-cusp* model (with a plateau at 0.4), and the *structure function model* (Métais and Lesieur, 1992) presented here, respectively; the latter seems to give the best agreement with the $k^{-5/3}$ Kolmogorov spectrum, with Kolmogorov constant $C_K \approx 1.4$. However, it gives excessive values close to $k_C$, due to the absence of cusp in the eddy viscosity. This test is satisfactory enough to justify the use of this model for the next two problems.

## 9.5 Separated Flow Behind a Backward-Facing Step

We consider an incompressible flow behind a backward-facing step in a plane channel. The simulation is done with the aid of a finite-volume method, with a third-order discretization scheme for the convective term (code TRIO, developed at the Commissariat à l'Energie Atomique by Grand et al., 1988). A preliminary simulation was done in two dimen-

sions (Silveira-Neto et al., 1991a), and showed Kelvin–Helmholtz type
eddies shed in the mixing layer behind the step. These eddies may
pair and travel downstream, or be trapped in the recirculation region
behind the step. Let $H$ and $W$ be the step and the channel outlet
heights, respectively. In the case of a *high step*, with the aspect ratio
of $W/H = 1.25$, the high Reynolds number three-dimensional simulations not employing any model (note that these simulations cannot be
properly referred to as direct numerical simulations, since the numerical
scheme used had a numerical viscosity making it possible to reach infinite Reynolds numbers) show how quasi two-dimensional billows form
behind the step and how they strain a set of intense longitudinal hairpin
vortices (of intensity about 40% of the spanwise maximum vorticity),
as shown in Figure 3. These vortices are analogous to those observed
in mixing layers in either laboratory settings (Breidenthal, 1981; Bernal
and Roshko, 1986; Lasheras and Choi, 1988) or numerical simulations
(Metcalfe et al., 1987; Comte et al., 1991). The calculation indicates
that these eddies merge downstream, in such a way that their spanwise
spacing scales with the layer thickness, following the pairing of large
spanwise vortices. When the structure function model is used, the same
vortex topology is found, although the small scales of the flow are more
chaotic. More details on these calculations are given in Silveira-Neto et
al. (1991 b,c).

    In order to assess the validity of these predictions, we have carried out
a numerical simulation behind a *low step*, corresponding to an aspect
ratio of $W/H = 2.5$, and for which experimental data are available
(Eaton and Johnston, 1980).

    Figure 4 (taken from Silveir-Neto et al., 1991b) shows, for a calcula-
tion with resolution of $91 \times 17 \times 17$ grid points, the mean velocity and
turbulence kinetic energy profiles at a downstream distance corresponding, approximately, to the reattachment region. The direct simulation
refers here to the calculation without any model, and it is clear that the
structure function model improves significantly the predictions of numerical simulation, as far as the statistics of the flow are concerned. In
the low-step simulations performed with a higher resolution, the vorticity plots reveal intense longitudinal vortices, but it is not clear whether
they are the debris of primary billows submitted to a *helical-pairing instability* (i.e., a spanwise staggered mode predicted by Pierrehumbert
and Widnall (1982) on the basis of a secondary instability analysis, and
which has been identified in *natural transition* numerical experiments by
Comte et al. (1991)), or have the same origin as in the case of the high

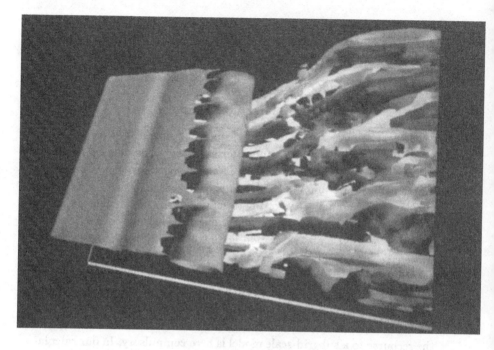

Fig. 3 Longitudinal hairpin vortices strained behind a backward-facing step; simulation without subgrid model (high step, resolution 200 × 30 × 30).

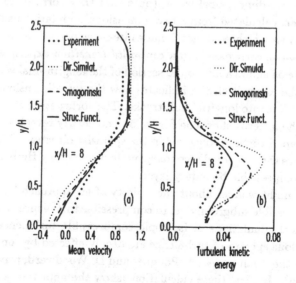

Fig. 4 Mean velocity and kinetic energy profiles calculated above the reattachment region behind the step.

step. The large-scale structure of the flow, seen in low-pressure plots, consists of highly distorted spanwise billows, which stretch longitudinal thinner vortices.

## 9.6 High-Supersonic Boundary Layer

The last application of the structure function model deals with the transition to turbulence in a supersonic boundary layer developing above an insulated flat plate at Mach 5. The numerical solution is based on a high-order finite-difference scheme proposed by Gottlieb and Turkel (1976); it also employs a coordinate stretching in the transverse direction and permeable boundary conditions (see Normand and Lesieur, 1992, for details). The resolution is $400 \times 40 \times 20$ in the streamwise, transverse, and spanwise directions, respectively. The Reynolds number, based on the upstream displacement thickness, is chosen equal to 10,000, in order to permit the growth of the instabilities in the domain. Such a Reynolds number is too high to allow for a direct numerical simulation (see, however, the work of Erlebacher and Hussaini, 1990), and the recourse to a subgrid-scale model is here compulsory. In our calculation, using the structure function model (with a local structure function calculated by averaging over the four grid points surrounding x and located in the plane parallel to the plate), the flow is forced upstream by a set of two-dimensional waves (note that they correspond to the *second mode*, calculated from a two-dimensional simulation using the same code). A small, three-dimensional white-noise perturbation, superposed on the wave, corresponds to a *controlled transition* experiment.

Figure 5a shows a horizontal section of the longitudinal velocity at the top of the layer: it clearly indicates how the two-dimensional upstream waves evolve into longitudinal streaks. The vortex structure corresponding to the latter is shown on Figure 5b and clearly reveals the formation of $\Lambda$-shaped vortices staggered in the spanwise direction. This structure resembles the staggered secondary mode proposed by Herbert (1988) for the incompressible boundary layer.

One might wonder about the validity of an approach based upon an incompressible subgrid model in compressible turbulence: indeed, direct numerical simulations of isotropic compressible turbulence show that the various spectra of turbulence are deeply affected by compressibility, even in the small scales (A. Pouquet and D. Woodward, private communication). In fact, these calculations show the appearance of shocklets

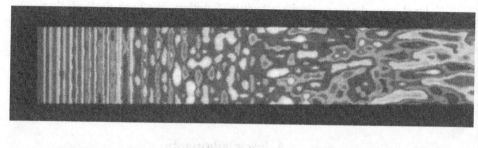

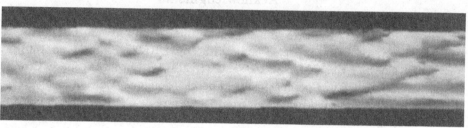

Fig. 5   Boundary-layer at Mach 5; horizontal cross-section of the longitudinal velocity on the top of the layer (a), top view of the vortex
structure in the transitional region (b).

whose interaction may generate vortex sheets and turbulence. However,
the turbulence thus produced is subsonic and might not be strongly affected by compressibility, once produced. As for the shocklets, they will
certainly not be very much affected by the subgrid model. Therefore,
the structure function model might do a good job even in these extreme
situations.

## 9.7 Conclusion and Discussion

We have surveyed the theory of spectral eddy coefficients derived from
the EDQNM two-point closure of isotropic turbulence and shown how
it can be employed for LES purposes. This theory gives satisfactory
results for LES of homogeneous isotropic or stably stratified turbulence,
and also for temporal mixing layers. We have also generalized the concept of spectral eddy viscosity to a flow where spatial intermittency and
inhomogeneities exist, introducing the concept of local kinetic energy
spectrum in physical space. The latter is calculated with the aid of a

local second-order velocity structure function. We have successfully applied LES based on this structure function model to isotropic turbulence, separated flows, and supersonic boundary layer above a flat plate. The two main advantages of the model are that it allows the vortex coherent structures of the flow to develop and accurately predicts of the flow statistics.

## Acknowledgments

The authors are indebted to Y. Fouillet, P. Comte, and E. David for developing the three-dimensional visualization software FLOSIAN. This work was supported by CNRS, by DRET (contract 88/150), by CNES/Avions Marcel Dassault (HERMES European Space Programme), and by Commissariat à l'Energie Atomique. Calculations were partially supported by a grant from the Centre de Calcul Vectoriel pour la Recherche.

## References

BATCHELOR, G.K. (1953) *The Theory of Homogeneous Turbulence.* Cambridge University Press, 197 pp.

BATCHELOR, G.K., CANUTO, V.M. AND CHASNOV, J.R. (1992) Homogeneous buoyancy-generated turbulence. *J. Fluid Mech.* **235**, 349–378.

BERNAL, L.P. AND ROSHKO, A. (1986) Streamwise vortex structure in plane mixing layers. *J. Fluid Mech.* **170**, 499–525.

BREIDENTHAL, R. (1981) Structure in turbulent mixing layers and wakes using a chemical reaction. *J. Fluid Mech.* **109**, 1–24.

CHOLLET, J.P. AND LESIEUR, M. (1981) Parameterization of small scales of three-dimensional isotropic turbulence utilizing spectral closures. *J. Atmos. Sci.* **38**, 2747–2757.

CHOLLET, J.P. (1983) Turbulence tridimensionnelle isotrope: Modélisation statistique des petites échelles et simulation numérique des grandes échelles. Doctorat thesis, University of Grenoble.

COMTE, P., LESIEUR, M. AND FOUILLET, Y. (1989) Coherent structures of mixing layers in large-eddy simulations. In *Topological Fluid Mechanics.* Ed. H.K. Moffatt and A. Tsinober, pp. 649–658. Cambridge University Press.

COMTE, P., FOUILLET, Y., GONZE, M.A., LESIEUR, M. AND NORMAND,

X. (1991) Generation of coherent structures in free shear layers. In *Turbulence and Coherent Structures*. Ed. O. Métais and M. Lesieur, pp. 45–73. Kluwer.

EATON, J.K. AND JOHNSTON, J.P. (1980) Stanford University Report MD–39.

ERLEBACHER, G. AND HUSSAINI, M.Y. (1990) Numerical experiments in supersonic boundary layer stability. *Phys. Fluids A2*, 94–104.

GOTTLIEB, D. AND TURKEL, E. (1976) Dissipative two-four methods for time-dependent problems. *Math. Comp.* **30**, 703–723.

GRAND, D., VILLAND, M. AND COULON, C. (1988) In *Proc. 3rd Int. Symp. on Refined Flow Modeling and Turbulence Measurements*, pp. 427–434. Nippon Toshi Center.

HERBERT, T. (1988) Secondary instability of boundary layers. *Ann. Rev. Fluid Mech.* **20**, 487–526.

HERRING, J.R., SCHERTZER, D., LESIEUR, M., NEWMAN, G.R., CHOLLET, J.P. AND LARCHEVÊQUE, M. (1982) A comparative assessment of spectral closures as applied to passive scalar diffusion. *J. Fluid Mech.* **124**, 411–437.

KRAICHNAN, R.H. (1966) Isotropic turbulence and inertial range structure. *Phys. Fluids.* **9**, 1728–1752.

KRAICHNAN, R.H. (1968) Small-scale structure convected by turbulence. *Phys. Fluids* **11**, 945–953.

KRAICHNAN, R.H. (1976) Eddy viscosity in two and three dimensions. *J. Atmos. Sci.* **33**, 1521–1536.

LASHERAS, J.C. AND CHOI, H. (1988) Three-dimensional instability of a plane free shear layer: An experimental study of the formation and evolution of streamwise vortices. *J. Fluid Mech.* **189**, 53–86.

LESIEUR, M. AND ROGALLO, R. (1989) Large-eddy simulation of passive scalar diffusion in isotropic turbulence. *Phys. Fluids A1*, 718–722.

LESIEUR, M., MÉTAIS, O. AND ROGALLO, R. (1989) Etude de la diffusion turbulente par simulation des grandes échelles. *C. R. Acad. Sci., Paris, Ser. II* **308**, 1395–1400.

LESIEUR, M. (1990) *Turbulence in Fluids* (2nd revised edition). Kluwer 412 pp.

MÉTAIS, O. AND LESIEUR, M. (1989) Large-eddy simulation of isotropic and stably-stratified turbulence. In *Advances in Turbulence Vol. 2*. Ed. H.H. Fernholz and H.E. Fiedler, pp. 371–376. Springer-Verlag.

MÉTAIS, O. AND LESIEUR, M. (1992) Spectral large-eddy simulations of isotropic and stably-stratified turbulence. *J. Fluid Mech.* **239**, 157–194.

METCALFE, R.W., ORSZAG, S.A., BRACHET, M.E., MENON, S. AND RILEY, J. (1987) Secondary instability of a temporally growing mixing layer. *J. Fluid Mech.* **184**, 207–243.

MOFFATT, H.K. (1991) Spiral structures in turbulent flows. In *Monte Verità Colloquium on Turbulence*. Ed. T. Dracos and A. Tsinober, in press.

NORMAND, X. AND LESIEUR, M. (1992) Numerical experiments on transition in the compressible boundary layer over an insulated flat plate. *Theor. Comp. Fluid Dynam.* **3**, 231–252.

ORSZAG, S.A. AND PATTERSON, G.S. (1972) Numerical simulations of turbulence. In *Statistical Models and Turbulence*, Vol. 12. Ed. M. Rosenblatt and C. Van Atta, pp. 127–147, Springer Verlag.

ORSZAG, S.A. (1977) Statistical theory of turbulence. In *Fluid Dynamics 1973, Les Houches Summer School of Theoretical Physics*. Ed. R. Balian and J.L. Peube, pp. 237–374. Gordon and Breach.

PIERREHUMBERT, R.T. AND WIDNALL, S.E. (1982) The two- and three-dimensional instabilities of the spatially periodic shear layer. *J. Fluid Mech.* **114**, 59–82.

SILVEIRA-NETO, A., GRAND, D. AND LESIEUR, M. (1991a) Simulation numérique bidimensionnelle d'un écoulement turbulent stratifié derrière une marche. *Int. J. Heat Mass Transfer* **34**, 1999–2011.

SILVEIRA-NETO, A., GRAND, D., MÉTAIS, O. AND LESIEUR, M. (1991b) Large-eddy simulation of the turbulent flow in the downstream region of a backward-facing step. *Phys. Rev. Lett.* **66**, 2320–2323.

SILVEIRA-NETO, A., GRAND, D., MÉTAIS, O. AND LESIEUR, M. (1991c) A numerical investigation of the coherent structures of turbulence behind a backward-facing step. *J. Fluid Mech.*, submitted.

SMAGORINSKY, J. (1963) General circulation experiments with the primitive equations. *Mon. Wea. Rev.* **91**, 99–164.

VINCENT, P. AND MÉNÉGUZZI, M. (1991) The spatial structure of homogeneous turbulence. In *Turbulence and Coherent Structures*. Ed. O. Métais and M. Lesieur, pp. 191–201.

# 10

---

# Anisotropic Representation
# of Subgrid-Scale Reynolds Stress
# in Large Eddy Simulation

## KIYOSI HORIUTI

## 10.1 Introduction

In this chapter, the Smagorinsky model (Smagorinsky, 1963), which has been commonly used in large eddy simulation (LES) of turbulent flows, is investigated. Various numerical simulations using the Smagorinsky model have yielded good results (Deardorff, 1970; Schumann, 1975; Moin and Kim, 1982; Horiuti, 1987; Piomelli et al., 1988). However, several issues on the model remain unsolved: (a) a poor correlation of the model value of the Reynolds stresses using the Smagorinsky model with the exact value computed from the direct numerical simulation (DNS) data; (b) universality of the Smagorinsky constant involved in the model; (c) development of the damping function with a wide applicability and theoretical foundations, in place of the ad hoc Van Driest type.

The aim of this chapter is to propose a new model which partially resolves these issues. This model is developed on the basis of a proper choice of energy scale in the subgrid-scale (SGS) eddy viscosity.

The basic approach adopted here is an incorporation of higher order terms in anisotropic representation (AR) (Horiuti, 1990) into the SGS Reynolds stresses, adding new terms to the conventional eddy viscosity. The same approach has already been adopted for Reynolds-averaged models of the $k$-$\epsilon$ type (Horiuti, 1990). It is well known that conventional $k$-$\epsilon$ models have several drawbacks (Tennekes and Lumley, 1972). First of all, models based upon isotropic eddy viscosity representation inevitably fail to predict the anisotropy of the Reynolds stresses, i.e., the inability to forecast accurately the normal stresses degrades model applicability. Second, numerical studies indicate that the conventional eddy viscosity magnitude is too large and excessively dissipative; thus the model is unable to provide a detailed prediction of the fine turbulence structures (Tennekes and Lumley, 1972). To overcome the first deficiency, second order AR models of the Reynolds stresses have been introduced by Leslie (1973), Yoshizawa (1984), and Speziale (1987). These models provide acceptable predictions of the anisotropy of turbulence intensities (Speziale, 1987; Nisizima and Yoshizawa, 1987). When second order AR is incorporated, however, the eddy viscosity, a proportionality coefficient between the Reynolds shear stress and the mean strain, remains unchanged. AR can be mathematically extended to higher orders. In Horiuti (1990), third order AR was used to develop supplementary eddy viscosity terms which effectively reduce its magnitude, therefore counteracting this drawback of the $k - \epsilon$ model. The most prominent finding was that as an energy scale in the eddy viscosity, the normal shear stress is preferable to the total turbulent energy. Besides, it was found that third order AR may be used as an alternative method to reduce the magnitude of the eddy viscosity in the buffer layer region by acting in a similar manner to the Van Driest (1956) damping function, commonly used in $k$-$\epsilon$ models. The present study extends the same approach to the SGS modeling in LES.

Large eddy simulations involve the modeling of various stress terms, which arise in filtering the Navier–Stokes (NS) equations. The raw variables $u_i$ are divided into filtered [or grid scale (GS)] components and the SGS components. GS variables $\overline{u}_i$ are defined as

$$\overline{u}_i(x_1, x_2, x_3) = \int_D \prod_{i=1}^{3} G_i(x_i - x_i') u_i(x_1', x_2', x_3') \, dx_1' \, dx_2' \, dx_3', \qquad (1)$$

where $u_i$ denotes the velocity component in the $i$th direction. This chapter deals with the plane Poiseuille flow, where $i = 1,2,3$ correspond to the $x$ (downstream), $y$ (normal to the walls), $z$ (spanwise) directions, respectively. $G_i$ is a filter function in the $i$th direction (Moin and Kim, 1982; Horiuti, 1987; Piomelli et al., 1988). Here the Gaussian filter is used as $G_i(i = 1, 3)$ in homogeneous directions and the top-hat filter is used as $G_2$ in the $y$ direction. The SGS component is defined as $u_i' = u_i - \overline{u}_i$. The length in wall units is denoted by $(\cdot)_+$, the horizontal average by $\langle \cdot \rangle$. The governing equations in LES for incompressible flow become

$$\frac{\partial \overline{u}_i}{\partial t} + \frac{\partial}{\partial x_j}(\overline{u}_i \overline{u}_j) = -\frac{\partial \tau_{ij}}{\partial x_j} - \frac{\partial \overline{p}}{\partial x_i} + \frac{1}{Re}\nabla^2 \overline{u}_i + 2\delta_{i1}, \qquad (2)$$

$$\frac{\partial \overline{u}_i}{\partial x_i} = 0. \qquad (3)$$

When raw variables are divided into GS and SGS components, the SGS stresses ($\tau_{ij}$) consist of three components, $\tau_{ij} = L_{ij} + C_{ij} + R_{ij}$: the Leonard terms, $L_{ij} = \overline{\overline{u}_i \overline{u}_j} - \overline{u}_i \overline{u}_j$; the cross-stress terms, $C_{ij} = \overline{\overline{u}_i u'_j + u'_i \overline{u}_j}$; and the SGS Reynolds stress terms, $R_{ij} = \overline{u'_i u'_j}$. Recent direct tests of LES models using a DNS database (Bardina, 1983; Clark et al., 1979; Horiuti, 1989a; Piomelli et al., 1988; Speziale et al., 1988) revealed that among these terms, $C_{ij}$, which have been neglected in most previous computations (Moin and Kim, 1982; Horiuti, 1987), have a significant contribution. The most commonly used model for $C_{ij}$ is that of Bardina (1983), which is based on the scale similarity hypothesis, approximating $C_{ij}$ and $R_{ij}$ as follows:

$$C_{ij} \sim C_{ij}{}^B = \overline{\overline{u}}_i(\overline{u}_j - \overline{\overline{u}}_j) + (\overline{u}_i - \overline{\overline{u}}_i)\overline{\overline{u}}_j, R_{ij} \sim (\overline{u}_i - \overline{\overline{u}}_i)(\overline{u}_j - \overline{\overline{u}}_j). \quad (4)$$

One of the most significant contributions of the Bardina model is a recovery of Galilean invariance (Speziale, 1985). In an *a priori* test (Piomelli et al., 1988; Horiuti, 1989a), it was revealed that the Bardina model is a fairly good approximation for $C_{ij}$, and the inclusion of the Bardina model significantly improves the correlation with the DNS data. In an *a posteriori* test (Piomelli et al., 1988; Horiuti, 1989a), the statistical values such as turbulence intensities were considerably improved.

## 10.2 Anisotropic Representation of the Reynolds Stress

Reynolds stresses up to the second order expansion in a scale parameter can be expressed as

$$
\begin{aligned}
\overline{u'_i u'_j} = {} & \delta_{ij}\left\{\frac{2}{3}k - \frac{1}{3}\frac{k^3}{\epsilon^2}\left[(C_{\tau 1} + C_{\tau 3})\frac{\partial \overline{u}_l}{\partial x_m} + C_{\tau 2}\frac{\partial \overline{u}_m}{\partial x_l}\right]\frac{\partial \overline{u}_l}{\partial x_m}\right\} \\
& - C_\nu\frac{k^2}{\epsilon}\left(\frac{\partial \overline{u}_i}{\partial x_j} + \frac{\partial \overline{u}_j}{\partial x_i}\right) + \frac{k^3}{\epsilon^2}\left[C_{\tau 1}\frac{\partial \overline{u}_i}{\partial x_l}\frac{\partial \overline{u}_j}{\partial x_l}\right. \\
& \left. + \frac{C_{\tau 2}}{2}\left(\frac{\partial \overline{u}_i}{\partial x_l}\frac{\partial \overline{u}_l}{\partial x_j} + \frac{\partial \overline{u}_j}{\partial x_l}\frac{\partial \overline{u}_l}{\partial x_i}\right) + C_{\tau 3}\frac{\partial \overline{u}_l}{\partial x_i}\frac{\partial \overline{u}_l}{\partial x_j}\right],
\end{aligned}
\qquad (5)
$$

where $k$ is the turbulent kinetic energy; $\epsilon$ is the dissipation rate of $k$; $C_\nu$, $C_{\tau 1}$, $C_{\tau 2}$, $C_{\tau 3}$ are constant coefficients; and $\delta_{ij}$ is the Kronecker delta symbol. After continuing to a third order expansion some noteworthy terms are produced (Horiuti, 1990), i.e., in the products of first order terms and second order terms:

$$-\frac{k^4}{\epsilon^3}\left(C_{A1}\frac{\partial\overline{u}_l}{\partial x_m}\frac{\partial\overline{u}_l}{\partial x_m}+C_{A2}\frac{\partial\overline{u}_l}{\partial x_m}\frac{\partial\overline{u}_m}{\partial x_l}\right)\left(\frac{\partial\overline{u}_i}{\partial x_j}+\frac{\partial\overline{u}_j}{\partial x_i}\right). \qquad (6)$$

If we combine (6) with (5), the eddy viscosity $\nu_e$, appearing as the coefficient in front of mean strain $\partial\overline{u}_i/\partial x_j+\partial\overline{u}_j/\partial x_i$, is rearranged to

$$\nu_e=\frac{3}{2}C_\nu\frac{k}{\epsilon}\left\{\frac{2}{3}k-\frac{2}{3}\frac{1}{C_\nu}\frac{k^3}{\epsilon^2}\left[\frac{C_{A1}+C_{A2}}{4}\left(\frac{\partial\overline{u}_l}{\partial x_m}+\frac{\partial\overline{u}_m}{\partial x_l}\right)^2\right.\right.$$

$$\left.\left.+\frac{C_{A1}-C_{A2}}{4}\left(\frac{\partial\overline{u}_l}{\partial x_m}-\frac{\partial\overline{u}_m}{\partial x_l}\right)^2\right]\right\}. \qquad (7)$$

The theoretical values of the model constants $C_{A1}$ and $C_{A2}$ are 0.119 and 0.0424, respectively. The second term in (7) is always positive and reduces the magnitude of the eddy viscosity. Note that the terms within the square brackets in Eq. (7) correspond to the second order AR turbulence intensities. The relationship between AR and algebraic stress model (ASM) (Rodi, 1976) will now be referred to. ASM approximates the transport equation for the Reynolds stress $\overline{u_i'u_j'}$ as

$$\overline{u_i'u_j'}\frac{P-\epsilon}{k}=P_{ij}-\frac{2}{3}\delta_{ij}\epsilon-C_1\frac{\epsilon}{k}(\overline{u_i'u_j'}-\frac{2}{3}\delta_{ij}k)-\gamma k\left(\frac{\partial\overline{u}_i}{\partial x_j}+\frac{\partial\overline{u}_j}{\partial x_i}\right)$$

$$-\alpha\left(P_{ij}-\frac{2}{3}\delta_{ij}P\right)-\beta\left(D_{ij}-\frac{2}{3}\delta_{ij}P\right), \qquad (8)$$

where

$$P_{ij}=-\overline{u_i'u_l'}\frac{\partial\overline{u}_j}{\partial x_l}-\overline{u_j'u_l'}\frac{\partial\overline{u}_i}{\partial x_l}, \qquad D_{ij}=-\overline{u_i'u_l'}\frac{\partial\overline{u}_l}{\partial x_j}-\overline{u_j'u_l'}\frac{\partial\overline{u}_l}{\partial x_i},$$

$$P=\frac{P_{mm}}{2}.$$

The left hand side of (8) is a model for convection and diffusion terms (Rodi, 1976), whereas the third to sixth terms in the right hand side are for pressure-strain terms (Launder et al., 1975). Usually these coupled equations are solved iteratively by setting the initial values of $\overline{u_i'u_j'}^{(0)}$ equal to $2\delta_{ij}k/3$ and successively inserting them into (8), with the superscript denoting the number of iterations. After the second iteration, an expression for $\overline{u_i' u_j'}$ is obtained,

$$
\begin{aligned}
\overline{u_i' u_j'}^{(2)} = {}& \delta_{ij} \left[ \frac{2}{3} k - \frac{2}{3} \frac{1}{C_1^2} \bar{\gamma}(1 - \alpha - \beta) \frac{k^3}{\epsilon^2} \left( \frac{\partial \overline{u}_l}{\partial x_m} + \frac{\partial \overline{u}_m}{\partial x_l} \right) \frac{\partial \overline{u}_l}{\partial x_m} \right] \\
& - \left\{ \left[ \frac{1}{C_1} \left( 1 + \frac{1}{C_1} \right) \bar{\gamma} \frac{k^2}{\epsilon} - \frac{1}{C_1^3} \bar{\gamma}^2 \frac{k^4}{\epsilon^3} \left( \frac{\partial \overline{u}_l}{\partial x_m} + \frac{\partial \overline{u}_m}{\partial x_l} \right) \frac{\partial \overline{u}_l}{\partial x_m} \right] \right. \\
& \left. \times \left( \frac{\partial \overline{u}_i}{\partial x_j} + \frac{\partial \overline{u}_j}{\partial x_i} \right) \right\} + \left\{ \frac{\gamma}{C_1^2} \frac{k^3}{\epsilon^2} \left[ 2(1 - \alpha) \frac{\partial \overline{u}_i}{\partial x_l} \frac{\partial \overline{u}_j}{\partial x_l} \right. \right. \\
& \left. \left. + (1 - \alpha - \beta) \left( \frac{\partial \overline{u}_i}{\partial x_l} \frac{\partial \overline{u}_l}{\partial x_j} + \frac{\partial \overline{u}_j}{\partial x_l} \frac{\partial \overline{u}_l}{\partial x_i} \right) + 2\beta \frac{\partial \overline{u}_l}{\partial x_i} \frac{\partial \overline{u}_l}{\partial x_j} \right] \right\},
\end{aligned}
\tag{9}
$$

where $\bar{\gamma} = \gamma + 2(1 - \alpha - \beta)/3$. Note that AR and ASM are similar, with the exception that (7) contains terms related to vorticity, possibly violating the principle of having the SGS Reynolds stresses indifferent under frame rotations (Speziale, 1983). This similarity has also been noted recently in Rubinstein and Barton (1990); however, no reference is made to the new eddy viscosity terms pointed out here. The origin of each term cannot be easily traced with the DIA approach, and thus ASM was utilized to do this. The eddy viscosity on the third iteration becomes

$$
\nu_e^{(3)} = \frac{\bar{\gamma}}{C_1} \frac{k^2}{\epsilon} - \frac{1}{C_1} \nu_e^{(2)} \frac{P^{(2)} - \epsilon}{\epsilon} - \frac{1}{C_1^2} (1 - \alpha - \beta)^2 \frac{k^2}{\epsilon^2} P^{(1)},
\tag{10}
$$

where

$$
P^{(2)} = -\overline{u_l' u_m'}^{(2)} \frac{\partial \overline{u}_l}{\partial x_m}.
$$

Both the second and third terms in (10) include third order AR. The second term results from the imbalance between production and dissipation, whereas the third term is from the products of $\delta_{il}$ terms in (8) with $\partial \overline{u}_j / \partial x_l$, and the $\delta_{jl}$ terms in (8) with $\partial \overline{u}_i / \partial x_l$, and is consequently related to the anisotropy of turbulence intensities. Based on these observations, it is assumed that third order AR arises from the deviation from equilibrium and also the anisotropy of turbulence intensities. For details, see Horiuti (1990).

Here, we refer to the relationship between the Leonard term, the Bardina model, and the second order AR. By using the Taylor expansion (Clark et al., 1979), $L_{ij}$ and $C_{ij}{}^B$ are expressed as follows:

$$L_{ij} = \left(\frac{\Delta^2}{24}\bar{u}_i\frac{\partial}{\partial x_k}\frac{\partial}{\partial x_k}\bar{u}_j + \frac{\Delta^2}{24}\bar{u}_j\frac{\partial}{\partial x_k}\frac{\partial}{\partial x_k}\bar{u}_i\right) + \frac{\Delta^2}{12}\frac{\partial\bar{u}_i}{\partial x_k}\frac{\partial\bar{u}_j}{\partial x_k} + O(\Delta^4),$$

$$\tag{11}$$

$$C_{ij}{}^B = -\left(\frac{\Delta^2}{24}\bar{u}_i\frac{\partial}{\partial x_k}\frac{\partial}{\partial x_k}\bar{u}_j + \frac{\Delta^2}{24}\bar{u}_j\frac{\partial}{\partial x_k}\frac{\partial}{\partial x_k}\bar{u}_i\right) + O(\Delta^4).$$

The first terms in the right hand side of each expansion have the same form with an opposite sign. The summation of $L_{ij}$ and $C_{ij}^B$ gives

$$L_{ij} + C_{ij}{}^B = \frac{\Delta^2}{12}\frac{\partial\bar{u}_i}{\partial x_k}\frac{\partial\bar{u}_j}{\partial x_k} + O(\Delta^4), \tag{12}$$

eliminating the terms that violate Galilean invariance (Speziale, 1985). It was revealed in an *a priori* test that the magnitude of (12) is not negligibly small (Horiuti, 1989a). The second order AR consists of three types of terms, namely

$$\Delta^2\frac{\partial\bar{u}_i}{\partial x_k}\frac{\partial\bar{u}_j}{\partial x_k}, \quad \Delta^2\frac{\partial\bar{u}_i}{\partial x_k}\frac{\partial\bar{u}_k}{\partial x_j}, \quad \Delta^2\frac{\partial\bar{u}_k}{\partial x_i}\frac{\partial\bar{u}_k}{\partial x_j}. \tag{13}$$

Therefore, $L_{ij} + C_{ij}{}^B$ provides an approximation for the second order AR, ignoring the last two terms in (13). Among the three terms in (13), the last term in particular is from $D_{ij}$ in the pressure-strain approximation terms in ASM. In this respect, the neglect of this term may result in a poor approximation of the pressure-strain terms, assuming that the model for the pressure-strain terms used in ASM is accurate. Inclusion of $D_{ij}$, however, violates the invariance of the SGS Reynolds stress under the frame rotation (Speziale, 1983), unless appropriate constants are chosen. Based upon his statistical theory, Leslie (1980) objects to the inclusion of $D_{ij}$ in (8).

## 10.3 A Priori Test of SGS Models

In LES, $k$ and $\epsilon$ in the preceding section are replaced by the SGS turbulent energy $K_G = \overline{u'_l u'_l}/2$ and $C_\epsilon K_G^{3/2}/\Delta$, respectively, giving the time scale as $\Delta/C_\epsilon K_G^{-1/2}$ and the eddy viscosity as

$$\nu_e = C_N\frac{\Delta}{K_G^{1/2}}E, \tag{14}$$

where $E$ denotes the energy scale and $C_\epsilon$ and $C_N$ are model constants. If

$K_G$ is chosen as $E$, assuming the local equilibrium between SGS energy production and dissipation, the conventional Smagorinsky model,

$$\nu_e = (C_S \Delta)^2 \left(\tfrac{1}{2} e_{ij} e_{ij}\right)^{1/2}, \qquad e_{ij} = \frac{\partial \overline{u}_i}{\partial x_j} + \frac{\partial \overline{u}_j}{\partial x_i}, \qquad (15a)$$

$$K_G = \frac{C_\nu}{C_\epsilon^3} \Delta^2 \left(\tfrac{1}{2} e_{ij} e_{ij}\right), \qquad (15b)$$

can be obtained.

In this section, *a priori* validation of the SGS models is made using the DNS database for turbulent channel flow. The Reynolds number ($Re$) for DNS was chosen equal to 360 based on the wall friction velocity and the channel width, with grid points of 128, 129, 128 in the $x, y, z$ directions, respectively, employing Fourier – Chebyshev polynomial expansions. The data are filtered using the Gaussian filter in the $x$ and $z$ directions and the top-hat filter in the $y$ direction, and divided into GS and SGS components, reducing the LES grid point numbers to 32, 65, and 32 in the $x, y,$ and $z$ directions, respectively. The $y$ distribution of the correlation coefficient (CC) between the model value for $\overline{u_1' u_2'}$ from the Smagorinsky model and the exact value from the DNS was generally very low and, in particular, became negative at $y \sim 0.04$ and 0.96 as in Horiuti (1989a). Besides, in the $y$ distribution of the $x$-$z$ plane average of $\overline{u_1' u_2'}$ (denoted by $\langle \overline{u_1' u_2'} \rangle$ in the following), the model value showed a prohibitively large peak at the locations where negative values of CC were found. This large peak is generally suppressed by multiplying the length scale $\Delta$ by the Van Driest damping function $(1 - \exp(-y_+/A_+))$ with $A_+ = 25$, implying that $K_G$ may not be a proper energy scale. Actually, when $K_G$ from the DNS data is used as $E$ and $K_G$ in (14), $\langle \overline{u_1' u_2'} \rangle$ shows a very large peak as can be found in Figure 1. We begin with a determination of a proper energy scale $E$, while retaining the exact value obtained from DNS data as $K_G$ in Eq. (14). It was suggested in Horiuti (1990) that a proper energy scale in the eddy viscosity of $k$-$\epsilon$ models is the normal shear stress rather than the total turbulent energy. Figure 2 shows CC between the model value of $\overline{u_1' u_2'}$ with $E = \overline{u_2' u_2'}$ from DNS in (14) and the exact value. Figure 3 is $\langle \overline{u_1' u_2'} \rangle$ for this case. Without introducing any ad hoc damping functions, the mean value is in a good agreement with the DNS data, while retaining a high correlation with the DNS. Therefore, the finding in Horiuti (1990) for the $k$-$\epsilon$ model is also confirmed by LES.

In actual LES computations, models that represent the normal shear stress by GS variables must be provided. One way to do this is to

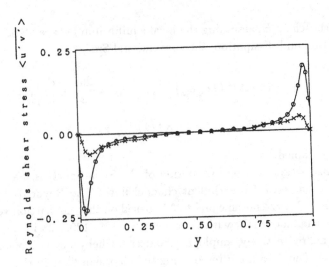

Fig. 1   The $y$ distribution of the exact value for $\langle \overline{u'_1 u'_2} \rangle$ ($\times$) and the model value from Eq. (14) using $K_G$ and $E = K_G$ computed from the DNS data ($\circ$).

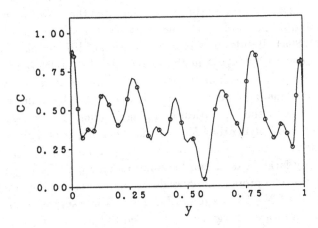

Fig. 2   The $y$ distribution of CC between the exact value for $\overline{u'_1 u'_2}$ and the model value from Eq. (14) using $K_G$ and $E = \overline{u'_2 u'_2}$ computed from the DNS data.

compute Eq. (7) directly for the eddy viscosity terms. First, we utilize Eq. (12) to approximate the terms within the square brackets in (6). In this model, the new term

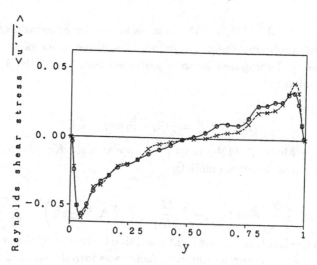

Fig. 3  The $y$ distribution of the exact value for $\langle \overline{u_1' u_2'} \rangle$ ($\times$) and the model value from Eq. (14) using $K_G$ and $E = \overline{u_2' u_2'}$ computed from the DNS data ($\circ$).

$$C_{A3}\Delta^2(\overline{\overline{u_l \overline{u}_l}} - \overline{\overline{u}}_l \overline{\overline{u}}_l) \tag{16}$$

is added to the conventional Smagorinsky model. These terms can be readily computed when the Gaussian filter is used with the spectral method. For an application of the finite-difference method to the approximation of these terms, see Horiuti (1989b). The $y$ distribution of $\langle u_1' u_2' \rangle$ of this model showed a large peak as in Figure 1, and the reduction of the eddy viscosity magnitude was not sufficient. To further suppress the model value, an inclusion of all the terms in (7) is necessary, together with the optimization of the model constants. This seems to be too demanding for LES because the universality of model constants is doubtful and they may depend on the flow field. Another approximation method for the normal shear stress is the Bardina model,

$$\overline{u_l' u_l'} = (\overline{u}_l - \overline{\overline{u}}_l)(\overline{u}_l - \overline{\overline{u}}_l), \tag{17}$$

which has proved to be fairly accurate in comparison with the DNS data (Horiuti, 1993). In the following, $E$ will be approximated by Eq. (17) with $l = 2$.

In the next step, $K_G$ in (14) must be properly represented by the GS variables. There are two approximation methods: One is to use the Bardina model (17), and the other is to use the Smagorinsky model (15b). In combining these approximations, three types of model are considered and compared in an *a posteriori* test. In Model 1, $K_G$ is approximated as

$$K_G \sim K_G^B = \frac{1}{2}(\overline{u}_l - \overline{\overline{u}}_l)(\overline{u}_l - \overline{\overline{u}}_l), \qquad (18)$$

whereas in Model 2, (15b) is used to approximate $K_G$. Model 3 is a combination of these two models,

$$\nu_e = C_\nu C_\epsilon \frac{3}{2}\frac{K_G}{\epsilon}E = C_\nu C_\epsilon \frac{K_G^2}{\epsilon}\frac{3E}{2K_G} \sim (C_S^N \Delta)^2 \left(\frac{1}{2}e_{ij}^2\right)^{1/2}\frac{3E}{2K_G^B}, \quad (19)$$

in which the last term on the right hand side is incorporated as a kind of damping factor. A similar damping factor was introduced by Yakhot et al. (1989), in which the GS turbulence intensities were used in place of the SGS turbulence intensities as the arguments in the damping factor. Therefore, when the computational meshes are refined, the effect of the damping factor never diminishes. Besides, no theoretical foundation for choosing the factor was given. The idea of making use of the preferential damping of the normal shear stress as the eddy viscosity magnitude reduction was also adopted by Launder and Tselepidakis (1990). Because of the insufficient accuracy of the models for the pressure-strain terms in the second order stress models, the result of the *a posteriori* test was not satisfactory. In the present study, the information in the finest resolved scale in the GS variables was used to extrapolate to the SGS to represent the effect of anisotropy. It should be noted here that, in deriving Model 1, no use is made of the local equilibrium hypothesis. The consequent defect of the model will be discussed in the following section. It also must be noted that the wall-limiting behavior of the SGS Reynolds stress is not satisfied by models 2 and 3 proposed here. Although most commonly used damping functions in LES do not satisfy the limiting behavior except for the one proposed by Piomelli et al. (1988) and Germano et al. (1991), it seems that incorrect wall behavior does not lead to seriously inaccurate computational results.

## 10.4 A *Posteriori* Test of the Proposed Models

This section presents numerical results of LES when the SGS models proposed in the preceding section are incorporated into the actual computation. $Re$ was chosen equal to 1280, and either 128, 129, and 128 grid points (Grid I) or 64, 62, and 64 grid points (Grid II) in the $x, y, z$ directions, respectively, were employed. It was not necessary to introduce damping functions in any of the three models.

This test revealed that although Model 1 yielded fairly good results with Grid I, it gave erroneous results with Grid II, in which the SGS model contribution was noticeably large. The mean velocity profile had a couple of bumps, and a significant deviation from the well-known logarithmic law was recognized. Local smoothing may eliminate these bumps, but such a procedure was not followed in this study. In the logarithmic region, the GS energy production term balances the dissipation term. It seems that a failure to use the local equilibrium hypothesis in Model 1 led to the erroneous results, because Models 2 and 3 yielded good results in this region with both Grids I and II. Figure 4 shows the mean velocity profile from Model 3. The von Kármán constant ($\kappa = 0.4$) and another constant ($B = 5.0$) are in good agreement with experimental measurement (Hussain and Reynolds, 1975). Without introducing any ad hoc damping function, a linear profile in the vicinity of the wall was obtained. Figure 5 is the GS turbulence intensities of the streamwise component. Note that the contributions from the Bardina model are included in the intensities. For a comparison, experimental measurements (Kreplin and Eckelmann, 1979) and numerical results (Horiuti, 1989b) obtained by using the Smagorinsky model combined with the Van Driest damping function are included in the figures. The peak positions of turbulence intensities with Models 2 and 3 are closer to the wall than in the previous result (Horiuti, 1989b) and are in better agreement with the experimental measurement. Overall agreement of Model 3 with the experimental measurement is good. The peak value from Model 2, however, seems to be too high. This is attributed to an excessively small eddy viscosity magnitude in the closest proximity of the wall. Therefore, Model 3 appears to be superior to the other two. Now a qualitative explanation for the difference in the optimized Smagorinsky constant in homogeneous flow ($C_S \sim 0.20$) (Clark et al., 1979), mixing layer ($C_S \sim 0.16$) (Mansour et al., 1978) and channel flow ($C_S \sim 0.10$) (Deardorff, 1970) can be given: the anisotropy of GS turbulence intensities becomes larger in this order, and this fact will be reflected in the SGS turbulent energy when the grid resolution is not sufficient. Because of the improper choice of energy scale in the Smagorinsky model, $C_S$ had to be adjusted depending on the flow. It may be possible that $C_S{}^N$ becomes universal independent of the flow, but $C_S$ may not be so.

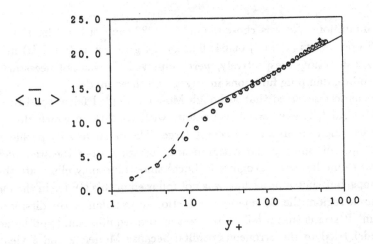

Fig. 4   Mean streamwise velocity profile $\langle \overline{u} \rangle$; o, computation from Model 3 using Grid II; $----$, $\langle \overline{u} \rangle = y_+$; $———$, $\langle \overline{u} \rangle = 1/0.4 \log y_+ + 5.0$.

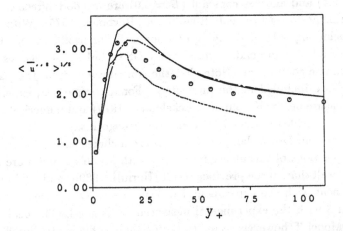

Fig. 5   Streamwise component of GS turbulence intensities in the vicinity of the lower wall (o) from Model 3 using Grid II; $———$, computational data using Model 2; $----$, experimental data (Kreplin et al., 1979); $———$, computational data using the Smagorinsky model combined with the Van Driest damping function.

As a matter of fact, the optimized value of $C_S{}^N$ in the channel flow was approximately 0.16, which is much higher than the optimized value of the Smagorinsky model.

## 10.5 Summary and Discussion

A new SGS Reynolds stress model has been proposed. The prominent feature of the model lies in the choice of the normal shear stress as the energy scale in the SGS eddy viscosity. In the present model, the previous discrepancy in the optimized values for the Smagorinsky model constant can be explained, and the use of the empirical damping functions can be avoided; in addition, a higher correlation with the DNS data can be maintained than in the Smagorinsky model. Unlike the commonly used damping function of Van Driest type, which globally reduces the eddy viscosity magnitude at the same level of $y_+$, this model can make a local reduction of the magnitude. The relationship between the Leonard term, the Bardina model, and the second order AR has been discussed. The validity of the proposed model is confirmed in an *a posteriori* test. For further details, see Horiuti (1993). Although the normal shear stress can be almost uniquely chosen in plane channel flow, further refinements are needed for its proper selection in complex geometry, such as in a corner flow of a backward-facing step, which will be left to future work.

## Acknowledgments

The author is grateful to Dr. A. Yakhot for valuable comments. This work was partially supported by Grant-in-Aid for Scientific Research Nos. 01613002 and 02302043 from the Ministry of Education of Japan. All computations were performed using University of Tokyo Computer Center's HITAC S-820 model 80 system, with the computational facilities provided under the University of Tokyo/Hitachi Ltd. Joint Research Program.

## References

BARDINA, J. (1983) Improved turbulence models based on large eddy simulation of homogeneous, incompressible turbulent flows. Ph.D. thesis, Stanford University.

CLARK, R.A., FERZIGER, J.H. AND REYNOLDS, W.C. (1979) Evaluation of subgrid-scale models using an accurately simulated turbulent flows. *J. Fluid Mech.* **91**, 1.

DEARDORFF, J.W. (1970) A numerical study of three-dimensional turbulent channel flow at large Reynolds numbers. *J. Fluid Mech.* **41**, 453.

GERMANO, M., PIOMELLI, U., MOIN, P. AND CABOT, W.H. (1991) A dynamic subgrid-scale eddy viscosity model. *Phys. Fluids A* **3**, 1760.

HORIUTI, K. (1987) Comparison of conservative and rotational forms in large eddy simulation of turbulent channel flow. *J. Comp. Phys.* **71**, 343.

HORIUTI, K. (1989a) The role of the Bardina model in large eddy simulation of turbulent channel flow. *Phys. Fluids A* **1**, 462.

HORIUTI, K. (1989b) Anisotropic representation of the Reynolds stress in large eddy simulation of turbulent channel flow. Proc. Int. Symp. on Comp. Fluid Dynamics, Nagoya, Nagoya, Aug. 1989, p. 233.

HORIUTI, K. (1990) Higher-order terms in the anisotropic representation of Reynolds stresses. *Phys. Fluids A* **2**, 10.

HORIUTI, K. (1993) A proper velocity scale for modeling subgrid-scale eddy viscosities in large eddy simulation. *Phys. Fluids A* **5**, 146.

HUSSAIN, A.K.M.F. AND REYNOLDS, W.C. (1975) Measurements in fully developed turbulent channel flow. *J. Fluid Eng.* **97**, 568.

KREPLIN, H. AND ECKELMANN, M. (1979) Behavior of the three fluctuating velocity components in the wall region of a turbulent channel flow. *Phys. Fluids* **22**, 1233.

LAUNDER, B.E., REECE, G.J. AND RODI, W. (1975) Progress in the development of a Reynolds-stress turbulence closure. *J. Fluid Mech.* **68**, 537.

LAUNDER, B.E. AND TSELEPIDAKIS, D.P. (1990) In *Near-Wall Turbulence*, Ed. S.J. Kline and N.H. Afgan, p. 818, Hemisphere.

LESLIE, D.C. (1973) *Developments in the Theory of Turbulence.* Clarendon Press.

LESLIE, D.C. (1980) Analysis of a strongly sheared, nearly homogeneous turbulent shear flow. *J. Fluid Mech.* **98**, 435.

MANSOUR, N.N., FERZIGER, J.H. AND REYNOLDS, W.C. (1978) Large-eddy simulation of a turbulent mixing layer. Report No. TF-11, Stanford University.

MOIN, P. AND KIM, J. (1982) Numerical investigation of turbulent channel flow. *J. Fluid Mech.* **118**, 341.

NISIZIMA, S. AND YOSHIZAWA, A. (1987) Turbulent channel and Couette flows using an anisotropic k-ε model. *AIAA J.* **25**, 414.

PIOMELLI, U., MOIN, P. AND FERZIGER, J.H. (1988) Model consistency in large eddy simulation of turbulent channel flows. *Phys. Fluids* **31**, 1884.

RODI, W. (1976) A new algebraic relation for calculating the Reynolds stresses. *Z. Angew. Math. Mech.* **56**, T219.

RUBINSTEIN, R. AND BARTON, J.M. (1990) Nonlinear Reynolds stress models and the renormalization group. *Phys. Fluids A* **2**, 1472.

SCHUMANN, U. (1975) Subgrid scale model for finite difference simulation of turbulent flows in plane channels and annuli. *J. Comp. Phys.* **18**, 376.

SMAGORINSKY, J. (1963) General circulation experiments with the primitive equations: Part I. The basic experiment. *Mon. Wea. Rev.* **91**, 99.

SPEZIALE, C.G. (1983) Closure models for rotating two-dimensional turbulence. *Geophys. Astrophys. Fluid Dynam.* **23**, 69.

SPEZIALE, C.G. (1985) Galilean invariance of subgrid-scale stress in the large eddy simulation of turbulence. *J. Fluid Mech.* **156**, 55.

SPEZIALE, C.G. (1987) On nonlinear $K - l$ and $k$-$\epsilon$ models of turbulence. *J. Fluid Mech.* **175**, 459.

SPEZIALE, C.G., ERLEBACHER, G., ZANG, T.A. AND HUSSAINI, M.Y. (1988) The subgrid-scale modelling of compressible turbulence. *Phys. Fluids* **31**, 940.

TENNEKES, H. AND LUMLEY, J.L. (1972) *A First Course in Turbulence.* MIT Press.

VAN DRIEST, E.R. (1956) On turbulent flow near a wall. *J. Aero. Sci.* **23**, 1007.

YAKHOT, A., ORSZAG, S.A., YAKHOT, V. AND ISRAELI, M. (1989) Renormalization group formulation of large-eddy simulations. *J. Sci. Comput.* **4**, 139.

YOSHIZAWA, A. (1984) Statistical analysis of the deviation of the Reynolds stress from its eddy-viscosity representation. *Phys. Fluids* **27**, 1377.

# 11

# Large Eddy Simulation
# of Transitional Flow

## THOMAS A. ZANG AND UGO PIOMELLI

## 11.1 Introduction

The transition from laminar to turbulent flow has many facets. The phenomena of particular interest to us pertain to the technological issues of the prediction, modeling and control of transition on aerospace vehicles. Accurate prediction and modeling are crucial to the forthcoming generation of advanced aerospace vehicles such as the High Speed Civil Transport and the National Aerospace Plane (Malik, Zang and Bushnell 1990). For design purposes one needs to know not only the location of transition onset, but also the extent and properties of the transitional zone. In most cases the peak skin friction and wall heat transfer occur near the end of the transitional zone. Accurate predictions of these peak values are crucial.

Transitional zone models can be developed both for the Reynolds–averaged Navier–Stokes (RANS) equations and for large eddy simulations. In the RANS approach, models are developed for the higher-order moments of the *ensemble-averaged* Navier–Stokes equations, whereas subgrid-scale (SGS) models for large eddy simulations are directed toward the small spatial scales of the *spatially averaged* Navier–Stokes equations. For engineering applications, of course, one wants an RANS transitional zone model, and there is a strong practical preference for a model which blends smoothly into an acceptable model for the fully turbulent zone. Virtually all the extant models for the transitional zone are modifications of RANS turbulence models. For recent reviews of RANS transitional models see Narasimha (1985), Arnal (1989), and Narasimha and Dey (1989).

The opportunities for calibrating RANS models of the compressible transitional zone are quite limited. Until recently experimental data have been the only benchmark. Surface measurements are essentially all that is currently available from supersonic flight and quiet tunnel measurements, and there are virtually no suitable measurements at hypersonic speeds. It is only at essentially incompressible speeds that detailed flow-field data are available.

Fortunately, numerical simulations of transition appear on the verge of providing an alternative source of reliable, detailed flow-field data for the transitional zone. The current status of this field is summarized by Kleiser and Zang (1991). To date, direct numerical simulation (DNS) of transition in wall-bounded flows has been employed primarily for studying the basic physics of laminar–turbulent transition. Such computations, based on the Navier–Stokes equations without recourse to any type of physical model, are extremely demanding computationally for the later stages of transition. In simple low-speed (essentially incompressible) flows it is now possible to compute reliably the entire transition process from laminar to fully developed turbulent flow in extremely simple situations (Gilbert 1988; Gilbert and Kleiser 1990). However, at present this is feasible only for forced rather than for natural transition, i.e., for flows in which transition develops from imposed, specific waves with finite amplitude (of the order of 1%) rather than from the natural disturbance background (which is composed of a broad spectrum of waves). Due to the massive computational effort required by this technique, however, direct simulation of the governing equations is presently limited to simple flows and low Reynolds numbers. The computational demands of the *compressible* transition problem are even more severe: the length scales are shorter, the transitional zones are longer, the equations themselves are more complex and, ultimately, shock wave and real gas effects must be taken into account. For the solution of problems of engineering interest, other, less computationally intensive methods are required, and these will involve some degree of modeling.

A less intensive numerical technique that has been successfully applied to the study of turbulent flows is large eddy simulation (LES). In LES only the large, energy-containing scales of the motion are computed directly; the effect of the small scales (subgrid scales), which appears in a residual (or subgrid-scale) stress term, is modeled. Since, in turbulent flows, the small scales tend to be more isotropic and homogeneous than the large scales and do not depend very strongly on the boundary conditions, their effect can be represented by fairly simple models. This approach offers greater hope of furnishing data on the transitional zone for compressible flow than does DNS. Indeed, a conceivable path to supplying transitional zone flow-field data for high-speed flow starts with validating LES for incompressible transitional flow against both experimental data and DNS results, continues with validating LES for simple compressible transitional flow against the very limited set of experimental and DNS data and concludes with applying LES to general compressible transitional problems. At present only the first tentative steps along this path have been taken.

Large eddy simulations have been successfully applied to a variety of incompressible turbulent, wall-bounded flows such as plane channel flow (Deardorff 1970; Moin and Kim 1982; Piomelli, Moin and Ferziger 1988), boundary layers (Schmitt, Richter and Friedrich 1985) and channel flow with transpiration (Piomelli, Moin and Ferziger 1991b), but only recently have efforts been made to study transition to turbulence using LES. Early work (Horiuti 1986; Dang and Deschamps 1987; Deschamps and Dang 1987) was characterized by the application of well-established SGS stress models to the simulation of laminar–turbulent transition. From the technological point of view, however, the issue is not whether LES can start with a laminar flow and end up with a turbulent flow, but rather whether it computes the transition *correctly*, i.e., with the proper duration and properties. Piomelli, Zang, Speziale and Hussaini (1990a) were the first to use the databases generated by well-resolved direct numerical simulations of transition to study the behavior of the SGS stress tensor during transition and to evaluate actual LES of the transitional zone. They observed that during the nonlinear stages, and in particular during the second-spike stage, the SGS dissipation (i.e., the energy transfer from large to small scales) is significantly smaller than in turbulent flow. They devised an intermittency-like modification of the Smagorinsky (1963) model which allowed accurate prediction of the early stages of transition in a flat-plate boundary layer (Piomelli et al. 1990a) and plane channel flow (Piomelli and Zang 1991). Subsequently, Piomelli, Zang, Speziale and Lund (1990b) calculated a transitional boundary-layer flow using an SGS stress model based on the renormalization group (RNG) theory of Yakhot and Orszag (1986). This predicts zero eddy viscosity as long as the magnitude of the strain-rate tensor is less than some threshold value. A recently devised dynamic SGS model was also successfully applied to plane channel flow by Germano, Piomelli, Moin and Cabot (1991).

This chapter describes these SGS models for incompressible transitional flow, reviews the *a priori* tests of the models against DNS of plane channel flow and presents *a posteriori* results of actual LES using the various models.

## 11.2 Formulation

In large eddy simulations of incompressible flow, the dependent variables (velocity $u_i$ and pressure $p$) are decomposed into a large-scale component (denoted by an overbar) and an SGS component. The large-scale field is defined by the filtering operation

$$\overline{f}(\mathbf{x}) = \int G(\mathbf{x}, \mathbf{x}') f(\mathbf{x}') \, d\mathbf{x}', \qquad (1)$$

where the integral is extended over the entire spatial domain and $G = \overline{G}_1 \overline{G}_2 \overline{G}_3$, $\overline{G}_i$ being the filter function in the $i$th direction. In the present work $x_1$ (or $x$) is the streamwise direction, $x_2$ (or $y$) is the wall-normal direction and $x_3$ (or $z$) is the spanwise direction; $u_1$, $u_2$ and $u_3$ (or $u$, $v$ and $w$) are the velocity components in the coordinate directions. The SGS component is denoted by a prime, and is defined as

$$f' = f - \overline{f}. \qquad (2)$$

In the present work, a sharp cutoff filter (Leonard 1974) has been applied in all directions.

The filtered Navier–Stokes and continuity equations, which describe the evolution of the large, energy-carrying eddies, can be obtained by applying the filtering operation to the incompressible Navier–Stokes and continuity equations to yield

$$\frac{\partial \overline{u}_i}{\partial t} + \frac{\partial}{\partial x_j}(\overline{u}_i \overline{u}_j) = -\frac{\partial \overline{p}}{\partial x_i} + \frac{1}{Re}\frac{\partial^2 \overline{u}_i}{\partial x_j \partial x_j} - \frac{\partial \tau_{ij}}{\partial x_j}, \qquad (3)$$

$$\frac{\partial \overline{u}_i}{\partial x_i} = 0, \qquad (4)$$

in which a reference length and velocity scale are used to make $\overline{u}_i$, $\overline{p}$, $x_i$, $t$ and the molecular kinematic viscosity $\nu$ dimensionless, and to define the Reynolds number $Re$; repeated indices denote summation. The effect of the subgrid scales appears in the SGS stress, $\tau_{ij} = \overline{u_i u_j} - \overline{u}_i \overline{u}_j$, which must be modeled. For example, in eddy viscosity models, the SGS stress is approximated as

$$\tau_{ij}^m = -2\nu_T \overline{S}_{ij} + \delta_{ij} q_{sgs}^2 / 3, \qquad (5)$$

where $\delta_{ij}$ is the Kronecker delta function, $q_{sgs}^2 = \tau_{kk}$ is the SGS energy (which is added to the pressure), $\overline{S}_{ij}$ is the large-scale strain-rate tensor,

$$\overline{S}_{ij} = \frac{1}{2}\left(\frac{\partial \overline{u}_i}{\partial x_j} + \frac{\partial \overline{u}_j}{\partial x_i}\right), \tag{6}$$

and $\nu_T$ is an eddy viscosity. The eddy viscosity originally proposed by Smagorinsky (1963), denoted here by $\nu_S$, is

$$\nu_S = C_S \Delta^2 \sqrt{2\overline{S}_{ij}\overline{S}_{ij}}, \tag{7}$$

in which $C_S$ is a constant and $\Delta$ is a length scale, usually taken as

$$\Delta = (\Delta x \, \Delta y \, \Delta z)^{1/3}. \tag{8}$$

We have departed here from the usual convention in which $C_S$ appears in Eq. (7) as $C_S^2$ because we wish to admit models in which $C_S$ is negative, at least locally.

To examine the effect of the SGS stress model on the resolved scales, consider the transport equation for (twice) the resolved kinetic energy $\overline{q}^2 = \overline{u}_i \overline{u}_i$:

$$\frac{\partial \overline{q}^2}{\partial t} + \frac{\partial}{\partial x_j}\left(\overline{q}^2 \overline{u}_j\right) = \frac{\partial}{\partial x_j}\left(-2\overline{p}\,\overline{u}_j - 2\overline{u}_i\tau_{ij} + \frac{1}{Re}\frac{\partial \overline{q}^2}{\partial x_j}\right)$$
$$-\frac{2}{Re}\frac{\partial \overline{u}_i}{\partial x_j}\frac{\partial \overline{u}_i}{\partial x_j} + 2\tau_{ij}\overline{S}_{ij}. \tag{9}$$

One-half of the last term on the right-hand side of Eq. (9) will be referred to as "subgrid-scale dissipation," $\epsilon_{SGS} = \tau_{ij}\overline{S}_{ij}$; it represents the energy transfer between resolved and subgrid scales. If it is negative, the subgrid scales remove energy from the resolved ones (forward scatter); if it is positive, they release energy to the resolved scales (backscatter). The backward and forward scatter components of $\epsilon_{SGS}$, respectively denoted by $\epsilon_+$ and $\epsilon_-$, are defined as

$$\epsilon_+ = \frac{1}{2}\left(\epsilon_{SGS} + |\epsilon_{SGS}|\right), \qquad \epsilon_- = \frac{1}{2}\left(\epsilon_{SGS} - |\epsilon_{SGS}|\right). \tag{10}$$

It is easy to see that eddy viscosity SGS stress models of the Smagorinsky type are absolutely dissipative, so long as $C_S > 0$, since they then assume that the eddy viscosity $\nu_T$ is positive, which gives

$$\tau_{ij}^m \overline{S}_{ij} = -2\nu_T \overline{S}_{ij}\overline{S}_{ij} < 0. \tag{11}$$

To study the behavior of the SGS stresses and dissipation in transitional flow we have used the database obtained from the direct simulation of transition in a plane channel at $Re = 8000$ (based on the channel halfwidth $\delta$ and the laminar centerline velocity $U_c$), with periodic boundary conditions in the streamwise and spanwise directions. Initial conditions consisted of the parabolic mean flow, on which a two-dimensional Tollmien-Schlichting (TS) mode of 2% amplitude and a three-dimensional TS mode of 0.02% amplitude were superimposed. The periodicity lengths in the streamwise and spanwise directions were $L_x = 2\pi$ and $L_z = 4\pi/3$, respectively. The initial conditions and Reynolds number matched those of the direct simulation described by Zang, Gilbert and Kleiser (1990). This finely resolved DNS used $216 \times 162 \times 216$ grid points, and the truncation errors in the primitive variables were less than 0.03% at all times. The spanwise symmetry of the initial conditions was exploited to reduce the computational effort by a factor of 2. For details on the particular numerical method, the physics of channel flow transition and results for closely related problems, see Zang et al. (1990).

We should note that this is a *temporal* transition problem, as opposed to the *spatial* transition problem of technological interest. The extension of LES from temporal to spatial transition problems is a necessary task, whose prime complication is the increased computing demands (by about an order of magnitude) for the spatial problem.

This simulation has been conducted to date up to a time of 330 (made dimensionless by $\delta$ and $U_c$). The results of the present DNS at $Re = 8000$ are qualitatively similar, in both the transitional and turbulent regions, to the pioneering simulation of Gilbert and Kleiser (1990) at $Re = 5000$ (see Zang et al. 1990). From $t = 0$ to $t = 165$ the evolution is well-described by a combination of linear stability theory and secondary instability theory (Herbert 1988). The strongly nonlinear, "one-spike stage" occurs between $t \approx 165$ and $t \approx 175$, the "multispike stages" last from $t \approx 175$ to $t \approx 185$, the "laminar breakdown" phase commences at $t \approx 185$, the peak wall shear stress occurs near $t = 220$, and fully developed turbulent flow ensues at about $t = 240$. The fully turbulent flow is characterized by a Reynolds number based on the wall shear velocity of $Re_\tau = 320$ and a Reynolds number based on mean centerline velocity of $Re_{CL} = 6200$, and it has a computational domain in wall units of $L_x^+ = 2010$ and $L_z^+ = 1340$. The "transitional zone," between $t = 165$ and $t = 240$, is the focus of the present work.

## 11.3 A Priori Analysis

Piomelli et al. (1990a) performed an *a priori* analysis of channel flow transition for $t \le 200$ using DNS data in which they focused on the early part of the transitional period. They found that the SGS stresses $\tau_{ij}$ and the SGS dissipation $\epsilon_{SGS}$ were significantly different from their counterparts in fully developed turbulent flow. They noted that there were substantial regions of the flow with negative SGS dissipation, indicating the presence of a reversed energy cascade. In other words, transitional flow exhibits backscatter.

The issue of backscatter has been addressed recently by Piomelli et al. (1991a), with an emphasis on turbulent flows, but with some results extracted from the transitional database described in the preceding section. To investigate the character of backscatter, the velocity fields obtained from the DNS of channel flow transition were filtered to yield the exact resolved and SGS velocities, and the exact SGS dissipation. Figure 1 shows the SGS dissipation at three representative times: $t = 170$ (the one-spike stage), $t = 200$ (after laminar breakdown), and $t = 220$ (near the peak wall shear stress). SGS dissipation and backscatter are normalized by the volume-averaged viscous dissipation $< \epsilon_v >_V$:

$$< \epsilon_v >_V = \frac{1}{2L_x L_z} \int_0^{L_x} \int_0^{L_z} \int_{-1}^{+1} \frac{2}{Re} S_{ij} S_{ij} \; dx \; dy \; dz. \qquad (12)$$

The amount of filtering is characterized by $\alpha$, the ratio of SGS kinetic energy, $< q_{SGS} >_V$, to total turbulent kinetic energy, $< q^2 >_V$, and by the ratio of filter width, $\Delta_i$, to grid size, $\Delta x_i$. Since filtering was applied only in the plane parallel to the wall, the ratios $\Delta_i / \Delta x_i$ reported here refer only to the streamwise and spanwise directions. Unless otherwise noted, all quantities are made dimensionless by the channel halfwidth $\delta$, the fluid density $\rho$ and the centerline velocity in laminar flow $U_c$.

Figure 1 indicates that the backscatter contribution to $\epsilon_{SGS}$ is much larger than the mean at all times (and also independent of filter width). Although the subgrid scales extract energy from the large scales in the mean, large values of $\epsilon_+$ and $\epsilon_-$ can be expected. The fraction of points in each plane which experience backscatter (shown in Figure 2) is close to 50%, virtually independent of time and distance from the wall.

Figure 3 shows the plane-averaged SGS shear stress $< \tau_{12} >$ at the same instants. Initially, the stress is very small, reflecting the absence of small scales in the flow. The positive stress which is observed for $t = 170$ at $y/\delta = -0.8$ is to a large extent responsible for the positive mean SGS dissipation, which also occurs around this location.

## 11.4 SGS Stress Models

The model considered by Piomelli et al. (1990a) and Piomelli and Zang (1991) for the SGS stress in transitional flow was the scaled Smagorinsky

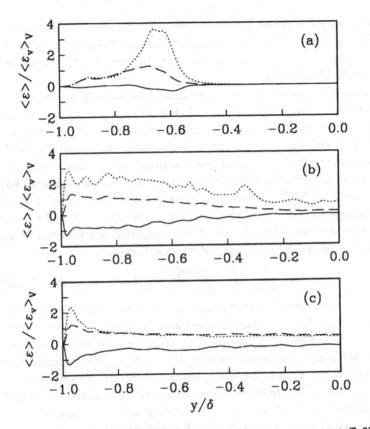

Fig. 1   Subgrid-scale dissipation normalized by $< \epsilon_v >_V$; cutoff filter.
———— Plane-averaged dissipation $< \epsilon_{SGS} >$; ········ root-mean-square fluctuation of $\epsilon_{SGS}$; - - - plane-averaged backscatter $< \epsilon_+ >$. (a) $t = 170$, $\Delta_i = 8\Delta x_i$, $\alpha = 0.01$; (b) $t = 200$, $\Delta_i = 6\Delta x_i$, $\alpha = 0.07$; (c) $t = 220$, $\Delta_i = 6\Delta x_i$, $\alpha = 0.11$.

model, which has the form of Eq. (5) with the eddy viscosity given by

$$\nu_T = \left[ \gamma \left( 1 - e^{-y^+/25} \right) \right]^2 C_S \Delta^2 \sqrt{2 \overline{S}_{ij} \overline{S}_{ij}}. \qquad (13)$$

A superscript $+$ indicates a quantity made dimensionless by the kinematic viscosity $\nu$ and the shear velocity $u_\tau = (\tau_w/\rho)^{1/2}$, where $\tau_w = \mu [\partial \overline{u}/\partial y]_w$ is the wall shear stress and $\rho$ the fluid density. The scaling factor $\gamma = (H_l - H)/(H_l - H_t)$ (in which $H$ is the shape factor, given by the ratio of the displacement thickness to the momentum thickness, and the subscripts $l$ and $t$ refer, respectively, to laminar and fully developed turbulent flow) was introduced to decrease the dissipation by the subgrid scales during the early stages of transition.

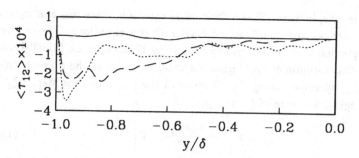

Fig. 2 Fraction of grid points at which $\epsilon_{SGS} > 0$; cutoff filter. ——— $t = 170$, $\Delta_i = 8\Delta x_i$, $\alpha = 0.01$; – – – $t = 200$, $\Delta_i = 6\Delta x_i$, $\alpha = 0.07$; ········ $t = 220$, $\Delta_i = 6\Delta x_i$, $\alpha = 0.11$.

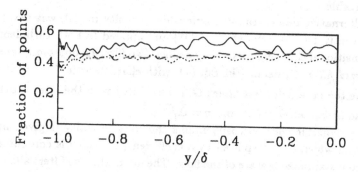

Fig. 3 Plane-averaged SGS stress $< \tau_{12} >$; cutoff filter. ——— $t = 170$, $\Delta_i = 8\Delta x_i$, $\alpha = 0.01$; – – – $t = 200$, $\Delta_i = 6\Delta x_i$, $\alpha = 0.07$; ········ $t = 220$, $\Delta_i = 6\Delta x_i$, $\alpha = 0.11$.

Piomelli et al. (1990b) modeled the SGS stress with an algebraic eddy viscosity model based on the RNG theory of Yakhot and Orszag (1986). With this approach the *total* viscosity, $\nu_{tot} = \nu + \nu_T$ [where $\nu_T$ is the eddy viscosity to be used in Eq. (5)] is given by the RNG formula

$$\nu_{tot} = \nu \left[ 1 + H \left( \frac{\nu_S^2 \nu_{tot}}{\nu^3} - C \right) \right]^{1/3} , \qquad (14)$$

where $H(x)$ is the ramp function

$$H(x) = \begin{cases} x & \text{for } x > 0, \\ 0 & \text{otherwise,} \end{cases} \qquad (15)$$

$C$ is a constant, $\nu$ is the molecular viscosity and $\nu_S$ is the Smagorin-sky (1963) eddy viscosity, given by Eq. (7).

If the argument of the ramp function $H(x)$ is greater than zero, Eq. (14) has the appearance of a cubic equation for the total viscosity. In reality, however, it is a functional equation, since $\nu_{tot}$ and $\nu_S$ are related through the renormalized Navier–Stokes equations. The numerical implementation of the model chosen by Piomelli et al. (1990b) was the explicit formula of evaluating the total viscosity at the present time step, $\nu_{tot}^n$, by substituting the viscosity at the previous time step, $\nu_{tot}^{n-1}$, in the right-hand side of (14) to yield

$$\nu_{tot}^n = \nu \left[1 + \frac{\nu_S^2 \nu_{tot}^{n-1}}{\nu^3} - C\right]^{1/3}. \tag{16}$$

The filter width $\Delta$ was given by Eq. (8). Recent results (Yakhot, Orszag, Yakhot and Israeli 1989) for LES of turbulent flows suggest that an alternative length scale formula specifically tuned for RNG models might be preferable.

An alternative model, the dynamic eddy viscosity model, was developed recently by Germano et al. (1991) and applied to both turbulent and transitional channel flows. They defined two filtering operators: one is the *grid filter*, $\overline{G}$, defined in Eq. (1), with characteristic filter width $\overline{\Delta}$, while the other, the *test filter*, $\widehat{G}$, is associated with the filter width $\widehat{\overline{\Delta}}$. In addition, let $\widehat{\overline{G}} = \overline{G}\widehat{G}$ and $\eta = \widehat{\overline{\Delta}}/\overline{\Delta} > 1$.

In the dynamic eddy viscosity model the eddy viscosity is given by Eq. (7), in which $C_S$ is replaced by a coefficient, $C$, which is computed from the instantaneous state of the flow. The use of the two filters allows extraction of information regarding the smallest resolved scales, which is then used to optimize the parameterization of the subgrid scales. The form of the coefficient $C$ used by Germano et al. (1991) is

$$C(y,t) = -\frac{1}{2} \frac{< \mathcal{L}_{kl}\overline{S}_{kl} >}{\widehat{\overline{\Delta}}^2 < |\widehat{\overline{S}}|\widehat{\overline{S}}_{mn}\overline{S}_{mn} > - \overline{\Delta}^2 < |\widehat{\overline{S}|\overline{S}_{pq}\overline{S}_{pq}} >}, \tag{17}$$

in which the resolved turbulent stress, $\mathcal{L}_{ij}$, is defined as

$$\mathcal{L}_{ij} = \widehat{\overline{u_i}\overline{u_j}} - \widehat{\overline{u}_i}\widehat{\overline{u}}_j, \tag{18}$$

and a hat indicates the application of the test filter and a planar average is denoted by $< \cdot >$. The coefficient $C$ used by Germano et al. (1991) is not completely localized; the average over planes parallel to the wall could, however, be relaxed to a more localized average, although some sort of average is necessary to avoid ill-conditioning arising from pointwise vanishing of the denominator in Eq. (17). The modeled SGS stress tensor is then given by

$$\tau_{ij}^m = \frac{< \mathcal{L}_{kl}\overline{S}_{kl} >}{\left(\widehat{\overline{\Delta}}/\overline{\Delta}\right)^2 < |\widehat{\overline{S}}|\widehat{\overline{S}}_{mn}\overline{S}_{mn} > - < |\overline{S}|\overline{S}_{pq}\overline{S}_{pq} >} |\overline{S}|\overline{S}_{ij}. \tag{19}$$

Note that the model gives zero SGS stress everywhere that $\mathcal{L}_{ij}$ vanishes (as long as the denominator remains finite). Such is the case in laminar flow. As Germano et al. (1991) observed, this model also produces the correct near-wall behavior of the SGS stress, and the modeled SGS dissipation, $\epsilon_{sgs}^m = \tau_{ij}^m \overline{S}_{ij}$, can be either positive or negative. Thus, the model does not rule out backscatter. The only adjustable parameter in the model is the ratio $\eta$.

## 11.5 LES Results

The three transitional SGS stress models described above have been applied to the channel flow transition problem described at the end of Section 2. The sharp cutoff filter was employed. The meshes for the LES used up to $48 \times 72 \times 48$ grid points (smaller meshes were used for the early part of the evolution – until $t = 165$ – during which the flow could be computed reliably with no SGS model; as increased resolution was required, the mesh was refined to its final value). The DNS imposed spanwise symmetry, whereas the LES calculations were free of this constraint; although the initial condition was spanwise symmetric, the LES results eventually became asymmetric due to the influence of round-off errors.

In the case of the scaled Smagorinsky model, the model constant $C_S$ was set equal to 0.01, and the values 2.5 and 1.7 were used for $H_l$ and $H_t$. The numerical results, at least in the fully developed turbulent regime, are not expected to depend much on the values of $H_l$ and $H_t$: turbulent statistics are insensitive to changes in $C_S$ of the order of 20% (Moin and Kim 1982). For the RNG model, the values recommended by Yakhot and Orszag (1986) were chosen for the model constants: $C = 100$ and $C_S = 0.016$. For the dynamic eddy viscosity model, the parameter $\eta$ was chosen to be 2. The results of Germano et al. (1991) suggested that this value was close to an optimal choice for this problem.

Figure 4 presents a comparison of wall shear stress, $\tau_w$, for the well-resolved DNS and the LES using the three different SGS models. In the following, time is normalized by $\delta/U_c$, velocities by $U_c$ and lengths by $\delta$; moreover, $U_i = <\overline{u}_i>$, and the resolved fluctuations are defined as $u_i'' = \overline{u}_i - U_i$.

Some remarks regarding the evolution of the wall shear stress in the DNS are in order. There is no perceptible change in the wall shear stress during the stages dominated by linear and secondary instability. It is only after the one-spike stage is under way that the wall shear stress rises

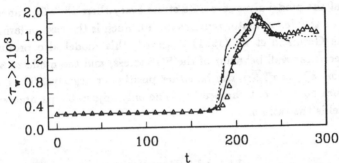

Fig. 4   Time evolution of the wall shear. ········· Scaled Smagorinsky model
(13) (Piomelli and Zang 1991); – – – RNG model (14); ——— dynamic eddy viscosity model (19) (Germano et al. 1991); △ fine direct
simulation (Zang et al. 1990).

above its laminar value. During the later part of the transitional zone,
the wall shear stress overshoots the turbulent value by roughly 20%.
Although these results were obtained for a highly idealized simulation of
forced transition, this sort of evolution of the wall shear stress is typical
of natural transition in a low-disturbance environment, and accurate
prediction of this overshoot as well as the length of the transitional zone
are significant indicators of the usefulness of the LES for transition.

During the linear and secondary instability stages all simulations agree
very well. Indeed, the eddy viscosity in all three LESs is negligible up
to this point ($t = 165$). All three exhibit a slightly premature rise in
wall shear stress near $t = 175$; this rise occurs earlier with the scaled
Smagorinsky and the RNG models than with the dynamic eddy viscosity
model. Both the scaled and the dynamic eddy viscosity models produce
an overshoot in the wall shear stress that agrees well with the actual
overshoot, and they also produce reasonable agreement for the time-
averaged wall shear stress in the fully turbulent state. The peak wall
shear stress computed with the scaled Smagorinsky model is within 3%
of the DNS prediction. The RNG model does not perform as well in
either of these respects.

In earlier work (Piomelli and Zang 1991), the LES results for the
scaled Smagorinsky model were compared with both the well-resolved
DNS results and those of a coarse-grid DNS which used the same number of points as the LES, but no model. The coarse-grid DNS results
were decidedly inferior to those with the SGS model: the prediction
of the peak wall stress from the coarse-grid DNS, e.g., was in error by
approximately 12%.

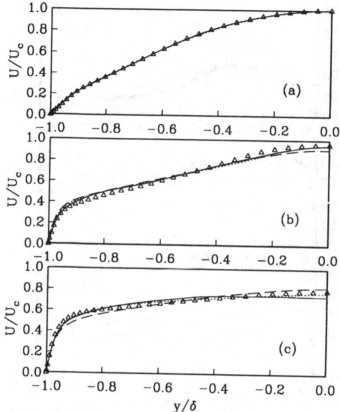

Fig. 5 Plane-averaged velocity. ········ Scaled Smagorinsky model (13) (Piomelli and Zang 1991); – – – RNG model (14); ——— dynamic eddy
viscosity model (19) (Germano et al. 1991); $\triangle$ fine direct simulation
(Zang et al. 1990). (a) $t = 175$; (b) $t = 200$; (c) $t = 220$.

In the transitional zone, the eddy viscosity begins to increase for all the
models; for example, in the calculation which used the scaled Smagorinsky model the eddy viscosity at $t = 220$ is approximately 40% larger
than it is in fully developed turbulent flow. This is due to the large
velocity fluctuations that occur during laminar breakdown, which cause
oscillations of the large-scale strain-rate tensor of greater amplitude than
in turbulent flow. Similar results were observed with all models.

Figure 5 displays the mean flow at selected times as predicted by
the DNS and the various LES calculations. The LES results obtained
with all models agree well with the DNS results. Streamwise velocity
fluctuations and Reynolds shear stress profiles are shown in Figures 6 and
7, respectively. Some differences between the predictions of the various

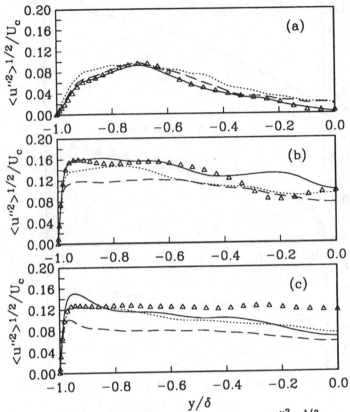

Fig. 6   Plane-averaged rms turbulent fluctuations $< u''^2 >^{1/2}$. ........
Scaled Smagorinsky model (13) (Piomelli and Zang 1991); – – – RNG
model (14); ——— dynamic eddy viscosity model (19) (Germano et
al. 1991); $\triangle$ fine direct simulation (Zang et al. 1990). (a) $t = 175$;
(b) $t = 200$; (c) $t = 220$.

models can be observed here. For $t = 170$ the scaled Smagorinsky model
is the least accurate; the errors are caused primarily by its prediction of
slightly premature transition. For $t \leq 200$ the dynamic eddy viscosity
model is significantly more accurate than the other two models studied.
At $t = 220$ the profiles of the rms fluctuations of $u$ that are predicted by
the LES calculations are peakier than those calculated by the DNS; the
level predicted by the RNG model is also lower than the DNS result.

During the late stages of transition the Reynolds stresses can be several times larger than their counterparts in turbulent flow, perhaps due
to the highly intermittent character of the late stages of the transition
process, which results in large velocity gradients and increased turbulent

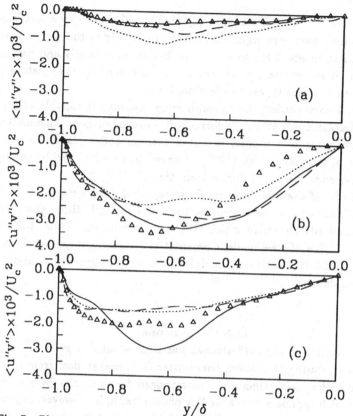

Fig. 7  Plane-averaged shear Reynolds stress, $< u''v'' >$. ········ Scaled Smagorinsky model (13) (Piomelli and Zang 1991); --- RNG model (14); ——— dynamic eddy viscosity model (19) (Germano et al. 1991); $\triangle$ fine direct simulation (Zang et al. 1990). (a) $t = 175$; (b) $t = 200$; (c) $t = 220$.

kinetic energy production. The large eddy simulations were carried out into the turbulent regime, and their results were found to compare well with experimental and numerical results (Dean 1978; Kim, Moin and Moser 1987). Little difference was observed, in turbulent flow, among the various models, with the dynamic eddy viscosity model giving slightly more accurate results (especially in the near-wall region) than the other two.

When the RNG model was used for LES of boundary-layer transition (Piomelli et al. 1990b), the eddy viscosity was found to depend very much on the length scale employed. If the length scale was changed by a factor of 2 due, e.g., to increased mesh resolution, the first term in the argument of the ramp function in Eq. (14) would change by a factor of 16. This may have very significant effects, especially at the early stages of transition, in which the model is switched on only locally and part of the time. If too coarse a mesh was used, moreover, Eq. (14) yielded a nonzero eddy viscosity, even in laminar flow.

As mentioned earlier, the dynamic eddy viscosity is capable of predicting backscatter. With the formulation presented here, backscatter is not localized: if it occurs, it occurs over an entire plane. In their calculation, Germano et al. (1991) observed backscatter (evidenced by a negative eddy viscosity) during transition (for $t \leq 185$).

The LESs of channel flow transition reported here were nearly two orders of magnitude less expensive (in terms of CPU time) than the highly resolved DNS with which they were compared. Thus, LES of transitional flow offers a realistic prospect of producing important information for RANS transition models in more complex geometries and for compressible flow.

## 11.6 Conclusions

The behavior of the SGS stresses has been studied *a priori*. It was found that, during transition, backscatter (i.e., energy flow from small to large scales) is significant. The standard Smagorinsky SGS stress model, which cannot account for this phenomenon, is, however, capable of predicting transition to turbulence fairly accurately, as long as the SGS dissipation predicted by the model is decreased by the introduction of some form of intermittency function.

Alternatively, a model based on RNG theory has been used which does not require an ad hoc intermittency function. With this model the SGS stresses are essentially zero throughout the linear and early nonlinear stages of transition. This model also is absolutely dissipative, but still captures most of the physical features of transition. Previous experience with this model, however, has shown it to be quite grid-dependent.

Finally, an SGS model which is not strictly dissipative was used. This model gives more accurate prediction of mean velocity and Reynolds stresses than the others, indicating that the use of models capable of predicting backscatter may be beneficial for the study of laminar–turbulent transition.

## Acknowledgments

This research was supported partly by the NASA Langley Research Center under Grant NAG-1-1089 and partly by the National Aeronautics and Space Administration under NASA Contract No. NAS1-18605 while the second author was in residence at the Institute for Computer Applications in Science and Engineering (ICASE), NASA Langley Research Center, Hampton, VA, 23681.

## References

ARNAL, D. (1989) Laminar-Turbulent Transition Problems at High Speeds. In *The Second Joint Europe/U.S. Short Course on Hypersonics.*

DANG, K.T., AND DESCHAMPS, V. (1987) Numerical simulation of transitional channel flow. In *Numerical Methods in Laminar and Turbulent Flows.* Ed. C. Taylor, W.G. Habashi, and M.M. Hafez, pp. 423–434. Pineridge.

DEAN, R.B. (1978) Reynolds number dependence of skin friction and other bulk flow variables in two-dimensional rectangular duct flow. *J. Fluids Eng.* 100, 215–223.

DEARDORFF, J.W. (1970) A numerical study of three-dimensional turbulent channel flow at large Reynolds numbers. *J. Fluid Mech.* 41, 453–480.

DESCHAMPS, V. AND DANG, K.T. (1987) Evaluation of subgrid-scale models for large-eddy simulation of transitional channel flows. In *6th Symp. on Turbulent Shear Flows*, Toulouse, France.

GERMANO, M., PIOMELLI, U., MOIN, P., AND CABOT, W.H. (1991) A dynamic subgrid-scale eddy viscosity model. *Phys. Fluids A* 3, 1760–1765.

GILBERT, N. (1988) Numerische Simulation der Transition von der laminaren in die turbulente Kanalströmung. DFVLR-FB 88-55.

GILBERT, N. AND KLEISER, L. (1990) Near-wall phenomena in transition to turbulence. In *Near-Wall Turbulence: 1988 Zoran Zaric Memorial Conference*, Ed. S.J. Kline and N.H. Afgan, pp. 7–27. Hemisphere.

HERBERT, T. (1988) Secondary instability of boundary layers. *Ann. Rev. Fluid Mech.* 20, 487–526.

HORIUTI, K. (1986) On the use of SGS modeling in the simulation of transition in plane channel flow. *J. Phys. Soc. Japan* **55**, 1528–1541.

KIM, J., MOIN, P., AND MOSER, R.D. (1987) Turbulence statistics in fully developed channel flow at low Reynolds number. *J. Fluid Mech.* **177**, 133–166.

KLEISER, L. AND ZANG, T.A. (1991) Numerical simulation of transition in wall-bounded shear flows. *Ann. Rev. Fluid Mech.* **23**, 495–537.

LEONARD, A. (1974) Energy cascade in large-eddy simulations of turbulent fluid flows. *Adv. Geophys.* **18A**, 237–248.

MALIK, M.R., ZANG, T.A., AND BUSHNELL, D.M. (1990) Boundary layer transition in hypersonic flows. AIAA Paper No. 90–5232.

MOIN, P. AND KIM, J. (1982) Numerical investigation of turbulent channel flow. *J. Fluid Mech.* **118**, 341–377.

NARASIMHA, R. (1985) The laminar–turbulent transition zone in the boundary layer. *Prog. Aerospace Sci.* **22**, 29–80.

NARASIMHA, R. AND DEY, J. (1989) Transition-zone models for 2-dimensional boundary layers: A review. *Sādhanā, Acad. Proc. Eng. Sci.* **14**, 93–120.

PIOMELLI, U., MOIN, P., AND FERZIGER, J.H. (1988) Model consistency in large eddy simulation of turbulent channel flows. *Phys. Fluids* **31**, 1884–1891.

PIOMELLI, U., ZANG, T.A., SPEZIALE, C.G., AND HUSSAINI, M.Y. (1990a) On the large eddy simulation of transitional wall-bounded flows. *Phys. Fluids A* **2**, 257–265.

PIOMELLI, U., ZANG, T.A., SPEZIALE, C.G., AND LUND, T.S. (1990b) Application of renormalization group theory to the large-eddy simulation of transitional boundary layers. In *Instability and Transition*, Vol. 2. Ed. M.Y. Hussaini and R.G. Voigt, pp. 480–496. Springer-Verlag.

PIOMELLI, U., CABOT, W.H., MOIN, P., AND LEE, S. (1991a) Subgrid-scale backscatter in turbulent and transitional flows. *Phys. Fluids A* **3**, 1766–1771.

PIOMELLI, U., MOIN, P., AND FERZIGER, J.H. (1991b) Large eddy simulation of the flow in a transpired channel. *AIAA J. Thermophys. Heat Transfer* **5**, 124–128.

PIOMELLI, U. AND ZANG, T.A. (1991) Large-eddy simulation of transitional channel flow. *Computer Phys. Comm.* **65**, 224–230.

SCHMITT, L., RICHTER, K., AND FRIEDRICH, R. (1985) A study of turbulent momentum and heat transport in a boundary layer using large eddy simulation technique. In *Finite Approximations in Fluid Mechanics*, Ed. E.H. Hirschel pp. 232–247. Vieweg.

SMAGORINSKY, J. (1963) General circulation experiments with the primitive equations. Part I. The basic experiment. *Mon. Wea. Rev.* **91**, 99–164.

YAKHOT, V. AND ORSZAG, S.A. (1986) Renormalization group analysis of turbulence. Part I. Basic theory. *J. Sci. Computing* **1**, 3–51.

YAKHOT, A., ORSZAG, S.A., YAKHOT, V., AND ISRAELI, M. (1989) Renormalization group formulation of large-eddy simulation. *J. Sci. Computing* **4**, 139–158.

ZANG, T.A., GILBERT, N., AND KLEISER, L. (1990) Direct numerical simulation of the transitional zone. In *Instability and Transition*, Vol. 2. Ed. M.Y. Hussaini and R.G. Voigt, pp. 283–299. Springer.

# COMPRESSIBLE AND REACTING FLOWS

# 12

## Direct Numerical Simulation and Large Eddy Simulation of Compressible Turbulence

GORDON ERLEBACHER AND MOHAMMED Y. HUSSAINI

### 12.1 Introduction

Compressible turbulence is currently the focus of attention in the fluid dynamics community, owing perhaps to the renewed interest in the design of supersonic and hypersonic transport vehicles. In the quest to understand the basic physical mechanisms that distinguish compressible from incompressible turbulence, the body of knowledge in incompressible turbulence accumulated over the past several decades serves as a springboard to new levels of understanding. Today, researchers are studying compressible turbulent flows using direct numerical simulation (DNS) (Passot and Pouquet 1987; Blaisdell 1990; Lele 1990; Lee, Lele and Moin 1991; Sarkar, Erlebacher and Hussaini 1991a; Kida and Orszag 1990 a,b, 1992) primarily. Some of these numerical results have been recently explained from a theoretical point of view (Tatsumi and Tokunaga 1974; Tokunaga and Tatsumi 1975; Erlebacher, Hussaini, Kreiss and Sarkar 1990a; Sarkar, Erlebacher, Hussaini and Kreiss 1991b). Past theoretical work on compressible turbulent flows concentrated mostly on modifications of incompressible scaling laws to include Mach number effects (Zakharov and Sagdeev 1970; Kadomtsev and Petviashvili 1973; Moiseev, Sagdeev, Tur and Yanovskii 1977; L'vov and Mikhailov 1978a,b) Unfortunately, it is unlikely that the near future will bring about experiments detailed enough to confirm or reject any of these theories. However, with the explosion in supercomputer memory and speed, it is possible that DNS may soon provide some clues to spur theoretical research along the right direction.

In his seminal paper on supersonic turbulence, Kovasznay (1957) described the decomposition of compressible turbulence into acoustic, vortical and entropy modes. This decomposition of the flow remains a beacon in guiding the interpretation of results in the fields of compressible turbulence and transition. Another way to characterize compressible and incompressible effects in homogeneous turbulence is to perform a Helmholtz decomposition on the velocity field (Moyal 1952; Passot and Pouquet 1987; Erlebacher et al. 1990a, Lee et al. 1991; Sarkar et al. 1991b). However, despite the progress made, much more work is required before the "incompressible" and "compressible" characteristics of the turbulent flows are fully classified. The cubic nonlinearities in the momentum equation and in the definition of kinetic energy add a degree of complexity to the interpretation of intermodal energy transfers. For example, we have become accustomed to a decomposition of the velocity field into Fourier components whose amplitude squared is correctly interpreted as the kinetic energy (or turbulence intensity) in the given mode. When the density fluctuations are taken into account in the definition of kinetic energy, it becomes necessary to make the distinction between the turbulence intensity, which involves only the velocity, and kinetic energy, which includes the effect of density fluctuations. Kida and Orszag (1990a, 1992) have defined an energy spectrum based on the Fourier decomposition of $\mathbf{u}\sqrt{\rho}$. Although this reduces the kinetic energy to a quadratic form, the resulting interpretation of modal energy is not clearly related to physical mechanisms. However, any proposed decomposition will enhance our knowledge of the intricate physical mechanisms that are surely present in compressible turbulent flows.

Direct numerical simulations have uncovered several regimes of isotropic turbulence which are now under study. An original classification was attempted by Passot and Pouquet (1987), who found that strong shocks are present when the fluctuating Mach number ($M_t$) is $O(1)$, while weak shocklets can be present even for small $M_t$. On the other hand, Feiereisen, Reynolds and Ferziger (1981) and Erlebacher, Hussaini, Speziale and Zang (1990b) performed DNS of compressible turbulence (with $M_t = 0.5$) which had *almost* incompressible statistical properties. The reason was brought out a few years later by Erlebacher et al. (1990a), who related the initial conditions of the flow to the characteristically different turbulent regimes. Since then, Kida and Orszag (1990a,b, 1992) have considered both forced and unforced isotropic turbulence, and work has begun on homogeneous shear flow turbulence (Blaisdell 1990; Sarkar et al. 1991a).

The ultimate goal of current research on the physics of turbulence (besides academic interest) is to develop practical turbulence models applicable to complex flows of practical interest. Direct numerical simulations of such flows of aerodynamic interest are not practical even on today's fastest computers. The turbulence models are therefore a necessity in engineering applications. There are two standard approaches to turbulence modeling. In the first approach, the Navier–Stokes equations are averaged over time. As a consequence, the terms which are modeled contain both small and large spatial scales. This approach is known as Reynolds stress modeling. The application of Reynolds stress models to compressible flows has a long history and is not discussed in this chapter. Comprehensive reviews can be found in Cebeci and Smith (1974) and Speziale (1991). An alternative approach to turbulence modeling is to decompose the flow field into its large-scale (energy-containing) eddies and the presumably universal small-scale eddies. While the dynamics of the large-scale eddies satisfy known evolution equations, there remain source terms, called subgrid-scale stress terms, which must be modeled. Subgrid-scale models are expected to be less sensitive to different large-scale flow conditions than their Reynolds stress counterparts if the assumption of small-scale universality proves to be valid. Of course, the price one pays for this increased generality is higher computational expense.

It is still too early to consider large eddy simulations (LES) of compressible flows as an engineering tool. However, once more reliable models are developed, LES databases can serve as an "exact" reference solution against which cheaper Reynolds stress models can be evaluated. This is pretty much the same relationship that exists between the existing DNS databases and the LES models under development.

In this chapter, we consider the advances in both the DNS and the LES of compressible turbulence. These subjects are treated together since the numerical techniques that are used to perform the simulations are identical for both types of problem. In fact the subgrid-scale models are often incorporated into the DNS codes. To date, LES of compressible turbulence is based less on theory than on extensions of incompressible models to the compressible regime. As we demonstrate, these models, in the absence of shocks and/or shocklets, are quite capable of predicting reasonably well the decay rates of the turbulence and of the thermodynamics fluctuation rms levels. We conclude with some current issues that should be addressed in the near future. A discussion of recent LES advances can be found in Ferziger (1984). More recent reviews of the achievements of DNS and LES appear in Reynolds (1989) and Hussaini, Speziale and Zang (1989).

## 12.2 Direct Numerical Simulation

Direct numerical simulations seek the solution to the full compressible Navier–Stokes equations without any modeling. For reference, the continuity, momentum and energy equations (expressed in nondimensional form) are

$$\frac{\partial \rho}{\partial t} + \frac{\partial(\rho u_k)}{\partial x_k} = 0, \tag{1}$$

$$\frac{\partial(\rho u_k)}{\partial t} + \frac{\partial(\rho u_k u_l)}{\partial x_l} = -\frac{\partial p}{\partial x_k} + \frac{\partial \sigma_{kl}}{\partial x_l}, \tag{2}$$

$$\frac{\partial(\rho h)}{\partial t} + \frac{\partial(\rho h u_k)}{\partial x_k} = \frac{\partial p}{\partial t} + u_k \frac{\partial p}{\partial x_k} + \frac{\partial}{\partial x_k}\left(\kappa \frac{\partial T}{\partial x_k}\right) + \Phi, \tag{3}$$

where $\rho$ is the mass density, $\mathbf{u}$ is the velocity vector, $p$ is the thermodynamic pressure, $\mu$ is the dynamic viscosity, $h$ is the enthalpy, $T$ is the absolute temperature and $\kappa$ is the thermal conductivity. The viscous stress $\sigma_{kl}$ and the viscous dissipation $\Phi$ are defined by

$$\sigma_{kl} = -\frac{2}{3}\mu \frac{\partial u_j}{\partial x_j}\delta_{kl} + \mu\left(\frac{\partial u_k}{\partial x_l} + \frac{\partial u_l}{\partial x_k}\right), \tag{4}$$

$$\Phi = -\frac{2}{3}\mu\left(\frac{\partial u_j}{\partial x_j}\right)^2 + \mu\left(\frac{\partial u_k}{\partial x_l} + \frac{\partial u_l}{\partial x_k}\right)\frac{\partial u_k}{\partial x_l}, \tag{5}$$

respectively. Equations (1)–(3) must be supplemented with

$$p = \rho RT, \tag{6}$$

for an ideal gas, where $R$ is the ideal gas constant. The dependence of the viscosity and thermal conductivity on the temperature must also be provided (i.e., relationships of the form $\mu = \mu(T)$ and $\kappa = \kappa(T)$ are needed and these depend on the gas under consideration). Typically Sutherland's law is appropriate for the viscosity when the medium is air. The Prandtl number is assumed constant by most investigators.

The majority of compressible turbulence direct numerical simulations are concerned with homogeneous turbulence. In all cases, the boundary conditions are periodic in all coordinate directions. Homogeneous shear flow simulations are rendered periodic through the Rogallo transformation (Rogallo 1981). For a long time, only spectral methods were considered accurate enough for the numerical solution to the full Navier–Stokes equations. This was due to the low dissipation and dispersion errors of the method, which are very desirable characteristics of the numerical algorithm when one is performing direct numerical simulations. Indeed, one wishes to protect the structure of the smallest turbulent eddies from numerical dissipation and dispersion. Recently, however, high order finite-difference methods (Lele 1990; Shu, Erlebacher, Zang, Whitaker and Osher 1992) are gaining acceptance in the DNS community. Tests at Langley confirm the high accuracy of sixth order compact schemes applied to turbulence calculations (Shu et al. 1992). Besides its accuracy, the sixth order compact scheme is two times faster than the spectral algorithm when compared on identical numerical grids. However, except for the recent work of Lee et al. (1991) almost all investigators involved with the DNS of homogeneous turbulence still opt for Fourier collocation schemes (Canuto, Hussaini, Quarteroni and Zang 1988) because of their very desirable numerical properties (low truncation errors and exponential convergence).

Passot and Pouquet (1987) use the nonconservative form of the momentum equations and the conservative enthalpy equation. All terms are treated explicitly, except for the viscous terms, which are partially implicit. Time advancement is a combination of Crank–Nicolson and third order Adams–Bashforth schemes. Erlebacher et al. (1990a, 1992) consider initial conditions with very low turbulent Mach number. Consequently, they require an efficient implicit scheme with very low dispersion and dissipation errors to advance the acoustic terms in time. To this effect, a two-step time-split scheme is used. The first step integrates (with an explicit third order Runga–Kutta scheme) the full equations with a set of linear acoustic terms subtracted off. In the second step, the remaining set of (linear) equations is solved exactly in Fourier space. Thus, large time steps do not sacrifice accuracy, since the proper acoustic behaviour is accounted for in the implicit step. The only errors incurred are splitting errors and errors due to the (small) variation of the average sound speed during the implicit stage. When time histories of statistical variables are required, the time step is decreased to resolve the sound waves.

Recently, Sarkar et al. (1991a) generated several direct numerical simulations of compressible homogeneous shear flow turbulence. Although a time-splitting algorithm similar to that of Erlebacher et al. (1990a) is possible, the simulations are currently fully explicit. The impact of the explicit scheme is not too high in the range of turbulent Mach numbers considered $(0.2 < M_t < 0.6)$.

Feiereisen et al. (1981) performed the first DNS of compressible homogeneous turbulence. They described potential problems which could result from an improper discretization of the convective terms. They advocate rewriting the convective terms in the form

$$\frac{1}{2}\left[\frac{\partial(\rho u_k u_l)}{\partial x_l} + \rho u_l \frac{\partial u_k}{\partial x_l} + u_k \frac{\partial(\rho u_l)}{\partial x_l}\right]. \tag{7}$$

As noted by Feiereisen et al. (1981), if this form is employed, then a wide variety of numerical algorithms (including Fourier collocation), when applied to the inviscid compressible equations, conserve energy in addition to mass and momentum in the absence of time differencing (and splitting) errors. Fornberg (1975) discusses the conditions under which the nonsymmetric form of the convective equations maintains numerical stability. Recently, numerical simulations of supersonic boundary layer flows undergoing transition from a laminar to a turbulent state have demonstrated the unstable nature of both the conservation form $(\partial/\partial x_j)(\rho u_i u_j)$ and the nonconservative forms $\rho u_j(\partial u_i/\partial x_j)$ using a spectral collocation algorithm. Stability was achieved with Eq. (7). Note that the alternative form

$$\frac{1}{2}\left[\frac{\partial(\rho u_k u_l)}{\partial x_l} + \rho u_k \frac{\partial u_l}{\partial x_l} + u_l \frac{\partial(\rho u_k)}{\partial x_l}\right] \tag{8}$$

is also subject to instabilities. The reasons for this have not yet been ascertained. Zang (1990) discusses the relative merits of the different forms that can be taken by the convective operator in incompressible flows. He concludes that aliasing errors decrease substantially when the skew-symmetric form is used instead of either the rotational form $(\omega \times \mathbf{u})$ or the nonconservative form $(\mathbf{u} \cdot \nabla \mathbf{u})$. However, he does not demonstrate an actual instability (using Eq. (7)) similar to one observed in recent simulations of transition of high speed flows. Blaisdell (1990) also discusses the aliasing properties of Eq. (7).

### 12.2.1 Initial Conditions

The choice of initial conditions in turbulent flow simulations is often more an art than a precise science. This is especially so for compressible turbulence simulations. Often, one can trace back a particular set of initial conditions to previous studies. In particular, incompressible turbulence initial conditions have been generalized to the compressible case without a full understanding of the consequences. For example, Feiereisen et al. (1981) and Erlebacher et al. (1992) choose a divergence-free initial velocity field and compute the initial pressure to ensure that the initial time derivative of the divergence of velocity will also be zero. Furthermore, the initial fluctuating density rms is set to zero. As explained by Erlebacher et al. (1990a) in the case of isotropic turbulence, these conditions lead to a very slow buildup of the compressibility effects on a convective $O(1)$ time scale. If one relaxes the condition that the initial time derivative of $u_i$ be zero, and specifies random fluctuations for pressure, the flow reaches a quasi-equilibrium on an $O(M_t)$ time scale instead. Sarkar et al. (1991b) have shown that in homogeneous shear turbulence, significant compressibility develops as the turbulence Mach number increases, even though the initial velocity field is divergence-free and the initial pressure is chosen to ensure that the initial time derivative of the divergence of velocity is zero.

More general initial conditions were considered by Passot and Pouquet (1987) and adopted by Erlebacher et al. (1992). These allow the generation of weak and strong shocks, or simply acoustic waves of prescribed strength superimposed on an essentially nonlinear incompressible flow field. These initial conditions are based on a Helmholtz decomposition of the velocity field

$$\mathbf{u} = \mathbf{u}^I + \mathbf{u}^C, \tag{9}$$

where $\mathbf{u}^I, \mathbf{u}^C$ are respectively the solenoidal and irrotational components of velocity. At $t = 0$, the reference turbulent Mach number $M_{t0}$, the Reynolds number $Re$, the Prandtl number $Pr$, the rms levels of $\mathbf{u}_0^C$, $\mathbf{u}_0^I$, $\rho_0$, $T_0$ and the autocorrelation spectrum for $\rho$, $T$, $\mathbf{u}^C$ and $\mathbf{u}^I$ are imposed. The ratio of irrotational kinetic energy to total kinetic energy is also a free parameter and is denoted by $\chi$. (The term "kinetic energy" actually refers to turbulent intensities.) A zero subscript refers to the initial state of the simulation. The energy spectrum of the density, temperature and velocity fluctuations are all given by

$$E(k) = k^4 e^{-2k^2/k_0^2}, \tag{10}$$

where $k_0$ corresponds to the spectrum peak. Strictly speaking the initially irrotational component of velocity should have an autocorrelation spectrum proportional to $k^2$ for low $k$ to guarantee analyticity of the velocity field (Batchelor 1953). However, Erlebacher et al. (1992) did not feel that this would qualitatively influence the numerical results.

### 12.2.2 Results

In this section we present some numerical results obtained over the past few years. Some theoretical background is also provided.

Passot and Pouquet (1987) performed two- and three-dimensional DNS of compressible isotropic turbulence on $256^2$ and $64^3$ grids and were able to produce both weak and strong shocks. The presence or absence of the shocks was controlled solely by the initial conditions, whose influence can never completely vanish in decaying isotropic flow. They found three separate regimes. The weak Mach number regime ($M_t < 0.3$) has two subcomponents. If $\delta\rho/\rho \leq M_t^2$, the flow remains quasi-incompressible for all time. However, if $\delta\rho/\rho > M_t^2$, even by a small amount, the flow switches to a state which is mostly irrotational. Under these conditions weak shocks (negligible entropy jumps) can appear which propagate at the velocity of sound across the domain. The third regime is the strong-shock regime characterized by $M_t = O(1)$. Passot and Pouquet considered the decay laws of their spectra compared with those predicted by various statistical theories. Results were inconclusive, mainly because the simulations are still not capable of producing an inertial range. They also found that, in the strong-shock regime, there is a strong influence of the baroclinic torque in the vicinity of the shocks (for two-dimensional flows). In these cases, $\nabla\rho$ and $\nabla p$ were found to be nearly perpendicular to each other.

The ratio of irrotational kinetic energy to compressible potential energy is defined (Sarkar et al. 1991b) as

$$F = \frac{(\gamma M_t)^2 \chi}{p^{C2}}, \tag{11}$$

where $p^C$ is the compressible perturbation pressure defined by

$$p^C = p - p^I - <p>, \tag{12}$$

$<>$ denotes a three-dimensional spatial average, and the pair $(\mathbf{u}^I, p^I)$ satisfies the incompressible Navier–Stokes equations. It was shown in Sarkar et al. (1991b) that, for isotropic turbulence, $F$ tends to unity after the initial acoustic transient dies out, implying thereby the tendency toward equipartition between the kinetic and potential energy of the compressible component. This trend appears to hold for homogeneous shear flow also (Sarkar et al. 1991a). Time histories of $F$ and $\chi$ are presented in Figs. 1 and 2, starting from four widely different sets of initial conditions, all chosen to avoid shocks on the acoustic time scale. The horizontal lines in Fig. 2 are the theoretical predictions for $\chi$ in the absence of viscous effects. Time histories on a longer (viscous) time scale are shown in Figs. 1b and 2b. Whereas $F$ still oscillates around unity on the viscous time scale, $\chi$ can either increase or decrease. There is no available theory which can predict the long time scale evolution trends for $\chi$.

Figures 3 and 4 show the time histories of the incompressible ($E^I$) and compressible ($E^C$) integrated turbulence intensities respectively from

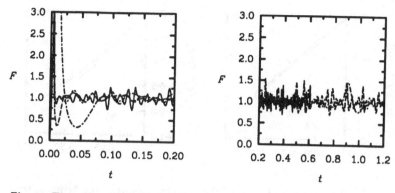

Fig. 1   Time history of $F$ for four sets of initial conditions. (a) Acoustic time scale, (b) viscous time scale.

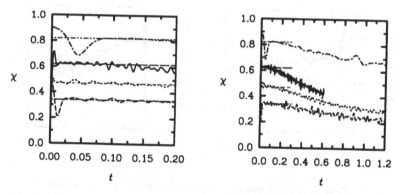

Fig. 2   Time history of the compressibility ratio $\chi$ for four sets of initial conditions. (a) Acoustic time scale, (b) viscous time scale.

the same four simulations. One notes the stronger oscillatory behavior in $E^C$, which is not surprising given its wavelike behavior. Also of interest is the stronger dependence of $E^C$ on initial conditions, even on a viscous time scale (compare Figs. 4a and 4b). Figure 5 compares the time histories of $\nabla \cdot \mathbf{u}$ without and with weak shocks in the flow. In the absence of shocks (or shocklets), the distribution of regions of expansion and compression is approximately equal (Fig. 5a). However, the presence of shocks leads to an asymmetry in the distribution of $\nabla \cdot \mathbf{u}$. Indeed, the preclusion of expansion shocks shows up as an asymmetry in Fig. 5b. The sharp negative peaks in the minimum divergence curve correspond to the crossing of two or more shocks. The width of the peaks is a function of both the shock speed and shock width. Lee et al. (1991) use the spatial average of velocity skewness to monitor the

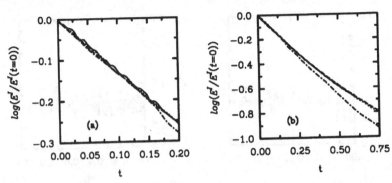

Fig. 3    Time history of solenoidal turbulent intensity. (a) Acoustic time scale, (b) viscous time scale.

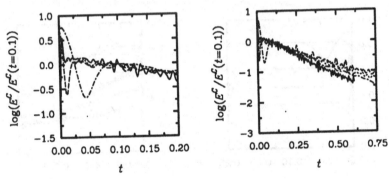

Fig. 4    Time history of compressible bulent intensity. (a) Acoustic time scale, (b) viscous time scale.

growth of shocks on the convective time scale. These shocks tend to occur when the turbulent Mach number is $O(1)$.

Erlebacher et al. (1990a) are concerned primarily with the weak shock regime treated by Passot and Pouquet (1987). They seek to explain and quantify rigorously the relationship between initial conditions and the subsequent evolution of turbulence. By a careful asymptotic analysis, Erlebacher et al. (1990a) found that the presence of weak shocks is linked to initial flow conditions, which are out of acoustic equilibrium. For example, this situation arises for the initial conditions $M_t = 0.04$, $p_{rms} = 10\%$ and $\chi = 0.1$, where $\chi$ is the ratio of irrotational to solenoidal kinetic energy.

The compressible simulations of Kida and Orszag (1990b) highlight some interesting differences with Passot and Pouquet (1987). For exam-

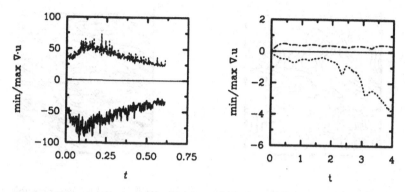

Fig. 5 Time history of $\nabla \cdot \mathbf{u}$ in a turbulent flow. (a) Without shocks, (b) with shocks.

ple, Kida and Orszag find that in three-dimensional decaying turbulence, $\nabla\rho$ and $\nabla p$ are almost aligned in strong shock regions. These alignment properties in two-dimensional and three-dimensional turbulence are not currently explained. In the three-dimensional simulations, they found that the vorticity budget is dominated (overall) by the vortex stretching and the dissipation terms. However, in the shock regions, the compression and baroclinic terms produce vorticity, while the dissipation term depletes it.

Sarkar et al. (1991a) performed DNS of compressible, homogeneous shear flow turbulence. Initial conditions are imposed as in our isotropic turbulence simulations. Results indicate that compressibility reduces the growth of the turbulent kinetic energy. After performing a Helmholtz decomposition of the velocity and computing the associated pressure fields, the authors found significant differences between the solenoidal and irrotational fields. The irrotational component acquires a wavelike character. This is most easily seen in the plots of $p^I$ and $p^C$, respectively shown in Figs. 6a and 6b. Preliminary visualizations of the magnitude of $\mathbf{u}^I$ and $\mathbf{u}^C$ have revealed the presence of streaks, although the character and orientation of the streaks are different for the two velocity fields. We are currently studying the properties of the streaks based on the invariants of the flow.

## 12.3 Large Eddy Simulation

Large eddy simulations incorporate subgrid-scale turbulence models into the Navier–Stokes equations in order to track accurately the small-scale phenomena that the numerical grid cannot resolve. One begins

Fig. 6 (a) Incompressible component of pressure $(p^I)$ and (b) compressible component of pressure $(p^C)$ in compressible homogeneous shear flow turbulence. Shear rate $S = 15$.

with a localized spatial average of the Navier–Stokes equations, which leads to unknown nonlinear terms that must be modeled. This approach is contrasted with Reynolds stress modeling, where one first averages the Navier–Stokes equations over time and then models the resulting unknown nonlinear terms. Under the (unproven) assumption that the character of turbulence in the small scales is somewhat universal, there is reason to believe that a good subgrid-scale model will be capable of treating a wider variety of complex geometrical configurations than their Reynolds-averaged counterparts. Large eddy simulations are still in their infancy and are still a long way from effectively proving to be reliable in realistic geometries. The greater computational expense of LES also makes it harder to run many simulations over a wide range of parameters, which is essential to gain familiarity and intuition, two necessary ingredients for progress.

The large eddy simulations reported in this chapter are based on the following Favré-filtered continuity, momentum and energy equations (Speziale, Erlebacher, Zang and Hussaini 1992):

$$\frac{\partial \overline{\rho}}{\partial t} + \frac{\partial \overline{\rho}\tilde{u}_i}{\partial x_i} = 0,$$

$$\frac{\partial}{\partial t}(\overline{\rho}\tilde{u}_i) + \frac{\partial \overline{\rho}\tilde{u}_i\tilde{u}_j}{\partial x_j} = -\frac{\partial \overline{p}}{\partial x_i} + \frac{\partial \overline{\sigma}_{ij}}{\partial x_j} + \frac{\partial \tau_{ij}}{\partial x_j}, \tag{13}$$

$$\frac{\partial}{\partial t}(C_v\overline{\rho}\tilde{T}) + \frac{\partial C_v\overline{\rho}\tilde{u}_i\tilde{T}}{\partial x_i} = -\overline{p\frac{\partial u_i}{\partial x_i}} + \overline{\sigma_{ij}\frac{\partial u_i}{\partial x_j}} + \frac{\partial}{\partial x_i}\left(\overline{\kappa\frac{\partial T}{\partial x_i}}\right) - \frac{\partial Q_i}{\partial x_i},$$

where $R$ is the ideal gas constant, $C_v$ is the specific heat at constant volume, $\tau_{ij}$ is the subgrid-scale stress tensor and $Q_i$ is the subgrid-scale heat flux. In Eqs. (13), an overbar represents a spatial filter whereas a tilde represents a mass-weighted or Favré filter, i.e.,

$$\tilde{\mathcal{F}} \equiv \frac{\overline{\rho \mathcal{F}}}{\bar{\rho}}, \tag{14}$$

where $\mathcal{F}$ is any flow variable. These equations are obtained by applying a Gaussian filter to the full Navier–Stokes equations.

### 12.3.1 Subgrid-Scale Models

The subgrid scale-stress tensor and subgrid-scale heat flux are modeled as follows (Speziale et al. 1992; Erlebacher et al. 1990b):

$$\tau_{ij} = -C_C \bar{\rho}(\widetilde{\tilde{u}_i \tilde{u}_j} - \tilde{u}_i \tilde{u}_j) + 2C_R \bar{\rho} \Delta_f^2 II_{\tilde{S}}^{1/2} (\tilde{S}_{ij} - \tfrac{1}{3}\tilde{S}_{kk}\delta_{ij})$$

$$- \tfrac{2}{3} C_I \bar{\rho} \Delta_f^2 II_{\tilde{S}} \delta_{ij}, \tag{15}$$

$$Q_i = C_v \bar{\rho} \left( \widetilde{\tilde{u}_i \tilde{T}} - \tilde{\tilde{u}}_i \tilde{T} - \frac{C_R}{Pr_T} \Delta_f^2 II_{\tilde{S}}^{1/2} \frac{\partial \tilde{T}}{\partial x_i} \right), \tag{16}$$

where $II_{\tilde{S}} = \tilde{S}_{ij}\tilde{S}_{ij}$, $\Delta_f$ is the spatial filter width and $C_R$ and $C_I$ are model constants. In the incompressible limit, the subgrid-scale stress model (15) reduces to the linear combination model of Bardina, Ferziger and Reynolds (1983). The modified Navier–Stokes equations are solved with the numerical method discussed in Erlebacher et al. (1990a) modified to account for the eddy viscosity term. Time advancement of the modeled terms is fully explicit; therefore, the extra terms pose no difficulty.

The model constants $C_C, C_I, C_R$ are related to the cross, Reynolds and isotropic subgrid-scale stresses respectively. We also define a turbulent Prandtl number $Pr_T$ in analogy with the standard Prandtl number.

Two subgrid-scale models that are gaining popularity include the dynamic model (Germano, Piomelli, Moin and Cabot 1990) and Lesieur's (1991) spectral viscosity model. Both models can be implemented in physical space. This requirement is necessary if LES concepts are to be useful in complicated geometrical configurations.

The objective of the dynamic model is to remove a major deficiency of the Smagorinsky model wherein the model constant is a function of the particular flow under study. It is evident that if the model constant must be recalibrated for a different class of flows, the usefulness of the model is compromised. The dynamic model assumes a prescribed functional form for the unresolved scales (generally a Smagorinsky model), and furthermore assumes that this form is independent of the filter width. This is approximately true if the cutoff range is in the inertial range.

(Without this latter hypothesis, LES simulations would generate grid-dependent results.) An algebraic identity derived by Germano (1990) relates the subgrid-scale stresses on two different coarse meshes. Substitution of the turbulence model into this identity provides the value of the "unknown" constant. It is shown by Germano (1990) that the model constant vanishes at solid boundaries, in laminar flow, and has the correct cubic dependence on the wall-normal coordinate near solid boundaries. In addition, the dynamic model allows backscatter (or inverse cascade) from the small to the larger scales in localized regions of the flow. The dynamic model has been extended to compressible homogeneous turbulence flows by Moin, Squires, Cabot, and Lee (1991). A summary of the compressible version of the model is given here. Its formulation closely follows the incompressible version found in Germano (1990).

Consider two filter operators $\overline{G}$ and $\hat{G}$. The former corresponds to the standard LES filter called the grid filter, while the latter is called the test filter. A composite filtering operator $\hat{\overline{G}} \equiv \overline{G}\hat{G}$ is also defined. Concentrating our attention on the averaged Reynolds stress $u_i u_j$ in the momentum equations, we obtain, after filtering the momentum equations with $\overline{G}$ and $\hat{\overline{G}}$, the unresolved stresses

$$\tau_{ij} = \overline{\rho}\,(\widetilde{u_i u_j} - \tilde{u}_i \tilde{u}_j) \tag{17}$$

and

$$T_{ij} = \hat{\overline{\rho}}\,(\boxed{u_i u_j} - \boxed{u_i}\,\boxed{u_j}). \tag{18}$$

Here, we have defined the Favré average on the two grids by

$$\tilde{u}_i = \frac{\overline{\rho u_i}}{\overline{\rho}} \tag{19}$$

and

$$\boxed{u_i} = \frac{\widehat{\rho u_i}}{\hat{\rho}}. \tag{20}$$

Following Germano et al. (1990), it is easy to derive the identity

$$L_{ij} = T_{ij} - \tilde{\tau}_{ij}, \tag{21}$$

where $L_{ij}$ has the interpretation of resolved turbulent stresses, i.e., those stresses whose scale lies between $\overline{\Delta}$ and $\hat{\overline{\Delta}}$.

If one chooses a compressible Smagorinsky model to approximate the tensors $T_{ij}$ and $\tau_{ij}$, one can derive a Smagorinsky constant similar to that found in Germano et al. (1990). Various types of tensor contractions allow for the separate modeling of the anisotropic and the isotropic components of the Reynolds stress without overly complicating the model.

Whether the properties of the dynamic model (established for simple in-compressible flows) hold for different types of compressible flows is one of many questions under investigation. Extensions of the dynamic model to handle the velocity–temperature correlations parallels the model for the subgrid Reynolds stresses.

The spectral viscosity model, developed by Lesieur (1991) has its origin in the eddy-damped quasi-normal Markovian (EDQNM) approximation to turbulence. Its original form is formulated in spectral space. A wave number $k_C$ is defined beyond which the turbulence is considered unresolved. Scales smaller than $k_C^{-1}$ are the subgrid scales. Subgrid-scale transfers of energy corresponding to wave triads with at least one wave number greater than the cutoff frequency $k_C$ are computed with the help of EDQNM. This in turns leads to an approximate model for the subgrid-scale transfers which can be incorporated into the evolution equations written in spectral space. The success of Lesieur's model rests on the ability of EDQNM to predict the correct energy scaling laws for various turbulent variables. The extension of EDQNM (Marion 1988) to compressible flows leads to the possibility of extending Lesieur's model to compressibility effects.

We reemphasize that to be useful in practical configurations, subgrid-scale models should be formulated in physical space. Both the dynamic grid model and Lesieur's model can be so formulated.

### 12.3.2 Results

Before performing the LES, *a priori* estimates of the model constants are obtained by an analysis of the DNS data. To this end, the direct simulation data are first injected onto a coarser grid ($16^3$ or $32^3$) and filtered. This reduced set of data corresponds to the results of an LES with an ideal subgrid-scale model. Under the assumption that our model is ideal, we effectively duplicate an LES without actually specifying the model. Using the coarse data, the various components of the subgrid-scale stresses are computed via Eq. (15). At the same time, the associated nonmodeled terms are computed on the fine DNS grid, followed by both coarsening and filtering. Model constants are then determined by comparing the coarsened exact stresses with the modeled stresses. This is accomplished in several ways, including least squares fits and comparisons of rms values of the different stress components. The analysis is performed on the tensor, vector and scalar levels. Full details are available in Erlebacher et al. (1992). Among the methods used to calculate the model constants, we find the most reliable to be the least squares fit of the full stress terms on the vector level. This means that the three components of $\tau_{ij,j}$ are compared individually. They are computed as a linear combination of the three individual subgrid-scale stresses with unknown coefficients. The least squares fit then allows the three constants to be determined. Of course, the cross-stress constant, $C_C$, must be unity for Galilean invariance of the model. Therefore, a procedure that determines all three constants was considered only valid if $C_C \approx 1$. The analysis produced results and correlations fully consistent with the results of Bardina et al. (1983). This is not surprising since the specified initial conditions led to turbulence, which remained quasi-incompressible for several large-eddy turnover times. A description of the initial conditions which lead to such situations is found in Erlebacher et al. (1990a).

Due to the low microscale Reynolds numbers of the DNS, there is no observable inertial range. As a consequence, the energy cascades directly from the large scales to the dissipation range. Since subgrid-scale models are expected to work best when the rate of energy flow is a constant, independent of the cutoff length scale, there is no reason to believe that the model constants will be independent of the LES grid resolution. This is borne out when comparing results of *a priori* analyses on $16^3$ and $32^3$ grids. The results are strongly grid dependent. However, we nonetheless felt sure that the sensitivity would not be as severe in a real LES. We chose the constants obtained with the $16^3$ grid because the Smagorinsky constant $C_R$ was closer to the value predicted by Bardina et al. (1983). Unless specified otherwise, the constants used in the LES simulations are $C_C = 1$, $C_R = 0.012$, $C_I = 0.066$. The turbulent Prandtl number is taken to be 0.7.

The subgrid-scale model given by Eqs. (15) was tested by Erlebacher et al. (1990b) and Zang, Dahlburg and Dahlburg (1991). Erlebacher et al. (1992b) compared the results of $96^3$ DNS of isotropic decaying compressible turbulence with the results of LES on $32^3$ grids. Both simulations had the following initial conditions: $Re = 250$, $\langle M_t^2 \rangle_0^{1/2} = 0.1$, $(\rho_{rms})_0 = 0$, $(T_{rms})_0 = 0.0626$ and a compressibility ratio $\chi_0 = 0.2$. Interestingly, this choice of parameters leads to an increase of $\chi$. In general, however, $\chi$ is a decreasing function of time. The reason for the anomaly is not currently known and awaits further investigation.

The initial conditions for the LES were spectrally interpolated from the DNS results after the first peak in the time history of total enstrophy. Simulations were performed with a variety of subgrid-scale widths to confirm that a width $\Delta_f = 2$ gives the best results. We also varied the various model constants to study the sensitivity of the results with respect to the magnitude of the various modeled terms. We found that there is a fair degree of sensitivity. This confirms the need for subgrid-scale models more sophisticated than the models which implicitly result from the truncation errors of low order numerical schemes. One such model is the dynamic model mentioned earlier.

The biggest surprise was the insensitivity of the results to variations of 50% of $C_I$. We have therefore omitted this term from our model on the grounds that it has no effect. Zang et al. (1991), on the other hand, multiply $C_I$ by 50, which results in a better match between the time histories of the rms density and the irrotational kinetic energy based on LES and DNS data. The initial conditions for this comparison were $\chi = 0.2$, $M_t = 0.1$. Unfortunately this improvement was not uniform when tried on simulation data based on runs with a range of initial conditions. Increasing $C_I$ sometimes led to worse results than simply setting it to zero. Moin et al. (1991) have found at least one case where *a priori* testing suggests that $C_I$ is indeed negligible, but DNS suggests otherwise. More investigation is clearly required. Therefore, it is still not clear whether this term should be part of the model. We do know, however, from the work of Sarkar et al. (1991b) that, within the framework of Reynolds stress modeling, the dilatational terms play an important role. It should therefore not come as a surprise if an extra term is required to model the isotropic components of the subgrid-scale Reynolds stress. More work is clearly in order to establish leading order corrections to the current model.

Results show that the subgrid-scale model is capable of correctly predicting the decay rates of kinetic energy (both compressible and solenoidal) and of the thermodynamic variables $(\rho, T, p)$. For illustration, Fig. 7 shows time histories of several turbulent quantities obtained from the LES with a filter width $\Delta_f = 2$. In all the graphs, the symbols refer to the DNS results (coarsened and filtered) at selected times, while the solid lines refer to the results of the LES. Figure 7a shows the good agreement of the decay rate of the incompressible and compressible total kinetic energy $E^I$ and $E^C$. The agreement for $E^I$ is not totally unexpected, since to leading order, the evolution of the solenoidal velocity and that of the irrotational velocity are decoupled for small $M_t$. However, it is not clear why the agreement for $E^C$ is so good since we are not specifically modeling compressibility, aside from mean density effects. If the general time evolution of terms directly related to com-

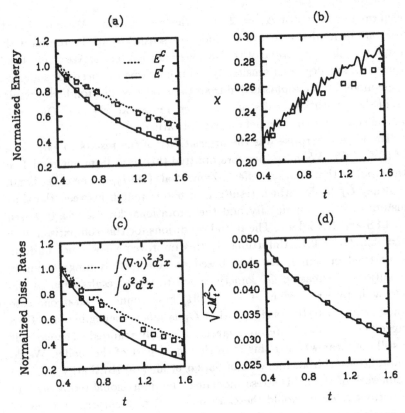

Fig. 7  Time histories of LES results on $32^3$ grid with $\Delta_f = 2$. (a) Solenoidal
and irrotational kinetic energy, (b) $\chi$, (c) $< (\nabla \cdot \mathbf{u})^2 >$ and $< \omega^2 >$,
(d) $M_t$.

pressibility is strongly linked to that of the incompressible turbulence,
the developement of subgrid-scale models should be greatly facilitated.

Large eddy simulations of compressible homogeneous shear flows
should help us find out if this good fortune is linked to the isotropic
nature of the flow or if it is more universal in character. A better measure of the model performance is given by the time history of $\chi$ (Fig.
7b). It is clear that the compressibility ratio is overpredicted. Therefore,
the decay rate of $E^C$ predicted by the LES is too low. In Fig. 7c, the
decay rates of the variance of dilatation and vorticity are shown. Again,
there is good agreement between DNS and LES. Good agreement of
the decay rates of kinetic energy (large scales) and dissipation energies
(small scales) is generally a good sign. Finally, time histories of the rms

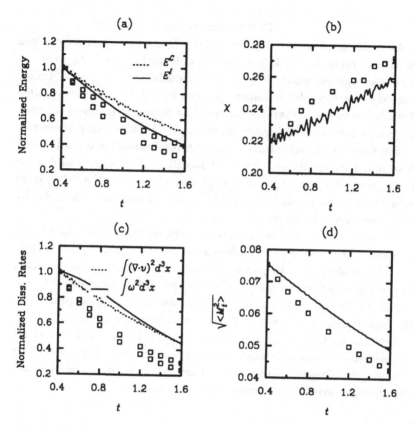

Fig. 8   Time histories of LES results on $32^3$ grid with $\Delta_f = 0$. (a) Solenoidal and irrotational kinetic energy, (b) $\chi$, (c) $< (\nabla \cdot \mathbf{u})^2 >$ and $< \omega^2 >$, (d) $M_t$.

of density, temperature and pressure are shown in Fig. 7d. Considering the (almost) arbitrary choice of $Pr_T$, the agreement is quite good. Tests were conducted with different filter widths and different model constant values. For the particular LES under study, the $C_R$ was optimal.

The effect of the model is most clearly demonstrated by performing a DNS on a $32^3$ grid with the same initial conditions. This is equivalent to a LES with a filter width $\Delta_f = 0$. Results are shown in Fig. 8 which exactly mirror the results presented in Fig. 7. Without the use of a filter, the mismatch between the DNS results (symbols) and the LES predictions (solid and dotted curves) is quite severe. Thus the agreement between LES and DNS observed in Fig. 7 is clearly the result of the proposed subgrid-scale model.

## 12.4 Conclusions

As emphasized in this brief overview, direct numerical simulations of several types of shock-free compressible turbulence have provided databases against which subgrid-scale models can be tested. Unfortunately, the Reynolds numbers that can be achieved in compressible flows are even lower than the already low Reynolds numbers inherent in today's best incompressible simulations. While we now know how to generate compressible turbulence with incompressible statistics, real flows may have significant compressibility effects (i.e., compressible mixing layers). Because of the low Reynolds numbers in the DNS, good subgrid-scale models at low Reynolds numbers may not automatically carry over to the larger ones.

A couple of LES models have been developed recently to address this problem. They make use of information on the cascade process from large to small scales. In particular, the dynamic subgrid-scale model has no adjustable constants (they are computed from the resolved component of the flow). By its very construction, it holds great promise at the higher Reynolds numbers of practical engineering interest.

Research on LES is still in its infancy, and there are many obstacles to overcome. Within the next decade, the first large eddy simulations of flow over a three-dimensional wing are expected. A major issue will then become quantitative verification of the results.

## References

BARDINA, J., FERZIGER, J.H., AND REYNOLDS, W.C. (1983) Improved turbulence models based on large-eddy simulation of homogeneous, incompressible turbulent flows. Stanford University Technical Report TF-19.

BATCHELOR, G.K. (1953) *The Theory of Homogeneous Turbulence*. Cambridge University Press.

BLAISDELL, G. (1990) Numerical simulation of compressible homogeneous turbulence. Ph.D Thesis, Stanford University.

CANUTO, C., HUSSAINI, M.Y., QUARTERONI, A. AND ZANG, T.A. (1988) *Spectral Methods in Fluid Dynamics*. Springer-Verlag.

CEBECI, T. AND SMITH, A.M.O. (1974) *Analysis of Turbulent Boundary Layers*. Academic Press.

ERLEBACHER, G, HUSSAINI, M.Y., KREISS, H.O. AND SARKAR, S. (1990a)

The Analysis and Simulation of Compressible Turbulence. *Theor. Comput. Fluid. Dynam.* **2**, 73–95.

ERLEBACHER, G., HUSSAINI, M.Y., SPEZIALE, C.G. AND ZANG, T.A. (1990b) On the large-eddy simulation of compressible turbulence. In proc. of the 12th Int. Conf. of Numerical Methods in Fluid Dynamics, Oxford University.

ERLEBACHER, G., HUSSAINI, M.Y., SPEZIALE, C.G AND ZANG, T.A. (1992) Toward the large-eddy simulation of compressible turbulent flows. *J. Fluid Mech.* **238**, pp. 155-185.

FEIEREISEN, W.J., REYNOLDS, W.C. AND FERZIGER, J.H. (1981) Numerical simulation of compressible, homogeneous, turbulent shear flow. Report TF-13, Stanford University, Dept. of Mechanical Engineering.

FERZIGER, J.H. (1984) Large eddy simulation: Its role in turbulence research. In *Theoretical Approaches to Turbulence.* Ed. D.L. Dwoyer, M.Y. Hussaini, and R.G. Voigt, Springer-Verlag.

FORNBERG, B. (1975) On a Fourier methods for the integration of hyperbolic equations. *Siam J. Numer. Anal.* **12** (4), 509-528.

GERMANO, M. (1990) Averaging invariance of the turbulent equations and similar subgrid-scale modeling. CTR Manuscript 116, Stanford University.

GERMANO, M., PIOMELLI, U., MOIN, P. AND CABOT, W.H. (1990) A dynamic subgrid-scale eddy viscosity model. In *Studying Turbulence Using Numerical Simulation Databases – III.* CTR Manuscript 124, Stanford University.

HUSSAINI, M.Y., SPEZIALE, C. AND ZANG, T.A. (1989) The potential and limitations of direct and large-eddy simulations, Comment 2. In *Wither Turbulence? Turbulence at the Crossroads.* Lecture Notes of Physics, vol. 357. Ed. J.L. Lumley, pp. 354-368.

KADOMTSEV, B.B. AND PETVIASHVILI, V.I. (1973) Acoustic turbulence. *Sov. Phys. Dokl.* **18**, 115-116.

KIDA, S. AND ORSZAG, S. (1990a) Energy and spectral dynamics in forced compressible furbulence. *J. Sci. Computing* **5**, No. 2, 85-125.

KIDA, S. AND ORSZAG, S. (1990b) Enstrophy budget in decaying compressible turbulence. *J. Sci. Computing* **5**, No. 1, 1-34.

KIDA, S. AND ORSZAG, S. (1992) Energy and spectral dynamics in decaying compressible turbulence. *J. Sci. Computing* **7**, No. 1, 1-34.

KOVASZNAY, L.S.G. (1957) Turbulence in supersonic flows. *J. Aero. Sciences* **20**, (10), 657-682.

LEE, S., LELE, S.K. AND MOIN, P. (1991) Eddy shocklets in decaying compressible turbulence. *Phys. Fluids* **3**, (4), 657–664.

LELE, S.K. (1990) Compact finite-difference schemes with spectral-like resolution. CTR Manuscript 107, Stanford University.

LESIEUR, M. (1991) *Turbulence in Fluids*. Kluwer.

L'VOV, V.S. AND MIKHAILOV, A.V. (1978a) Sound and hydrodynamic turbulence in a compressible liquid. *Sov. Phys. J. Exp. Theor. Phys.* **47**, 756–762.

L'VOV, V.S. AND MIKHAILOV, A.V. (1978b) Scattering and interaction of sound with sound in a turbulent medium. *Sov. Phys. J. Exp. Theor. Phys* **47**, 840–847.

MARION, J-D. (1988) Etude spectrale d'une turbulence isotrope compressible. Ph.D. thesis, Ecole Centrale de Lyon.

MOIN, P., SQUIRES, K., CABOT, W. AND LEE, S. (1991) A dynamic subgrid-model for compressible turbulence and scalar transport. CTR Manuscript 124, Stanford University.

MOISEEV, S.S., SAGDEEV, R.Z., TUR, A.V. AND YANOVSKII, V.V. (1977) Structure of acoustic-vortical turbulence. *Sov. Phys. Dokl.* **22**, 582–584.

MOYAL, J.E. (1952) The spectra of turbulence in a compressible fluid; Eddy turbulence and random noise. *Proc. Cambridge Phil. Soc.* **48**, part 1, 329–344.

PASSOT, A. AND POUQUET, A. (1987) Numerical simulation of compressible homogeneous flows in the turbulent regime. *J. Fluid Mech.* **181** 441–466.

REYNOLDS, W.C. (1989) The potential and limitations of direct and large-eddy simulations. In *Wither Turbulence? Turbulence at the Crossroads*. Lecture Notes of Physics, vol. 357. Ed. J.L. Lumley, pp. 313–343.

ROGALLO, S. (1981) Numerical experiments in homogeneous turbulence. NASA TM-81315.

SARKAR, S., ERLEBACHER, G. AND HUSSAINI, M.Y. (1991a) Direct simulation of compressible turbulence in a shear flow. *Theor. Comput. Fluid Dynam.* **2**, 291–305.

SARKAR, S., ERLEBACHER, G., HUSSAINI, M.Y. AND KREISS, H.O. (1991b) The analysis and modeling of dilatational terms in compressible turbulence. *J. Fluid Mech.* **227**, 473–494.

SHU, C-W, ERLEBACHER, G., ZANG, T.A., WHITAKER, D. AND OSHER, S. (1992) High order ENO schemes applied to two- and three-dimensional compressible Euler and Navier–Stokes equations. *J. Appl. Num. Math.* **9**, 45–71.

SPEZIALE, C.G. (1991) Analytical methods for the development of Reynold-stress closures in turbulence. *Ann. Rev. Fluid Mech.* **23**, 107–157.

SPEZIALE, C.G., ERLEBACHER, G., ZANG, T.A. AND HUSSAINI, M.Y. (1988) The subgrid scale modeling of compressible turbulence. *Phys. Fluids* **31**, 940–942.

TATSUMI, T. AND TOKUNAGA, H. (1974) One-dimensional shock-turbulence in a compressible fluid. *J. Fluid Mech.* **65**, 581.

TOKUNAGA, H. AND TATSUMI, T. (1975) Interaction of plane nonlinear waves in a compressible fluid and two-dimensional shock-turbulence. *J. Phys. Soc. Japan* **38**, 1167–1179.

YOSHIZAWA, A. (1986) Statistical theory for compressible turbulent shear flows, with the application to subgrid modeling. *Phys. Fluids* **29**, 2152–2164.

ZAKHAROV, V.E. AND SAGDEEV, R.Z. (1970) Spectrum of acoustic turbulence. *Sov. Phys. Dokl.* **15**, 439–441.

ZANG, T.A. (1991) On the rotation and skew-symmetric forms for incompressible flow simulations. *Appl. Numer. Mathematics* **7**, 27–41.

ZANG, T.A, DAHLBURG, R.B. AND DAHLBURG, J.P. (1991) Direct and large-eddy simulations of compressible, isotropic Navier–Stokes turbulence. NRL Memorandum Report 6799.

# 13

---

# Large Eddy Simulation of Mixing Layers

### SAAD A. RAGAB AND SHAW-CHING SHEEN

## 13.1 Introduction

The nonlinearity of the governing equations of fluid mechanics is a primary source of many of the difficulties (and excitements) that a fluid analyst faces. It is the nonlinearity that leads to the generation of turbulent motions and precludes the development of satisfactory methods for their prediction. Even in the presence of coherent structures, turbulent flows are random fields and comprise a disparity of temporal and spatial scales. This disparity of scales increases quite rapidly as a characteristic Reynolds number increases.

Starting with certain initial conditions, direct numerical simulation (DNS) resolves all scales of motion from the largest scales imposed by external effects (a pipe diameter or an airfoil chord) down to the smallest scales which are responsible for the dissipation of turbulence kinetic energy into thermal energy. Therefore, this approach is limited to low Reynolds numbers. Remarkable progress in the understanding of the physics and structure of incompressible turbulent flows has been achieved using DNS (Reynolds 1990). Direct numerical simulations have been successfully used in the analysis of isotropic turbulence (Orszag and Patterson, 1972), homogeneous shear flows (Rogers and Moin, 1987), fully developed channel flow (Kim et al., 1987), and zero-pressure-gradient boundary layer flow (Spalart, 1988). Considering the projected increase in the computing power in the next two decades, we may not be too optimistic to hope for an order of magnitude increase in the Reynolds number of such simulations. Applications of DNS to compressible flows have been limited to isotropic and homogeneous turbulence (Passot and Pouquet, 1987, 1988; Blaisdell et al., 1988; Erlebacher et al., 1990).

Direct numerical simulations are not meant to be an engineering analysis or a design tool. The design engineer has to depend on predictive techniques that provide information about low-order statsticial correlations of turbulent flows. Therefore, turbulence modeling (or the closure problem, which is a result of the nonlinearity of the Navier–Stokes equations) remains the fundamental issue for predictive methods of turbulent flows. Notwithstanding the low Reynolds number limitations, DNS will continue to provide insight and guidance for the development of turbulence models.

In large eddy simulation (LES), scales imposed by external boundaries or immersed bodies are numerically resolved in time and space. The computational grid spacings should be fine enough that some scales in the inertial range of the spectrum are resolvable. A turbulence model is used to model the effect of subgrid-scale (SGS) motions in terms of the resolvable part of the flow. LES is envisaged not only as a research tool for the understanding of the physics of turbulence and its structure but also as a predictive method for flows of technological interest. Thus far LES has been applied mostly to incompressible flows and simple geometries. Extensions to complex geometries of technological interest and to compressible turbulence are needed.

### 13.1.1 Filtering

Filtering is a process by which scales smaller than a certain cutoff value are eliminated from the total flow, and hence it defines the resolvable part of the flow. In turbulent channel flows, Piomelli et al. (1988) found strong dependence of the structure of the SGS field on the type of filter used. Because filters define the resolvable scales, a consistency between the filter and the SGS model should be observed.

### 13.1.2 SGS Models

The premise of LES is that only small scales, which tend to be isotropic and hence more universal in nature, need to be modeled. Furthermore, the small scales carry a small portion of the total turbulent energy, and thus one anticipates that the SGS models are less complicated than those required for the Reynolds-averaged equations.

Depending on the type of filter used (Leonard, 1974), the filtered Navier–Stokes equations may contain three terms that need to be modeled in terms of the resolved fields. The three terms are referred to as the Leonard stress, cross-stress, and Reynolds stress terms. The Leonard term needs no modeling becuase it can be computed directly by filtering products of the resolved fields (Ferziger, 1977). The majority of LES results obtained to date used an eddy viscosity model for the remaining terms. An improved model referred to as the linear combination model has been developed and tested by Bardina et al. (1983) (see also Speziale, 1985). The eddy viscosity coefficient is usually determined from the Smagorinsky model (Smagorinsky, 1963). Erlebacher et al. (1990) and Speziale et al. (1988) presented a generalization of the linear combination model to compressible flows.

Sophisticated SGS models may be necessary if LES is to be performed on a coarse mesh. In this case, the vast literature on isotropic and homogeneous turbulence will prove very useful in developing such models. Recently, efforts have been undertaken by Yoshizawa (1986, 1990) and Hartke et al. (1988) to extend the direct interaction approximation of Kraichnan (Leslie, 1973) to compressible turbulence. Another promising approach of developing SGS models is the renormalization group theory of Yakhot et al. (1989). SGS models based on these theories need to be implemented and tested for complex turbulent flows.

A new SGS eddy viscosity model has been developed by Germano et al. (1991). In this model the Smagorinsky constant is computed dynamically as the calculation progresses rater than input *a priori*. It performs very well in transitional flow and near-wall regions. It has been extended to compressible flows by Moin et al. (1991).

### 13.1.3 Numerical Methods

Both spectral and finite-difference methods have been applied to LES in simple domains. For practical engineering problems, there is a need to adapt spectral-element methods (Korczak and Patera, 1986) and high-order (fourth- and sixth-order) finite-difference methods (Gottlieb and Turkel, 1976; Lele 1989; Tang et al., 1989a) to general curvilinear coordinates. Furthermore, finite-difference methods of low order (e.g., second order) should also be evaluated because most engineering fluid dynamics computations have been performed using these methods, and it would be advantageous to the engineering community if LES could be performed using second-order finite-difference or finite-volume methods. However, one should be careful with low-order methods because of their built-in low-pass filter, which may confuse the issue of resolved scales versus SGS motions.

A more serious problem in developing numerical methods for compressible LES is the interaction of shock waves with turbulence and shock-generated turbulence. Resolving the internal viscous structures of shock waves is not feasible with LES because such simulations are intended for high Reynolds number flows. A shock wave must be treated and captured as a surface of discontinuity. Spectral and high-order finite-difference methods cannot be used without modifications across such a discontinuity; they produce hig-frequency oscillations and thus contaminate the turbulence field. On the other hand, first-order methods are nonoscillatory, but they severely damp high frequencies; therefore, genuine turbulence generated by interactions of shock waves would be completely lost. The strategy followed in the development of high-resolution nonoscillatory schemes (TVD, total variation diminishing) should be exploited to supplement high-order finite-difference schemes or spectral methods (e.g., Hirsch, 1990; Swanson and Turkel, 1990).

As a final note on numerical methods, we emphasize the importance of considering schemes that take advantage of parallel machines such as the connection machine because the projected increase in computing speed (the teraflop range) will be achieved on such machines. Explicit methods can run very efficiently on these machines.

### 13.1.4 Boundary and Initial Conditions

Far-field boundary conditions have to be treated very carefully so that spurious pressure reflections do not excite the internal turbulent flow. A demonstration of the seriousness of such reflections in the development of supersonic mixing layers is given by Ragab and Sheen (1990). They found that consistency of the simulation results with linear stability predictions is necessary but not sufficient for the correctness of the results at large times or distances. Inflow boundary conditions are important for spatial simulations of developing shear layers. Sandham and Reynolds (1987) demonstrated that some randomizing of the inflow conditions may be necessary if simulations representative of natural turbulent flows are to be obtained. Wall boundary conditions are simple to apply in LES; however, they introduce difficulties with SGS models if the computations are to be continued to the wall because the small scales are not isotropic near the wall.

Reynolds (1990) states that initial conditions are very important for homogeneous isotropic turbulence, where the decay history is determined by the initial state. Inhomogeneous flows, such as the channel flow, establish their own steady-state spectrum and hence the intial conditions are not important.

### 13.1.5 The Mixing Layer

In this chapter we are interested in the numerical simulations of mixing layers. The nonlinear development of instability waves and low Reynolds number DNS in supersonic mixing layers have been performed using finite-difference representations. Lele (1989) simulated a developing mixing layer using a compact finite-difference scheme with spectral-like accuracy. Time advancement was carried out by a compact-storage third-order Runge–Kutta scheme. Simulations were conducted for both spatially evolving and temporally evolving mixing layers. The Reynolds number based on the velocity difference across the layer and the initial vorticity thickness was in the range 100–500. Tang et al. (1989a,b) also presented direct simulation of two-dimensional (2D) spatially evolving mixing layers using a modified MacCormack scheme developed by Gottlieb and Turkel (1976) and used in boundary layer research by Bayliss et al. (1985, 1986). The scheme is second-order accurate in time and fourth-order accurate in space for the convective terms. A numerical investigation of 2D and 3D disturbances in unconfined temporal mixing layers is given by Sandham and Reynolds (1989). Their work is the most recent and most complete investigation of the subject.

Large eddy simulations of incompressible mixing layers have been obtained by Cain et al. (19891) using spectral methods and by Maruyama (1988) using a finite-difference method. The SGS model is the Smagorinsky eddy viscosity model. The effects of the initial field on the development of the mixing layer is thoroughly investigated. More recently, Leith (1990) formulated an SGS model that accounts for both dissipation and backscatter, and applied it to compressible mixing layers at low convective Mach numbers.

### 13.1.6 Objective

Our objective is to develop finite-difference methods for LES with emphasis on compressible turbulent shear flows. A preliminary investigation of a MacCormack-type scheme is presented in this chapter. The flow congfiguration is a low convective Mach number mixing layer.

## 13.2 Equations of Large Eddy Simulations

The compressible Navier–Stokes equations are

$$\frac{\partial \rho}{\partial t} + \frac{\partial (\rho u_j)}{\partial x_j} = 0, \tag{1}$$

$$\frac{\partial \rho u_i}{\partial t} + \frac{\partial (\rho u_i u_j + p\delta_{ij} - \tau_{ji})}{\partial x_j} = 0, \qquad (2)$$

$$\frac{\partial \rho E}{\partial t} + \frac{\partial \left[ (\rho E + p)u_j - u_i \tau_{ji} + q_j \right]}{\partial x_j} = 0, \qquad (3)$$

where $\rho, u_j$, and $p$ are the density, velocity vector, and pressure, respectively.

The total specific energy is given by

$$E = e + \tfrac{1}{2} u_j u_j, \qquad (4)$$

and the viscous stress tensor $\tau_{ij}$ and heat flux vector $q_j$ are given by

$$\tau_{ij} = \lambda \frac{\partial u_k}{\partial x_k} \delta_{ij} + \mu \left[ \frac{\partial u_i}{\partial x_j} + \frac{\partial u_j}{\partial x_i} \right] \qquad (5)$$

and

$$q_j = -\kappa \frac{\partial T}{\partial x_j}, \qquad (6)$$

respectively. A perfect gas with constant specific heat capacities, $C_v$ and $C_p$, is assumed; hence the equations of state are

$$e = C_v T \qquad (7)$$

and

$$p = \rho R T. \qquad (8)$$

The resolvable flow field is defined by

$$\overline{f}(x,t) = \int_D G(x - \xi) f(x,t) \, d\xi, \qquad (9)$$

where $D$ is the flow domain and $G$ is a filter function that satisfies the normalization condition

$$\int_D G(x - \xi) \, d\xi = 1. \qquad (10)$$

It is assumed that the function $G$ is such that filtering and differentiation with respect to space and time commute, that is,

$$\frac{\overline{\partial f}}{\partial t} = \frac{\partial \overline{f}}{\partial t} \qquad (11)$$

and

$$\frac{\overline{\partial f}}{\partial x_j} = \frac{\partial \overline{f}}{\partial x_j}. \qquad (12)$$

Following Erlebacher et al. (1990), we introduce the Favre-filtered field,

$$\tilde{f} = \frac{\overline{\rho f}}{\overline{\rho}}, \qquad (13)$$

and decompose the total flow field into a resolvable field and an SGS

field,
$$f = \tilde{f} + f'. \tag{14}$$
Note that, in general,
$$\tilde{\tilde{f}} \neq \tilde{f} \tag{15}$$
and
$$\tilde{f}' \neq 0. \tag{16}$$
Filtering the mass and momentum equations, we obtain
$$\frac{\partial \bar{\rho}}{\partial t} + \frac{\partial (\bar{\rho} \tilde{u}_j)}{\partial x_j} = 0, \tag{17}$$
$$\frac{\partial \bar{\rho} \tilde{u}_i}{\partial t} + \frac{\partial (\bar{\rho} \tilde{u}_i \tilde{u}_j + \bar{p} \delta_{ij} + R_{ij} + L_{ij} + C_{ij} - \bar{\tau}_{ji})}{\partial x_j} = 0, \tag{18}$$
where
$$R_{ij} = \rho \widetilde{u'_i u'_j}, \tag{19}$$
$$L_{ij} = \bar{\rho} (\widetilde{\tilde{u}_i \tilde{u}_j} - \tilde{u}_i \tilde{u}_j), \tag{20}$$
and
$$C_{ij} = \bar{\rho} (\widetilde{\tilde{u}_i u'_j} + \widetilde{\tilde{u}_j u'_i}) \tag{21}$$
are referred to as the Reynolds stress, Leonard stress, and cross-stress terms. The filtered pressure is given by
$$\bar{p} = \bar{\rho} R \tilde{T}. \tag{22}$$
Filtering the total energy equation, we obtain
$$\frac{\partial (\bar{\rho} \tilde{E} + q)}{\partial t} + \frac{\partial \left[ (\bar{\rho} \tilde{E} + \bar{p}) \tilde{u}_j + K_j + Q_j - \overline{u_i \tau_{ji}} + \overline{q_j} \right]}{\partial x_j} = 0, \tag{23}$$
where
$$\tilde{E} = \tilde{e} + \tfrac{1}{2} \tilde{u}_i \tilde{u}_i, \tag{24}$$
$$q = \tfrac{1}{2}(R_{kk} + L_{kk} + C_{kk}), \tag{25}$$
$$K_j = \tfrac{1}{2}\bar{\rho}(\widetilde{\tilde{u}_i \tilde{u}_i \tilde{u}_j} - \tilde{u}_i \tilde{u}_i \tilde{u}_j + \widetilde{u'_i u'_i \tilde{u}_j} + \widetilde{\tilde{u}_i \tilde{u}_i u'_j} + \widetilde{u'_i u'_i u'_j} + 2\widetilde{u'_i \tilde{u}_i \tilde{u}_j} + 2\widetilde{u'_i \tilde{u}_i u'_j}), \tag{26}$$
and
$$Q_j = C_p \bar{\rho} (\widetilde{T' u'_j} + \widetilde{\tilde{T} \tilde{u}_j} - \tilde{T} \tilde{u}_j + \widetilde{T' \tilde{u}_j}). \tag{27}$$

## 13.3 An SGS Model

Here, we summarize the SGS model of Erlebacher et al. (1990) (see also Speziale et al., 1988). This model is an extension to compressible flows of a linear combination model developed by Bardina et al. (1983) for incompressible flows. The Leonard stress term needs no modeling; it can be computed directly once the filter function $G$ is specified, while the cross-stress term is modeled by a scale similarity model,

$$C_{ij} = \overline{\rho}(\tilde{u}_i \tilde{u}_j - \tilde{\tilde{u}}_i \tilde{\tilde{u}}_j). \tag{28}$$

Therefore, the sum of the two terms can be written as

$$L_{ij} + C_{ij} = \overline{\rho}(\widetilde{\tilde{u}_i \tilde{u}_j} - \tilde{\tilde{u}}_i \tilde{\tilde{u}}_j). \tag{29}$$

The Reynolds stress term is modeled by an eddy viscosity model in which the eddy viscosity coefficient is given by the generalized Smagorinsky formula,

$$R_{ij} = -2C_R \Delta^2 \overline{\rho} \sqrt{\Pi} (\widetilde{S}_{ij} - \tfrac{1}{3} \widetilde{S}_{kk} \delta_{ij}) + \tfrac{2}{3} C_I \Delta^2 \overline{\rho} \Pi \delta_{ij}, \tag{30}$$

where

$$\widetilde{S}_{ij} = \frac{1}{2} \left( \frac{\partial \tilde{u}_i}{\partial x_j} + \frac{\partial \tilde{u}_j}{\partial x_i} \right) \tag{31}$$

is the rate of strain tensor of the resolvable field, and

$$\Pi = \widetilde{S}_{ij} \widetilde{S}_{ij}. \tag{32}$$

In Eq. (30), $\Delta$ is the filter width.

The temperature–velocity correlation terms are modeled by a gradient-transport model,

$$\widetilde{T' u'_j} = -C_T \Delta^2 \sqrt{\Pi} \frac{\partial \tilde{T}}{\partial x_j}, \tag{33}$$

and the cross term is modeled by a scale similarity mode,

$$\widetilde{T' \tilde{u}_j} + \widetilde{\tilde{T} u'_j} = \widetilde{\tilde{T} \tilde{u}} - \tilde{\tilde{T}} \tilde{\tilde{u}}_j. \tag{34}$$

Substituting Eqs. (33) and (34) into Eq. (27), we obtian

$$Q_j = C_p \overline{\rho} \left( -C_T \Delta^2 \sqrt{\Pi} \frac{\partial \tilde{T}}{\partial x_j} + \widetilde{\tilde{T} \tilde{u}_j} - \tilde{\tilde{T}} \tilde{\tilde{u}}_j \right), \tag{35}$$

where the constant $C_T$ is

$$C_T = \frac{C_R}{Pr_T} \tag{36}$$

and $Pr_T$ is a turbulent Prandtl number.

We model the third and last term in Eq. (26) as

$$\overline{\rho} \widetilde{u'_i \tilde{u}_i u'_j} \approx \overline{\rho} \widetilde{u'_i u'_j} \tilde{u}_i = R_{ij} \tilde{u}_i, \tag{37}$$

where $R_{ij}$ is given by Eq. (30). We note that the first two terms of Eq. (26) can be computed directly, whereas a scale similarity model may be used for the remaining three terms.

In the present preliminary investigation, we neglected all the terms in Eq. (26) except the last term, for which we used Eq. (37). Also, we neglected the $q$ term in Eq. (23) and the Leonard and cross terms. The effect of these terms will be investigated in a forthcoming paper using a box filter in the physical domain (Ragab et al. 1992). Because free shear layers are treated in this chapter, the viscous and the heat conduction terms are also neglected.

The parameter $\Delta^2$ in our calculations is defined by

$$\Delta^2 = \left[ (\Delta x)^2 + (\Delta y)^2 + (\Delta)^2 \right] / 3.$$

The numerical values of $C_R$ and $C_I$ depend on how $\Delta^2$ is defined. In the work of Erlebacher et al. (1990), $\Delta^2$ is defined by $4\delta^2$, where $\delta$ is the step size. In this chapter we use $C_R = 0.05$ and $C_I = 0.01$. The value of $C_I$ may not be very important.

## 13.4 The Numerical Method

The numerical scheme is a modified MacCormack scheme developed by Gottlieb and Turkel (1976) and used in boundary layer research by Bayliss et al. (1985, 1986) and Lesieur et al. (1991), and in supersonic mixing layers by Tang et al. (1989a,b).

The scheme is explicit second-order accurate in time and fourth-order in space. For the 1D equation $q_t + f_x = 0$, the predictor step is

$$\overline{q_j} = q_j^n - \frac{\Delta t}{6\Delta x} \left[ 7(f_{j+1} - f_j) - (f_{j+2} - f_{j+1}) \right], \tag{38}$$

and the corrector step is

$$q_j^{n+1} = \tfrac{1}{2}(q_j^n + \overline{q}_j) - \frac{\Delta t}{12\Delta x} \left[ 7(\overline{f}_j - \overline{f}_{j-1}) - (\overline{f}_{j-1} - \overline{f}_{j-2}) \right]. \tag{39}$$

In operator form the two steps (38) and (39) can be written as

$$q^{n+1} = L_x^{FB} q^n. \tag{40}$$

The $FB$ superscript refers to forward predictor (38) and backward corrector (39). There is an alternative version which employs a forward predictor and a backward corrector. We denote such a scheme by $L_x^{BF}$.

For multidimensional equations, an operator splitting technique (Strang, 1968) is used. Discretization in three dimensions is achieved by using 1D operators in a symmetrical fashion. Thus

$$q^{n+4} = L_x^- L_y^- L_z^- L_z^+ L_y^+ L_x^+ q^n, \tag{41}$$

where

$$L_x^- = L_x^{BF} L_x^{FB}, \tag{42}$$

$$L_x^+ = L_x^{FB} L_x^{BF}, \tag{43}$$

with similar definitions for the $y$ and $z$ operators.

Assuming a harmonic solution of wavelength $\lambda = 2\pi/\beta$, we obtain the amplification factor for the linear equation $u_t + cu_x = 0$ as

$$g = 1 - \frac{i\nu(-\sin 2\beta + 8 \sin \beta)}{6}$$
$$+ \frac{\nu^2(-7 \cos 2\beta + 64 \cos \beta - 57)}{36}, \tag{44}$$

where $\nu = c\Delta t/\Delta x$ is the Courant (CFL) number. The scheme is stable if $\nu < \frac{2}{3}$. The magnitude of $g$ and its relative phase are shown in Fig. 1 as functions of $2\Delta x/\lambda$ for different values of the CFL number. The results show that waves whose wavelengths are shorter than $6\Delta x$ are very much damped by the artificial viscosity inherent in the numerical scheme. Therefore, about two-thirds of the spectrum in the high-wavenumber range will be influenced by the artificial damping to some degree.

## 13.5 Two-Dimensional Simulations

Temporal simulations are obtained for compressible mixing layers at a convective Mach number of 0.4 and density ratio of 1. The initial mean velocity profile is assumed to be a hyperbolic tangent profile, $u = U_0 \tanh(y/\delta)$, where $U_0$ and $\delta$ are used as reference velocity and length throughout the simulations. The mean temperature profile is obtained from the Busemann–Crocco energy equation, and the mean pressure is assumed to be uniform. Disturbance fields are then superimposed on these mean profiles. The velocity disturbance field is a white-noise divergence-free field. No density or pressure disturbances are used.

According to the linear stability theory, the most amplified 2D wave has a wave number of 0.4, and hence the corresponding wavelength is $\lambda = 5\pi$. To be able to predict the pairing phenomenon, we choose a computational box that extends $4\lambda$ in the streamwise direction, and over the range $-30 \le y \le 30$ in the transverse direction. A uniform grid of (257,301) points is used. Periodic boundary conditions are imposed in the $x$-direction, and zero-derivative boundary conditions are imposed in the $y$-direction.

A Fourier analysis in the $x$-direction is performed at selected time steps, and the modal energy,

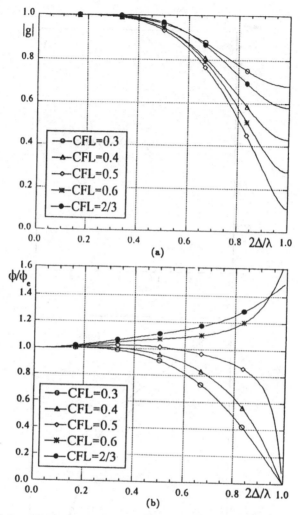

Fig. 1 (a) Amplification factor of the fourth-order MacCormack scheme for the model equation $u_t + cu_x = 0$. CFL $= c\Delta t/\Delta x$. (b) Phase error of the fourth-order MacCormack scheme for the model equation $u_t + cu_x = 0$. CFL $= c\Delta t/\Delta x$.

$$E_k(t) = \int_{ymin}^{ymax} (\hat{u}\hat{u}^* + \hat{v}\hat{v}^*)_k \, dy, \qquad (45)$$

is computed, where $k$ is the wave number.

The initial spectrum of $E_k$ is shown in Fig. 2a; a white-noise spectrum is evident. As time progresses, the most amplified wave (m = 4) extracts energy from the mean flow and dominates the large-scale structure of the mixing layer. This is clearly evident in Fig. 3a, which shows vor-

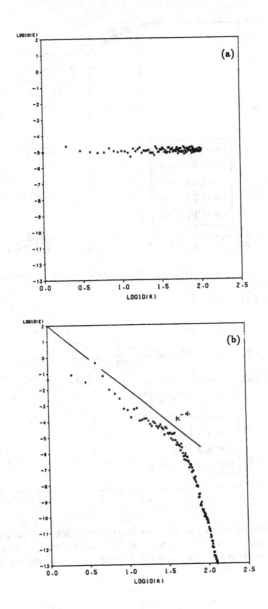

Fig. 2   One-dimensional kinetic energy spectrum of a two-dimensional mixing layer [Eq.(45)]. Convective Mach = 0.4, temperature ratio = 1.0. (a) Time = 0, (b) time = $51.2(\delta/U_0)$, (c) time = $83.2(\delta/U_0)$,(d) time = $128(\delta/U_0)$.

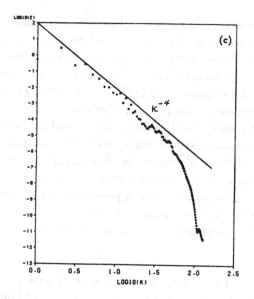

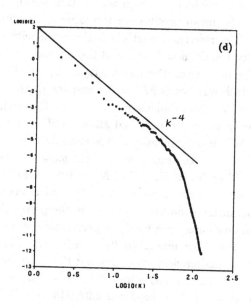

Fig. 2 (continued)

ticity contours at time $t = 51.2(\delta/U_0)$. The corresponding modal energy spectrum is shown in Fig. 2b. As this mode saturates, the first subharmonic mode ($m = 2$) grows and becomes the most energetic mode. This mode is responsible for the first pairing that is shown by the vorticity contours at time $t = 83.2(\delta/U_0)$ in Fig. 3b. The energy spectrum at this time is shown in Fig. 2c, which shows a peak at $m = 2$. At a much later time, $t = 128(\delta/U_0)$, the second subharmonic ($m = 1$) becomes the most energetic mode, and a second pairing appears in the vorticity contours as shown in Fig. 3c and Fig. 2d of the modal energy. We also note that the spectra shown in Figs. 2b,c,d suggest that the energy-containing eddies follow a power law of exponent of ($-4$). Such a power law has been reported by Lesieur et al. (1988) in their 2D large eddy simulations of an incompressible mixing layer.

## 13.6 Three-Dimensional Simulations
### *13.6.1 Computational Domain and Initial Conditions*

The temporal problem is treated here so that adequate resolution can be realized. The mean profiles are the same as the profiles of the 2D simulations. The reference length is half the vorticity thickness of the initial velocity profile, $\delta = \delta_{w0}/2$, and the reference velocity is half the velocity difference across the mixing layer, $U_0 = (U_1 - U_2)/2$. The convective Mach number is $M^+ = 0.4$ and the density ratio is 1. The computational box is of dimensions $L_x = L_z = 2(2\pi/\alpha)$, where $\alpha$ is the wave number of the most amplified 2D wave of the hyperbolic tangent profile. At $M^+ = 0.4$, linear stability analysis gives $\alpha = 0.4$. In the transverse direction, the box extends from $y_{min} = -24$ to $y_{max} = 24$. A uniform grid of $N_x = N_z = 73$, $N_y = 129$ points is used. Periodic boundary conditions are imposed in the $x$- and $z$-directions, and zero-derivative boundary conditions are used in the $y$-direction.

The initial fields are given by the superposition of the hyperbolic tangent profile, the three modes $(\alpha,0),(\alpha,\pm\beta)$, and $(\alpha/2,\pm\beta)$ of linear analysis, and a random field. We note that the 2D subharmonic mode is not excited by the initial conditions. The rms of the stream-wise velocity component of the three modes is $0.02, 0.02$, and $0.005$, respectively, and their phases are $\pi/2$, $\pi$, and $0$, respectively. We will refer to these modes as the (2,0), (2,2), and (1,2) modes. The mode numbers $(m,n)$ are related to the wave numbers $(k_x,k_z)$ by $k_x = 0.2m$ and $k_z = 0.2n$. The velocity field of the random disturbance is divergence free when

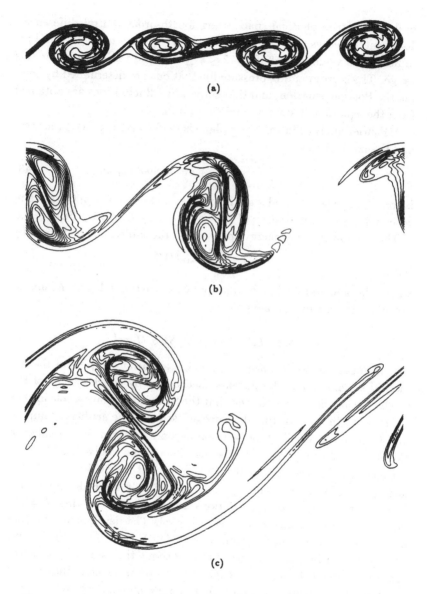

Fig. 3   Vorticity contours of two-dimensional simulations. Convective
Mach = 0.4, temperature ratio = 1. (a) Time = $51.2(\delta/U_0)$, (b)
time $83.2(\delta/U_0)$, (c) time = $128(\delta/U_0)$.

evaluated in the physical space using fourth-order central differences. Its strength decays as $e^{-(y/4)^2}$. Its 3D energy spectrum function $E(k)$ is uniform at a value of $10^{-6}$ if $0.412 \leq k \leq 7.09$ and vanishes outside this range. The corresponding pressure fluctuations are determined by solving the Poisson equation, and the temperature fluctuations are obtained from the equation of state assuming uniform density.

A Fourier analysis in the $x - z$ plane is performed at every time step, and the modal kinetic energy,

$$E_{mn}(t) = \int_{ymin}^{ymax} (\hat{u}\hat{u}^* + \hat{v}\hat{v}^* + \hat{w}\hat{w}^*)_{mn} \, dy, \qquad (46)$$

is computed, where $m$ and $n$ are respective mode numbers in the streamwise and spanwise directions of the Favre-averaged velocity field.

The one-dimensional energy spectrum in the $x$-direction is defined by

$$E_m(t) = \sum_n E_{mn}(t)/2, \qquad (47)$$

where the summation is over all spanwise wave numbers. A similar definition is used for the z-direction.

### 13.6.2 LES Without an SGS Model

It is important to demonstrate that the artificial damping of the numerical scheme is not the primary mechanism by which energy of the resolved scales is dissipated, and that the SGS model plays a role in that dissipation. In fact, for the MacCormack scheme, the artificial damping is not enough to stabilize the solution beyond a certain time because of the buildup of energy at high wave numbers of the resolved field.

Computations are conducted with and without the SGS model. In the latter case, the equations being solved are actually the Euler equations. Without the model, the calculations are terminated at time $t = 66$ because a negative temperature is predicted. The development in time of the modal energy of the modes (2,0), (1,0), (2,2), (1,2), (0,2), and (0,1) is shown in Fig. 4. The results without the model show larger growth rates and hence more energetic disturbances near the time of negative temperature. We note that this event happens well beyond the time of maturity of the spanwise rollers, which is completed at time $t = 35$. Therefore, the failure of the calculations without the SGS model is due to strong nonlinear interactions between different modes and the production of smaller scales. This can be better demonstrated with the 1D energy spectrum at time $t = 66$, just before termination, which is shown in Fig. 5. The buildup of energy at the high wave numbers is

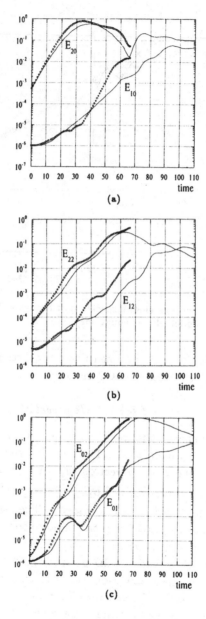

Fig. 4   Time development of modal energy. (a) Fundamental (2,0) subharmonic (1,0). (b) Modes (2,2) and (1,2). (c) Modes (0,2) and (0,1). Smagorinsky, no SGS model (Euler equations).

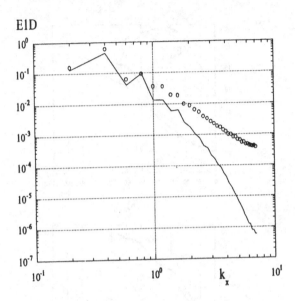

EID

Fig. 5　One-dimensional energy spectrum at time $= 66\delta/U_0$. Smagorinsky, no SGS model (Euler equations).

evident, and the difference between the two curves must be attributed to the dissipative effect of the SGS model.

### 13.6.3 Large-Scale Structures

In Fig. 6a, we show contours of the spanwise vorticity component $w_z$ in one $x$-$y$ plane at time $t = 32.71$. The rollers have been fully formed. Figure 6b shows the velocity vectors $(u, v)$ in the same plane and the contours of $w_{xy}$, which is defined by $\sqrt{w_x^2 + w_y^2}$. The stretching by the strain field in the braid region is evident. The stretching mechanism makes the streamwise vorticity in the braid region grow into elongated, thin vortex tubes and wrap around the spanwise rollers. This can be seen in Fig. 7a, which shows a 3D perspective of a surface of constant $w_{xy}$ (periodicity in the $x$- and $z$-directions is used to extend the domain by 50% of each direction for visualization purposes). Projections of this surface onto a vertical $(x - y)$ and a horizontal $(x - z)$ plane are shown in Figs. 7b and c, respectively. In these figures, the ellipsoid-shaped surfaces represent the streamwise vorticity component contributed by the undulated spanwise rollers, which are shown by the constant pressure surface in Fig. 7d. Regions of high spanwise vorticity form a stag-

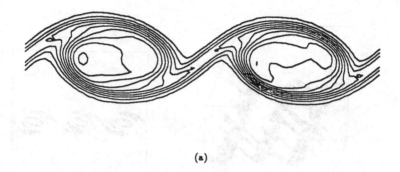

(a)

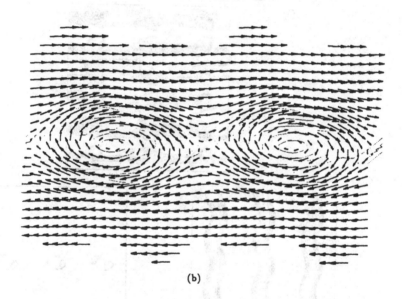

(b)

Fig. 6  (a) Vorticity contours of spanwise component $w_z$.  Plane $z =$
0.861$L_z$, time $= 32.71\delta/U_0$, $w_{max}= -0.038$, $w_{min} = -0.71$.  (b)
Velocity field, plane $z = 0.861$, $L_z$, time $= 32.71\delta/U_0$.

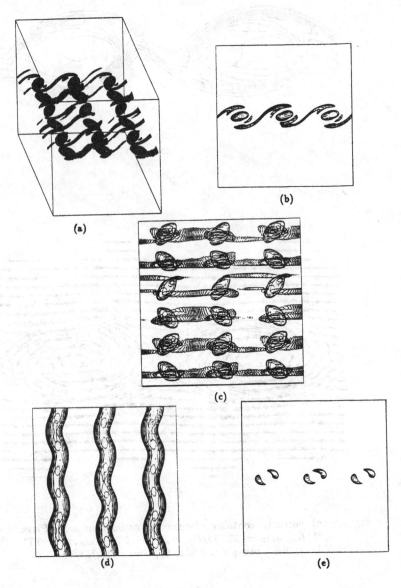

Fig. 7   (a) Constant vorticity surface, perspective view, $w_{xy}$ component. Time $= 40.78\delta/U_0$, $w_{xy} = 0.25$. (b) Projection of constant $w_{xy}$ surface on the $x$-$y$ plane. Time $= 40.78\delta/U_0$, $w_{xy} = 0.22$. (c) Projection of constant $w_{xy}$ surface on the $x$-$z$ plane. Time $= 40.78\delta/U_0$, $w_{xy} = 0.22$. (d) Projection of constant pressure surface on the $x$-$z$ plane. Time $= 40.78\delta/U_0$, $(P/\rho_o RT_o)) = 0.90$. (e) Projection of constant $w_z$ surface on the $x$-$y$ plane. Time $= 40.78\delta/U_0$, $w_z = -0.85$.

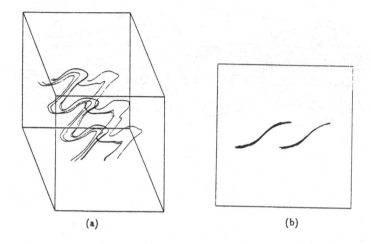

Fig. 8  (a) Perspective view of vortex lines through streamwise vortex
tubes. Time = $40.78\delta/U_0$. (b) Projection of vortex lines on $x$-$y$
plane. Time = $40.78\delta/U_0$.

gered pattern of cup-shaped surfaces as shown in Fig. 7e, and discussed
by Moser and Rogers (1991). Few vortex lines through pionts of high
streamwise vorticity are shown in Fig. 8a, and their projections on the
$x$-$y$ plane are shown in Fig. 8b. These vortex lines are as those described
by Bernal and Roshko (1986).

The stretching mechanism continues to strengthen the streamwise vorticity in the three rows of counterrotating streamwise vortices in a vertical plane that cuts through the "mean" center of a spanwise roller. The
$w_{xy}$ vorticity contours are shown in Fig. 9a and the velocity vectors are
shown in Fig. 9b. The four vortices on top result from the intersection
with streamwise vortices in the braid region that ends on the top of the
spanwise roller, as shown in Fig. 7b, while the four vortices in the lower
row are associated with the braid region that ends on the bottom, as
in Fig. 7b. The four vortices in the middle row are the intersection
with the undulated spanwise roller. The induced velocity by the middle
row of vortices at the centers of the top row will make the two central
vortices move away from each other, and hence it brings the outer pair
of vortices closer. The opposite movement happens to the lower row of
vortices. The new locations of the vortices in the same plane at a later
time $t$ = 64 are shown in Fig. 10. Now, the projection onto the $x$-$z$

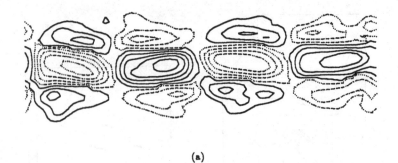

(a)

(b)

Fig. 9  Vorticity contours of $w_{xy}$. Plane $x = \frac{1}{4}L_x$. Time $= 48.65\delta/U_0$, $w_{max} = 00.689$, $w_{min} = -0.689$. Negative values are shown by solid lines. (b) Velocity field and contours of $w_{xy}$. Plane $x = \frac{1}{4}L_x$. Time $= 48.65\delta/U_0$.

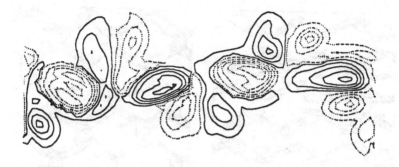

Fig. 10   Vorticity contours of $w_{xy}$. Plane $x = \frac{1}{4}L_x$. Time = $64.016/U_0$, $w_{max} = 1.418$, $w_{min} = -1.418$. Negative values are shown by solid lines.

plane of a surface of constant $w_{xy}$ appears as in Fig. 11a. Following Sandham and Reynolds (1989), we use surfaces of constant low pressure to show the location of regions of high vorticity. The projection onto the $x$-$z$ plane of a surface of constant pressure is shown in Fig. 11b. Note that Fig. 11b includes the effects of total vorticity, whereas Fig. 11a is for the $w_{xy}$ component only. This explains the continuity of the spanwise undulated structure of the pressure, where the structure in Fig. 11a breaks where $w_{xy}$ becomes small.

Perspectives of surfaces of constant vorticity $w_{xy}$ and pressure are shown in Figs. 11c and 11d, respectively. These figures are consistent with the description of the translative mode of Pierrehumbert and Widnall (1982). This mode is computed by Sandham and Reynolds (1991) using DNS.

The projections onto the $x$-$z$ plane of the $w_{xy}$ and pressure surfaces at time $t = 71$ are shown in Figs. 12a and 12b, respectively. Notice the breaking of the original spanwise structures and the dominance of the streamwise structures, which also bend and produce a spanwise component of vorticity, as shown in Fig. 12c. Soon after the breaking

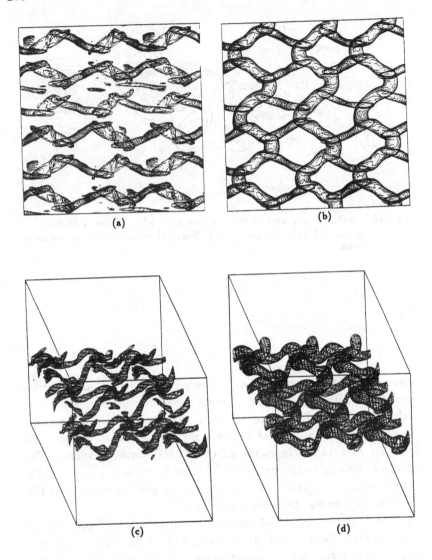

Fig. 11  (a) Projection of constant $w_{xy}$ surface on $x$–$z$ plane. Time = $64.01\delta/U_0$, $w_{xy} = 1.0$. (b) Projection of constant pressure surface on the $x$-$z$ plane. Time = $64.01\delta/U_0$, $(P/(\rho_0 R T_0)) = 0.91$. (c) Constant vorticity surface, perspective view, $w_{xy}$ component. Time = $64.01\delta/U_0$, $w_{xy} = 1.15$. (d) Constant pressure surface, perspective view. Time = $64.01\delta/U_0$, $(P/(\rho_0 R T_0)) = 0.905$.

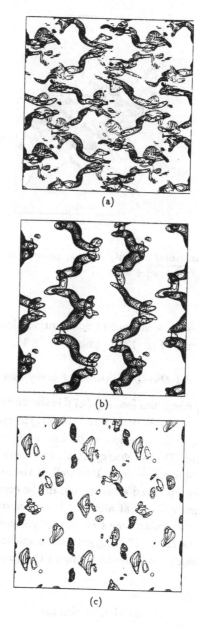

Fig. 12  (a) Projection of constant $w_{xy}$ surface on $x$-$z$ plane.  Time $=$ $71.25\delta/U_0$, $w_{xy} = 1.35$.  (b) Projection of constant pressure surface on the $x$-$z$ plane.  Time $= 71.25\delta/U_0$, $(P/(\rho_0 R T_0)) = 0.88$. (c)Projection of constant pressure surface on the $x$-$z$ plane.  Time $= 71.25\,\delta/U_0, w_z = -1.2$.

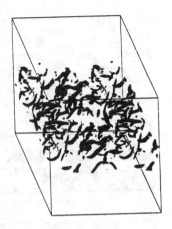

Fig. 13   Constant total vorticity surface, perspective view. Time = 78.41
$\delta/U_0$, $\sqrt{w_x^2 + w_y^2 + w_z^2}$= 1.75.

of the spanwise roller, a surface of constant vorticity $|w|$ becomes very irregular, as shown in Fig. 13 at time $t = 78.4$.

### 13.6.4 One-Dimensional Energy Spectrum

The 1D energy spectrum [see Eq. (47)] is shown in Fig. 14 at the times $t = 66, 71, 91$, and 110. At early times $t < 71$, the energy is concentrated in the $m = 2$ $(k_x = 0.4)$ mode and its harmonic $m = 4$ $(k_x=0.8)$. For times greater than 71, the energy at these modes decreases while the energy of the subharmonic $m = 1$ as well as the small scales increases. Beyond $t = 91$ we observed a saturation in the energy spectrum with a slight drop at time $t = 100$ at almost all wave numbers.

The energy spectrum at time $t = 71$ and the corresponding vorticity and pressure plots (Fig. 12) suggest that transition to small scales of the translative mode starts by breaking in the spanwise rollers.

### 13.7 Conclusions

Large eddy simulations of temporally developing mixing layers at a convective Mach number of 0.4 have been obtained. The SGS model used is a Smagorinsky eddy viscosity. The numerical method is an explicit

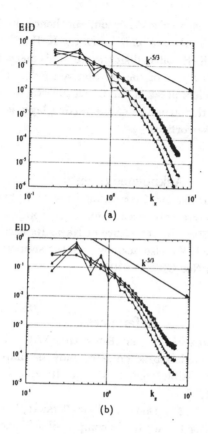

Fig. 14    (a) One-dimensional energy spectrum, $E_x$ at time t = 66. $\Delta$; t = 71, $\Delta$; t = 91, o; t = 110, •. (b) One-dimensional energy spectrum $E_z$ at time t = 66, $\Delta$; t  71, $\Delta$; t = 91, o; t = 110, •.

MacCormack method that is fourth-order in space and second-order in time.

Two-dimensional simulations initialized by white-noise velocity disturbance develop a power-law 1D energy spectrum with exponent −4, in agreement with published results for incompressible mixing layers (Lesieur et al., 1988).

Three dimensional simulations are obtained for the translative mode of Pierrehumbert and Widnall (1982) and Sandham and Reynolds (1991). The strain field in the region between the Kelvin–Helmholtz (K–H) rollers stretches and intensifies the streamwise vorticity, producing vortices initially lying in a vertical plane (streamwise–transverse plane). Then, the undulated K–N rollers induce velocities that tilt the streamwise vortices into the spanwise direction, and hence produce spanwise vorticity components in the region between the K–H rollers. Simultaneously, the undulated K–H rollers disintegrate. This marks the beginning of transition to small scales of the translative mode. The mechanism by which the K–H rollers break is still under investigation. At the later stages of transition, the structure of the mixing layer is dominated by small-scale streamwise vortices.

## Acknowledgments

We thank Professor Sankar, Georgia Institute of Technology, for providing us with a copy of his direct simulation computer code. This work has been supported by the Office of Naval Research under Grant No. N00014-91-J-1233. We also acknowledge the support of the NCSA, Illinois, under Grant No. CBT920010N.

## References

BARDINA, J., FERZIGER, J. H. AND REYNOLDS, W.C. (1983) Improved subgrid-scale models base on large-eddy simulation of homogeneous, incompressible, turbulent flows. Report No. TF-19, Stanford University.

BAYLISS, A., PARIKH, P., MASTRELLO, L., AND TURKEL, E. (1985) A fourth order scheme for the unsteady compressible Navier–Stokes equations. AIAA Paper No. 85-1694.

BAYLISS, A., MASTRELLO, L., PARIKH, P. AND TURKEL, E. (1986) Numerical simulation of boundary layer excitation by surface heating and cooling. *AIAA J.* **24**, 1095–1101.

BERNAL, L.P. AND ROSHKO, A. (1986) Streamwise vortex structure in plane mixing layers. *J. Fluid Mech.* **170**, 499–525.

BLAISDELL, G.A., MANSOUR, N.N., AND REYNOLDS, W.C. (1988) Numerical simulation of compressible homogeneous turbulence. Presented at Workshop on the Physics of Compressible Turbulence Mixing, Princeton, NJ.

CAIN, A.B., REYNOLDS, W.C., AND FERZIGER, J.H. (1981) A three-dimensional simulation of transition and early turbulence in a time-developing mixing layer, Report No. TF-14, Stanford University, Department of Mechanical Engineering.

ERLEBACHER, G., HUSSAINI, M.Y., SPEZIALE, C.G., AND ZANG, T.A.

(1990) Toward the large-eddy simulation of compressible turbulent flows. ICASE Report No. 87-20.

FERZIGER, J.H. (1977) Large eddy numerical simulations of turbulent flows. *AIAA J.* **15**(9), 1261-1267.

GOTTLIEB, D. AND TURKEL, E. (1976) Dissipative two–four methods for time-dependent problems. *Math. Comp.* **30**, 703–723.

GERMANO, M., PIOMELLI, U., MOIN, P. AND CABOT, W. (1991) A dynamic subgrid-scale eddy viscosity model. *Phys. Fluids A* **3**, 1760–1765. See also *Phys. Fluids A* **3**, 3128(E).

HARTKE, G.J., CANUTO, V.M., AND ALONSO, C.T. (1988) A direct interaction approximation treatment of turbulence in a compressible fluid. Part I. Formalism. *Phys. Fluids A* **31**, 1034–1050.

HIRSCH, C. (1990) *Numerical Computation of Internal and External Flows, Vol. 2: Computational Methods for Inviscid and Viscous Flows.* Wiley.

KIM, J., MOIN, P., MOSER, R.D. (1987) Turbulence statistics in fully-developed channel flows at low Reynolds number. *J. Fluid Mech.* **177**, 133–136.

KORCZAK, K.Z., AND PATERA, A.T. (1986) Isoparametric spectral element method for solution of the Navier–Stokes equations in complex geometry. *J. Comp. Phys.* **26**, 361–382.

LEITH, C.E. (1990) Stochastic backscatter in a subgrid-scale model: Plane shear mixing layer. *Phys. Fluids A* **2**, 297–299.

LELE, S. (1989) Direct numerical simulation of compressible free shear flows. AIAA Paper 89-0374.

LEONARD, A. (1974) Energy cascade in lare-eddy simulations of turbulent fluid flows. *Adv. Geophysics* **18**, 237–248.

LESIEUR, M., STAQUET, C., LE ROY, P., AND COMTE, P. (1988) The mixing layer and its coherence examined from the point of view of two-dimensional turbulence. *J. Fluid Mech.* **192**, 511–534.

LESIEUR, M., COMTE, P., AND NORMAND, X. (1991) Direct and large-eddy simulation of transitioning and turbulent shear flows, AIAA Paper No. 91-0335.

LESLIE, D.C. (1973) *Developments in the Theory of Turbulence,* Clarendon.

MARUYAMA, Y. (1988) A numerical simulation of plane turbulent shear layer. *Trans. Japan Soc. Aero. Space Sci.* **31**, 79–93.

MOIN, P., SQUIRES, K., CABOT, W., AND LEE, S. (1991) A dynamic subgrid-scale model for compressible turbulence and scalar transport. *Phys. Fluids A* **3**, 2746–2757.

MOSER, R.D., AND ROGERS, M.M. (1991) Mixing transition and the cascade to small scales in a plane mixing layer. *Phys. Fluids A* **3**, 1128–1134.

ORSZAG, S.A., AND PATTERSON, G.S. (1972) Numerical simulation of three-dimensional homogeneous isotropic turbulence. *Phys. Rev. Lett.* **28**, 76–79.

PASSOT, T., AND POUQUET, A. (1987) Numerical simulation of compressible homogeneous flows in the the turbulent regime. *J. Fluid Mech.* **181**, 441–446.

PASSOT, T., AND POUQUET, A. (1988) Hyperviscosity for compressible flows using spectral methods. *J. Compt. Phys.* **75**, 300–313.

PIERREHUMBERT, R.T., AND WIDNALL, S.E. (1982) The two- and three-dimensional instabilities of a spatially periodic shear layer. *J. Fluid Mech.* **114**, 59–82.

PIOMELLI, U., MOIN, P., AND FERZIGER, J.H. (1988) Model consistency in large eddy simulation of turbulent channel flows. *Phys. Fluids* **31**, 1884–1891.

RAGAB, S.A., AND SHEEN, S. (1990) Numerical simulation of a compressible mixing layer. AIAA Paper No. 90-1669.

RAGAB, S.A., SHEEN, S., AND SREEDHAR, M. (1992) An investigation of finite-difference method for large-eddy simulation of a mixing layer. AIAA Paper No. 92-0554.

REYNOLDS, W.C. (1990) The potential and limitations of direct and large eddy simulations. *In Whither Turbulence.* Ed. J.L. Lumley, Springer-Verlag.

SANDHAM, N.D., AND REYNOLDS, W.C. (1987) Some inlet plane effects on the numerical simulated spatially-developing mixing layer. Springer-Verlag, Proc. Sixth Symp. Turb. Shear Flows, pp. 441–454.

SANDHAM, N.D., AND REYNOLDS, W.C. (1989) A numerical investigation of the compressible mixing layer. Report No. TF-45, Department of Mechanical Engineering, Stanford University.

SANDHAM, N.D., AND REYNOLDS, W.C. (1991) Three-dimensional simulations of large eddies in the compressible mixing layer. *J. Fluid Mech.* **224**, 133–158.

SMAGORINSKY, J. (1963) General circulation experiments with the primitive equations. Part I. The basic experiment. *Mon. Wea. Rev.* **91**, 99–164.

SPALART, P.R. (1988) Direct numerical simulation of a turbulent boundary layer up to $Re = 1410$. *J. Fluid Mech.* **187**, 61–98.

SPEZIALE, C.G., ERLEBACHER, G., ZANG, T.A., AND HUSSAINI, M.Y. (1988) The subgrid-scale modeling of compressible turbulence. *Phys. Fluids A* **31**, 940–942.

STRANG, G. (1968) On the construction and comparison of difference schemes. *SIAM J. Num. Anal.* **5**, 506–517.

SWANSON, R.C. AND TURKEL, E. (1990) On central-difference and upwind schemes. ICASE Report No. 90-44.

TANG, W., KOMERATH, N., AND SANKAR, L. (1989a) Numerical simulation of the growth of instabilities in supersonic free shear layers. AIAA Paper No. 89-0376.

TANG, W., SANKAR, L., AND KOMERATH, N. (1989b) Mixing enhancement in supersonic free shear layers. AIAA Paper No. 89-0981.

YAKHOT, A., ORSZAG, S.A., YAKHOT, V., AND ISRAELI, M. (1989) Renormalization group formulation of large eddy simulations. *J. Sci. Comput.* **4**, 139–158.

YOSHIZAWA, A. (1986) Statistical theory for compressible turbulent shear flows, with the application to subgrid modeling. *Phys. Fluids A* **29**, 2152–2164.

YOSHIZAWA, A. (1990) Three-equation modeling of inhomogeneous compressible turbulence based on a two-scale DI approximation. *Phys. Fluids A* **2**, 838–850.

# 14

# A Linear-Eddy Mixing Model
# for Large Eddy Simulation
# of Turbulent Combustion

SURESH MENON, PATRICK A. MCMURTRY,

AND ALAN R. KERSTEIN

## 14.1 Introduction

To improve predictive capabilities for modern combustion processes, sophisticated models that faithfully represent the physics of turbulent mixing and reaction are required. Models that are presently used to study combustion systems, such as those based on gradient diffusion assumptions, omit important aspects of the subgrid mixing process. More sophisticated methods based on probability density function (PDF) transport equations treat the reaction process exactly, but the molecular mixing submodels that are employed are not fully satisfactory. Although the utility of these models should not be overlooked, there is clearly a need to develop alternative methods for calculating turbulent mixing processes.

In modeling the scalar mixing process, one encounters difficulties not present in modeling momentum transport. These difficulties can be primarily attributed to the interactions among turbulent stirring, molecular diffusion, and chemical reaction at the smallest scales of the flow. A reliable model of subgrid mixing and reaction should therefore include and distinguish among these distinctly different physical processes. To accomplish this, it appears that a comprehensive description of the scalar microfield is needed. This is fundamentally different from momentum transport modeling, in which the main influence of the small scales is to provide dissipation for the large-scale structures. Thus, while the effect of subgrid stresses on the momentum transport can be reasonably treated with various eddy viscosity models, a similar characterization of the subgrid scalar field in terms of an eddy diffusivity is neither sufficient nor correct, since an accurate description of the small-scale dynamics is critical to the overall mixing and the combustion process. As a result of these features of the turbulent mixing process, progress in the development of subgrid mixing models for use in large eddy simulation (LES) has been limited.

These fundamental issues are relevant to both the solution of the Reynolds-averaged flow field (where only the mean motion is of interest) and the solution of the unsteady flow field simulated using LES techniques. In this chapter, the development of a subgrid modeling approach specifically directed at scalar mixing is addressed. The model is based on Kerstein's (1988, 1991) *linear eddy model*. The emphasis here is placed on premixed combustion applications, although the general formulation has also been applied to turbulent diffusion flames (McMurtry et al., 1991, 1992).

In the following section, some of the issues associated with turbulent mixing are addressed along with issues associated with premixed flame propagation. In Section 14.3 the basic ideas behind the linear eddy model, upon which the subgrid model described here is developed, are discussed. The subgrid model formulation is presented in Section 14.4, while model results are presented in Section 14.5.

## 14.2 Mixing Mechanisms and Modeling Issues

Mixing in a typical turbulent flow may be envisioned as a two-step process consisting of (1) turbulent stirring and length-scale reduction of initially unmixed species and (2) molecular diffusion. In many classical approaches to modeling turbulent mixing at high Reynolds numbers, it is assumed that the first step is rate limiting and that the second step can therefore be omitted from consideration. Though adequate for many purposes, this picture does not account for important mixing properties observed in turbulent shear flows, and it omits other mixing properties that are more subtle but nevertheless affect combustion processes.

The key experimental observations with regard to turbulent shear flow mixing are, first, the observed sensitivity of the observed mixing rate to the molecular diffusivity for a given Reynolds number and, second, a spatial structure of the mixing field that cannot be explained solely in terms of large-scale gradient diffusion (for a review, see Broadwell and Mungal, 1991). The first observation indicates that molecular diffusion accounts for a significant portion of the total mixing time at Reynolds numbers of practical interest. The second observation indicates the interplay of large-scale and fine-scale processes even in the limit of infinite Reynolds number.

Other, more subtle mixing effects that affect combustion include differential molecular diffusion and compositional variations within fluid that is nominally fully mixed. Differential diffusion can influence the thermochemical state within a flame and thus impact ignition, extinction, and other important properties. The high sensitivity of local chemistry to the relative abundance of fuel and air (e.g., the propensity to form soot in fuel-rich zones) requires accurate modeling of small deviations from complete mixing. Such deviations may occur over a range of length scales, requiring considerable time to be dissipated, and in some instances persisting at the combustor exit.

Most parameterizations of the mixing process used in simulations of turbulent flows do not contain sufficient physics to predict accurately the amount of molecular mixing that occurs. For example, at the small scales, most conventional models make no distinction between turbulent advection and molecular diffusion. This distinction is crucial if an accurate representation of the mixing process is to be made. In combustion applications, the difficulties in treating mixing are magnified due to the extreme range of length scales present. The small-scale mixing process must be parameterized over spatial regions that may be meters in extent, while the actual mixing and reaction occur at length scales in the diffusion range of the order of millimeters and smaller.

The above discussion is relevant to nonpremixed combustion, in which fuel and oxidizer must come into molecular contact before combustion can occur. When premixed combustion is considered, molecular diffusion of the primary species (i.e., the fuel and oxidizer) is no longer a critical issue since they are already mixed. However, there are issues particular to LES of premixed combustion that require accurate modeling of the processes occurring at the small scales.

### 14.2.1 Premixed Combustion Issues

In premixed combustion, the amount of heat release per unit area of flame is determined by the local flame speed and by the specific chemical energy available in the fuel. If a finite-rate chemical mechanism for premixed combustion is employed in an LES, the numerical simulation must implicitly compute the local flame speed. Unfortunately, this is difficult to achieve in practice. The flame speed depends upon the dissipation mechanism and therefore the internal structure of the flame sheet. Because the number of grid points is limited in LES, the flame sheet cannot be resolved adequately. Also, all numerical schemes involve some form of artificial dissipation, either explicitly added to stabilize the computations or implicitly present due to the differencing algorithm. As a result, the computed flame structure will be numerically diffused and the temporal–spatial distribution of the heat release could be overwhelmed by numerical diffusion.

The problems associated with employing a classical finite-rate kinetics model can be circumvented by using a *thin flame model*. In such an approach it is assumed that the flame thickness is small compared with the smallest turbulent length scale, and if the changes in the reaction-diffusion structure due to turbulent straining are also small, then the reaction zone can be considered to be asymptotically thin. Within the thin flame approximation, a model equation for premixed combustion is considered in which the local flame speed explicitly appears. If the local flame speed $u_F$ is known, a progress variable $G$ can be defined that is governed by the equation (Kerstein et al., 1988; Menon and Jou, 1991)

$$\frac{\partial \rho G}{\partial t} + \frac{\partial}{\partial x_i} \rho u_i G = \rho u_F |\nabla G|, \tag{1}$$

where $\rho$ is the density and $u_i$ is the fluid velocity. The flame is located along a specified isosurface $G = G_o$ separating the fuel mixture $G < G_o$ and the fully burned combustion product $G > G_o$. Equation (1) describes the convection of the flame by the local fluid velocity and the flame propagation into the unburned mixture through a Huygens-type mechanism, $u_F |\nabla G|$. For laminar premixed combustion, the local flame speed $u_F$ is the laminar flame speed $S_L$, which contains the information on the chemical kinetics and the molecular dissipation.

If Eq. (1) is applied to turbulent flows without the use of a subgrid model that explicitly represents the fine scales, the local flame speed $u_F$ is taken to be the local turbulent flame speed $u_T$, where $u_T$ is a prescribed function of local turbulence intensity and the laminar flame speed $S_L$ (here treated as a constant chemical property, though in reality it is sensitive to the strain field affecting the flame). The implementation of the thin flame model as a part of the LES transport equations therefore explicitly requires the specification of the subgrid turbulence kinetic energy to determine the turbulent flame speed.

This approach is subject to the same limitations as subgrid closures for other molecular properties. In the present case, the closure involves assigning $u_F$ in Eq. (1) as an effective property representing the joint influence of the underlying molecular process, in this instance laminar flame propagation, and turbulent convection at the unresolved scales. This is analogous to turbulent viscosity and turbulent diffusivity assumptions adopted in LES models of turbulent transport and mixing, respectively. As in those applications, the nominal influence of the molecular process vanishes in the limit of high turbulence intensity. Namely, $u_F$ scales as the root-mean-square (rms) fluctuation $u'$ of the unresolved velocity field, plus corrections involving $S_L$ that vanish in this limit.

As noted earlier, closures of this type are problematic when the combustion process is sensitive to aspects of the coupling between the molecular processes and the flow field. Coupling of chemistry and pressure variations is important in some premixed combustion applications (Menon and Jou, 1991). Apart from the question of couplings, the determination of a general functional relationship between the turbulent flame speed, the laminar flame speed, and the turbulence intensity which is valid for all types of fuel and flow conditions appears to be difficult (Kerstein and Ashurst, 1992).

The subgrid model described in this chapter provides an approach whereby, for premixed combustion, the issue of specification of a turbulent flame speed is no longer necessary. Furthermore, as described below, the current approach also enables the resolution of fine-scale geometrical structure of the flame surface in a fundamentally correct manner.

In the next section, the linear-eddy model is formulated in the context of application to molecular mixing in turbulent flow, and it is shown how that formulation, used in previous work, is modified to address premixed flame propagation.

## 14.3 The Linear-Eddy Model

The subgrid mixing model presented in Section 4 is based on the linear-eddy model, developed by Kerstein (1988, 1991). The original formulation of the model was motivated by the need to compute the dependence of the scalar field statistics on Reynolds, Schmidt, and Damköhler numbers. The key feature of the linear eddy approach is the separate treatment of the two primary physical mechanisms that govern the scalar (or chemical species) mixing process: turbulent stirring and molecular diffusion. For nonpremixed combustion, within the framework of the thin flame model, molecular diffusion is replaced by a flame propagation mechanism (Eq. (1)). When heat release occurs the effects of thermal expansion must also be modeled.

To maintain the distinction among these mechanisms at all scales of the flow, it is necessary to resolve *all* relevant length scales. This is not computationally feasible in three, or even two, dimensions. To achieve this resolution, evolution of the scalar field is simulated in one spatial dimension in the linear eddy model. Within the one-dimensional domain, molecular diffusion is treated explicitly, and turbulent stirring is modeled by a stochastic rearrangement process applied to the scalar field along the linear domain. The implementation of molecular diffusion, flame propagation, thermal expansion, and turbulent mixing within the linear eddy model is described below.

### 14.3.1 Molecular Diffusion Process

For scalar mixing or nonpremixed combustion, molecular diffusion and chemical reaction are implemented by the numerical time integration of the diffusion equation along the linear domain using standard finite-difference schemes. The equation is of the form

$$\frac{\partial Y_k}{\partial t} = D_k \frac{\partial^2 Y_k}{\partial x^2} + W_k, \tag{2}$$

where $D_k$ is the $k$th species diffusion coefficient, $Y_k$ is the $k$th species mass fraction, and $W_k$ denotes the reaction rate for the $k$th species. The exact specification of the linear domain $x$ within the context of LES modeling is described later in Section 4.1.

Although Eq. (2) can in general be solved for $k$ species, the computational effort can be quite extensive, particularly when multistep, finite-rate chemical reactions are considered. To reduce the overall computational effort, the effect of finite-rate kinetics was recently incorporated by McMurtry et al. (1991, 1992) using a reduced-mechanism approach. An $H_2$–air combustion system was modeled and the formation of $NO_x$ was also studied using a reduced Zeldovich mechanism. In this approach, the $H_2$–air diffusion flame process was solved by the solution of equations governing a conserved scalar $\xi$ and a progress variable $\eta$ of the form given by Eq. (2). The source term for the equation for the progress variable was determined using a lookup table.

Although Fickian diffusion, as in Eq. (2), has been assumed in applications to date, the model can accommodate multicomponent diffusion or any other evolution process which can be formulated as a partial differential equation in one spatial dimension. In particular, it can accommodate the flame propagation process, as discussed in Section 3.3.

### 14.3.2 Turbulent Mixing Process

The key feature of the model is the manner in which turbulent convection is treated. This is implemented in a stochastic manner by random, instantaneous rearrangements of the scalar field along the line. Each event involves spatial redistribution of the species field within a specified segment of the spatial domain. The size of the selected segment represents the eddy size, and the distribution of eddy sizes is obtained by applying Kolmogorov scaling laws. In this model, rearrangement of a segment of size $l$ represents the action of an eddy of size $l$.

The rearrangement events are specified by two parameters: $\lambda$, which is a rate parameter with dimensions $[L^{-1}t^{-1}]$, and segment size $l$, selected randomly from a length-scale PDF $f(l)$. The values of these parameters are determined by recognizing that the rearrangement events induce a random walk of a marker particle on the linear domain. Equating the diffusivity of the random process with scalings for the turbulent diffusivity provides the necessary relationships to determine $\lambda$ and $f(l)$. For a high Reynolds number turbulent flow described by a Kolmogorov cascade, the result is (Kerstein 1991)

$$f(l) = \frac{5}{3} \frac{l^{-8/3}}{\eta^{-5/3} - L^{-5/3}}, \qquad (3)$$

$$\lambda = \frac{54}{5} \frac{\nu Re_L}{L^3} \left(\frac{L}{\eta}\right)^{5/3}, \qquad (4)$$

where $\eta$ is the Kolmogorov scale and $L$ is the integral scale. Here, $Re_L$ is operationally defined as $(L/\eta)^{4/3}$. The numerical coefficient in Eq. (4) is specific to the rule for redistributing the scalar field during each rearrangement event.

Once the time, location, and region for a rearrangement event have been determined, the event is implemented as a mapping of the scalar field within the specified domain to a new distribution within the same domain. The effect of the mapping that is employed, termed the *triplet map*, is illustrated in Fig. 1. The scalar field within the chosen size-$l$ segment is first compressed by a factor of 3, thus tripling the scalar gradient within the segment. The original scalar field within the segment $l$ is then replaced by three copies of this compressed field with the middle copy reversed. The resulting scalar field is shown in Fig. 1b. The implementation of this mapping on the discretized computational domain is

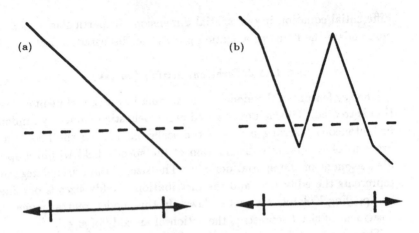

Fig. 1    a) Original scalar field before application of rearrangement. b) Scalar field after rearrangement process. The scalar rearrangement (turbulent stirring) process is carried out by the use of the "triplet" map. The triplet map involves selecting a segment of the linear domain for rearrangement, making three compressed copies of the scalar field in that segment, replacing the original field by the three copies, and inverting the center copy.

defined so as to recover this rule in the continuum limit, while satisfying species conservation exactly in the discrete implementation.

The triplet map can be understood on both an intuitive and a theoretical basis. It was shown by Kerstein (1991) that the triplet mapping reproduces the qualitative effect of a single eddy on the scalar gradient field. More details of the theoretical considerations behind the triplet map are given in that reference.

Depending on the application, one or more empirical inputs or parameter adjustments are required for comparison of the model with measured quantities. Spectral and related scaling exponents are obtained directly, with no empiricism (Kerstein 1991). For other applications, coefficients relating the physical values of $Re$, $Sc$, and $L$ to their model counterparts must be determined empirically. Since the coefficients are found to be of the order of unity in all instances, all such coefficients are set to unity in computations presented here. The present objective is to illustrate the principles involved rather than to reproduce particular measurements.

### 14.3.3 Flame Propagation Model

For premixed combustion, within the framework of the thin flame approximation, the molecular diffusion process is replaced by a flame propagation model. In this approach, the thin flamelets propagate with their *laminar* speeds while simultaneously undergoing turbulent mixing. When employed as a subgrid model, this formulation avoids the specification of a turbulent flame speed since the underlying laminar process is modeled explicitly, along with the influence of turbulent stirring at all length scales not resolved by the LES.

Flame propagation in the thin flamelet model is governed by a Huygens propagation mechanism,

$$\frac{\partial G}{\partial t} = u_F |\nabla G|, \tag{5}$$

on the linear domain, where $u_F$ is now taken to be the laminar flame speed $S_L$. Equation (5) omits the convective term appearing in Eq. (1) because convective effects are represented by the concurrent rearrangement process. As noted before, $G > G_o$ denotes the product (completely burned), and $G < G_o$ denotes the fuel (completely unburned); the flame is located at points of transition between these two states. The existence of the flame is determined by using a Heaviside function such that $T = T_{fuel}$ for $G < G_o$ and $T = T_{product}$ for $G > G_o$, where $T_{product} > T_{fuel}$. Here, $T$ is the temperature and the thin flames are located wherever the $G = G_o$ surface occurs. In general, $G_o$ can be arbitrarily chosen within the [0, 1] interval, provided that the locations where $G = G_o$ initially correspond to the initial configuration of the physical flame sheet. In the present study, $G_o = 0.5$.

Interpretation of flame evolution simulated in this fashion as a one-dimensional cut through a locally isotropic turbulent flame brush has been validated by comparison of simulated and measured scaling properties governing the geometrical structure of the wrinkled flame sheet and the overall propagation rate $u_T$, which is now an output of the model rather than an input (Menon and Kerstein, to appear).

### 14.3.4 Thermal Expansion Process

To include the effect of heat release, an additional small-scale process is implemented. This process involves volumetric expansion within the linear domain at regular time intervals to account for thermal expansion occurring since the previous expansion event. This process is briefly described here using premixed combustion as an example. Details of the algorithm are presented elsewhere in the context of an application to nonpremixed combustion (Kerstein 1991).

At any instant, the spatial distribution of flamelets corresponds to the points on the computational domain at which $G = G_o$. Some of the linear-eddy cells that contained unburned fuel $(G < G_o)$ at a temperature $T_{fuel}$ at the conclusion of the previous expansion event can now be in a burned state $(G > G_o)$ at the product temperature $T_{product}$. Since the product is at a higher temperature, the number of cells in the burned state must be increased accordingly. This is carried out by creating additional burned cells adjacent to each newly burned cell, such that the volume change conserves mass based on the new mass density.

This procedure increases the total number of cells in the computational domain, corresponding to the overall volume increase that occurs physically. In stand-alone model formulations representing spatially developing flows, this extension of the computational domain is the model analog of flow induced by thermal expansion in the direction corresponding to the computational line.

In the subgrid formulation of Section 14.4, the linear-eddy computation is interpreted as a representative sample of the thermochemical field, characterizing intensive quantities such as temperature and species mole fractions but not extensive quantities such as volume. In this case, the expansion process is implemented as outlined above but the computational domain is truncated to maintain a constant number of cells. Volume expansion is implemented within the LES. Thermochemical input for the equation of state within each LES cell is obtained by averaging over the linear-eddy domain corresponding to that cell. In this context, volume expansion is subsumed in the overall solution of the momentum and continuity equations, as in any variable-density LES implementation.

### 14.3.5 Summary

The linear-eddy model treats the processes of turbulent stirring, molecular diffusion, and chemical reaction within a turbulent reacting flow in a physically sound manner. For nonreactive scalar mixing and non-premixed combustion, molecular diffusion is simulated by solving Fick's law or an appropriate multicomponent generalization along the linear domain. Finite-rate chemistry is likewise implemented by finite-difference solution of the differential equations representing the chemical mechanism. For premixed combustion, this diffusion-reaction process is replaced by the flame propagation model, Eq. (5).

Turbulent stirring is modeled by randomly occurring rearrangement events, which interrupt the diffusion or flame propagation process. The frequency of these stirring events is governed by a parameter $\lambda$ as determined by the physical scaling laws described above.

The linear-eddy model was originally developed as a stand-alone mixing model to study scalar field microstructure in homogeneous turbulence and to interpret observed features of the mixing process in spatially developing flows. In the former application, linear-eddy computations of a statistically steady scalar configuration obtained by imposing a uniform scalar gradient in a homogeneous turbulent flow field reproduced key features of the scalar power spectrum. These features included effects of Reynolds and Schmidt numbers and scaling properties of higher order scalar statistics (Kerstein 1991).

Applications to spatially developing flows have reproduced the following measured properties: (1) three distinct scaling regimes governing turbulent plume growth and spatially resolved scalar fluctuation statistics within such plumes (Kerstein 1988, 1992); (2) the spatially resolved cross-correlation of diffusive scalars in a three-stream mixing configuration (Kerstein 1992); (3) Damköhler number dependencies of reactant concentrations in a two-stream configuration (Kerstein 1992); (4) spatially resolved scalar fluctuation statistics in free shear flows and the dependence of local and overall mixing on Reynolds and Schmidt numbers (Kerstein 1989, 1990; Broadwell and Mungal 1991); and (5) scalar fluctuation statistics reflecting differential molecular diffusion effects (Kerstein 1990). In these different studies, configuration-specific aspects are reflected in the initial and boundary conditions of the computations and in the model analogs of such quantities as the Reynolds, Schmidt, and Damköhler numbers.

The success of the linear eddy model in predicting the scalar field statistics in these stand-alone applications lends strong support for its potential use as a subgrid model within an LES. Such a formulation is discussed next.


## 14.4 The Linear-Eddy Subgrid Model

The usual implementation of large eddy simulations involves filtering out the small scales of motion and explicitly computing the larger-scale motions which can be resolved on a computational grid. The modeling process is then concerned with modeling only the unresolved, small-scale motions and their interactions with the large scales. Subgrid models for momentum transport have been based primarily on ideas of a subgrid eddy viscosity. The eddy viscosity approach is reasonable to the extent that the small-scale motions provide a mechanism for energy dissipation and transfer. For turbulent mixing and reaction processes, however, it is not possible to characterize the overall statistical state of the scalar field in such a manner. This is a result of the subtle interactions between turbulent stirring, molecular diffusion, and chemical reaction at the smallest scales of the flow. A reliable subgrid model should account for these distinct physical mechanisms at all unresolved scales including molecular.

The linear eddy model described in Section 14.3 has been demonstrated to be an effective tool for studying scalar field mixing processes by retaining the distinction among the relevant physical mechanisms at all length scales. Here the use of the model is extended for application as a subgrid model within LES. Two separate processes are implemented to incorporate linear eddy as a subgrid model. The first process involves performing separate linear-eddy calculations in each LES grid cell. The second process involves transport of subgrid scalar information across LES cell boundaries as prescribed by the large-scale resolvable velocity field. Because the small scales have much higher frequencies than the large scales, a number of scalar rearrangement events, along with the flame propagation (or Fickian diffusion for nonpremixed combustion) process at the subgrid level, must be performed in each linear-eddy calculation between successive time steps in the LES. For large computational domains, this process can be demanding on computer resources, but it is well suited for efficient implementation on massively parallel computers. This feature, in conjunction with its economy relative to direct numerical simulation (discussed in Section 14.4.1), indicates the likelihood that the present approach will become the method of choice for a class of turbulent reacting flow problems not accessible by direct numerical simulation and not adequately modeled by approaches lacking explicit spatial structure at subgrid length scales.

## 14.4.1 Linear-Eddy Calculations in Each LES Grid Cell

In each LES grid cell of characteristic cell size $\Delta$, the linear eddy algorithm described in Section 3 is carried out. This process represents turbulent mixing and flame propagation (or molecular diffusion) at the unresolved scales. The linear eddy grid spacing $\delta$ in each LES cell $\Delta$ is determined on the basis of scaling laws so that the smallest eddy within each LES grid cell is numerically resolved. With the Reynolds number of the flow as an input parameter, classical scaling arguments lead to an approximation for the range of eddy sizes in the subgrid range. In the subgrid formulation, the Reynolds number is defined in terms of the LES grid size and the subgrid turbulent rms velocity fluctuation $u'$, i.e., $Re = u'\Delta/\nu$. High Reynolds number scaling laws give $\Delta/\eta = Re^{3/4}$, where $\eta$ is the smallest eddy size in the given LES grid of size $\Delta$. With the cell Reynolds number specified, the smallest eddy size, $\eta$, in each LES grid is then determined. Assuming that approximately six cells are needed to resolve the smallest eddy (i.e., $\delta = \eta/6$), the subgrid resolution within each LES grid is $N_l = \Delta/\delta = 6\Delta/\eta$. For diffusion flame applications, it may be necessary to refine the linear eddy spacing even further to account for difficulties in resolving internal flame structure.

Within each large eddy cell, a single linear eddy calculation involving $N_l$ cells of size $\delta$ (and assumed volume $\delta^3$) with periodic end conditions is implemented. *No directional dependence* is associated with the individual linear eddies. The linear eddy represents a characteristic statistical state of the scalar field within the subgrid, and it is instructive to consider the linear domain as a time-varying space curve aligned with the local subgrid scalar gradient. This interpretation is consistent with the action of the rearrangement events, which always result in an increase in scalar gradient. The range of length scales can therefore be accounted for with a one-dimensional line of scalar information. Therefore, if the range of length scales in a large eddy grid cell is $N$, the scalar information can be represented with an array of size $N$, whether in one, two, or three dimensions. Conventional simulations in two dimensions would require an array of size $N^2$, and in three dimensions, $N^3$. The potential economy of this method relative to direct simulation of the three-dimensional convection-diffusion-reaction equations is apparent.

To carry out the subgrid turbulent mixing process within each LES cell, certain parameters must be determined. These are the frequency parameter $\lambda$ and parameters determining the eddy-size PDF $f(l)$. Equations (3) and (4) for $f(l)$ and $\lambda$, respectively, are still valid, with the integral length scale $L$ replaced by $\Delta$ in the subgrid model application. The linear eddy calculation in the subgrid is thus completely specified.

## 14.4.2 Mean and Turbulent Transport Across Grid Cells

In addition to the rearrangement events (turbulent convection) and molecular diffusion that are treated in each LES cell, the communication of scalar information across the LES grid cells must be carried out. These communication events reflect the LES-resolved convective transport corresponding to the resolved velocity field ($\tilde{u}$) and additional transport due to turbulent fluctuations in each LES cell. Here, the tilde denotes Favre filtering, which is used to obtain the LES-resolved equations in compressible flows (e.g., Speziale et al., 1988). It is important to include fluctuation effects, since they can cause transport locally orthogonal to the instantaneous streamlines of the resolved velocity field, or even upstream with respect to that field.

In the present implementation of the linear eddy subgrid model, convective transport of the scalar field across LES grid cells is accomplished by *splicing events*. Splicing involves (1) computation of volume transfers between pairs of adjacent LES cells, (2) for each transfer, identification of the donor and receiver cells involved based on the local LES-resolved velocity vector or velocity fluctuation vector, (3) selection of a portion of the donor-cell linear-eddy domain to be transferred, (4) insertion of a copy of the selected portion at a randomly selected location of the receiver-cell linear-eddy domain, and (5) upon completion of all transfers, removal of excess portions of the linear-eddy domains, so that there is no net change in domain sizes at the completion of steps 4 and 5. Splicing events take place at a frequency governed by the LES time step. This time step is much greater than the time step for the finite-difference solution of the diffusion–reaction equations within the linear-eddy calculations or the mean time interval between linear-eddy rearrangement events. The details of the splicing process, and its rationale based on the underlying physical mechanisms, are as follows.

Given a characteristic velocity field $u_i$ defined at the cell center in each LES grid cell, the volume flux across any cell boundary of area $dS_k$ in a time interval $\Delta t_{les}$ is given as

$$dV_k = u_k \, \Delta t_{les} \, dS_k. \tag{6}$$

Here, subscript $k$ indicates that the volume flux is computed in each of the three spatial directions (i.e., $x$, $y$, and $z$) in a three-dimensional flow. Both the LES "mean" velocity field ($\tilde{u}$) and the turbulent fluctuations in the LES cells will contribute to the volume flux. For now, we characterize the turbulent fluctuations in the LES cells by specifying the turbulence intensity $u' = \sqrt{2k}$, where $k = \frac{1}{2}\widetilde{u_i''^2}$ is the subgrid turbulence kinetic energy.

During each splicing event, three volume transfers are invoked for each pair of adjacent LES cells $i - 1$ and $i$. The first transfer is based on the volume flux across their common interface as prescribed by the LES mean velocity field, $\tilde{u}$. The other two transfers are equal-and-opposite exchanges based on the average of the turbulent velocity fluctuation $u'$ in the two cells.

Note the following features of the algorithm: (1) The first transfer involves a net volume transfer between the cells. Nevertheless, the net volume transfer into or out of any cell based on its interaction with all neighbors is zero owing to the continuity condition that is enforced within the LES algorithm. (2) The subgrid turbulent fluctuation can cause transport both orthogonal and opposite to the mean motion. Although there is no net volume transfer resulting from the equal-and-opposite exchange, there may be net transport of individual species if there are large-scale gradients of species concentration and/or density. Thus, the algorithm accommodates countergradient transport at the LES level (and in fact at the linear-eddy level as well, due to the nonlocality of the stirring process at that level).

To summarize, a given cell both donates to and receives from a given neighbor during a splicing event because there are three distinct splicings across each LES cell interface: the splicing corresponding to the LES-resolved flux and the equal-and-opposite splicings corresponding to the subgrid turbulent velocity fluctuations. A schematic representation of this process is given in Fig. 2. The portion of the linear-eddy domain of the donor cell that is chosen for the splicing and the insertion point in the linear-eddy domain of the receiver cell are both chosen randomly.

Each volume transfer that is computed is normalized by the receiver-cell volume to obtain the fraction $\phi$ of the receiver-cell volume that, at the end of the splicing process, is occupied by material originating in the donor cell. This fraction determines the portion of the donor-cell linear-eddy array copied into the receiver cell, based on the interpretation of the linear-eddy array as a representative statistical sample of spatial structure of the thermochemical field in the cell. In particular, array elements are regarded as having equal volume weighting, although an actual volume per element need not be assigned. This is consistent with the interpretation of the array as a representative linear cut through the LES cell.

On this basis, the number $N_{i,j}$ of array elements of donor cell $i$ that are inserted into the array of receiver cell $j$ (which has $N_j$ elements initially) must satisfy $N_{i,j}/N_j' = \phi_{i,j}$, where $\phi_{i,j}$ is the normalized volume transfer from cell $i$ to cell $j$ and $N_j'$ is the size of the receiver-cell array after all insertions into that array have been completed. (Here,

Suresh Menon et al.

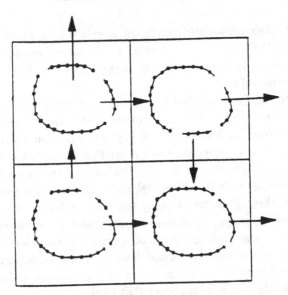

Fig. 2    Schematic illustration of linear eddy subgrid model. The one-dimensional elements within each grid cell represent the on-going linear eddy calculations that are performed in each cell. At each large eddy times step, linear eddy elements are exchanged across neighboring grid cells, accounting for large-scale transport. The arrows indicate the components of convective flux across the grid cells, which determines the amount of scalar information contained in the linear eddies that is exchanged at each large eddy time step.

it has been assumed that each linear-eddy array has the same absolute spatial resolution. Resolution variation from cell to cell can also be treated, but the algorithm becomes more complicated.) $N'_j$ is equal to $N_j$ plus the sum $\sum_i N_{i,j}$ of all donations to cell $j$ during the splicing step. The foregoing relations give $N'_j = N_j/(1 - \sum_i \phi_{i,j})$, and thus $N_{i,j} = N_j \phi_{i,j}/(1 - \sum_i \phi_{i,j})$.

Note that $N'_j$ is obtained by adding to $N_j$ the donations to cell $j$, but that donations from cell $j$ to other cells are not subtracted. This is because donation is a copying process that does not involve removal of the copied elements from the donor array. This ostensible replication of material does not violate conservation laws because the linear-eddy arrays are treated as representative statistical samples of the concentration field and, accordingly, they are used to determine intensive quantities but not extensive quantities.

The increase in the size of the linear-eddy arrays due to this copying process causes no difficulty in principle, but it does increase the computational cost. Therefore, as a final step in the splicing process, a randomly selected portion of each linear-eddy array is removed, such that the remaining portion contains only the required number of elements. Because the selection is random, the remaining portion is a representative statistical sample of the concentration field.

The treatment of scalar transport as a two-step process of copying followed by removal of excess elements causes individual realizations to violate species conservation and, for variable-density flows, total mass conservation. The impact of this artifact on the predictive capability of the model should be assessed on a case-by-case basis. A strictly conservative algorithm can be formulated, but its advantages may not outweigh the added complexity.

Another artifact is the occasional creation of spurious chemical states due to the fact that splicing can instantaneously bring initially separated linear-eddy cells into direct contact. For instance, this can result in cold fuel meeting cold air in nonpremixed configurations, including configurations in which such nonburning fuel–air interfaces do not occur physically. This artifact is mitigated by the rarity of splicing events relative to the subgrid-scale events that account for most of the mixing during the simulation.

In summary, the linear eddy subgrid model implemented in each LES cell (Section 4.1) represents the physical processes occurring at the small scales, and the splicing process described in this section accounts for the transport due to the large-scale motions and the local subgrid turbulent fluctuations. Species transport equations are not solved along with the LES equations for the mass, momentum, and energy, since the large-scale convection is accounted for by the splicing process and the small-scale mixing and diffusion is implemented in the linear-eddy calculations. The large-scale flow affects the scalar field through the splicing process and through the closure model for the subgrid stresses which determine the value of $u'$ in each LES cell. Note that $u'$ plays the dual role of determining the fluctuation contribution to large-scale transport as implemented by splicing and determining the value of the grid Reynolds number that is used in Eq. (4) for the rearrangement frequency parameter $\lambda$.

The scalar properties affect the large-scale flow through the equation of state, as discussed in Section 3.4. The linear-eddy computations embody all thermochemical properties of the flow. Averages and other statistical procedures, whether for updating the equation of state or for comparison with measured properties, are applied directly to the linear-eddy domains in the respective LES cells.

*14.4.3 Implementation of the Subgrid Model in an LES Code*

In this section, we outline the approach by which the subgrid model for scalar transport can be fully coupled to the Navier–Stokes solver used for LES. Since the scalar transport processes, as well as mixing and chemical reaction, are relegated to the individual linear-eddy computations and the splicing process, the LES simulation does not involve the explicit computation of scalar transport. The number of equations in the LES solver will, as a result, remain the same whether or not scalar processes are being modeled. The LES equations for compressible flows consist of a conservation equation for density ($\bar{\rho}$), conservation equations for momentum ($\widetilde{\rho\mathbf{u}}$), and a conservation equation for total energy ($\widetilde{\rho E}$). These equations are obtained using an appropriate filter function, for example, the Favre filter (e.g. Speziale et al., 1988). To close the momentum equations, the subgrid stresses require modeling. In addition, all velocity-coupled correlations appearing in the energy equation must also be modeled. The present subgrid approach does not attempt to address these closure issues. It is assumed that available subgrid models for the subgrid stresses and other velocity-coupled correlations can be employed. However, some of the terms in the filtered (LES) equations are affected by the evolution of the subgrid-scale scalar fields, as discussed in Sections 3.4 and 4.2. This coupling, mainly reflecting heat release, is effected through the determination of LES-filtered temperature $\tilde{T}$ and species mass fractions $\tilde{Y}_k$ by averaging over the respective linear-eddy domains.

The LES method with the flame propagation subgrid model described above is currently being applied to premixed combustion in a ramjet dump combustor. The following section describes the implementation of the linear eddy model in a simpler formulation in which the resolved velocity field and the fluctuation intensity $u'$ are specified rather than computed from dynamical equations. This formulation highlights the novel aspects of the subgrid mixing model implementation in the context of a premixed combustion application.

## 14.5 Application to Premixed Combustion

Several simplifications are adopted in the following demonstration of the linear eddy subgrid mixing model. Both $\Delta$ and the subgrid rms fluctuation $u'$ are specified as constants over the entire domain. This allows us to fix the subgrid Reynolds number and the linear eddy cell size $\delta$ in every LES grid cell. This simplification reduces computational requirements while allowing the features of the subgrid mixing model to be demonstrated.

In a general LES code, the LES grid will be nonuniform and hence $\Delta$ will be different at various locations. Also, the subgrid Reynolds number will vary throughout the domain, since both $\Delta$ and $u'$ will be nonuniform. As a result, the linear eddy cell resolution $N_l$ and the stirring time scale $\tau_s$ will also vary in the computational domain. However, the general implementation of the subgrid mixing model exemplified here is the same as it would be in a full Navier–Stokes solver. The extension of the present approach to account for nonuniform grid and subgrid turbulent fluctuations can be accomplished without affecting the major conclusions of the present discussion.

The linear-eddy subgrid model for flame propagation has been studied for a physical configuration that models a vortex roll-up in a turbulent mixing layer. A time-dependent analytic velocity field that describes a vortex roll-up in a shear layer was used. This analytical field was used in an earlier study by Cetegen and Sirignano (1988) and satisfies the incompressible equation of continuity. The analytic form of the velocity field is given as

$$u(x,y,t) = \left(\frac{\Gamma}{2a}\right)\frac{\sinh\left(2\pi y/a\right)}{\cosh\left(2\pi y/a\right) - \cos\left(2\pi x/a\right)} - \left(\frac{\Gamma}{2\pi}\right)\frac{y}{x^2+y^2}\exp\left[-(x^2+y^2)/4\nu t\right], \tag{7}$$

$$v(x,y,t) = (-\Gamma/2a)\frac{\sin\left(2\pi x/a\right)}{\cosh\left(2\pi y/a\right) - \cos\left(2\pi x/a\right)} - \left(\frac{\Gamma}{2\pi}\right)\frac{x}{x^2+y^2}\exp\left[-(x^2+y^2)/4\nu t\right]. \tag{8}$$

This velocity field was computed at each time step in the computational domain. The domain was chosen as a segment $[-\pi, \pi]$ in both the $x$- and $y$-directions and an LES grid resolution of $32 \times 32$ was employed for all the calculations. The constant $\Gamma$ and the kinematic viscosity $\nu$ were set to values of $2a$ and 0.0005, respectively. Here, $a$ is the spacing between the vortices in the shear layer. We choose $a = 2\pi$, implying that there is only one vortex present in the computational domain.

For the following demonstration simulations, the effect of heat release was studied for a fixed value $T_{product}/T_{fuel} = 5$. Two simulations (Cases 1 and 2) with different values of $u'$ were carried out with the same laminar flame speed $S_L$ to study the effect of variable turbulence intensity in the LES cells. The effect of changing fuel type can be modeled by changing the laminar flame speed $S_L$. The turbulence intensity $u'$ is determined by specifying the subgrid Reynolds number (based on $\Delta$ and $u'$). Simulations with $Re = 50$ and $Re = 100$ were performed, corresponding to $u'/S_L$ =1.5 for case 1 and 3.0 for case 2.

As noted earlier, we assumed that six linear-eddy cells are required to resolve the smallest eddy size ( the Kolmogorov scale). For the lower value of $u'/S_L$ (i.e., case 1) a linear eddy resolution of 100 elements in each LES cell was sufficient, and for the higher $u'/S_L$ (i.e., case 2) a resolution of 200 elements was required. The most reliable method for determining the resolution requirements is to increase resolution until the computed results become insensitive to resolution. This method was employed in previous linear-eddy applications but not in the present demonstration.

The time step $\Delta t_{les}$ used to advance the mean velocity field on the large eddy grid was taken to be 0.01. This is approximately one-third of the CFL condition stability limit. The sensitivity of the results to the large eddy time step will be addressed in a rigorous manner in the future.

Between successive LES updates, linear eddy calculations for flame propagation, thermal expansion, and turbulent stirring (as described in Section 14.3.1) are carried out in each LES cell. In conjunction with each LES update, the splicing algorithm described in Section 14.3.2 is used to update each linear-eddy array to account for the mean (based on the specified velocity field) and turbulent (based on the specified $u'$) fluxes across each of the four cell surfaces.

To begin each simulation, the $G$ field in each linear-eddy array is set equal to a constant value depending on the nominal $y$ location of the corresponding LES grid cell, with the $G$ value linearly decreasing from unity at $y = -\pi$ to zero at $y = \pi$. (The initial field can be chosen arbitrarily provided that the level surface $G = G_o$, where in this case $G_o = 0.5$, corresponds to the desired initial location of the flame surface.)

As the subgrid flame propagation process evolves according to Eq. (2), the $G$ field increases wherever its gradient is nonzero. The gradient is initially zero within each linear-eddy array, but splicing causes $G$ variations within individual arrays and thus initiates the propagation process. Due to turbulent stirring that is implemented concurrently, a single flamelet (corresponding to a single crossing of the level $G = G_o$) will spawn multiple flamelets. Physically, this can be interpreted as an effect of the wrinkling of the flame surface due to the turbulent stirring process. A measure of the wrinkling effect can also be obtained by interpreting the spacing between the flamelets within an LES cell as an estimate of (twice) the radius of curvature of the wrinkles on the flame surface.

The total number of the flamelets gives an estimate of the turbulent flame speed $u_T$ of the flame brush within the linear-eddy array, analogous to the subgrid viscosity and subgrid diffusivity. Thus, the subgrid model as implemented not only models the small-scale processes correctly, but also provides a means to estimate the subgrid turbulent flame speed, a quantity often treated as an input in other formulations. Also, since both $S_L$ and $u'$ are inputs to the subgrid model, the functional relationship $u_T = u_T(S_L, u')$, which is of great interest, can be determined. Earlier, a simplified version of this subgrid model was implemented in a stand-alone mode to study this functional relationship (Menon and Kerstein, 1992). One of the results of that study was that for the case of no heat release, the turbulent flame speed $u_T$ scales linearly with the turbulence intensity $u'$, consistent with scaling analysis of the turbulent flame propagation process. It is shown by the results described below that a similar conclusion can be obtained with heat release included.

To present the results of the simulations, a progress variable $\bar{g}$ is defined in each LES grid cell to represent the fraction of burned cells $(G > G_o)$ in the corresponding linear-eddy array. Thus, when $\bar{g} = 0$, all linear eddy elements are unburned and when $\bar{g} = 1$, all elements are burned Any intermediate value of $\bar{g}$ represents partial conversion. This definition of the progress variable is the same as used by Bray (1980) in earlier premixed combustion studies.

Figure 3 shows scatter plots of the subgrid turbulent flame speed, $u_T$, normalized by $u'$ as a function of the progress variable at two instants during the simulation for Cases 1 and 2. A linear dependence of the turbulent flame speed on the subgrid turbulent intensity is evident since the simulations with different values of $u'$ appear to collapse on each other. Each point in these scatter plots represents the subgrid state in an LES grid cell at the chosen instant. It is evident that the simulations exhibit the parabolic dependence $\bar{g}(1 - \bar{g})$ of the reaction rate on the progress variable that follows from elementary geometrical considerations concerning the relationship between the burned volume fraction and the flame surface area (Bray, 1980). In contrast to other formulations, this relationship is an outcome of the model rather than an input to the model. Furthermore, the model can be used to investigate

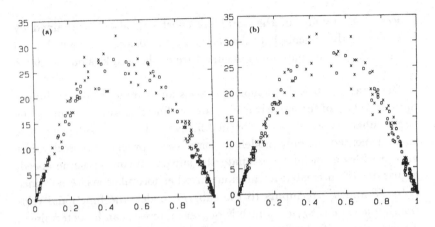

Fig. 3   Scatter plot of $u_T/u'$ vs. progress variable for Cases 1 and 2. Linear dependence of flame speed is indicated by collapse of normalized $u'$ values. a) $t=t_1$. b) $t=2t_1$.

the mechanistic origin of any random or systematic deviations from this relationship that may occur.

Most of the points in the scatter plots are clustered near the two extreme values of $\overline{g}$ reflecting the fact that many of the LES cells are in either the fully unburned ($\overline{g} \sim 0$) or the fully burned ($\overline{g} \sim 1$) state. As time evolves and the flame propagates into the fuel, an increasing proportion of the LES cells become fully burned. This can be seen more clearly in Fig. 4, which shows the cumulative distribution of the progress variable at the same two times as in Fig. 3. At the earlier time, nearly 40% of the LES cells are fully burned, but at the later time nearly 60% of the LES cells are fully burned.

The effect of increasing $u'$ is to increase the magnitude of the turbulent flame speed, which results in a faster propagation of the turbulent flame brush through the fuel mixture. This results in an increase in the number of cells that are in the burned state at a given instant, as can be seen in Fig. 4.

Figures 5a and 5b show the contours of the progress variable, for the earlier and later time, respectively, for Case 1, and Figs. 5c and 5d show the corresponding contours for Case 2. The propagation of the flame into the fuel is seen in these figures, and comparison of Figs. 5a and 5b with 5c and 5d shows that the increase in $u'$ results in the faster propagation of the flame. Note that the mean velocity field, as given by Eqs. (7) and (8), models a clockwise rotating vortex field centered in the middle of

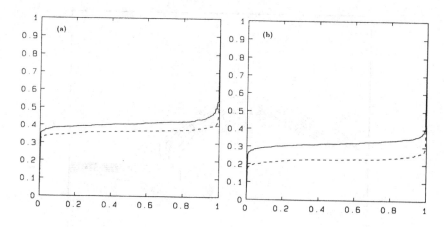

Fig. 4 Cumulative distribution of progress variable at same times as Fig. 3. a) $t=t_1$. b) $t=2t_1$.

these contour plots. This results in a significant negative $v$ velocity near $y = 0$ and $x = \pi$ and can result in a transport of the fuel in the negative $y$-direction by the splicing mechanism. However, the counterpropagating flame brush overcomes this downward convective flux.

It is evident in Figs. 5a and 5c that the width of the turbulent flame brush at a given instant, as reflected in the spread of the progress-variable contours, is slightly higher for the lower of the two $u'$ values. Physically, the opposite trend, if any, would be expected. This may be an artifact of the selection of an LES time step that is too large. Note that an increase of $u'$ has two effects on the propagation process. First, stirring and therefore the overall conversion rate within each linear-eddy domain is accelerated. Second, the turbulent fluctuation contribution to the splicing process is enhanced. The first effect causes partially burned domains to achieve complete conversion more rapidly at large $u'$, thereby reducing the population of partially burned domains at any instant. The second effect accelerates the ignition of otherwise unburned domains, thereby contributing to the spread of the turbulent flame brush. If the time scale for conversion of individual domains is less than the LES time step, then the second effect may be artificially inhibited, though the first effect is fully represented. These considerations indicate the importance of adequately resolving the LES time step as well as the linear-eddy length and time scales in order to obtain a physically valid simulation.

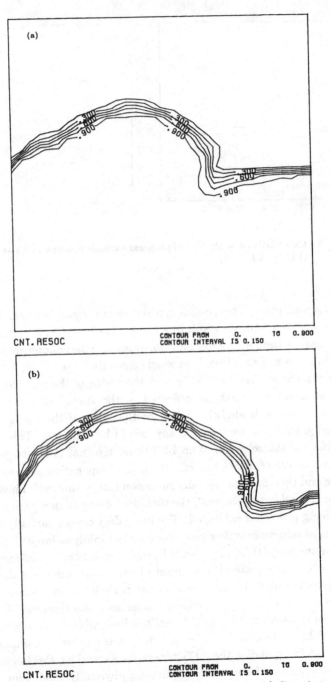

Fig. 5   Contours of progress variable. a) Case 1, $t=t_1$. b) Case 1, $t=2t_1$.
c) Case 2, $t=t_1$. d) Case 2, $t=2t_1$.

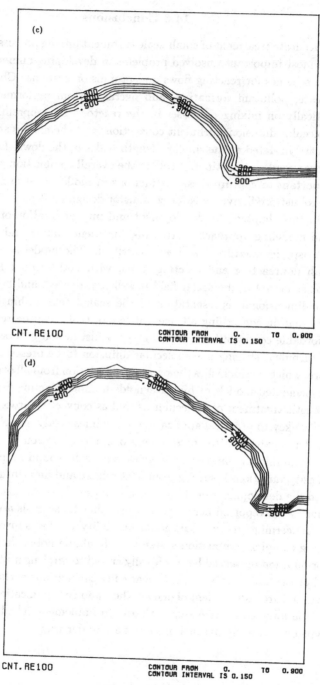

Fig. 5 (continued)

## 14.6 Conclusions

Accurate treatment of small-scale turbulent mixing processes is among the most important unsolved problems in developing numerical predictive schemes for reacting flows and propulsion systems. Chemical heat release, pollutant formation, and overall system performance depend critically on mixing processes in the reactor. In nonpremixed flames, molecular diffusion, turbulent convection, and chemical reaction are intimately related at the smallest length scales in the flows. Each of these processes plays a distinctive role in the overall combustion process. It is important to capture these distinctions in models that are intended to be robust predictive tools for combustor design.

In this chapter, the development and implementation of a new subgrid modeling approach to treating turbulent mixing and reaction for multispecies mixtures has been described. The model is applicable to both nonreacting and reacting flows, with and without heat release. The structure of the scalar field is well represented, and because of the one-dimensional representation of the scalar field within the subgrid, information concerning all relevant length scales can be retained in an affordable computation. This allows the distinction between the effects of turbulent stirring and molecular diffusion to be preserved. This feature, which distinguishes the present formulation from all other currently implemented models of high Reynolds number reacting flows, allows a realistic treatment of molecular as well as convective processes.

The key to practical application of the linear eddy model as a subgrid model is extension to three dimensions, incorporation of mesh generation suitable for complex geometries, physically sound coupling between density variations resulting from heat release and momentum transport, and the development of efficient computational algorithms. The preliminary work reported here demonstrates that these goals are obtainable.

Concerning the computational feasibility of this approach for handling complex combustion systems, it should be noted that the subgrid formulation presented here is ideally suited to implementation on parallel processing machines. The linear eddy calculations performed in each grid cell are independent of one another, and communication occurs only at the large eddy time steps. Efforts to implement calculations of this type on massively parallel machines are under way.

## Acknowledgments

The work was supported in part by NASA-Ames Research Center under Contract NAS2-13354 to Quest Integrated, Inc (SM), the Division of Engineering and Geosciences, Office of Basic Energy Sciences, U.S. Department of Energy (AK), and the Advanced Combustion Engineering Research Center (PM). Funds for the Research Center are received from the National Science Foundation, the state of Utah, 26 industrial participants, and the U.S. Department of Energy.

## References

BRAY, K.N.C. (1980) Turbulent flows with premixed reactants. In *Turbulent Reacting Flows.* Ed. P.A. Libby and F.A. Williams, pp. 115–184. Springer-Verlag, Berlin.

BROADWELL, J.E., AND MUNGAL, M.G. (1991) Large-scale structures and molecular mixing. *Phys. Fluids A* **3**, pp. 1193–1206.

CETEGEN, B.M., AND SIRIGNANO, W.A. (1988) Study of molecular mixing and a finite rate chemical reaction in a mixing layer. *22nd Symp. (Int.) on Combustion*, The Combustion Institute, pp. 489–494.

KERSTEIN, A.R. (1988) Linear-eddy model of turbulent scalar transport and mixing. *Comb. Sci. Tech.* **60**, pp. 391–421.

KERSTEIN, A.R (1989) Linear-eddy modeling of turbulent transport. Part 2: Application to shear layer mixing. *Comb. Flame* **75**, pp. 397–413.

KERSTEIN, A.R. (1990) Linear-eddy modeling of turbulent transport. Part 3: Mixing and differential molecular diffusion in round jets. *J. Fluid Mech.* **216**, pp. 411–435.

KERSTEIN, A.R. (1991) Linear-eddy modeling of turbulent transport. Part 6: Microstructure of diffusive scalar fields. *J. Fluid Mech.* **216**, pp. 361–394.

KERSTEIN, A.R. (1992) Linear-eddy modeling of turbulent transport. Part 7. Finite-rate chemistry and multi-stream mixing. *J. Fluid Mech.* **240**, pp. 289–313.

KERSTEIN, A.R., AND ASHURST, W.T. (1992) Propagation rate of growing interfaces in stirred fluids. *Phys. Rev. Let.* **68**, pp. 934–937.

KERSTEIN, A.R., ASHURST, W.T., AND WILLIAMS, F.A. (1988) Field equa-

tion for interface propagation in an unsteady homogeneous flow
field. *Phys. Rev. A* **37**, pp. 2728–2731.

McMurtry, P.A., Menon, S., and Kerstein A R. (1991) A linear eddy
subgrid model for turbulent combustion: Application to $H_2$–air
combustion. Presented at the 4th Int. Conf. on Numerical Combustion, St. Petersburg, FL, Dec 2–4.

McMurtry, P.A., Menon, S., and Kerstein A R. (1992) A new subgrid
model for turbulent mixing and reactions. AIAA Paper No. 92-
0234.

Menon, S., and Kerstein, A.R. (1992) Stochastic simulation of the
structure and propagation rate of turbulent premixed flames. *Proceedings of the 24th Symp. (Int.) on Combustion*, in press.

Menon, S., and Jou, W.-H. (1991) Large-eddy simulations of
combustion instability in an axisymmetric ramjet combustor.
*Comb. Sci. Tech.* **75**, pp. 53-72.

Menon, S., McMurtry, P.A., and Kerstein, A.R. (1991) A linear-eddy
flamelet subgrid model for large-eddy simulations of turbulent premixed combustion. Presented at the 4th Int. Conf. on Numerical
Combustion, St. Petersburg, FL, Dec 2–4.

Menon, S., McMurtry, P.A., Kerstein, A.R., and Chen, J.-Y. (1992)
A mixing model to predict NOx production during rapid mixing
in a dual-stage combustor. AIAA Paper No. 92-0233.

Speziale, C.G., Erlebacher, G., Zang, T.A., and Hussaini, M.Y.
(1988) On the subgrid-scale modeling of compressible turbulence.
*Phys. Fluids* **31**, pp. 940–942.

# 15

## Direct Numerical Simulation and Large Eddy Simulation of Reacting Homogeneous Turbulence

CYRUS K. MADNIA AND PEYMAN GIVI

### 15.1 Introduction

In reviews of recent computational work on turbulent reacting flows, Oran and Boris (1987), Givi (1989), and Drummond (1991) indicate the need for further development in both the methodology and the implementation of direct numerical simulation (DNS) and large eddy simulation (LES). According to these reviews, previous and ongoing efforts to develop advanced numerical algorithms for simulating reacting turbulent fields have been successful, thus warranting their continued utilization. With the broad knowledge gained to date, it is now widely accepted that anticipated developments in advanced computational facilities will not be sufficient to relax the restriction of DNS to flows having small to moderate variations of the characteristic length and time scales (Rogallo and Moin, 1984; Schumann and Friedrich, 1986, 1987; Givi, 1989; Reynolds, 1990). Nevertheless, DNS can be (and has been extensively) used to enhance our understanding of the physics of chemically reacting turbulent flows by providing specific information concerning the detailed structure of the flow and by furnishing a quantitative basis for evaluating the performance of turbulence closures. In this sense, DNS has provided an effective tool for such studies and has established useful guidelines for future investigations.

LES appears to provide a good alternative to DNS for computing flows having ranges of parameters similar to those encountered in practical systems (Ferziger, 1981, 1982; Schumann and Friedrich, 1987; Hussaini et al., 1990; Reynolds, 1990). This approach has a particular advantage over analogous Reynolds-averaging procedures in that only the effects of small-scale turbulence have to be modeled. However, despite their success in the analysis of both incompressible (Ferziger, 1981) and compressible (Erlebacher et al., 1987) flows, they have seldom been employed for calculations of reacting turbulence. This is primarily due to the appearance of additional unclosed terms in the transport equation governing the evolution of the filtered quantities in LES. In these equations, the correlations among the scalar variables must be accurately modeled in order to provide a realistic estimate of the reactant conversion rate.

A methodology which has proven useful for predictions of turbulent reacting flows is based on the probability density function (PDF) or the joint PDF of the scalar variables which characterize the field (Pope, 1985, 1990). A particular advantage of the method is the fundamental property of the probability density, namely that it provides the complete statistical information about the behavior of the variable. This, in comparison with moment methods, provides a unique advantage in that once the distribution of the PDF is known, the mean and higher order moments of the reaction rate (or any other functions of the scalar field) can be directly determined without a need for further closure assumptions. The implementation of PDF methods in conjunction with the Reynolds- and/or Favré-averaged form of the equations of turbulent combustion has proven very successful (for a recent review see Pope, 1990). During the past 15 years, these methods have been used for simulating a variety of reacting flow configurations, and in many cases, the predicted results have been indicative of the higher accuracy of these methods in comparison with conventional moment methods. On the basis of this indication, it is safe to state that from the statistical point of view, PDF methods are the most attractive choice for the description of chemically reacting turbulent flows.

A similar scenario may be surmised to be the case when the PDF methods are used in the context of subgrid closures for LES. In this case, the PDF (or the joint PDF) of the scalar variables within the ensemble of grids (identifying the large scales of the flow) may uniquely determine the subgrid average value of the reaction rate (or other functions of the scalar variables). This PDF, in conjunction with an appropriate model for the hydrodynamic fluctuations within the subgrid, may facilitate the implementation of LES for the analysis of reacting turbulence. Our main goal in this work is to initiate an assessment of the use of PDF methods for their possible future implementation in LES, as well as to further our investigation of the PDF methods in modeling turbulent combustion problems. The approach followed is similar to those previously used in *a priori* analyses. Namely, the data obtained by DNS are utilized to construct the appropriate PDFs that can be used for either turbulence modeling or subgrid closures. We are currently in a preliminary stage of our work, especially in tasks directed at subgrid modeling.

In order to establish a framework for the discussion of the results, a review of PDF methods and subgrid closures is provided in Section 15.2. This rather brief review is presented solely to provide a reference for the remainder of the chapter. The reader is referred to other edited works and survey articles which discuss these statistical methods in greater detail. Also, the discussion on PDF methods focuses on the treatment of scalar quantities alone without including the treatment of scalar-velocity PDFs. The specification of the problem under current scrutiny is provided in Section 15.3, in which the relevance of some of the characteristic parameters is expressed, together with a description of the geometrical configuration and the initialization procedure. The results of simulations are presented in Section 15.4, in which we emphasize the use of stochastic methods for modeling the mixing phenomenon. In Section 15.5 we summarize our findings and also provide some speculations and suggestions for future work.

## 15.2 Background Review
### *15.2.1 PDF Methods*

In the majority of previous work on the computational treatment of turbulent combustion, statistical methods have generally been employed. These methods involve the representation of physical variables in a stochastic sense and, in conjunction with appropriate transport equations, predict the approximate behavior of the flow field. In the framework of such methods, the physical quantities are treated as *random* variables and the transport equations describing the evolution of these quantities are used in such a way as to yield their stochastic variations. At present, two general types of strategy are identified which can be utilized for statistical treatment of turbulent combustion problems (Libby and Williams, 1980): (1) the moment method and (2) the PDF method. In the first approach, "averages" (or "means") of the relevant physical variables are introduced, and the governing transport equations are subsequently averaged. These averages are generally defined in such a way as to yield the desired mathematical properties and are usually introduced in the form of time, ensemble, or Favré averages. The PDF approach relies directly on the PDF of these variables by which the desired moments can be obtained.

Moment methods usually involve the determination of the statistical means by virtue of solving transport equations for the various unclosed moments of the equations. These equations provide, in a sense, a description of the variables at one level higher than the equations for the mean values (first moments). This results in explicit transport equations for the second order moments. These equations, however, are still not in a closed form due to the appearance of the unclosed third order correlations which require further modeling. Regardless of the degree of truncation of the higher moments, the closure problem always remains and transport equations at any level require modeling at higher orders.

The closure based on the second approach, the PDF method, has proven very useful in the theoretical description of turbulent reactive flows (Hawthorne et al., 1949). The philosophy of this approach is to consider the transport of the PDFs rather than the finite moments of the statistical variables. Representing the field involving $\sigma$ scalars by $\Phi = (\phi^1, \phi^2, \ldots, \phi^\sigma)$, and the reaction conversion rate of scalar $\alpha$ by $\omega_\alpha(\Phi)$, the mean of this rate (denoted by $\langle\ \rangle$) can be written as

$$\langle \omega_\alpha \rangle = \int_{-\infty}^{+\infty} \mathcal{P}_1(\Xi) \omega_\alpha(\Xi)\, d\Xi. \tag{1}$$

In this equation, $\Xi$ represents the scalar space and $\mathcal{P}_1(\Xi)$ is the PDF of the scalar variable at a point, defined so that

$$\mathcal{P}_1(\Xi) d\Xi = \text{Probability } (\Xi \leq \Phi \leq \Xi + d\Xi), \tag{2}$$

where $d\Xi = (d\xi^1, d\xi^2, \ldots, d\xi^\sigma)$ denotes an elemental hypervolume at $\Xi$.

The scalar PDF, $\mathcal{P}_1(\Xi)$, defined in Eqs. (1)–(2), is in its simplest form in that it contains information only about the scalar variables, $\Phi$. Also, it governs the probability distribution only at a single point. However, it contains much more information than is required to determine the mean value of any functions of the random variables $\Phi$. Therefore, its evaluation is understandably more difficult than direct evaluation of $\langle \omega_\alpha \rangle$ (if it could be done). Nevertheless, since $\mathcal{P}_1(\Xi)$ includes all the statistical information about the scalars, its determination is in many ways more useful than that of the mean values.

A systematic way of evaluating a PDF is to obtain and solve a transport equation governing its evolution. A transport equation for $\mathcal{P}_1(\Xi, \Xi = \xi^1, \xi^2, \ldots, \xi^\sigma)$ can be constructed by relating the PDF to $\Phi(\underline{x}, t)$ in terms of Dirac delta functions (Pope, 1985),

$$\mathcal{P}_1(\Xi) = \langle \varrho(\underline{x}, t) \rangle,$$ (3)

where $\varrho$ is the single-point fine-grained density defined by

$$\varrho = \prod_{\alpha=1}^{\alpha=\sigma} \delta(\xi^\alpha - \phi^\alpha).$$ (4)

From the appropriate conservation equations for the scalar field and Eqs. (1)–(4) one can obtain a transport equation describing the evolution of the PDF. Limiting the formulation to that of a constant-density Fickian diffusion flow, we have

$$\frac{\partial \mathcal{P}_1}{\partial t} + \langle \underline{V} \cdot \nabla \varrho \rangle = -\frac{\partial}{\partial \xi^\alpha}(\omega_\alpha \mathcal{P}_1) + \Gamma \langle \nabla_{\underline{x}}^2 \phi^\alpha \frac{\partial \varrho}{\partial \xi^\alpha} \rangle.$$ (5)

Here, $\underline{V}$ is the velocity field and $\Gamma$ is the diffusion coefficient of the species. From this equation, it can be seen that the effects of reactivity and diffusivity (the terms on the right-hand side) are to transport the PDF into the composition space ($\Xi$), whereas the convective transport is a physical space phenomenon. Furthermore, it is shown that the influences of $\omega_\alpha(\Phi)$ in the composition space appear in closed form. However, models are needed for the closures of the molecular mixing term (second term on the right-hand side) and the turbulent convection term (second term on the left-hand side).

Since the term that includes the interaction between the scalar variables through the reaction rate $\omega_\alpha$ appears in a closed form in Eq. (5), in what follows we limit the discussion to the transport of a single reacting scalar (i.e., $\sigma = 1$) which is being convected by the velocity field. In this frame, the diffusive transport may be expressed as

$$\langle \nabla_{\underline{x}}^2 \phi \frac{\partial \varrho}{\partial \xi} \rangle = \frac{\partial}{\partial \xi} \lim_{\underline{x}_2 \to \underline{x}_1} \nabla_{\underline{x}_2}^2 [E(\xi_2|\xi_1)\mathcal{P}_1(\xi_1)].$$ (6)

Here, $E(\xi_2|\xi_1)$ represents the expected value of the scalar at $\underline{x}_2$ conditioned on its value at $\underline{x}_1$. This equation explicitly reveals a fundamental difficulty associated with the PDF formulations, namely that the transport equation for a one-point PDF ($\mathcal{P}_1(\xi, \underline{x}, t)$) requires information about the second point, $\underline{x}_2$. In general, $n$-point PDFs, $\mathcal{P}_n(\xi_1, \xi_2, \ldots, \xi_n, \underline{x}_1, \underline{x}_2, \ldots, \underline{x}_n, t)$, require information about the $(n+1)$th point, $\underline{x}_{n+1}$. This is further demonstrated by considering the transport of $\mathcal{P}_n$ in the compositional space (Ievlev, 1973; Kuo and O'Brien, 1981),

$$\frac{\partial \mathcal{P}_n}{\partial t} + \sum_{q=1}^{n} \frac{\partial}{\partial \xi_{(q)}} [\omega(\xi_{(q)})\mathcal{P}_n] = \sum_{q=1}^{n} \frac{\partial}{\partial \xi_{(q)}} (C^{(q)}\mathcal{P}_n),$$ (7)

where

$$C^{(q)} = \Gamma \lim_{\underline{x}_{n+1} \to \underline{x}_{(q)}} \nabla^2_{\underline{x}_{n+1}} [E(\xi_{n+1}|\xi_1, \xi_2, \ldots, \xi_n, \underline{x}_1, \underline{x}_2, \ldots, \underline{x}_n)] \qquad (8)$$

and $E(\xi_{n+1}|\xi_1, \xi_2, \ldots, \xi_n, \underline{x}_1, \underline{x}_2, \ldots, \underline{x}_n)$ is the expected value of $\xi_{n+1}$ at the point $\underline{x}_{n+1}$ conditioned on the values of $\xi_1, \xi_2, \ldots, \xi_n$ at $\underline{x}_1, \underline{x}_2, \ldots, \underline{x}_n$. The dependence of $C^{(q)}$ on the $(n+1)$th point in the transport equation for $\mathcal{P}_n$ indicates the need for a closure model which relates $\mathcal{P}_{n+1}$ to $\mathcal{P}_n$.

Despite the attractiveness of PDF methods in obtaining a closed representation of the reaction term, there is another difficulty with the implementation of the method. This difficulty is caused by the increase of the dimensionality of the PDF as the number of scalars and/or the number of statistical points increase. This imposes a severe limit on the maximum number $n$ in the equation for $\mathcal{P}_n$. The majority of previous work on PDFs has closed the equations at the first level $(n = 1)$ and only recently has closure at the $n = 2$ level been attempted. Both these closures are reviewed by Givi (1989).

### 15.2.2 LES and Subgrid Closures

Despite the present capability of modern supercomputers to allow calculations with more than 1 million grid points, the range of length and time scales that can be resolved by DNS is substantially smaller than those of turbulent flows of practical interest. In DNS the largest computable scales are limited by the size of the computational domain and the turnover time of the large-scale structures; the smallest resolvable features are limited by the molecular length and the time scales of the viscous dissipation and chemical reactions. DNS, therefore, in comparison with turbulence modeling, finds its greatest application in basic research problems in which the scales of the excited modes remain within this band of computationally resolved grid sizes. This implies that, for an accurate simulation the magnitudes of the viscosity and the diffusivity must be large enough to damp out the unresolved scales, and the size of the computational time step must be kept small enough to capture the correct temporal evolution of the turbulent flame. In this context, since the instantaneous behavior of the flow variables is directly simulated at all spatial points at every instant, the computed data can actually be used to provide a quantitative basis for evaluating the validity of turbulence models and for assessing the performance of subgrid closures in LES.

The limitations associated with DNS may be alleviated, to an extent, by prefiltering the transport equations (Aldama, 1990), a practice implemented in LES. This is effectively equivalent to eliminating the scales smaller than those resolvable within a given mesh. In this way, the variables can be represented on the number of grid points that are available for the simulations. This filtering procedure facilitates the simulations of flows with larger parameter ranges on a coarser grid. The disadvantage is that some modeling is required for the closure of the scales excluded by filtering. The selection between DNS and LES (and turbulence modeling for that matter) is dependent on the type of flow being considered, on the range of physical parameters that characterize the turbulent field, and the degree of desired accuracy.

A straightforward implementation of LES involves the decomposition of the transport variables into large-scale and subgrid-scale components. The former is related to the large-scale eddies in the turbulent field, whereas the latter is the component containing the small-scale fluctuations. The pre-filtering of the variable $\Phi(\underline{x}, t)$ is performed by means of the convolution integral (Ferziger, 1977; Aldama, 1990),

$$\widetilde{\Phi}(\underline{x}, t) = \int \int \int_{\Delta} F\ell(\underline{x} - \underline{x}')\Phi(\underline{x}', t)\, d\underline{x}', \qquad (9)$$

where $F\ell$ is an appropriate filter function with a characteristic length $\Delta$, and the integration is over the entire flow field. The remaining portion of $\Phi$ from the filtered quantity is defined as the subgrid-scale field and is represented by

$$\Phi''(\underline{x}, t) = \Phi(\underline{x}, t) - \widetilde{\Phi}(\underline{x}, t). \qquad (10)$$

The magnitudes of prefiltered averaged $\widetilde{\Phi}$ and its subgrid component $\Phi''$ are obviously dependent on the filter type (function $F\ell$) and its size $\Delta$ (for reviews see Ferziger, 1981, 1982). The simplest choice would be a box filter with

$$F\ell(\underline{x} - \underline{x}') = \begin{cases} 1, & \text{if } |\underline{x} - \underline{x}'| \leq \Delta \\ 0, & \text{otherwise,} \end{cases} \qquad (11)$$

where $\Delta = (\Delta_x, \Delta_y, \Delta_z)$ represents the dimensions of the box, and the length in each direction should be larger than the original grid spacing in that direction. Other types of filters have also been suggested in the literature, each with computational and physical advantages as well as limitations (see Schumann and Friedrich, 1986, 1987; Aldama, 1990).

Implementation of LES as a practical tool involves a combination of DNS for the filtered portion of the transport variable $\widetilde{\Phi}$, and modeling of the subgrid-scale portion $\Phi''$. The transport equations for the large-scale components are obtained by filtering the instantaneous transport equations by means of implementing the filter defined by Eq. (9). In the resulting filtered equations, analogous to those in Reynolds-averaged ones, the equations representing the large-scale field contain unclosed terms involving the fluctuation of the subgrid components. The methodology practiced to date in dealing with these fluctuations is very similar to that employed in Reynolds-averaging procedures. One may be able to develop and solve transport equations for the moments of fluctuations (Ferziger, 1981). These equations, analogously, contain some higher order moments which need to be modeled. There are more of these terms than of the corresponding ones that appear in Reynolds-averaged transport, because some items that are zero in the Reynolds-averaging approach are nonzero when filtering is used. Correspondingly, the task required for modeling the subgrid fluctuations would be somewhat more complicated than that in turbulence modeling. Nevertheless, since the small scales of turbulence are believed to exhibit a more "universal" character, it is anticipated that the approach based on subgrid modeling would be more promising than the procedures based on ensemble averaging closures.

Within the past two decades, there have been many attempts to fine-tune the models by optimizing the closure for subgrid fluctuations and the type of filtering to identify the large-scale components of the variables. In most of these efforts, the closure of hydrodynamic fluctuations has been the subject of major concern. There have also been some results for the passive scalar simulation and modeling of velocity–scalar correlations. However, no attempts have been made to treat reacting scalars and the treatment of scalar–scalar interactions. This is probably due to the consensus among the researchers in this field that until an accurate subgrid model is constructed to represent the evolution of nonreacting scalar variables, the extension to reactive flow simulations will be a difficult task.

With the new developments and progress in PDF methods and the advantages offered by such schemes over moment methods, one may anticipate that these methods may also prove useful in LES. In this case, the influences of the chemical reactions and the subgrid scalar–scalar correlations can be included by considering the PDFs of the fluctuating scalars within the computational grid. The advantage of using these methods for the subgrid closure is apparent, since once these PDFs are known, any statistical quantity related to scalar fluctuations can be subsequently determined. This determination, although an ambitious task, is not impossible. The obvious and simplest choice is to follow an approach based on guessed PDF methods. Similar to Reynolds averaging, the first two subgrid moments of the scalar variables can be solved by LES and then the shape of the PDF can be parameterized based on these two moments. This parameterized approximate distribution can be appraised by a comparison with the "exact" shape constructed via DNS results. Similar to common practice in *a priori* assessments (Erlebacher et al., 1987), the PDF distribution can be specified by performing two sets of calculations, one with a coarse mesh, the other with a fine mesh. The results obtained from the fine-mesh simulations can be used to construct the PDF distribution, which in turn may be used in extensive subsequent simulations on the coarse mesh. In this setting, calculations with large transport parameters which otherwise could not be simulated on the coarse grids are possible.

A more direct but substantially more complicated procedure involves solving the transport equations for the PDFs of the subgrid scalars rather than assuming their form. The advantages of this approach, like its counterpart in turbulence modeling, is that the effect of scalar–scalar correlations appears in a closed form. However, models are needed for the molecular diffusion within the subgrid, and one has to resort to mixing models (a subject of current intense research) for providing the closure. The approach based on guessed PDF methods is feasible and is within the reach of present-day computers. The approach based on solving the transport equation for the subgrid-scale PDF might require extensive computational resources (Pope, 1990). These models must initially be developed in a simple flow. A homogeneous flow is a suitable configuration for this purpose, and will be discussed in the next section. After the establishment of a successful model, it may be utilized for simulating more complicated flows. The extent of this utilization is dependent on the available computational resources and on the performance of the model in rigorous trials.

## 15.3 Description of the Problem

The subject of our DNS is a two-dimensional homogeneous box flow under the influence of a binary chemical reaction of the type A + B → Products. To impose homogeneity, periodic boundary conditions are employed in all directions of the flow, identifying the box as a homogeneous member of a turbulent environment. This periodicity allows the mapping of all the aerothermochemical variables from the physical domain into a Fourier domain, thus allowing the implementation of spectral methods for numerical simulations. The flow field is assumed to be homogeneous and isotropic, and is initialized by a procedure similar to that of Passot and Pouquet (1987). This involves the superposition of a "random" velocity to a zero mean velocity and also includes random initial density, temperature, and Mach number fluctuations. These fluctuations have certain energy spectra, and the ratio between the compressible and the incompressible kinetic energy can be varied to assess the effects of compressibility. The specification of the fluctuating field with a random field is to mimic a "probabilistic" turbulent field in the context of "deterministic" DNS. This approach is similar to that followed in previous direct simulations of turbulent reacting flows (Riley et al., 1986; Givi and McMurtry, 1988; McMurtry and Givi, 1989). In a laboratory flow, these fluctuations appear as the result of interactions with the surrounding universe. Such interactions do not exist in the isolated homogeneous flow considered in our simulations. Therefore, in order to introduce the "noise," which plays a central role in laboratory turbulence, these perturbations are randomly superimposed to initialize the fluctuating field for DNS. The generated turbulence field is of decaying nature, i.e., there is no artificial forcing mechanism to feed energy to the small wave numbers.

The scalar fields are defined to be square waves with the two species out of phase and at stoichiometric conditions. The species field is assumed to be dynamically passive, in that turbulence influences the consequent transport of the scalar field with the neglect of reverse influences. In the nonreacting case, the trace of only one of these reactants is considered whereas in the reacting case, the transport of an appropriate Shvab–Zeldovich variable is assumed to portray the reactants' conversion rate. This is possible by assuming an infinitely fast chemical reaction rate, and by assuming that the reactants have identical thermochemical properties. In this framework, the effects of nonequilibrium chemistry are neglected; the inclusion of such effects is postponed for future investigations.

The computational package employed in simulations is based on the modification of a computer code developed by Erlebacher et al., (1987). This code is based on a spectral collocation algorithm with Fourier trial functions. All the variables are spectrally approximated on $N^2$ collocation points, where $N$ represents the number of collocation points in each of the directions. The spatially homogeneous flow evolves in time, and at each time step $N^2$ defines the sample data size for the purpose of statistical analysis. A third order accurate Runge–Kutta finite-difference scheme is employed for temporal discretization. For a trustworthy simulation, the magnitudes of the Reynolds and Peclet numbers must be kept at moderate levels and the size of the time step should be kept sufficiently small. The code is capable of simulating both two- and three-dimensional flows and we have performed both such simulations. In the presentation of our results in the next section, however, we limit our discussion to that of a two-dimensional flow.

## 15.4 Results

Computations are performed on a domain with a normalized dimension of $2\pi$ in each of the directions of the flow. With the available computational resources, a resolution of $256 \times 256$ collocation grid is attainable. With this resolution, the magnitude of the Taylor microscale Reynolds numbers that could be simulated is in the range $Re_\lambda \approx 20 - 30$. Simulations are performed with a wide spectrum of compressibility levels; here we report only the results obtained by the use of two extreme cases: one with a low compressibility level and the other with a relatively high level. In the former, the initial value of the normalized density rms is very small, i.e. $\rho_{rms} = \sqrt{\langle \rho'^2 \rangle}/\rho_0 \approx 0$, whereas in the latter, this value is of order unity. The compressibility level was monitored by adjusting the initial values of the following parameters (Passot and Pouquet, 1987): (1) the rms of the Mach number, $M_t = \sqrt{\langle M'^2 \rangle}$, and (2) the ratio of the compressible energy to that of the total kinetic energy, $\chi$. In what follows we refer to pseudoincompressible simulations in which the values $M_t = 0.2$, $\chi = 0.01$ are used initially, and to compressible simulations in which at the initial time $M_t = 0.6$, $\chi = 0.2$.

The compressibility effects can be assessed by both flow visualization and by considering the global behavior in an integral sense. In the former, the contour plots of the relevant variables constructed from the DNS results show the qualitative behavior, whereas in the latter, the ensemble averages of the simulated results portray the quantitative response. To demonstrate this, in Figures 1 and 2 we present the contour plots of the density for the pseudoincompressible case and for the compressible case, respectively. A prominent difference between the two cases is the formation of steep gradients in the high-compressible case which are not observed in the pseudoincompressible simulations. The regions of high gradients are referred to as shocklets and, consistent with

Fig. 1   Plot of density contours for the pseudoincompressible case. $t = 6.142$.

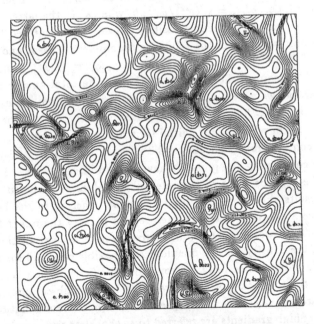

Fig. 2   Plot of density contours for the compressible case. $t = 0.686$.

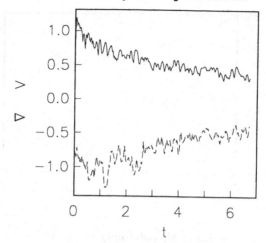

Fig. 3   Temporal variation of the minimum and maximum values of velocity divergence for the pseudoincompressible case.

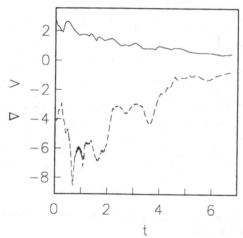

Fig. 4   Temporal variation of the minimum and maximum values of velocity divergence for the compressible case.

the previous simulations of Passot and Pouquet (1987), appear when the initial levels of density and Mach number fluctuations are high.

The effects of increased compressibility can be quantitatively demonstrated by examining the temporal variations of the maxima and minima of the divergence of the velocity, the fluctuating Mach number, and the ratio of the compressible to the total kinetic energy. These are presented in Figures 3–6. In Figures 3 and 4 the temporal variations of the minimum and maximum values of the divergence of the instantaneous

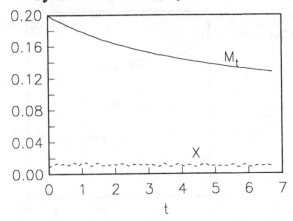

Fig. 5   Temporal evolution of $M_t$ (solid line) and $\chi$ (dashed line) for the pseudoincompressible case.

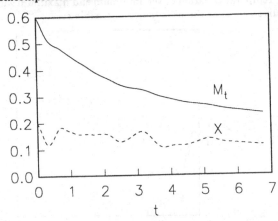

Fig. 6   Temporal evolution of $M_t$ (solid line) and $\chi$ (dashed line) for the compressible case.

velocity ($\nabla \cdot \underline{V}$) are presented for the pseudoincompressible and for the compressible cases, respectively. Note that in the low-compressible flow, the minimum and maximum values are approximately mirror images of each other (with respect to $\nabla \cdot \underline{V} = 0$). In the high-compressible case, however, the maximum and minimum divergence curves are asymmetric. The minimum value of divergence occurs at the normalized time of $t \approx 0.7$. This time corresponds to the onset of formation of the shocklets, and indicates severe compression. It must be mentioned that the capture of these shocklets by means of global numerical methods (without any artificial dissipation) is rather difficult. With $256 \times 256$ collocation points, this was the strongest shock that we were able to capture, and a lower resolution would result in a significant oscillation in the magnitude of the velocity divergence after the appearance of the shock. One must be careful not to confuse these numerical oscillations with compression and expansion waves.

The effects of compressibility are further quantified in Figures 5–6. For the pseudoincompressible flow, Figure 5, the magnitude of $M_t$ decreases monotonically, whereas in the compressible case, Figure 6, the decrease in $M_t$ is interrupted by plateaus. The locations of these plateaus coincide with those corresponding to a rise in the compressible kinetic energy and the local minima of the velocity divergence. Therefore, these locations correspond to the times at which shocklets can be formed in the flow. In the pseudoincompressible flow the magnitude of $\chi$ remains fairly constant, indicating that the initial low level of compressibility remains low at all times.

With the development of the flow the species field would consequently evolve. To visualize the flow, the contour plots of species $A$ are presented in Figure 7. Parts (a) and (b) of this figure correspond to the zero and the infinitely fast reaction rate cases, respectively. The figure exhibits the effects of random motion on the distortion of the scalar field and the mixing of the two initially segregated reactants. The effects of chemical reaction are to increase the steepness of the scalar gradients and to reduce the instantaneous values of the reactants, as indicated by a comparison between parts (a) and (b). The quantitative behavior of the scalar development is depicted in a statistical sense by examining the evolution of the PDFs of the conserved Shvab–Zeldovich variable $\mathcal{J}$. This variable is normalized and defined within the region $[0, 1]$. Correspondingly, its PDF, $\mathcal{P}_1(\mathcal{J})$, is always bounded in this region. The temporal variation of $\mathcal{P}_1(\mathcal{J})$ is presented in Figure 8. It is shown in this figure that at the initial time, the PDF is approximately composed of two delta functions at $\mathcal{J} = 0, 1$, indicating the two initially segregated reactants $A$ and $B$. At later times, it evolves through an inverse-like diffusion in the composition space. The heights of the delta functions decrease and the PDF is redistributed at other $\mathcal{J}$ values in the range $[0, 1]$. At even later times, the PDF becomes concentrated around the mean value. Proceeding further in time results in a sharper peak at this mean concentration, and the PDF can be approximated by a Gaussian distribution. This Gaussian-like behavior has been observed in previous simulations of Givi and McMurtry (1988) and Eswaran and Pope

(1988), and also has been corroborated in a number of experimental investigations.

An interesting character of these PDFs is that throughout their evolution, the simulated results compare remarkably well with that of a beta density. This is also shown in Figure 8 by a comparison between the beta density and the DNS-generated PDFs. The beta density is parameterized with the same first two moments as those of the DNS. Therefore, in all the figures the results are presented with respect to a time scale ($t^*$) proportional to the decay of the variance of the scalar $J$. Higher order moments of the DNS data for the variable $J$ also show good agreements in comparison with those predicted by the beta density. This is demonstrated in Figures 9 and 10, which show the normalized kurtosis ($\mu_4$) and superskewness ($\mu_6$) of the random variable $J$ and the reactant $A$, respectively. At time zero, these moments are close to unity and monotonically increase as mixing proceeds. For the Shvab–Zeldovich variable $J$, the magnitudes of the kurtosis and superskewness resulting from the beta density asymptotically approach the limiting values of 3 and 15, respectively. These correspond to the normalized fourth and sixth moments of a Gaussian distribution as the variance of $J$ tends to zero (as $t^* \rightarrow \infty$). The DNS-generated results are very close to those of the beta distribution throughout the simulations. However, the limiting value for variance of $J$ approaching zero cannot be obtained in the simulations due to obvious numerical difficulties. In the reacting case, the moments of the scalar $A$ are consistently higher than those of the conserved scalar, but portray similar trends in both beta density and DNS-generated results.

The trends shown above are also observed in the compressible simulations. The profiles of the PDFs and those of the higher order moments and their comparisons with the corresponding parameterized beta density are presented in Figures 11–13. Again, the comparison is very good. The main difference between these results and those of pseudoincompressible simulations is the time scale of the decay of the variance. This is not illustrated in the figures clearly, since the time is scaled by the rate of variance decay. This time scale cannot be incorporated into the PDF description in the format employed here. The single-point nature of these PDFs does not allow for any information about the length scales (or any other scales requiring two-point statistics) of turbulence. For a systematic inclusion of the length scale into the PDF and quantitative description of the differences between the results in the two cases, one is required to consider the evolution of the PDFs at two points (at

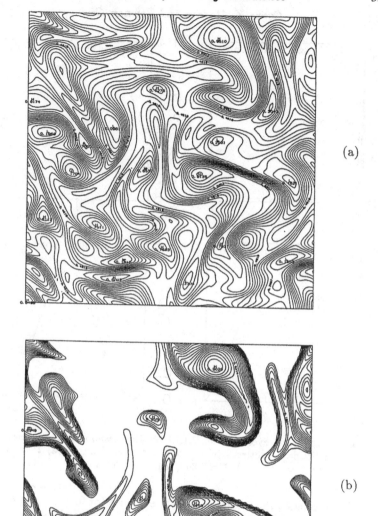

Fig. 7 Plot of species A concentration contours at $t^* = 1.0715$. (a) Nonreacting case; (b) reacting case.

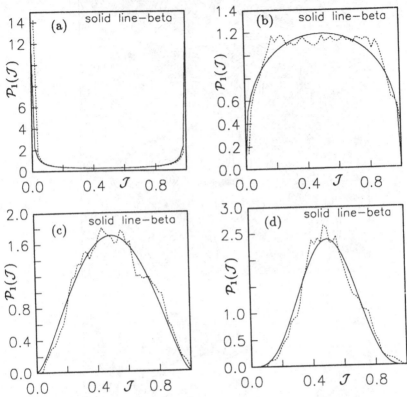

Fig. 8    Temporal evolution of $\mathcal{P}_1(\mathcal{J})$ for the pseudoincompressible case. DNS data (dotted line), beta density (solid line). (a) $t^* = 0.035$, (b) $t^* = 0.549$, (c) $t^* = 0.800$, (d) $t^* = 1.0715$.

least). Employing the results of DNS to evaluate (or to generate) models for more than one-point-level closures (Eswaran and O'Brien, 1989) has proven to be a challenging task, and is the subject of current research (Frankel et al., 1992).

The results of these simulations indicate that the approximation of a Gaussian PDF for the final stages of mixing of a conserved scalar is well justified. This corroborates the results of laboratory experiments (e.g., Miyawaki et al., 1974; Tavoularis and Corrsin, 1981) and those of other numerical simulations (Eswaran and Pope, 1988; Givi and McMurtry, 1988). The numerical experiments performed here, however, are similar to most of the laboratory investigations in that they include the results of only one experimental realization. Future numerical experiments using different initializations and/or with better numerical resolution and

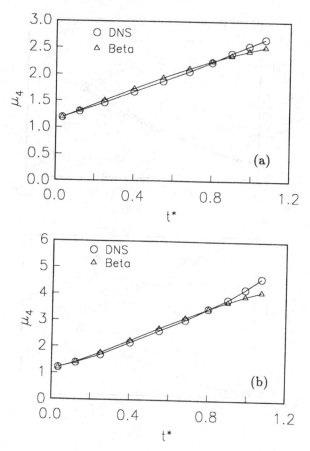

Fig. 9 Temporal variation of the kurtosis, $\mu_4$, for the pseudoincompressible case. (a) Variable $\mathcal{J}$, (b) species A in reacting case.

including three-dimensional effects would be logical extensions of this work.

In addition to the beta distribution approximations, the DNS results also compare very well with the mapping closure recently introduced by Kraichnan (see Chen et al., 1989). This closure has the property of allowing the relaxation of the PDF of a conserved scalar property to an asymptotic Gaussian distribution. Pope (1991) has utilized this closure for the prediction of PDF evolution in an isotropic homogeneous turbulent flow with an initial double delta distribution (similar to the condition imposed here). He showed that in the context of a one-point PDF, the results obtained by this closure compare favorably with the DNS results. However, again there is no length scale information built

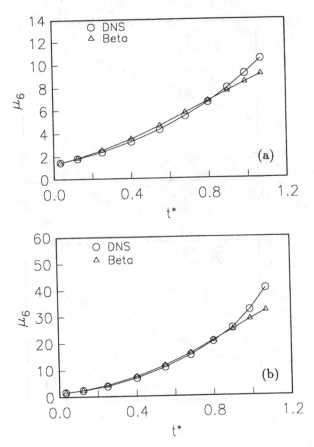

Fig. 10 Temporal variation of the superskewness, $\mu_6$, for the pseudoincompressible case. (a) Variable $\mathcal{J}$, (b) species A in reacting case.

into the model and only the evolution of the PDF without any information regarding the time scale of evolution can be predicted. Madnia et al. (1992) have extended this model to predict the decay rate of a reacting scalar in homogeneous turbulence. The generated results obtained by applying this model to the prediction of the decay rate of a reacting scalar for the cases considered in DNS are shown in Figures 14 and 15. In these figures, the DNS-generated results and those predicted by a beta density are also shown. The two PDFs (by the mapping closure and the beta density) are constructed in such a way as to yield the same first two normalized moments of the Shvab–Zeldovich variable as those of DNS. In this context, the predicted results of the decay by both models show good agreement. The agreement for the mapping closure

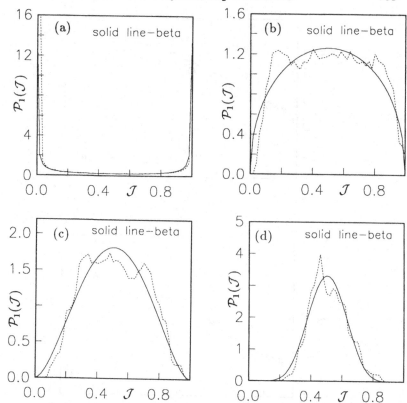

Fig. 11  Temporal evolution of $\mathcal{P}_1(\mathcal{J})$ for the compressible case. DNS data (dotted line), beta density (solid line). (a) $t^* = 0.014$, (b) $t^* = 0.587$, (c) $t^* = 0.848$, (d) $t^* = 1.365$.

model is rather interesting considering that this closure was developed for an isotropic three-dimensional turbulent flow. This is probably due to matching of the normalized first two moments. A better test of the model would be its utilization at the two-point level and its comparison with DNS-generated results. Also, the agreement between the mapping closure results and those of DNS worsens somewhat as the compressibility level is increased. This again is understandable in that Kraichnan's model was developed for a purely incompressible flow without including any compressibility effects.

With the construction of the DNS database, the results are used to construct the PDFs of the scalar quantities within the subgrid. Since the major riddle in PDF modeling is associated with the closure of the diffusion term and not the chemical reaction term, we limit the analysis

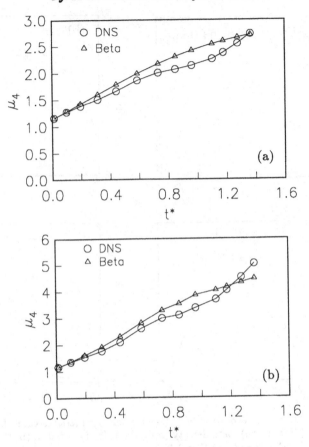

Fig. 12 Temporal variation of the kurtosis, $\mu_4$, for the compressible case. (a) Variable $\mathcal{J}$, (b) species A in reacting case.

to that of the generated PDFs of the conserved Shvab–Zeldovich variable. The construction of the PDFs is implemented by placing a coarse $(n^2)$ mesh over the physical space occupied by the fine mesh $(N^2)$. This mesh can be considered the one to be potentially used in LES. In this way, it is assumed that the LES predicts the averaged results at the centers of the $n^2$ mesh, and the fluctuations within the coarse grids are to be modeled by an appropriate PDF closure. The actual LES in this case would correspond effectively to simulations with a homogeneous box filter in which the values of the filtered means are constant. This is somewhat analogous to the procedure followed by Schumann (1989). The main difference is the mechanism by which the fluctuations are considered. Schumann (1989) simply neglected such effects but with

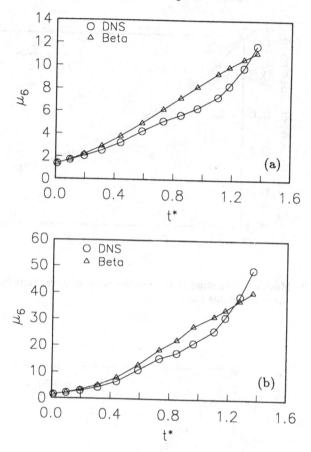

Fig. 13  Temporal variation of the superskewness, $\mu_6$, for the compressible case. (a) Variable $\mathcal{J}$, (b) species A in reacting case.

consideration of the PDF it may be possible to account for such effects, albeit statistically. Within each cell of the coarse mesh, we have the values of the scalar quantities at $N_\phi = (N/n)^2$ equally spaced points. This means that there is an ensemble of $N_\phi$ sample points at each location to construct the PDFs. This construction is implemented by statistical sampling of $N_\phi$ number of data points. The domain of the scalar property $\phi \in [\phi_{min}, \phi_{max}]$ is divided into a number of bins of size $\Delta\phi$, and the number of scalar variables within each of these bins is counted. The PDF of the variable $\phi$, denoted by $\mathcal{P}(\xi)$, is calculated in the interval $\xi \leq \phi \leq \xi + d\xi$, $d\xi \equiv d\phi$, from the definition of the probability

$$\mathcal{P}(\xi)\, d\xi = \text{Probability}(\xi \leq \phi \leq \xi + d\xi), \qquad (2')$$

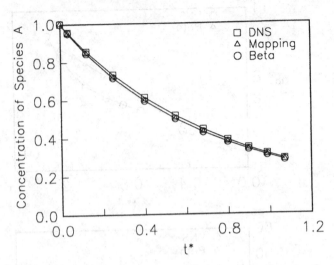

Fig. 14  **Mean concentration of the reacting species A vs. time for the pseudoincompressible case.**

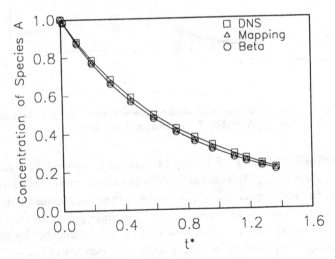

Fig. 15  **Mean concentration of the reacting species A vs. time for the compressible case.**

which translates into

$$\mathcal{P}(\xi)\,d\xi \equiv \text{Probability density} \tag{12}$$
$$= \frac{\text{Number of elements in the interval } \xi \leq \phi \leq \xi + d\xi}{N_\phi}$$

The value of $N_\phi$ is a measure of the width of the filter, in that it determines the scale at which the fluctuations of the scalar field are not accounted for deterministically, but are included in a probabilistic manner by the PDFs. Consequently, the shapes of these distributions are dependent on the magnitude of the sample size $N_\phi$. If this value is too small, then there will be a large scatter in the data. If it is too large, then the PDFs correspond to those appropriate for Reynolds averaging and not LES. In this primeval implementation, we have not yet performed an actual LES (using a PDF subgrid closure). Rather, we have focused our efforts on evaluating the performance of the presumed PDFs within the subgrids. These PDFs can again be quantified by the magnitudes of their moments to provide a reasonable *a priori* assessment of the closure for future actual LES. With the $256^2$ available total data, there is a limit on the magnitude of the ensemble data for an *a priori* assessment. Our experience indicated that an ensemble of about $8^2$ was the lowest acceptable limit for a meaningful statistical analysis. Therefore, the domain was broken into $32 \times 32$ squares. Within each of these squares, the results obtained by statistical analysis were compared directly with those attained by means of an assumed beta distribution.

For the simple kinetics mechanism considered, two parameters that are very useful in this *a priori* analysis are the subgrid-averaged product concentrations $\tilde{P}$ and the unmixedness $\Psi^2$. These parameters can be calculated directly from the DNS results and can also be modeled by means of the presumed beta distribution. The beta density of the Shvab–Zeldovich variable is defined (Papoulis, 1965) by

$$\mathcal{P}_1(\mathcal{J}) = \frac{1}{B(\alpha,\beta)}\mathcal{J}^{\alpha-1}(1-\mathcal{J})^{\beta-1}, \tag{13}$$

where B denotes the beta function and the parameters $\alpha$ and $\beta$ are dependent on the first two subgrid moments of the random variable. Based on this distribution, all the moments of the species field can be directly determined. For a stoichiometric mixture, the implementation is rather straightforward and the results can be directly obtained by a procedure similar to that followed in generating Figures 7, 9, and 10. Within the subgrid, however, the mixture is usually nonstoichiometric. In this case, the procedure is somewhat more complex. The complete derivation is provided by Madnia et al. (1992) and will not be repeated here; only the final results are presented:

$$\tilde{P} = 1 - \Theta_1 - \Theta_2,$$

(14)

$$\Psi^2 = -\Theta_1\Theta_2.$$

Here the variables $\Theta_1$ and $\Theta_2$ are related to the parameters of the beta density through

$$\Theta_1 = B(\alpha,\beta)\frac{\mathcal{J}_{st}^{\alpha}(1-\mathcal{J}_{st})^{\beta-1}}{(\alpha+\beta)} + \frac{1}{1-\mathcal{J}_{st}}\left(\frac{\alpha}{\alpha+\beta}\mathcal{J}_{st}\right)\mathcal{I}_{\mathcal{J}_{st}}(\alpha,\beta),$$

(15)

$$\Theta_2 = B(\alpha,\beta)\frac{\mathcal{J}_{st}^{\alpha-1}(1-\mathcal{J}_{st})^{\beta}}{(\alpha+\beta)} + \left(1 - \frac{\alpha}{\mathcal{J}_{st}(\alpha+\beta)}\right)\mathcal{I}_{\mathcal{J}_{st}}(\alpha,\beta).$$

In (15) $\mathcal{I}_{\mathcal{J}_{st}}(\alpha,\beta)$ is the incomplete beta function and $\mathcal{J}_{st}$ denotes the stoichiometric value of the Shvab–Zeldovich variable. To assess the performance of the model for subgrid PDF, in Plates 2 and 3 the contour plots of the subgrid product concentration and the unmixedness estimated by Eqs. (14)–(15) are compared with those obtained directly via DNS results. The comparisons made in these Plates reveal a good agreement between the results predicted by the subgrid model and those generated by the DNS data. This agreement, albeit primitive, provides a reasonable justification for recommending Eqs. (14)–(15) as working relations in practical LES applications. However, these relations are advocated only for statistical predictions at this level. Due to a rather small sample size within the subgrid, the comparison of higher order statistical quantities generated by the model with those of DNS data cannot be made in an accurate manner. Also, the implementation of the model requires the accurate input of the first two moments of the Shvab–Zeldovich variable within the subgrid. Here, these moments were supplied via DNS, but in an actual LES application they must be provided by an appropriate subgrid closure (e.g. the closure suggested by Antonopoulos-Domis, 1981).

The specification of these parameters is substantially simpler than that of the reacting scalars, even with the assumption of an infinitely fast reaction. With such specification, Eqs. (14)–(15) are recommended, at least in the absence of better alternatives, for a reasonably accurate and inexpensive prediction of the limiting bounds of the reactant conversion rate within the subgrid. Finally, the results are valid only in the limit of an infinitely fast chemical reaction. The generalization to include finite-rate kinetics may not be very straightforward, since the first two moments (including the covariance of the scalar fluctuations) must be known (or modeled) *a priori*. The versatility of the beta distribution based on the input of these parameters has yet to be determined. By the same token, the extension to more complex kinetics with the inclusion of nonequilibrium effects, which are very important for practical applications, remains a challenging task. An improvement of the procedure obviously involves the solution of a transport equation for the PDF (rather than its parameterization with its moments). However, the computational burden may be prohibitive due to the increased dimensionality of the subgrid PDF transport equations. An estimate of the computational requirements indicates that the cost associated with the implementation of the LES procedure involving the stochastic solution of the PDF transport equation may be of the same order as that of DNS on the fine grids, unless the ratio of the fine to coarse grid is large. In this regard, the trade-off between LES and DNS depends on other factors such as the physical complexity and computational resources.

## 15.5 Concluding Remarks

A spectral collocation algorithm was employed to simulate the phenomena of mixing and reaction in a decaying two-dimensional homogeneous turbulent flow involving a stoichiometric mixture with two initially unpremixed reactants $A$ and $B$. The evolution of the species field in a one-step reaction of the type $A + B \rightarrow$ Products was the subject of the investigation. Calculations were performed for zero-rate and infinitely fast rate chemistry in both pseudoincompressible and relatively high compressible flows. The effects of compressibility are delineated by the formation of shocklets and can be quantified by a global examination of the data. Mixing and the reaction conversion rate (in the reacting case) are characterized by examining the evolution of a Shvab–Zeldovich conserved scalar quantity extracted from the DNS. The results show that the PDF of this scalar variable evolves from an initial double-delta function distribution to an asymptotic shape which can be approximated by a Gaussian distribution. During this evolution, a beta density qualitatively describes the DNS-generated PDFs in both pseudoincompressible and compressible simulations. This is quantitatively demonstrated by a good agreement between the values of the higher order moments of the

PDFs calculated from the DNS data and those predicted by the beta approximation. The generated results also compare very well with those predicted by the mapping closure of Kraichnan. The decay of the mean of the reacting scalar determined by DNS compares very well with that predicted by this model and that obtained via the use of the beta density. This trend is observed in both pseudoincompressible and compressible simulations. The main difference is the time scale of the decay of the variance of the PDFs.

The beta density seems also to predict the behavior of the fluctuating field within the subgrid for a conserved scalar quantity. Therefore, it may provide a reasonable means of parameterizing the subgrid fluctuations in a stochastic sense. Based on this parameterized density, a simple analytical relation is suggested for estimating the limiting bound of reactant conversion rate averaged within the subgrid. The predicted results based on this model agree well with DNS-generated data. Therefore, this relation is recommended for LES of reacting flows involving irreversible fast reactions. The implementation of the procedure, however, requires the input of the first two moments of the Shvab–Zeldovich mixture fraction within the subgrid. These must be provided by modeled LES transport equations. Also, the generalization to include finite-rate kinetics may not be very straightforward since these moments must be modeled *a priori*. An improvement of the procedure involving the solution of a transport equation for the PDF is possible, but may not be practical due to the increased dimensionality of the modeled subgrid PDF transport equations. From the computational point of view, it is recommended that the applicability of such a procedure in (yet) simpler flows be assessed before it is implemented in the simulation of more complex flows.

In comparing the DNS results with those of predictions, the first two moments of the Shvab–Zeldovich variable must be matched. This is because single-point PDFs do not include information about the frequency scales of turbulence. Therefore, the evolution of the length scale, or any other parameter requiring two-point information, must be introduced in an ad hoc manner (here, such information is provided via DNS). Future validation studies of the PDFs by DNS need to consider the evolution of multipoint (at least two-point) statistics. Frankel et al., (1992) have already employed a two-point formulation in the statistical description of homogeneous turbulent flows. The preliminary results obtained by such formulations are in good agreement with experimental data. However, a comparison between the predicted results and the DNS data, similar to the ones reported here for single-point PDFs is recommended. Also, the analytical solution obtained for the mapping closure corresponds to that of an exact initial double-delta distribution. This solution infers an infinite dissipation rate at $t = 0$, before relaxation to finite values for $t > 0$. Future comparisons with DNS data should also include smooth initial PDFs with finite scalar dissipation. Finally, future work should also deal with finite-rate chemistry with the inclusion of nonequilibrium effects and also with the extension of PDF methods for the subgrid closures in the LES of such flows. In a recent article, Madnia et al. (1992) provide some guidelines in this direction.

## Acknowledgments

We are indebted to Dr. Gordon Erlebacher for providing us with the initial version of the DNS computer code. This research was sponsored in part by the NASA-Langley Research Center under Grant NAG-1-1122, by the Office of Naval Research under Grant N00014-90-J-4013, and by the National Science Foundation under Grants CTS-9253488 and CTS-9012832. Computational resources were provided by Numerical Aerodynamic Simulation (NAS) at NASA Ames Research Center and by the National Center for Supercomputing Applications (NCSA) at the University of Illinois.

## References

ALDAMA, A.A. (1990) *Filtering Techniques for Turbulent Flow Simulations.* Lecture Notes in Engineering, Vol. 56, Ed. C.A. Brebbia and S.A. Orszag, Springer.

ANTONOPOULOS-DOMIS, A. (1981) Large-eddy simulation of a passive scalar in isotropic turbulence. *J. Fluid Mech.* 104, 55–79.

CHEN, H., CHEN, S. AND KRAICHNAN, R.H. (1989) Probability distribution of a stochastically advected scalar field. *Phys. Rev. Lett.* 62, 2657–2660.

DRUMMOND, J.P. (1991) Supersonic reacting internal flow fields. In *Numerical Approaches to Combustion Modeling.* Progress in Astronautics and Aeronautics, Vol. 135. Ed. E.S. Oran and J.P. Boris, pp. 365–420. AIAA.

ERLEBACHER, G., HUSSAINI, M.Y., SPEZIALE, C.G. AND ZANG, T.A. (1987) Toward the large eddy simulations of compressible turbulent flows.

NASA CR 178273, ICASE Report 87-20, NASA-Langley Research Center.

ESWARAN, V. AND O'BRIEN, E.E. (1989) Simulations of scalar mixing in grid turbulence using an eddy-damped closure model. *Phys. Fluids* A **1** (3), 537–548.

ESWARAN, V. AND POPE, S.B. (1988) Direct numerical simulations of the turbulent mixing of a passive scalar. *Phys. Fluids* **31**(3), 506–520.

FERZIGER, J.H. (1977) Large eddy numerical simulations of turbulent flows. *AIAA J.* **15** (9), 1261–1267.

FERZIGER, J.H. (1981) Higher-level simulations of turbulent flows. Report TF-16, Stanford University, Dept. of Mechanical Engineering.

FERZIGER, J.H. (1982) Turbulent flow simulation: A large eddy simulator's viewpoint. In *Recent Contributions to Fluid Mechanics*. Ed. W. Haase, pp. 69–77. Springer.

FRANKEL, S. H., JIANG, T.-L. AND GIVI, P. (1992) Modeling of isotropic reacting turbulence by a hybrid mapping-EDQNM closure. *AIChE J.* **38** (4), 535–543.

GIVI, P. AND McMURTRY, P.A. (1988) Nonpremixed reaction in homogeneous turbulence: Direct numerical simulations. *AIChE J.* **34** (6), 1039–1042.

GIVI, P. (1989) Model-free simulations of turbulent reactive flows. *Prog. Energy Comb. Sci.* **15**, 1–107.

HAWTHORNE, W.R., WEDELL, D.S. AND HOTTEL, H.C. (1949) Mixing and combustion in turbulent gas jets. In Proc. Third Symp. on Combustion, Flames and Explosion Phenomena, pp. 266–288. The Combustion Institute.

HUSSAINI, M.Y., SPEZIALE, C.G. AND ZANG, T.A. (1990) The potential and limitations of direct and large-eddy simulations. In *Whither Turbulence? Turbulence at the Crossroads*. Lecture Notes in Physics, Vol. 357. Ed. J.L. Lumley, pp. 354–368. Springer.

IEVLEV, V.M. (1973) *Dokl. Akad. S.S.S.R.* **208**, 1044 (also *Sov. Phys.-Dokl.* **18**, 117).

KUO, Y.Y. AND O'BRIEN, E.E. (1981) Two-point probability density function closure applied to a diffusive-reactive system. *Phys. Fluids* **24** (2), 194–201.

LIBBY, P.A. AND WILLIAMS, F.A. (eds.), (1980) *Turbulent Reacting Flows*. Topics in Applied Physics, Vol. 44. Springer.

MADNIA, C.K., FRANKEL, S.H. AND GIVI, P. (1992) Reactant conversion in homogeneous turbulence: Mathematical modeling, computational validations, and practical applications. *Theoret. Comput. Fluid Dynamics* 4, 104–134.

McMurtry, P.A. and Givi, P. (1989) Direct numerical simulations of mixing and reaction in a nonpremixed homogeneous turbulent flow. *Comb. Flame* 77, 171–185.

Miyawaki, O., Tsujikawa, H. and Uraguchi, Y. (1974) Turbulent mixing in multi-nozzle injector tubular mixer. *J. Chem. Eng. Japan* 7, 52–74.

Oran, E.S. and Boris, J.P. (1987) *Numerical Simulations of Reactive Flow*. Elsevier.

Papoulis, A. (1965) *Probability, Random Variables, and Stochastic Processes*. McGraw-Hill.

Passot, T. and Pouquet, A. (1987) Numerical simulation of compressible homogeneous flows in the turbulent regime. *J. Fluid Mech.* 181, 441–466.

Pope, S.B. (1985) PDF methods for turbulent reactive flows. *Prog. Energy Comb. Sci.* 11, 119–192.

Pope, S.B. (1990) Computations of turbulent combustion: Progress and challenges. In Proc. Twenty-Fourth Symp. on Combustion, Flames and Explosion Phenomena, pp. 591–612.

Pope, S.B. (1991) Mapping closures for turbulent mixing and reaction. *Theoret. Comput. Fluid Dynamics* 2, 255–270.

Reynolds, W.C. (1990) The potential and limitations of direct and large eddy simulations. *Whither Turbulence? Turbulence at the Crossroads*, Lecture Notes in Physics, Vol. 357. Ed. J.L. Lumley, pp. 313–343. Springer.

Riley, J.J., Metcalfe, R.W. and Orszag, S.A. (1986) Direct numerical simulations of chemically reacting turbulent mixing layer. *Phys. Fluids* 29 (2), 406–422.

Rogallo, R.S. and Moin, P. (1984) Numerical simulation of turbulent flows. *Ann. Rev. Fluid Mech.* 16, 99–137.

Schumann, U. and Friedrich, R. (eds.), (1986) *Direct and Large Eddy Simulation of Turbulence*, Proceedings of the EUROMECH Colloquium No. 199, Munchen, FRG, Sep. 30–Oct. 2, 1985.

Schumann, U. and Friedrich, R. (1987) On direct and large eddy simulation of turbulence. In *Advances in Turbulence*. Ed. G. Comte-Bellot and J. Mathieu, pp. 88–104. Springer.

Schumann, U. (1989) Large-eddy simulation of turbulent diffusion with chemical reactions in the convective boundary layer. *Atmos. Environ.* 23 (8), 1713–1726.

TAVOULARIS, S. AND CORRSIN, S. (1981) Experiments in nearly homogeneous turbulent shear flow with a uniform mean temperature gradient. *J. Fluid Mech.* **104**, 311–347.

# PART THREE

## LARGE EDDY SIMULATION IN GEOPHYSICS

# 16

## Large Eddy Simulation in Geophysical Turbulence Parameterization: An Overview

### JOHN C. WYNGAARD AND CHIN-HOH MOENG

## 16.1 Introduction

Today scientists are working particularly hard, and with new resources, to develop improved models of the atmosphere and the ocean. One motivation is the increasing interest in global-change issues, which has led to proposals for building comprehensive numerical models of the global climate system. These would involve coupled submodels of the general circulations of the atmosphere and the ocean. The latter are called general circulation models, or GCMs.

GCMs rely heavily on subgrid-scale parameterizations. Their grid square is typically no less than about 100 km on a side, and more often a few hundred kilometers. Phenomena that occur on smaller scales, such as clouds, mixing, and surface exchange, must be represented approximately – *parameterized*, in the language of the community. Many of these processes involve turbulence.

Geophysical turbulence parameterization not only faces the challenge of representing the effects of the nonlinear, three-dimensional, stochastic fields we call turbulence, but must also include the effects of stratification, phase change, radiation, and rotation on that turbulence. Furthermore, geophysical turbulence parameterizations must be concise. A better subgrid scheme that doubles the run time for the parent model probably will not be used.

Turbulence parameterization is also an important task in engineering fluid mechanics. In geophysical flows, however, the measurements required to develop parameterizations are much more difficult and expensive. Geophysical turbulence researchers have substantially less experimental access to turbulence than do their counterparts in engineering. One result seems to be that geophysical turbulence researchers rely more on numerical models. They use models to generate simpler models – something that the history of turbulence research suggests would better be done through experiment.

Large eddy simulation (LES) entered planetary boundary layer (PBL) research two decades ago through Deardorff's efforts. He did not call it "large eddy simulation," but rather "three-dimensional numerical study of turbulence." (The term "large-eddy simulation" – which is more precise and, fortunately, now in common use – came from the engineering community when it adopted the technique several years later.) Deardorff's first simulations were of turbulent channel flow (Deardorff, 1970a), using a total of 6720 grid points. Today's largest computers allow $10^3$ times that number!

Deardorff credited Smagorinsky et al. (1965) and Lilly (1967) with establishing the basic foundations of his subgrid-scale closure. In Chapter 1 of this volume, Smagorinsky reviews the early history of the developments that led to that closure. The adaptation of LES within the engineering community, which benefited from this pioneering work by Smagorinsky, Lilly, and Deardorff, is reviewed in Chapter 6 of this volume by Piomelli.

Deardorff soon turned to LES of the atmospheric boundary layer. His early findings (e.g., Deardorff, 1970b) changed the way we perceive the structure and scaling of the convective case. He used it to study the dynamics of PBL turbulence in unprecedented detail (Deardorff, 1974) and later studied cloud-topped boundary layers (Deardorff, 1980).

Since Deardorff's first studies, several other groups have carried out LES of the planetary boundary layer. In Chapter 18 of this volume Schumann discusses an application of LES to boundary layer flow over a wavy surface; in Chapter 20, McWilliams et al. describe the application of LES to the upper ocean. With the larger number of grid points possible on today's machines, one can achieve finer grid spacing and better resolution of energy-containing range dynamics than Deardorff did. There are still some problems, such as inadequate subgrid-scale parameterizations near the bottom and top of the boundary layer. Nonetheless, LES has clearly demonstrated its potential in geophysical turbulence parameterization.

## 16.2 LES "Databases" and Turbulence Modeling

Early geophysical turbulence models were based on eddy diffusivity closures, but second-order models are generally considered the leading-edge approach today. As direct statements about the Reynolds fluxes in the large-scale equations, they specifically address some of the physics of the maintenance of those fluxes. Furthermore, they can usually be written in a form compatible with the large-scale equations in the parent model. To provide a background for our discussion of the potential of LES in the development of turbulence models, let us discuss some of the historical mileposts in the community's turn to second-order closure.

### 16.2.1 Second-Order Closure: The Early Years

The unusual transport properties of the turbulence in the convective boundary layer of the atmosphere have been evident for some time. Deardorff (1966) mentioned that the results of the Great Plains Turbulence Field Program in the mid-1950s showed clearly that the lapse rate at 100 m and higher, when the heat flux was unquestionably upward at these levels, was positive, so that the eddy diffusivity was negative. Since then, numerous atmospheric field studies have shown a characteristic daytime mean potential temperature profile that decreases with height near the surface, attains very small gradients by mid-layer, and maintains a slightly positive gradient above that. The turbulent temperature flux remains positive, so the eddy diffusivity for temperature has a mid-layer singularity.

Deardorff (1966) used the conservation equation for temperature variance to discern reasons for this singularity and the region of "countergradient" temperature flux above it. He showed that the mid-layer change of sign of the temperature gradient corresponds to a change in the source of temperature variance from gradient production to turbulent transport (i.e., third-moment flux divergence). Later Deardorff (1972) used the conservation equation for temperature flux, with its third-moment term neglected, to estimate the magnitude of the stable lapse rate in the countergradient region.

There were other early demonstrations of the ability of simple second-order closures to explain important features of the atmospheric boundary layer. Donaldson (1973) was one of the first to apply the same model to both laboratory and geophysical flows and to explore the effects of stratification. Mellor (1973) and Lewellen and Teske (1973) reproduced the known Monin–Obukhov similarity functions with second-order models. André et al. (1978) modeled the 24-hour evolution of the PBL. The models of Brost and Wyngaard (1978) and Nieuwstadt (1984) predicted the structure of nocturnal boundary layers that agreed well with observations. Mellor and Yamada (1982) described a second-order model applicable to a wide range of geophysical problems.

During this period it also became evident, however, that the closure approximations within such second-order models were not uniformly valid in atmospheric turbulence. Wyngaard (1973) showed, for example, that surface layer data tended not to support the usual second-order-closure approximation of downgradient diffusion for turbulent transport. Zeman and Lumley (1976) argued that most second-order closures could not reproduce the entrainment process at the base of the capping inversion. It was discovered that most models did not satisfy realizability requirements such as positivity of variances (Schumann, 1977), something that was probably known much earlier to the model programmers.

The increasing activity in turbulence modeling in the 1970s, particularly within the engineering community, also generated some outspoken criticism. For example, in reviewing the state of turbulence research Liepmann (1979) wrote, "Computational approaches to turbulence tend to follow two distinctly different paths: a rigorous approach to an approximate problem or an approximate approach to a rigorous problem. . . . Problems of technological importance are always approached by approximate methods, and a large body of turbulence modeling has been established under prodding from industrial users. . . . Turbulent modeling is still on the rise owing to rapid development of computers coupled with the industrial need for management of turbulent flows. I am convinced that much of this huge effort will be of passing interest only. . . . much of this work is never subjected to any kind of critical or comparative judgment. . . . The only encouraging prospect is that current progress in understanding turbulence will restrict the freedom of such modeling and guide these efforts toward a more reliable discipline."

Today, some dozen years after these words appeared, Liepmann's "encouraging prospect" has indeed materialized. Progress in understanding turbulence has been steady, thanks in large part to turbulence simulation, and is guiding the development of more reliable turbulence models.

### 16.2.2 Reflections on the Early Years

In discussing the performance of second-order models of atmospheric turbulence, Lumley (1983) pointed out that "we are dealing with a calibrated surrogate for turbulence. . . . We would thus expect that the models would work satisfactorily in situations not too far removed geometrically, or in parameter values, from the benchmark situations used to calibrate the model. . . . some of the successes have been in flows dominated by inertia or mean buoyancy. Thus emboldened, the modelers have been overenthusiastic in promoting their models for other complex situations, often without considering at depth the difficult questions that arise. Consequently, there is some disillusionment with the models, a feeling that they embody too many *ad hoc* assumptions, and that they are unreliable as a result. This reaction is probably justified, but it would be a shame if it resulted in a cessation of efforts to put a little more physics and mathematics into the models."

The disillusionment Lumley referred to did not lead to a cessation of efforts to improve the models, as he feared. We believe the pronounced lack of what Lumley calls "benchmark situations used to calibrate the model" has seriously *limited* these efforts, however. The model developers of the 1970s and 1980s used what data they could find, and quite often these data were from laboratory shear flows or the lowest portions of the atmospheric surface layer. The models were soon being used as predictive tools for a far wider range of conditions, including the neutral, unstable, and stably stratified PBLs from the surface to boundary layer top. The results generally looked plausible, but in most instances could not be tested critically due to lack of data.

The stratification of geophysical turbulence further complicates its modeling. Only some of the effects of stratification on boundary layer turbulence appear explicitly in the Reynolds stress and flux equations. We have learned that stratification also has profound effects on the *structure* of boundary layer turbulence (Panofsky and Dutton, 1984; Wyngaard, 1988). Early hopes that a single set of closures could treat a wide range of stratification have faded.

In a symposium on the state of turbulence research Lumley (1990) stated: "That [turbulence models have not predicted anything] is true. However, I believe it is foolhardy to expect them to. These models are simply an embodiment of experience; they are something constructed to behave like turbulence, in situations where it has been observed, as a design tool. A model cannot, except by accident, contain more than is put into it. One tries, of course, to make models as universal as possible, building into them behavior designed to avoid violation of as many commandments as one can manage. In this sense, a good model can be reasonably forgiving, continuing to produce reasonable looking results far beyond the parameter range for which it was constructed. That does not mean the results are right. If, in this parameter range, a physical phenomenon is important that was not built into the model, the results will be wrong. You should never expect a model to predict something you did not foresee. Use it to get a better numerical value for something you can already estimate on the back of an envelope."

The geophysical turbulence community's hopes for the predictive value of second-order models were gradually found to be misplaced as the nature of these models became better understood and accepted. We believe the community would now agree that in order to build models capable of representing new regimes in geophysical turbulence, it is necessary to have experience with that regime – to have observed that turbulence.

### 16.2.3 Testing Turbulence Models

There are two means of testing turbulence models. Their predicted fields can be compared with observations; this is a way of evaluating the overall performance of the model. One could call this *system testing*. One can also test the fidelity of a model's individual components, e.g., the individual closures used within a second-order model. This could be called *component testing*. Any turbulence model is a nonlinear system, so the relationships between these two are not completely clear.

Turbulence modeling is based on the faith that the system performance can exceed that of its individual components. It is hoped that a crude assumption about some detail within a model system will allow the system to run without comparably poor *overall* performance. Generally speaking, this faith has been sustained by our experience.

The history of geophysical turbulence modeling suggests the following pattern of events is typical. A model is built, using benchmark situations in its design and calibration. It is then used in other situations, some quite far removed from the benchmark cases. System performance problems are discovered; they are studied, traced to weak model components, and the model is improved. This cycle can involve extensive revisions and a great deal of effort. In discussing his group's efforts to develop a second-order model of the PBL displaying countergradient fluxes, Lumley (1990) said they published a total of 49 papers on the subject between 1970 and 1989.

Due to the dearth of geophysical turbulence data, sometimes this evaluation process occurs almost entirely on the modeling plane. For example, Garwood et al. (1985) argued that rotational effects on oceanic mixed-layer turbulence, through their role in the equations for components of the turbulent kinetic energy, are responsible for the disparity in mixed-layer depths between the eastern and western regions of the equatorial Pacific Ocean. Galperin et al. (1989) reexamined the question with a more detailed second-order-closure model. Their results did not support the Garwood et al. findings. They did not find a dramatic effect of rotation on the depth of the mixed layer; according to their model, the stable stratification that is typical of the Pacific attenuated the effect. Due to the difficulty of obtaining upper-ocean turbulence data of the necessary quality and detail, the issue has not been observationally resolved.

### 16.2.4 LES and Model Testing

The problem of obtaining the geophysical turbulence data necessary for turbulence modeling can, we believe, be relieved somewhat through LES. Let us illustrate how an LES "database" can be used for both system testing and component testing.

Moeng and Wyngaard (1986, 1989) used LES to evaluate some contemporary parameterizations of the pressure covariance, turbulent transport, and dissipation-rate terms in second-order models. In the first study they resolved the fluctuating pressure field found through LES of a convective boundary layer into turbulence–turbulence, mean shear, buoyancy, Coriolis, and subgrid-scale contributions: $p = p_T + p_S + p_B + p_C + p_{SG}$. The first four of these are defined through

$$\frac{1}{\rho_0}\nabla^2 p = -\left(\frac{\partial u_i}{\partial x_j}\frac{\partial u_j}{\partial x_i} - \overline{\frac{\partial u_i}{\partial x_j}\frac{\partial u_j}{\partial x_i}}\right) - 2\frac{\partial U_i}{\partial x_j}\frac{\partial u_j}{\partial x_i} + \frac{g}{T}\frac{\partial \theta}{\partial x_3}$$
$$+ f\left(\frac{\partial u_2}{\partial x_1} - \frac{\partial u_1}{\partial x_2}\right). \tag{1}$$

They found that $p_B$ and $p_T$ dominated. Focusing on the pressure gradient–scalar-concentration covariance appearing in the scalar flux budget, they found that to a good approximation

$$\frac{1}{\rho_0}\overline{c\frac{\partial p_B}{\partial z}} \simeq \frac{1}{2}\frac{g}{T}\overline{\theta c}, \tag{2}$$

where $c$ is the fluctuating scalar, $g/T$ is the buoyancy parameter, $z$ is the vertical coordinate, and $\theta$ is the temperature fluctuation. Isotropic tensor modeling (Lumley, 1978) predicts such a relation (with a proportionality factor of $\frac{1}{3}$), but it was not known how well the prediction would apply to the PBL.

A Rotta-type parameterization (Rotta, 1951) is commonly used for the turbulent–turbulent part of the pressure gradient–scalar covariance:

$$\frac{1}{\rho_0} \overline{c \frac{\partial p_T}{\partial z}} = \frac{\overline{wc}}{\tau}. \tag{3}$$

Here $\tau$ is a turbulent time scale. They found that this parameterization can be made to fit the LES results, but that the time scale $\tau$ depends on the nature of the scalar diffusion process. In top-down diffusion (scalar flux maximum at the PBL top, zero at the bottom) $\tau$ is larger and has a different vertical profile shape than in the bottom-up case.

Second-order modelers more often split $\overline{c \partial p / \partial z}$ into two terms, writing

$$\overline{c \frac{\partial p}{\partial z}} = \frac{\partial \overline{pc}}{\partial z} - \overline{p \frac{\partial c}{\partial z}}, \tag{4}$$

and lumping the first term on the right with the turbulent transport term. Isotropic tensor modeling predicts that this first term vanishes, but the LES results indicate that it can be quite significant. The LES results also show that the Rotta parameterization applied to the turbulent–turbulent contribution to the second term on the right (the scalar gradient–pressure covariance) is less effective than when applied to the pressure gradient–scalar covariance [i.e., (3)].

In a second study, Moeng and Wyngaard (1989) used LES results to evaluate some of the turbulent transport and dissipation-rate closures used in contemporary second-order models. The gradient-diffusion approximation for turbulent transport fared poorly, due in large part to the direct influence of buoyancy. They showed that this causes poor predictions of the vertical profiles of some turbulence profiles. They found that the length scales used in closures for mechanical and thermal dissipation rates are typically a factor of 2–3 too small, leading to underprediction of turbulent kinetic energy levels.

Moeng and Wyngaard also found that the flux and variance budgets for conservative scalars are substantially different for top-down and bottom-up diffusion in the convective PBL. They surmised that in order to capture these differences in a second-order model, it would be necessary to model the turbulent transport, pressure covariance, and molecular destruction terms differently in the two cases.

## 16.3 An Illustrative Example: Modeling Scalar Transport

Let us consider an old but important problem, the vertical transport of a conservative scalar through the convective boundary layer. The

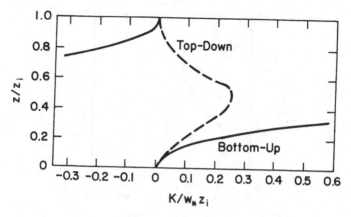

Fig. 1   Vertical profiles of the eddy diffusivities of bottom-up (solid curve) and top-down (dashed curve) scalars from large eddy simulations.

transport of species – e.g., trace chemical constituents – is of increasing interest in global-change science.

Through LES we found (Wyngaard and Brost, 1984; Moeng and Wyngaard, 1984) that the mid-layer singularity of eddy diffusivity discussed in Section 2 occurs only in the bottom-up process; in the top-down process it is well behaved. Figure 1 shows these eddy diffusivity profiles. It follows (Wyngaard and Brost, 1984) that there can also be a singularity in eddy diffusivity in the general case of combined top-down and bottom-up diffusion.

The difference between the top-down and bottom-up diffusivity profiles creates what we have termed "transport asymmetry" (Wyngaard and Weil, 1991). This refers to the different transport properties and scalar statistics in top-down and bottom-up diffusion. Let us examine its implications for modeling scalar transport.

The Reynolds equation for $\overline{cw}$, the vertical flux of a conservative scalar, is

$$\frac{\partial \overline{cw}}{\partial t} = -\overline{w^2}\frac{\partial C}{\partial z} - \frac{\partial \overline{w^2 c}}{\partial z} + \frac{g}{T}\overline{\theta c} - \frac{1}{\rho_0}\overline{c\frac{\partial p}{\partial z}}. \tag{5}$$

We have used the Boussinesq approximation in writing the buoyancy term. In order to use this equation in a second-moment model, one needs to parameterize the turbulent transport (second on the right side) and pressure (fourth) terms. Micrometeorological experiments show clearly that in the surface layer the pressure term is the principal sink of scalar flux (Wyngaard et al., 1971). The principal sources are the first term,

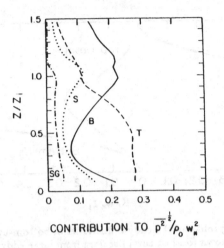

Fig. 2　Profiles of the turbulence–turbulence ($T$), buoyancy ($B$), mean shear ($S$), and subgrid-scale ($SG$) contributions to the rms pressure field, obtained from large eddy simulations and nondimensionalized with $\rho_0 w_*^2$.

representing gradient production, and the buoyancy term. In mid-PBL, the principal sources are buoyancy and transport; the mean-gradient term becomes a sink.

In a convective atmospheric PBL, the rms contributions to the total fluctuating pressure field, as given in (1), behave as shown in Fig. 2, according to Moeng and Wyngaard (1986). We see that at a minimum one needs to parameterize the buoyant and turbulent–turbulent parts of pressure. The mean-shear contribution is normally important as well, and in some flows, such as the upper-ocean PBL, it appears that one also needs to parameterize the Coriolis contribution (Garwood et al., 1985; Galperin et al., 1989).

Moeng and Wyngaard (1986) suggested the combination

$$\frac{1}{\rho_0}\left(\overline{c\frac{\partial p_T}{\partial z}} + \overline{c\frac{\partial p_B}{\partial z}}\right) = \frac{\overline{cw}}{\tau} + \frac{1}{2}\frac{g}{T}\overline{\theta c}, \qquad (6)$$

where $\tau$ is a time scale. They tested this against the calculated behavior of the covariance for temperature and for three passive, conservative scalars, each having different combinations of top-down and bottom-up diffusion. They found it to be an effective parameterization if the time scale $\tau$ is tailored properly; $\tau$ has different profiles in the top-down and bottom-up cases.

Holtslag and Moeng (1991) used (6) to close the buoyancy flux budget. That approach leads to an interesting result, as we will now show for the more general case of the bottom-up scalar flux. The scalar flux budget (5) is, in quasi-steady conditions,

$$0 = -\overline{w^2}\frac{\partial C}{\partial z} + T + B + P, \tag{7}$$

where $T$, $B$, and $P$ are the transport, buoyant, and pressure terms, respectively. Figure 3 shows the vertical profiles of the terms in the bottom-up case as determined from LES results. We see that

$$T \sim P + \text{constant}, \quad \text{constant} \sim 2w_*^2 c_*/z_i, \tag{8}$$

as Holtslag and Moeng (1991) first pointed out for the special case of the temperature flux budget (which is nearly a bottom-up process). Here $w_*$ is the convective velocity scale $(gQ_0 z_i/T)^{1/3}$, where $Q_0$ is the surface temperature flux and $w_* c_* = \overline{cw}_0$ the surface scalar flux. We use the linearity of the scalar flux profile in quasi-steady conditions to rewrite (8) as

$$T \sim P + 2\frac{w_*\overline{cw}_0}{z_i} = P - 2w_*\frac{\partial \overline{cw}}{\partial z}. \tag{9}$$

If we also use the parameterization (6) in the form

$$P \sim P_T + P_B \sim -\frac{\overline{cw}}{\tau} - \frac{B}{2}, \tag{10}$$

then (7) becomes

$$\overline{cw} = -\frac{\overline{w^2}\tau}{2}\frac{\partial C}{\partial z} - w_*\tau\frac{\partial \overline{cw}}{\partial z}. \tag{11}$$

As Holtslag and Moeng pointed out, for the special case of temperature, (11) bears a resemblance to previous results (e.g., Deardorff, 1972) that "correct" the eddy diffusivity relation for the observed countergradient heat flux in the convective PBL. Equation (11) is also similar to a result derived by Wyngaard and Weil (1991) by entirely different means. Let us now review that derivation.

Lumley (1975) derived a general, kinematic expression for the flux of a passive, conservative scalar in turbulence. In one dimension his result reads

$$\text{constituent flux} = \sum_{n=0}^{\infty} \frac{(-1)^n}{(n+1)!}\frac{\partial^n}{\partial z_0^n}\left[C(z_0,0)\overline{\dot{\zeta}^{n+1}}\right]. \tag{12}$$

Here $z_0$ is an arbitrary space point, $C(z_0,0)$ is the concentration there at time 0, $\zeta$ is the displacement between initial position and that at time $t$, and a dot signifies a time derivative. Wyngaard and Weil (1991) required that this be invariant to the addition of a constant to $C$, and applied it to skewed but homogeneous turbulence – the simplest possible

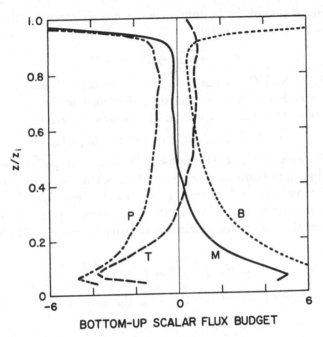

Fig. 3   Profiles of the terms in the bottom-up scalar flux budget, nondimensionalized with $w_*^2 c_*/z_i$.

approximation to a convective boundary layer. This gives

$$\overline{cw} = -\frac{1}{2}\dot{\overline{\zeta^2}}\frac{\partial \overline{C}}{\partial z} + \frac{1}{6}\dot{\overline{\zeta^3}}\frac{\partial^2 \overline{C}}{\partial z^2} - \frac{1}{24}\dot{\overline{\zeta^4}}\frac{\partial^3 \overline{C}}{\partial z^3} + \cdots . \qquad (13)$$

From the classical literature on dispersion, we recognize that for $t \gg T_L$, where $T_L$ is the Lagrangian integral time scale,

$$\dot{\overline{\zeta^2}} = 2\overline{w^2}T_L = 2K, \qquad (14)$$

and $K$ is the eddy diffusivity. If we treat $\dot{\overline{\zeta^3}}$ analogously, relating it to $\overline{w^3}$, we then have

$$\overline{cw} = -K\frac{\partial \overline{C}}{\partial z} + \frac{SK\sigma_w T_L}{2}\frac{\partial^2 \overline{C}}{\partial z^2} + D\frac{\partial^3 \overline{C}}{\partial z^3} + \cdots . \qquad (15)$$

Here $S = \overline{w^3}/(\overline{w^2})^{3/2}$ is the skewness of $w$. If we consider this to be an equation for $\partial C/\partial z$, we can write its solution as another series expansion:

$$\frac{\partial \overline{C}}{\partial z} = -\frac{\overline{cw}}{K} - \left(\frac{S\sigma_w T_L}{2K}\right)\frac{\partial \overline{cw}}{\partial z} - \left(\frac{(S\sigma_w T_L)^2}{4K} + \frac{D}{K^2}\right)\frac{\partial^2 \overline{cw}}{\partial z^2} + \cdots . \qquad (16)$$

Now consider the shape of the $\overline{cw}$ profile. In the horizontally homogeneous vertical diffusion of a scalar field from a large, plane source, the mean concentration balance is

$$\frac{\partial \overline{C}}{\partial t} + \frac{\partial \overline{cw}}{\partial z} = 0. \tag{17}$$

If instead we have a balance between $z$-independent mean advection and flux divergence,

$$U\frac{\partial \overline{C}}{\partial x} + \frac{\partial \overline{cw}}{\partial z} = 0. \tag{18}$$

In each of these cases, which are reasonable models of typical situations within a grid square of a large-scale meteorological model, the $\overline{cw}$ profile is linear, which truncates the expansion (16). If the flow has depth $z_i$, (16) then becomes

$$\overline{cw} = -K\frac{\partial C}{\partial z} - \frac{S\sigma_w T_L}{2}\frac{\partial \overline{cw}}{\partial z}. \tag{19}$$

If we write (19) as

$$\overline{cw}\left(1 + \frac{S\sigma_w T_L}{2\overline{cw}}\frac{\partial \overline{cw}}{\partial z}\right) = -K\frac{\partial C}{\partial z}, \tag{20}$$

we see that the effective eddy diffusivity $K^{eff}$ is

$$K^{eff} = K\left(1 + \frac{S\sigma_w T_L}{2\overline{cw}}\frac{\partial \overline{cw}}{\partial z}\right)^{-1}. \tag{21}$$

Equation (21) shows transport asymmetry: in top-down diffusion (which we can define generally as when the right-hand term in the denominator is positive), $K^{eff} < K$, and in bottom-up diffusion (when the term is negative) $K^{eff} > K$.

In convective turbulence the Lagrangian integral time scale $T_L$ behaves initially as if it is of the order of the Eulerian time scale $\tau \sim z_i/w_*$ (Weil, 1990). [Formally, $T_L$ is zero if the motions associated with the $w$-field are bounded, as in a boundary layer flow (Tennekes and Lumley, 1972); this means that the autocorrelation function for $w$ has a long, oscillating tail.] Thus, we can associate $\tau$ with $T_L$ and $\overline{w^2}\tau$ with the eddy diffusivity $K$. $S$, the skewness of $w$, is $O(1)$ in the convective boundary layer. Thus, we see that our budget result (11) and the kinematic result (19) have the same form in the special case of bottom-up diffusion.

Equation (19) is based on a small-time-scale expansion that would not be expected to be valid in convection. Wyngaard and Weil (1991) suggested that the apparent applicability of (19) stems not from its roots in this kinematic model, but from the heuristic model of Wyngaard (1987) for the asymmetry in top-down and bottom-up diffusion, whose results can be written in the same form. The heuristic model shows how the time changes inherent in these two diffusion processes (due to the flux divergence in each) can interact with the updraft–downdraft asymmetry (traceable to the skewness of the $w$-field) to give transport asymmetry.

Equation (19) [or, equivalently, Eq. (11)] is a very simple parameterization that reflects the enhanced diffusivity for scalars in bottom-up diffusion and the reduced diffusivity in the top-down case. Such situations exist commonly in the lower atmosphere and, presumably, in the upper ocean. We can also write, using (15), an equivalent expression in terms of $\overline{C}$ derivatives,

$$\overline{cw} = -K\frac{\partial \overline{C}}{\partial z} + K'\frac{\partial^2 \overline{C}}{\partial z^2}, \qquad (22)$$

where $K' \sim SK\sigma_w T_L$. It is interesting that such an expression, even with $K$ and $K'$ constant, shows transport asymmetry. We are working now to incorporate these physics in the simple PBL representations in large-scale models.

## 16.4 Looking to the Future

LES is finding many new applications in small-scale meteorology, as demonstrated by Cotton et al. in Chapter 17 of this volume. Its use in oceanography is more recent, but Chapter 19 by Holloway and Chapter 22 by Cane point to important issues that LES can address in the small-scale fluid mechanics of the ocean. Müller, in Chapter 21, reviews the understanding of diapycnal mixing of buoyancy in the ocean interior as a guide to the application of LES to that problem. LES is also being used to advantage in the study of turbulent flow and transport in surface waters (Bedford and Yeo, Chapter 23).

While these new applications of LES are being pursued, there is also continuing development work on the technique, some of it quite fundamental. Aldama (Chapter 24) discusses some new concepts in the basic filtering operations that underlie LES. Bedford and Yeo also discuss filtering operations with an eye toward clarifying the use of averaged equations in turbulent flow and transport modeling.

Thus, fundamental work on LES formulation continues, even though the technique has been under development for over 20 years. These studies are also touching on some basic issues in the treatment of subgrid-scale physics. At first glance this might be surprising, particularly since one of the strengths of LES has been perceived to be its relative insensitivity to the details of this subgrid-scale modeling. It has been discovered, however, that in boundary layer flows LES can show significant departures from the observed profiles in the near-surface region (Mason and Thomson, 1992). Those authors find that the inclusion of stochastic variability in the model for subgrid-scale stress leads to a marked improvement in the near-wall flow simulation. In Chapter 5 Leith also discusses the physics of this phenomenon, sometimes called "stochastic backscatter." This should serve to remind us that subgrid-scale models can be *locally* quite important, and their continued improvement will probably be necessary as LES is applied to new situations.

Finally, we must emphasize the importance of experiment in guiding and shaping the development of LES and its use in geophysical fluid mechanics. We have discussed the difficulty of making measurements of geophysical turbulence, but we cannot do without them. As our "ground truth," they alone can tell us whether our computer simulations of nature are to be trusted, and if not, where they must be improved.

## Acknowledgments

This work was carried out in the PBL Model Evaluation and Development Project at the National Center for Atmospheric Research. We are grateful to project members A. Andreń (Uppsala University), B. Holtslag (Royal Netherlands Meteorological Institute), J. McWilliams, and J. Tribbia for helpful comments.

The National Center for Atmospheric Research is sponsored by the National Science Foundation.

## References

ANDRÉ, J.C., DE MOOR, G., THERRY, G. AND DU VACHAT, R. (1978) Modeling the 24-hour evolution of the mean and turbulent structures of the planetary boundary layer. *J. Atmos. Sci.* **35**, 1861–1883.

BROST, R.A. AND WYNGAARD, J.C. (1978) A model study of the stably stratified planetary boundary layer. *J. Atmos. Sci.* **35**, 1427–1440.

DEARDORFF, J.W. (1966) The counter-gradient heat flux in the atmosphere and in the laboratory. *J. Atmos. Sci.* **23**, 503–506.

DEARDORFF, J.W. (1970a) A numerical study of three-dimensional turbulent channel flow at large Reynolds numbers. *J. Fluid Mech.* **41**, 453–480.

DEARDORFF, J.W. (1970b) Convective velocity and temperature scales for the unstable planetary boundary layer and Rayleigh convection. *J. Atmos. Sci.* **27**, 1211–1213.

DEARDORFF, J.W. (1972) Theoretical expression for the countergradient vertical heat flux. *J. Geophys. Res.* **77**, 5900–5904.

DEARDORFF, J.W. (1974) Three-dimensional numerical study of turbulence in an entraining mixed layer. *Bound.-Layer Meteorol.* **7**, 199–226.

DEARDORFF, J.W. (1980) Stratocumulus-capped mixed layers derived from a three-dimensional model. *Bound.-Layer Meteorol.* **18**, 495–527.

DONALDSON, C. DU P. (1973) Construction of a dynamic model of the production of atmospheric turbulence and the dispersal of atmospheric pollutants. In *Workshop on Micrometeorology*. Ed. D.A. Haugen, pp. 313–390. Amer. Meteorol. Soc.

GALPERIN, B., ROSATI, A., KANTHA, L.H. AND MELLOR, G.L. (1989) Modeling rotating stratified turbulent flows with application to oceanic mixed layers. *J. Phys. Oceanogr.* **19**, 901–916.

GARWOOD, R.W., MULLER, P. AND GALLACHER, P.C. (1985) Wind direction and equilibrium mixed layer depth in the tropical Pacific Ocean. *J. Phys. Oceanogr.* **15**, 1332–1338.

HOLTSLAG, A.A.M. AND MOENG, C.-H. (1991) Eddy diffusivity and counter-gradient transport in the convective atmospheric boundary layer. *J. Atmos. Sci.* **48**, 1690–1698.

LEWELLEN, W.S. AND TESKE, M.E. (1973) Prediction of the Monin–Obukhov similarity functions from an invariant model of turbulence. *J. Atmos. Sci.* **30**, 1340–1345.

LIEPMANN, H.W. (1979) The rise and fall of ideas in turbulence. *Am. Scientist* **67**, 221–228.

LILLY, D.K. (1967) *Proceedings of the IBM Scientific Computing Symposium on Environmental Sciences*. IBM Form No. 320-1951, 195–202.

LUMLEY, J.L. (1975) Modeling turbulent flux of passive scalar quantities in inhomogeneous flows. *Phys. Fluids* **18**, 619–621.

LUMLEY, J.L. (1978) Computational modeling of turbulent flows. *Adv. Appl. Mech.* **18**, 123–176.

LUMLEY, J.L. (1983) Atmospheric modeling. Mechanical Engineering Transactions, The Institution of Engineers, ME8: 153–159.

LUMLEY, J.L. (1990) The utility and drawbacks of traditional approaches. Comment 1. *Whither Turbulence? Turbulence at the*

*Crossroads.* Lecture Notes in Physics, vol. 357. Ed. J. L. Lumley, pp. 49–57. Springer-Verlag.

MASON, P.J. AND THOMSON, D.J. (1992) Stochastic backscatter in large-eddy simulations of boundary layers. *J. Fluid Mech.* **242**, 51–78.

MELLOR, G.L. (1973) Analytic prediction of the properties of stratified planetary surface layers. *J. Atmos. Sci.* **30**, 1061–1069.

MELLOR, G.L. AND YAMADA, T. (1982) Development of a turbulence closure model for geophysical flow problems. *Rev. Geophys. Space Phys.* **20**, 851–875.

MOENG, C.-H. AND WYNGAARD, J.C. (1984) Statistics of conservative scalars in the convective boundary layer. *J. Atmos. Sci.* **41**, 3161–3169.

MOENG, C.-H. AND WYNGAARD, J.C. (1986) An analysis of closures for pressure-scalar covariances in the convective boundary layer. *J. Atmos. Sci.* **43**, 2499–2513.

MOENG, C.-H. AND WYNGAARD, J.C. (1989) Evaluation of turbulent transport and dissipation closures in second-order modeling. *J. Atmos. Sci.* **46**, 2311–2330.

NIEUWSTADT, F.T.M. (1984) The turbulent structure of the stable nocturnal boundary layer. *J. Atmos. Sci.* **41**, 2202–2216.

PANOFSKY, H.A. AND DUTTON, J.A. (1984) *Atmospheric Turbulence.* Wiley, 397 pp.

ROTTA, J.C. (1951) Statistische theorie nichthomogener turbulenz. *Z. Phys.* **129**, 547–572.

SCHUMANN, U. (1977) Realizability of Reynolds stress turbulence models. *Phys. Fluids* **20**, 721–725.

SMAGORINSKY, J., MANABE, S. AND HOLLOWAY, J.L. (1965) Numerical results from a nine-level general circulation model of the atmosphere. *Mon. Wea. Rev.* **93**, 727–768.

TENNEKES, H. AND LUMLEY, J.L. (1972) *A First Course in Turbulence.* MIT Press, 300 pp.

WEIL, J.C. (1990) A diagnosis of the asymmetry in top-down and bottom-up diffusion using a Lagrangian stochastic model. *J. Atmos. Sci.* **47**, 501–515.

WYNGAARD, J.C. (1973) On surface-layer turbulence. In *Workshop on Micrometeorology.* Ed. D. A. Haugen, pp. 101–149. Amer. Meteorol. Soc.

WYNGAARD, J.C. (1987) A physical mechanism for the asymmetry in top-down and bottom-up diffusion. *J. Atmos. Sci.* **44**, 1083–1087.

WYNGAARD, J.C. (1988) Structure of the PBL. In *Lectures on Air Pollu-*

*tion Modeling.* Ed. A. Venkatram and J. C. Wyngaard, pp. 9–61. Amer. Meteorol. Soc.

WYNGAARD, J.C., COTÉ, O.R. AND IZUMI, Y. (1971) Local free convection, similarity, and the budgets of shear stress and heat flux. *J. Atmos. Sci.* **28**, 1171–1182.

WYNGAARD, J.C. AND BROST, R.A. (1984) Top-down and bottom-up diffusion of a scalar in the convective boundary layer. *J. Atmos. Sci.* **41**, 102–112.

WYNGAARD, J.C. AND WEIL, J.C. (1991) Transport asymmetry in skewed turbulence. *Phys. Fluids A* **3**, 155–162.

ZEMAN, O. AND LUMLEY, J.L. (1976) Modeling buoyancy driven mixed layers. *J. Atmos. Sci.* **33**, 1974–1988.

# ATMOSPHERIC SCIENCES

# Using the Regional Atmospheric Modeling System in the Large Eddy Simulation Mode: From Inhomogeneous Surfaces to Cirrus Clouds

WILLIAM R. COTTON, ROBERT L. WALKO,

KEELEY R. COSTIGAN, PIOTR J. FLATAU,

AND ROGER A. PIELKE

## 17.1 Introduction

Beginning with the pioneering work of Deardorff (1972a,b; 1973; 1974), large eddy simulation (LES) has revealed a great deal of new insight into the behavior of the convective boundary layer over horizontally homogeneous terrain. In recent years LES has been extended to simulations of the convective boundary layer over inhomogeneous surfaces (Hadfield, 1988; Hadfield et al., 1991; Krettenauer and Schumann, 1989; Walko et al., 1990) and to stably stratified boundary layers as well (Mason and Derbyshire, 1990).

In this paper we describe applications of a particular, multipurpose modeling system, the Regional Atmospheric Modeling System (RAMS) developed at Colorado State University, to LES over complicated land surfaces and upper tropospheric clouds. We also describe the computational environment in which we are performing large eddy simulations.

## 17.2 Summary of Rams

The Colorado State University Regional Atmospheric Modeling System (RAMS) is an advanced, multipurpose system capable of simulating a wide variety of meteorological phenomena over a broad range of spatial and temporal scales. RAMS is a completely new computer code that represents the merger of the cloud modeling program developed under the direction of Dr. William R. Cotton (Tripoli and Cotton, 1982) and the mesoscale modeling program directed by Dr. Roger A. Pielke (Mahrer and Pielke, 1977). Versatility in RAMS is gained through the use of a pre-processor code that allows the user to configure a unique FORTRAN model code that can run efficiently on vector processors.

Some of the general features of RAMS are as follows:

• Pressure can be evaluated either hydrostatically (Tremback et al., 1985) or nonhydrostatically using a time-split compressible scheme (Tripoli and Cotton, 1982; 1989).

• Turbulence is parameterized using simple sub-grid scale, first-order Smagorinsky-type closure schemes with Richardson number dependence (Hill, 1974), Deardorff's scheme based on predicting turbulent kinetic energy (TKE) (Deardorff, 1980), or an ensemble-averaged TKE scheme based on Mellor and Yamada (1982).

• The lower boundary is either a flat, Cartesian surface or a terrain-following surface (Gal-Chen and Somerville, 1975) with a surface boundary condition based on diabatic similarity theory (Louis, 1979).

• Surface fluxes or surface temperatures can be specified, or predicted using a multi-level soil model (Tremback and Kessler, 1985). Vegetation influences on the atmosphere can be represented using the procedure to be reported in Lee (1991).

• Lateral boundary conditions are either cyclic, open-radiative type (Klemp and Wilhelmson, 1978; Orlanski, 1976; Klemp and Lilly, 1978), or time-specified conditions nudged to observed fields.

• Top boundary conditions are either a wall, a radiative condition (Klemp and Durran, 1983) or a Rayleigh-friction wave absorbing layer.

• The computational grid may be Cartesian or polar-stereographic in the horizontal directions with spacing ranging from tens of meters (*e.g.*, Hadfield, 1988; Hadfield et al., 1991), to 50 km or greater (Tremback et al., 1985; Cram et al., 1991), and may be of constant spacing or stretched vertically.

• An interactive grid-nesting procedure can be activated that is based on techniques developed by Clark and Farley (1984) but is generalized to include such features as moving nests.

• Cloud processes can be explicitly simulated including a variety of cloud microphysical fields (Cotton et al., 1982; 1986) or, on coarser regional grids, cumulus clouds can be parameterized.

• Shortwave and longwave radiation flux divergences are parameterized either for clear air (Mahrer and Pielke, 1977) or for a cloudy, partly-cloudy atmosphere (Chen and Cotton, 1983).

## 17.3 LES over Simple Hilly Terrain in a Cyclic Domain

In this particular application of RAMS we examine the influence of a simple hilly surface on the behavior of the convective boundary layer. This work is a direct extension of Hadfield's (1988; Hadfield et al., 1991) simulation of the response of the convective boundary layer over an inhomogeneously heated surface. A more complete report of the present work is given in Walko et al. (1991).

For this application, RAMS is configured with a computational domain extending 10 km by 6 km horizontally and 3 km vertically. The horizontal resolution is 125 m, and the vertical resolution begins at 30 m near the ground, stretching gradually to 80 m at a height near 400 m, and remaining at 80 m above. Subgrid diffusion is parameterized following Deardorff (1980). A Rayleigh friction absorbing layer is activated in the upper 750 m of the domain to prevent reflections of gravity waves off the top boundary which is a rigid lid. The hilly terrain is sinusoidal in one horizontal direction and invariant in the other, thus consisting of a series of parallel ridges and valleys. In the simulations reported here, 5 terrain cycles are used. The terrain height is approximately 15% of the CBL depth. The lateral boundaries in both directions are cyclic. The simulations are run with a constant surface heat flux of 250 W m$^{-2}$. Initially, no horizontal wind is present, and the potential temperature is constant from the ground to 1 km in height, increasing above that at a rate of 10 K km$^{-1}$. The lowest grid level in the model is perturbed slightly with a random temperature variation to trigger the development of a three-dimensional structure in the eddy field. Simulations with both flat and hilly terrain were conducted for a period of 150 minutes, and statistics to be presented here were averaged over the final 15 minutes of that period. Averages taken over several previous 15-minute periods confirmed that the statistical fields became steady after 60 to 90 minutes into the simulations.

In a LES with horizontally homogeneous forcing, time-averaged fields contain no expected horizontal variations, and it is practical to horizontally-average quantities of interest in order to obtain statistically significant vertical profiles. Such an average groups together like properties of the CBL (*i.e.*, those from equivalent locations in the atmosphere relative to the ground, CBL top, etc.) while preserving their vertical variability. However, this is not necessarily the case when horizontal variations in terrain height are imposed. Over sloping terrain, field variables on a horizontal surface do not share the property of being located a given

height above the ground, and to average them over that horizontal surface would mix together various properties unique to each fractional
position between the ground and the CBL top. A logical alternative is
to average with respect to the horizontal coordinates on each of a family
of surfaces defined to be located a fixed fraction of the vertical distance
from the ground to the CBL top. Such a surface is defined by a constant
value of $z_*$, defined by

$$z_* = z_i(z - z_s)/(z_i - z_s) \qquad (1)$$

where $z$ is the vertical coordinate, $z_i$ is the CBL height, and $z_s$ is the
ground height, all relative to a reference height such as sea level. We shall
denote an average of a quantity $\phi$ on this surface by $< \phi >_{x,y,t}$, where
x and y are the horizontal coordinates, and t denotes time averaging. A
deviation from the average over the surface is denoted by $(\phi)_{x,y}$.

A similar argument applies to the vertical velocity, which at the
ground becomes zero when the ground is horizontal. With hilly terrain, it is useful to define a modified vertical velocity which is always
zero at the ground even when up-slope or down-slope flow exists. It
is also useful for this modified velocity to equate with the true vertical
velocity at the CBL top, provided the CBL top is nearly horizontal (as
it is in the present study). A modified vertical velocity having these
properties is defined as

$$\omega = w - u\frac{z - z_i}{z_s - z_i}\frac{\partial z_s}{\partial x}, \qquad (2)$$

where the coefficient of u is the local slope of a surface of constant $z_*$.
The quantity $\omega$ represents the vertical velocity of a parcel relative to the
local $z_*$ surface, while $w$ is the vertical velocity relative to the vertical
coordinate $z$. These two differ for any parcel having a component of
horizontal motion parallel to the height gradient of a constant $z_*$ surface.
A zero value of $\omega$ implies no change in fractional position between the
bottom and top of the CBL. Thus, $\omega$ is the appropriate variable for
describing vertical transport through the CBL.

The presence of a well-defined horizontal variation in forcing of the
CBL makes useful a second type of average, termed a phase average
(Hadfield, 1988; Hadfield et al., 1991). This is denoted by $< \phi >_{y,p,t}$,
where $p$ refers to the phase angle of the terrain height, which has its
variation in the x-direction. This average combines the five hill-valley
terrain cycles into one, thus reducing the variation of $\phi$ with respect
to the entire x-span of the domain to a variation across only a single
terrain cycle. Averaging with respect to the y-coordinate (and time) is

performed as before since no terrain variation occurs in that direction. It is to be understood that a phase average is still performed on a surface of constant $z_*$. The phase average avoids filtering out features in the CBL eddies having the same horizontal scale as the terrain, and allows determination as to whether such features are of significant amplitude.

In comparing results from the flat and hilly simulations, we first consider those averages which employ the operator $<>_{y,x,t}$. We have applied this operator to all first and second order moments involving perturbations of the velocity components $(u,v,\omega)$ and to potential temperature $\theta$, and to the triple correlation of modified vertical velocity $\omega$. Thus, we have evaluated expressions of the forms $< (\alpha)_{y,x} >_{y,x,t}$, $< (\alpha)_{y,x}(\beta)_{y,x} >_{y,x,t}$, and $< (\omega)^3_{y,x} >_{y,x,t}$, where $\alpha$ and $\beta$ represent any velocity component or potential temperature. A simple summary of the comparisons of these quantities between the flat and hilly simulations can be made: *Very little difference occurs.* This indicates that the mean vertical structure of the boundary layer, without regard to horizontal variations, is not significantly affected by the ridges and valleys. This somewhat surprising result, although negative, is of special interest in applications such as the designing of a boundary layer parameterization for use in lower resolution numerical models.

We next consider results obtained by applying the averaging operator $<>_{y,p,t}$ to the moments discussed above. Given sufficiently long time averaging, such averages would be expected to yield horizontally homogeneous results for the flat terrain simulation, and would therefore not produce any useful information beyond the $<>_{y,x,t}$ averages obtained earlier. Over the hilly terrain, however, there is the potential that the topography induces signals into the eddy field having the same horizontal scale as the topography, and phase-averaged quantities would reveal such signals.

Figure 1 is a plot of $< \omega >_{y,p,t}$, which is a function of $x$, over one terrain cycle, and of $z$. This field shows that mean vertical motion is upward over the hills and downward over the valleys. Near the ground, a positive value of $\omega$ indicates that a parcel is increasing its height above the ground. In contrast, the $< w >_{y,p,t}$ field, shown in Figure 2, has positive values everywhere on the ground, except where the ground is horizontal. While this variable reveals the mean up-slope flow between the ridge and valley axes, it is not a useful indicator of changing distance between a parcel and the ground surface. Figure 3 shows the mean horizontal velocity perpendicular to the ridge and valley axes, given by $< u >_{y,p,t}$, which together with $< w >_{y,p,t}$ satisfies the two-dimensional

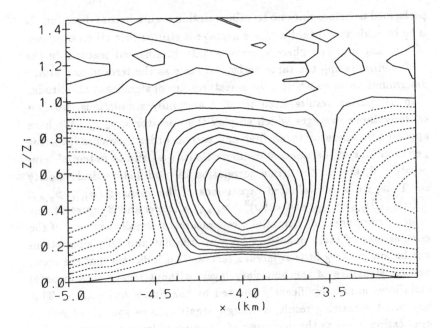

Fig. 1 Cross section of $< \omega >_{y,p,t}$ for the hilly case as a function of normalized height $z/z_i$ and $x$ (in km) over one terrain cycle. The contour interval is $0.04\, w_*$, with dashed lines in this and all other contour plots indicating negative values. The sinusoidal terrain height appears in correct vertical scale in the lower part of the plot.

continuity equation. Air in the lower CBL flows upslope to converge near the ridge top, and a compensating reversal of the flow occurs at higher levels. The lower branch of the horizontal motion reaches twice the peak speed of the upper branch.

The entire circulation is a prominent feature of the CBL eddy motions, with peak vertical and horizontal velocities reaching $0.3\, w_*$ and $0.5\, w_*$, respectively. The resolved kinetic energy of the circulation, defined by (Hadfield, 1988; Hadfield et al., 1991)

$$E_C = \int_{z_s}^{z_i} [< (u)_{y,x} >^2_{y,p,t} + < (w)_{y,x} >^2_{y,p,t}]dz, \qquad (3)$$

accounts for more than 12% of the total resolved kinetic energy in the CBL, which is given by

$$E_T = \int_x \int_{z_s}^{z_i} [< (u)^2_{y,x} >_{y,x,t} + < (v)^2_{y,x} >_{y,x,t} + < (w)^2_{y,x} >_{y,x,t}]dz\, dx. \qquad (4)$$

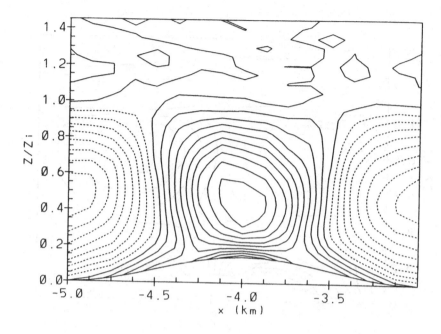

Fig. 2 As in Figure 1 but for $< w >_{y,p,t}$.

In spite of the mean upward motion over the higher terrain and mean downward motion elsewhere, as revealed in Figure 1, the local, instantaneous motion often is reversed in both regions. The mean probability of occurrence of positive $\omega$ as a function of $x$ and $z$, denoted by $< P(\omega) >_{y,p,t}$, is shown in Figure 4. A very strong horizontal variation of this quantity is indicated, with a maximum probability of 72% occurring just above the surface at the ridge top, and a minimum probability less than 15% near $0.6 z_i$ over the valley. This wide range is a further indication of the prominence of the mean circulation relative to the total eddy motion within the CBL.

Figure 5 shows the mean potential temperature perturbation from the horizontal average, given by $< (\theta)_{y,x} >_{y,p,t}$. Within the boundary layer, a broad region of above-average temperature occurs over the more elevated half of the surface terrain, coinciding approximately with the region of mean updraft in Figure 1. Similarly, the cooler region coincides with the subsident region. This is a partial indication that the mean circulation is thermally direct. The warmest perturbation (below the CBL top) from the layer-averaged mean is $1.6\,\theta_*$, located just above the

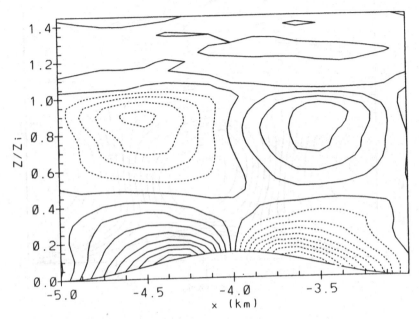

Fig. 3  As in Figure 1 but for $< u >_{y,p,t}$. The contour interval is $0.06\,w_*$.

ridge top, where the circulating air is most likely to just have spent the longest time being warmed by the surface. The coolest perturbation, $-0.6\,\theta_*$, occurs just above the valley floor, where air is most likely to have been out of contact with the ground for the longest time. At greater heights within the CBL, the horizontal temperature gradient weakens. In the layer capping the CBL, a strong, local cool perturbation occurs above the topographic ridge, and a warm perturbation occurs above the valley, both of which counteract the respective positive and negative buoyant affects of the layer below.

A plot of mean subgrid scale turbulent kinetic energy, given by $< e >_{y,p,t}$, is shown in Figure 6. Maximum values are found near the surface, where $e$ is most rapidly generated by buoyancy in a shallow superadiabatic layer, and values are somewhat higher over the higher terrain than over the valley. As in the case for $< \theta >_{y,p,t}$, this horizontal variation results from the mean circulation advecting $e$ through the low level generation region, from which the column over the high terrain is immediately downstream.

The sum of the covariance $< (\omega)_{y,x}(\theta)_{y,x} >_{y,p,t}$, which is a measure

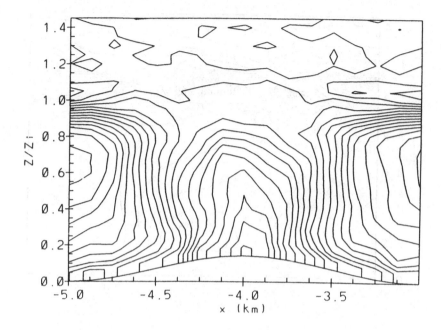

Fig. 4  As in Figure 1 but for the $y, p, t$-averaged probability of positive
$\omega$. The contour interval is 0.03, with the highest contour 0.72 near
the hilltop surface, and the minimum contour 0.15 over the valley at
$z = 0.6\, z_i$.

of the resolved vertical eddy $\theta$ flux, and the subgrid vertical $\theta$ flux is
shown in Figure 7. The $<>_x$ average of this field over the terrain cycle
yields the well-known linear vertical profile, ranging from its normalized
surface value of $1.0\, w_* \,\theta_*$ to its (negative) minimum at $z = z_i$. The figure
shows, however, that the mean upward heat transport is above average
over the higher terrain, exceeding the surface value by more than 20%
a short distance above the ground, while below-average upward heat
transport occurs above the valley. Most of the warm entrainment aloft,
indicated by the negative values at the top of the CBL, also occurs above
the higher terrain.

The above quantities which were phase-averaged to the same horizontal scale as the terrain showed that a prominent signal is induced in
the CBL eddies at that scale. To gain a broader perspective of how all
scales of CBL eddy motion are affected by the hills, we perform Fourier
analyses of $\omega$ for both the hilly and flat terrain cases. Specifically, we
compute the quantities $< F_x(\omega) >_{y,t}$ and $< F_y(\omega) >_{x,t}$, where $F_x(\omega)$ is

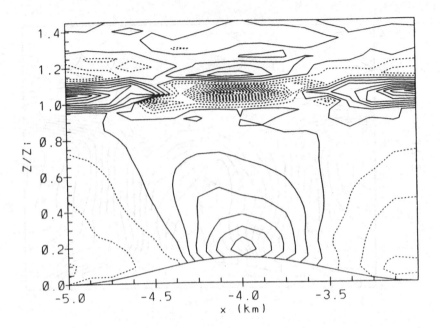

Fig. 5   As in Figure 1 but for $< (\theta)_{y,x} >_{y,p,t}$. The contour interval is
0.06 $w_*$.

the Fourier transform in the $x$-direction of $\omega$, and is a function of $y$, $z$, $t$,
and wave number $k_x$. Subsequent averaging over $y$ and $t$, as indicated,
yields a function of $z$ and $k_x$ only. A similar interpretation applies to
$< F_y(\omega) >_{x,t}$, which is also a function of $z$ and $k_y$ only.

Figure 8 shows $< F_x(\omega) >_{y,t}$ for the flat terrain case. The peak amplitude occurs at middle levels in the CBL at wave number 3, whose
wavelength is 3.3 $km$ or 2.5 $h_*$. Wave numbers 4 and 5 (having wavelengths of 2.0 $h_*$ and 1.6 $h_*$, respectively), are of only slightly weaker
amplitude. The highest wave number is 40, since there are 80 model grid
points spanning the $x$-direction. Figure 9 shows $< F_y(\omega) >_{x,t}$, which
has only 24 wave numbers since there are 48 model grid points in the
$y$-direction. Here, wave number 2 (whose wavelength is 2.3 $h_*$) achieves
the greatest amplitude, with other wave numbers from 0 to 3 all having
only slightly weaker amplitudes. We point out that wave number 5 in
the $x$-direction and wave number 3 in the $y$-direction, which are each the
highest wave number to have nearly the maximum amplitude, are of the
same wavelength, equal to 1.5 $h_*$, and wave number 8 in the $x$-direction

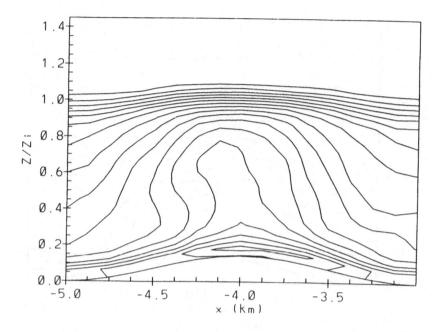

Fig. 6   As in Figure 1 but for $< e >_{y,p,t}$. The contour interval is $0.01\, w_*^2$.

and wave number 5 in the $y$-direction, which are the highest to have over half the peak amplitudes, are also nearly identical in wavelength. Thus, very similar spectra are exhibited in the two directions.

A very different situation arises in the hilly case, for which $< F_x(\omega) >_{y,t}$ is plotted in Figure 10. Here, a sharp peak occurs at wave number 5, corresponding to the 5 terrain cycles. The amplitude of wave number 5 is 40% stronger than in the flat terrain case. Wave numbers 3 and 4, from which most of the energy was apparently drawn, lost a third of their amplitude from the flat terrain case. Other waves were little affected by the terrain, except for wave number 6 which actually gained about 15% in amplitude. Figure 11 shows $< F_y(\omega) >_{x,t}$ for the hilly case. A strong peak in amplitude now occurs at wave number 0, representing a 40% increase from wave number 0 in the flat case. This echoes the mean upward motion over ridges and the mean subsidence over valleys which, as seen in Figure 6, appears in the $<>_y$ average over their entire length. Somewhat surprisingly, wave numbers 1 and 2 appear little affected by the hills, while wave number 3 was reduced in amplitude by over 20%. It is evident from the Fourier analyses that the dominant scales of the

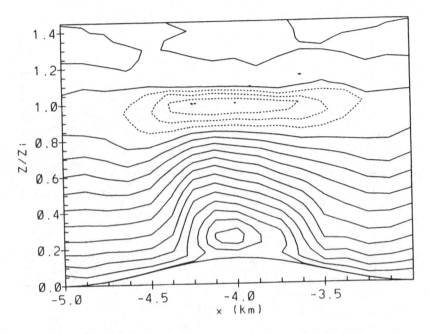

Fig. 7   As in Figure 1 but for $< (\omega)_{y,x}(\theta)_{y,x} >_{y,p,t}$. The contour interval is
0.01 $w_* \theta_*$.

CBL eddy motions are strongly forced toward the horizontal scale of the
terrain.

In summary, the results of simulating a convective boundary layer
over a series of ridges indicate that prominent solenoidal circulations
exist which consist of upslope flow near the surface, counter flow aloft,
upward motion over the hills, and subsidence over the valleys. For the
hill and valley configuration used, the solenoidal circulation accounts for
more than 12% of the total resolved eddy kinetic energy in the boundary layer. A Fourier analysis of the horizontal scales of vertical motion
in the boundary layer eddies showed the spectrum to be strongly influenced by terrain features. With hills rising to only 15% of the CBL
top from the valley floors, a 40% increase in amplitude resulted in the
harmonic matching the horizontal terrain scale. Most of this energy was
drawn from neighboring harmonics. In spite of the significant circulation
induced by the hilly terrain, mean vertical profiles of several meteorological variables, obtained by horizontally averaging across all terrain
features, were found to be almost identical between the hilly and flat

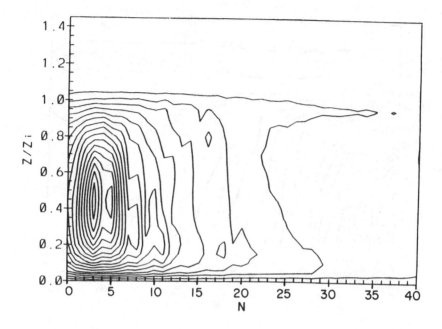

Fig. 8   Mean Fourier transform of vertical velocity, given by $< F_x(\omega) >_{y,t}$, for flat terrain case as a function of normalized height and wave number $N$. Contour interval is $0.01\,w_*$.

terrain simulations. This indicates that the hills and valleys had little influence on the global properties of the boundary layer, although they had large local influences.

## 17.4 LES with Nested Grids in Complex Terrain

Large eddy simulation has mainly been confined to the simulation of idealized physical situations. In this application we apply RAMS to a specific observational case study that took place during the Phoenix II experiment (Lilly and Schneider, 1990). Because the experimental area is immediately to the east of the foothills of the Rocky Mountains, larger-scale terrain-forced flows interact with boundary layer circulations. We have therefore decided to configure RAMS with the two-way interactive grids in which the courser resolution grids simulate the mountain slope flows while the finer resolution grids are used to simulate the behavior of the convective boundary layer. The results presented here represent

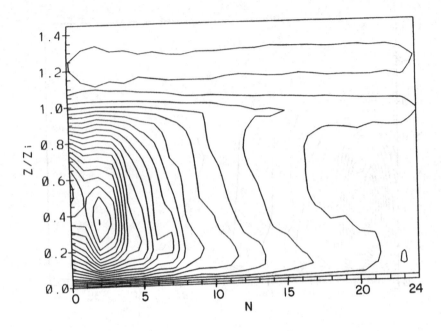

Fig. 9  As in Figure 8 but for $< F_y(\omega) >_{x,t}$.

preliminary work to the planned three-dimensional simulations. The model is configured as a two-dimensional model to test out the potential impacts of nested gridding on the LES and examine the effects of initial conditions (i.e. winds and thermodynamic fields) and terrain mapping on the behavior of the convective boundary layer in two-dimensions. While the two-dimensional results are not accurate quantitative representations of the three dimensional structure of the PBL, they provide us some insight into the expected behavior of the convective boundary layer over complicated terrain and, moreover, provide guidance in the final design of the computationally more costly three-dimensional LES experiments.

The model runs are intended to simulate 22 June 1984 in the area of the Boulder Atmospheric Observatory (BAO) tower on the high plains east of Boulder, CO. This date and location were chosen to coincide with the analysis of the Phoenix II data set being conducted by Lilly and Schneider (1990). The Phoenix II experiment was designed to study the motion field of the atmospheric boundary layer primarily with dual doppler radar. Aircraft, tower and PAM micronet surface measurements

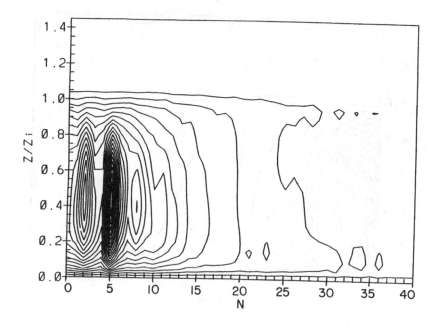

Fig. 10  As in Figure 8 but for the hilly case.

were also available for comparison and verification. The analysis by
Lilly and Schneider indicates that the boundary layer east of Boulder,
CO is not a purely convective boundary layer but is influenced by the
mountains to the west.

The weather conditions on 22 June 1984 can be summarized as basically sunny skies with relatively dry conditions and strong solar heating.
Estimates of the boundary layer depth throughout the day by aircraft
observations and observations of the depth of mixing of released chaff
varied with time and space and showed a lack of a well-defined top to the
boundary layer. The temperature profiles show that the mixed layer is
capped by a transition layer with only very weak static stability. The low
level winds were observed to be light easterlies. Aloft, the winds were
westerly and increased with height, providing significant wind shear.

The features of RAMS used in the preliminary two-dimensional experiments are as follows: the nonhydrostatic option was employed along
with a terrain following coordinate system. The LES grid contained 242
grid points with 190 m resolution in the horizontal (see Figure 12).

The intermediate grid consisted of 134 grid points and 760 m hori-

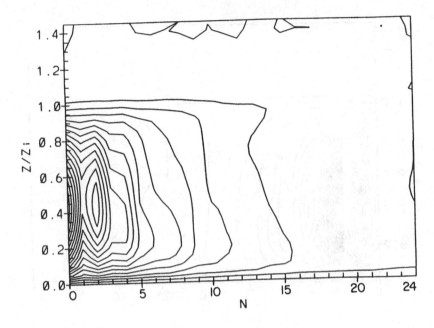

Fig. 11  As in Figure 8 but for $< F_y(\omega) >_{x,t}$ for the hilly case.  Contour interval is $0.015\,w_*$.

zontal resolution, while the largest grid had 58 grid points and 3060 m resolution. The LES grid is nested entirely within the second grid on the plains and in the vicinity of the Phoenix II study and extends west of Boulder, into the foothills. The western boundary of the middle grid is located west of the continental divide and the domain of the largest grid reaches more than 50 km west of the second grid boundary. The main purpose of the largest grid is to avoid lateral boundary problems on grid 2. All the grids have 55 points in the vertical with 200 m resolution up to 7 km and a stretched grid above, extending to beyond 15 km. This resolution is relatively course compared with other LES but is justified by the large depth of the boundary layer and size of the eddies observed on that day. The top boundary of the domain is a wall with a Rayleigh friction layer.

Radiation and soil processes were parameterized along with the surface layer processes. The subgrid parameterization used for the LES grid is the Deardorff (1980) TKE scheme. On the larger grids the Mellor and Yamada (1982) level 2.5 scheme was used for the vertical diffusion while

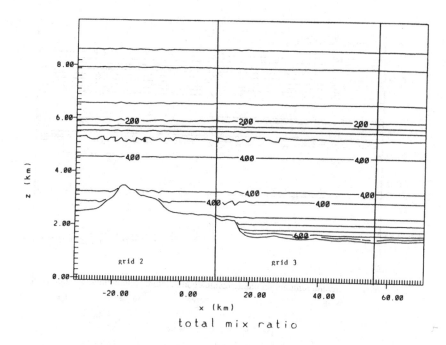

total mix ratio

Fig. 12 Total water mixing ratio field on grid 2 at the initial time (0000 UTC). The contour interval is 0.5 g $kg^{-1}$. The location of the grid 3 boundaries is indicated.

a deformation-based parameterization was employed for the horizontal diffusion. The subgrid TKE was not communicated between grids. The model was initialized horizontally homogeneously at 0000 UTC with the Denver, Colorado thermodynamic sounding (Figures 12 and 13) and the initial wind profile is taken to be similar to the observed winds at Denver close to sunrise on 22 June. Condensation is not modeled and atmospheric moisture is treated as a passive tracer in this simulation.

The model is run with the two largest grids for 12 hours of simulation to set up a realistic nocturnal boundary layer. Figures 14 and 15 show the potential temperature and u component of the winds on grid 2 at 1200 UTC which is near the time of sunrise. Cold air has pooled on the plains and flatter areas of the elevated terrain. The water mixing ratio field (not shown) also shows pooling in low lying areas at this time. The easterly winds on the east side of the divide are remnants of the low level easterlies which were used as initial conditions. Evidence of a mountain wave can also be seen in these figures.

At 12 hours into the simulation (1200 UTC), the LES grid is added.

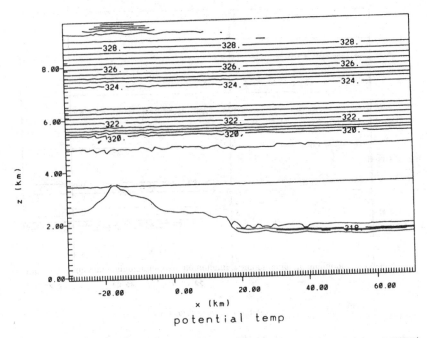

Fig. 13  Potential temperature field on grid 2 at the initial time (0000 UTC).
The contour interval is 0.5 K.

As solar heating increases through the morning, eddies develop first over
the highest terrain. Later eddies also develop over the plains with the
"youngest" eddies on the eastern plains. The daytime boundary layer
over the higher terrain develops separately from the boundary layer over
the plains. Because of the wind shear present on this day, the elevated
boundary layer is advected eastward over the growing boundary layer of
the plains.

Figure 16 gives the approximate location of Boulder and the BAO
tower within the LES grid (grid 3). The potential temperature field
is shown here for the LES grid at 1720 UTC. Evidence of developing
eddies of the growing plains boundary layer can be seen in this figure.
An indication of the advection of the elevated boundary layer is given
by the 319.0 K contour line above Boulder. The top of the boundary
layer is not well defined as was noted in the Phoenix II observations.

Also at 1720 UTC on grid 2, the u component of the wind (Figure
17) indicates eddies in the boundary layer with the largest (and oldest)
eddies over the highest terrain. General westerlies above low level east-

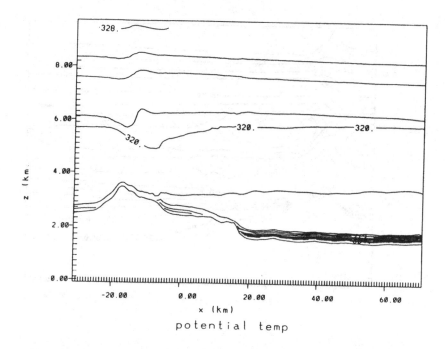

Fig. 14  Potential temperature field on grid 2 near sunrise (1200 UTC). The contour interval is 2.0 K.

erlies maintain the vertical wind shear. Because moisture is a passive tracer in this simulation, it is interesting to look at the moisture field at this same time (Figure 18). Again, the eddies over the elevated terrain are larger than the eddies over the plains and the larger ones are carrying moisture to higher levels. Some of the moisture of the elevated boundary layer can also be seen advecting eastward.

Figure 19 shows a time series of the subgrid TKE field at 30 minute intervals on the LES grid. Figure 19a is at 1650 UTC when the TKE associated with the elevated boundary layer is beginning to move eastward. The turbulence seems to weaken as it moves away from the elevated terrain. At 1720 (Figure 19b), notice that the subgrid TKE field is indicating that the elevated boundary layer is advecting eastward not as a smooth layer but in a more wave like manner. Gravity waves appear to be triggered by the convection in the elevated boundary layer and several researchers have suggested that gravity waves influence the character of the eddies in the atmospheric boundary layer (Clark *et. al.*, 1986; LeMone, 1990; and Moeng, 1990). In Figure 19c, the plains

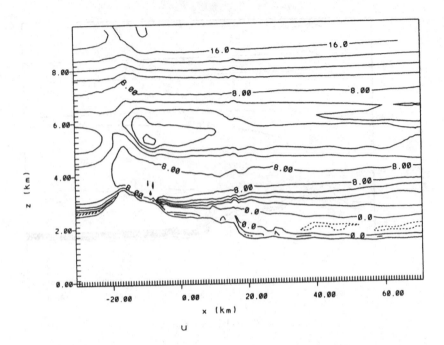

Fig. 15  The U wind component on grid 2 near sunrise (1200 UTC). The contour interval is 2 m $s^{-1}$.

boundary layer eddies are deeper while the upper level TKE continues to advect eastward. By 1820 UTC (Figure 19d), the plains boundary layer has grown deep enough to interact with the upper level boundary layer.

Figure 20 gives the total water mixing ratio on the LES grid at 1820 UTC. This figure indicates that the eddies are nearly 10 km across in this simulation. Although the eddies in the simulation are similar in character to the observed eddies, they are much larger. The observations indicate eddies of 1 or 2 km across. The simulated eddies are expected to be smaller in our upcoming three dimensional simulations where the circulations can be 3-dimensional and energy can cascade down scale.

In summary these preliminary two-dimensional simulations suggest that rather complicated interactions between the mountain-generated convective boundary layer and the plains boundary layer are possible. Because the two-dimensional model produced eddies that appear too large in scale, coupling between the upper and lower, plains boundary layers is probably too active. We shall have to wait for results from our

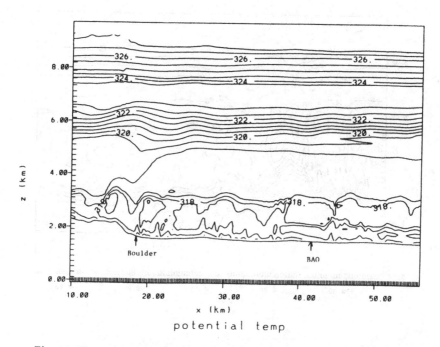

Fig. 16  Potential temperature field on grid 3 at 1720 UTC (1020 MST). The contour interval is 0.5 K. The location of Boulder and the BAO tower are indicated.

planned three-dimensional channel numerical experiments that we are now initiating.

## 17.5 LES of Cirrus Clouds

We are also in the process of commencing LES of cirrus clouds. Cirrus clouds present a unique challenge to LES because they develop in stably stratified air in the upper troposphere. Energy input to turbulence generation does not come from a lower boundary as in the planetary boundary layer but rather arises internally through radiative-convective interactions or externally through vertical shear of the horizontal winds and breaking of gravity waves. Recent analysis of aircraft data obtained during the First ISCCP Regional Experiment (FIRE) and the Global Atmospheric Sampling Program (GASP) revealed that the turbulence is stronger in magnitude inside cirrus clouds than in the non-cloudy regions (Flatau et al., 1990). Moreover, the turbulence spectra suggest

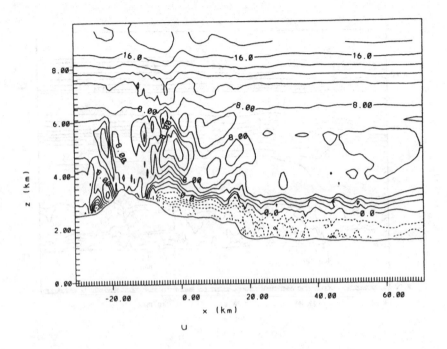

Fig. 17 The U wind component on grid 2 at 1720 UTC (1020 MST). The contour interval is 2 m $s^{-1}$.

that vertical velocity variance is much smaller in magnitude than horizontal velocity variances at all levels except very near cloud base. This suggests that radiative-convective interactions are important in exciting three-dimensional turbulence near the base of cirrus clouds but the bulk of the cloud layer is controlled by more two-dimensional turbulence having large eddies in the horizontal but small vertical dimensions. This places quite a challenge on the design of the LES model since it should have nearly equal-spaced horizontal and vertical grids on the order of ten meters or so in order to simulate the lower cloud three-dimensional turbulence but cover a horizontal domain large enough to depict the horizontal, two-dimensional turbulent eddies having scales of 5–10 km.

Plans are to perform exploratory LES experiments using a typical LES grid moving with the mean wind speed at cirrus levels using a Jacobian of transformation and cyclic boundary conditions. We also plan to perform LES simulations using our interactive nested grid procedures in which the courser grids are used to simulation mesoscale variations in cirrus

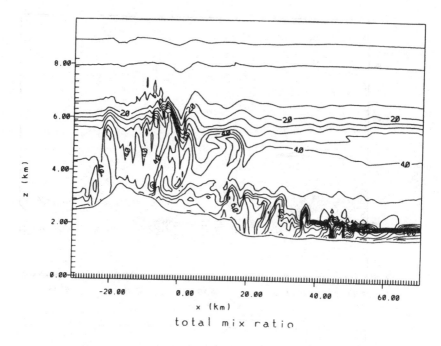

Fig. 18  The total water mixing ratio at 1720 UTC (1020 MST). The contour interval is 0.5 g $kg^{-1}$.

cloud forcing and an LES grid (8 km× 8 km × 2 km box) is allowed to float through the mesoscale grid.

## 17.6 Computational Resources

While the most computationally demanding LES calculations are performed on super-computers such as the NASA AMES Cray-II, much of our LES work is done on super-workstations. All the simulations described in sections 3.0 and 4.0 have been performed on a Stardent 1540 4-processor system. RAMS runs approximately one-fifteenth the speed of a single-processor Cray-X-MP on a single-processor Stardent 1540. We also have a 4-processor Stardent 3040 on which RAMS runs on a single processor at approximately one-sixth the speed of a single processor Cray X-MP. Moreover, RAMS runs at nearly one-fifth the speed of a single processor Cray X-MP on a single processor IBM RS-6000/320 RISC workstation. These workstations provide an economical, convenient platform for performing modestly computationally intense LESs

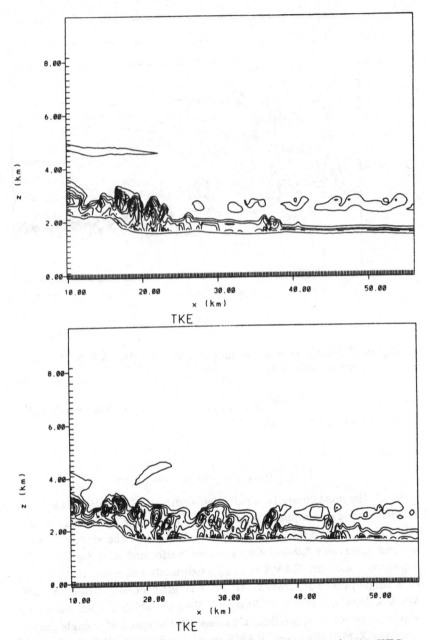

Fig. 19 Subgrid turbulent kinetic energy fields on grid 3 at a) 1650 UTC (0950 MST), b) 1720 UTC (1020 MST), c) 1750 UTC (1050 MST), d) 1820 UTC (1120 MST). The contour interval is 0.4 $m^2 s^{-2}$.

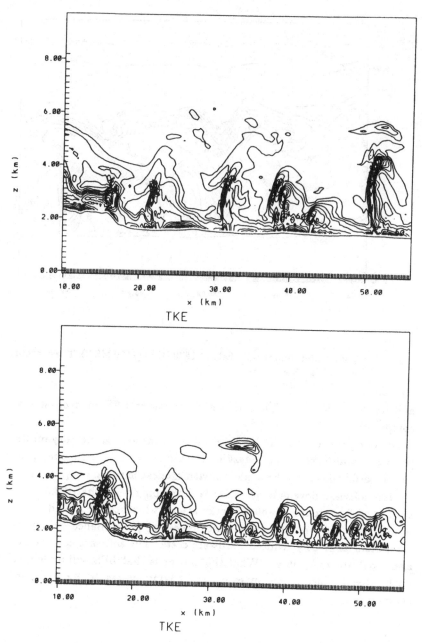

TKE

TKE

Fig. 19  (Continued )

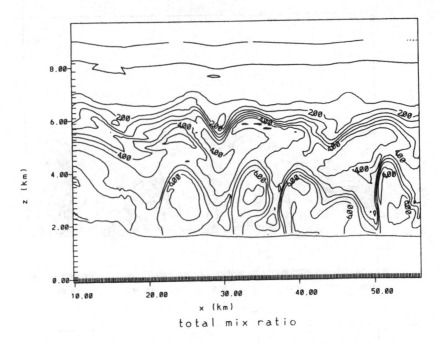

total mix ratio

Fig. 20 Total water mixing ratio field at 1820 UTC (1120 MST). The contour interval is 0.5 g $kg^{-1}$.

and for preliminary design and testing of larger LES runs on super-computers.

Greater performance can be obtained from the shared memory multiprocessor Stardents by multi-tasking the model so that single Cray processor performance can be achieved with modest hardware expense and modest software development. IBM is developing a nonshared memory cluster of 8 or more RS-6000 workstations which could provide performance levels comparable to multiprocessing performance on super-computers at greatly reduced hardware costs. The software costs, however, could be quite large. What this means is that LES will no longer be limited to just a few investigators having access to large super-computers.

## 17.7 Summary

In summary we have illustrated just a few examples of some of the problems that can be attacked using a versatile modeling system such as RAMS. Moreover, we have shown that much can be done in LES research on super-workstations thus providing a convenient, fast turn-around, economical alternative to super-computers.

## Acknowledgments

This research was supported by Army Research Office contract No. DAAL03-86-K-0175 and the Air Force Office of Scientific Research under contact No. AFOSR-88-0143. Computations were performed on the Department of Atmospheric Science at Colorado State University Star-dent 1540. The Stardent 1540 was purchased under support of ARO. The model used in this research, the CSU RAMS, was developed under support of the ARO and NSF.

## References

CHEN, C. AND COTTON, W.R. (1983) A one-dimensional simulation of the stratocumulus-capped mixed layer. *Boundary-Layer Meteorol.* 25, 289.

CLARK, T.L. AND R.D. FARLEY, R.D. (1984) Severe downslope windstorm calculations in two and three spatial dimensions using anelastic interactive grid nesting: A possible mechanism for gustiness. *J. Atmos. Sci.* 42, 329.

CLARK, T.L., HAUF, T. AND KUETTNER, J.P. (1986) Convectively forced internal gravity waves: Results from two-dimensional numerical experiments. *Quart. J. R. Met. Soc.*, 112, 899.

COTTON, W.R., STEPHENS, M.A., NEHRKORN, T. AND TRIPOLI, G.J. (1982) The Colorado State University three-dimensional cloud/mesoscale model – 1982. Part II: An ice phase parameterization. *J. de Rech. Atmos.* 16, 295.

COTTON, W.R., TRIPOLI, G.J., RAUBER, R.M. AND MULVIHILL, E.A. (1986) Numerical simulation of the effects of varying ice crystal nucleation rates and aggregation processes on orographic snowfall. *J. Clim. Appl. Meteor.* 25, 1658.

CRAM, J.M., PIELKE, R.A. AND COTTON, W.R. (1992) Numerical simulation and analysis of a prefrontal squall line. Part I: Observations and basic simulation results. *J. Atmos. Sci.* 49, 189-208.

DEARDORFF, J.W. (1972a) Three-dimensional numerical modeling of the planetary boundary layer. In *Workshop on Micrometeorology*, Aug. 14-18, American Meteorological Society, Boston, MA.

DEARDORFF, J.W. (1972b) Numerical investigation of neutral and unstable planetary boundary layers. *J. Atmos. Sci.* 29, 91.

DEARDORFF, J.W. (1973) Three-dimensional numerical modeling of the planetary boundary layer. In *Workshop on Micrometeorology*, edited by D.A. Haugen (American Meteorological Society, Boston), p. 429.

DEARDORFF, J.W. (1974) Three-dimensional numerical study of the height and mean structure of a heated planetary boundary layer. *Boundary Layer Meteorol.* 7, 81.

DEARDORFF, J.W. (1980) Stratocumulus-capped mixed layers derived from a three-dimensional model. *Boundary-Layer Meteorol.* 18, 495.

FLATAU, P.J., GULTEPE, I., NASTROM, G., COTTON, W.R. AND HEYMSFIELD, A.J. (1990) Cirrus cloud spectra and layers observed during the FIRE and GASP projects. In *Proceedings, Conference on Cloud Physics*, July 23-27, 1990, San Francisco, CA.

GAL-CHEN, T. AND SOMERVILLE, R.C.J. (1975) On the use of a coordinate transformation for the solution of the Navier-Stokes equations. *J. Comp. Physics* 17, 209.

HADFIELD, M.G. (1988) The response of the atmospheric convective boundary layer to surface inhomogeneities. Colorado State University Atmospheric Science Paper no. 433.

HADFIELD, M.G., COTTON, W.R., AND PIELKE, R.A. (1991) Large-eddy simulations of thermally-forced circulations in the convective boundary layer. Part I: A small-scale circulation with zero wind. *Boundary-Layer Meteorol.* 57, 79-114.

HILL, G.E. (1974) Factors controlling the size and spacing of cumulus clouds as revealed by numerical experiments. *J. Atmos. Sci.* 31, 646.

KLEMP, J.B. AND DURRAN, D.R. (1983) An upper boundary condition permitting internal gravity wave radiation in numerical mesoscale models. *Mon. Wea. Rev.* 111, 430.

KLEMP, J.B. AND WILHELMSON, R.B. (1978) The simulation of three-dimensional convective storm dynamics. *J. Atmos. Sci.* 35, 1070.

KLEMP, J.B., AND LILLY, D.K. (1978) Numerical simulation of hydrostatic mountain waves. *J. Atmos. Sci.* 35, 78.

KRETTENAUER, K. AND SCHUMANN, U. (1989) Direct numerical simulation of thermal convection over a wavy surface. *Meteorol. Atmos. Phys.* 41, 165.

LEE, T.J. (1991) The impact of vegetation on severe storms. *Ph.D. dissertation.* Colorado State University, Dept. of Atmospheric Science, Fort Collins, Colorado, in preparation.

LEMONE, M.A. (1990) Some observations of vertical velocity skewness in the convective planetary boundary layer. *J. Atmos. Sci.* 47, 1163.

LILLY, D.K. AND SCHNEIDER, J.M. (1990) Dual doppler measurement of momentum flux: results from the Phoenix II study of the convective boundary layer. Preprints of Ninth Symp. on Turbulence and Diffusion. 30 April to 3 May 1990, Roskilde, Denmark.

LOUIS, J.F. (1979) A parametric model of vertical eddy fluxes in the atmosphere. *Boundary-Layer Meteorol.* 17, 187.

MAHRER, Y. AND PIELKE, R.A. (1977) A numerical study of the airflow over irregular terrain. *Beitrage zur Physik der Atmosphare* 50, 98.

MASON, P.J., AND DERBYSHIRE, S.H. (1990) Large-eddy simulation of the stably-stratified atmospheric boundary layer. *Boundary-Layer Meteorol.* 53, 117.

MELLOR, G.L. AND YAMADA, T. (1982) Development of a turbulence closure model for geophysical fluid problems. *Rev. of Geophys. Space Phys.* 20, 851.

MOENG, C.-H., AND WYNGAARD, J.C. (1984) Statistics of conservative scalars in the convective boundary layer. *J. Atmos. Sci.* 41, 2052.

ORLANSKI, I. (1976) A simple boundary condition for unbounded hyperbolic flows. *J. Comp. Phys.* 21, 251.

TREMBACK, C.J. AND KESSLER, R. (1985) A surface temperature and moisture parameterization for use in mesoscale numerical models. Preprints, 7th Conference on Numerical Weather Prediction, 17-20 June 1985, Montreal, Canada, AMS.

TREMBACK, C.J., TRIPOLI, G.J. AND COTTON, W.R. (1985) A regional scale atmospheric numerical model including explicit moist physics and a hydrostatic time-split scheme. Preprints, 7th Conference on Numerical Weather Prediction, June 17-20, 1985, Montreal, Quebec, AMS.

TRIPOLI, G.J., AND COTTON, W.R. (1982) The Colorado State University three-dimensional cloud/mesoscale model - 1982. Part I: General theoretical framework and sensitivity experiments. *J. de Rech. Atmos. Sci.* 16, 185.

TRIPOLI, G., AND COTTON, W.R. (1989) A numerical study of an observed orogenic mesoscale convective system. Part 1. Simulated genesis and comparison with observations. *Mon. Wea. Rev.* 117, 273.

WALKO, R.L., COTTON, W.R. AND PIELKE, R.A. (1990) Large eddy simulation of the CBL over hilly terrain. Preprints of Ninth Symp. on Turbulence and Diffusion, Roskilde, Denmark.

WALKO, R.L., COTTON, W.R. AND PIELKE, R.A. (1992) Large eddy simulations of the effects of hilly terrain on the convective boundary layer. *Boundary-Layer Meteor.* 58, 133-150.

# 18

# Large Eddy Simulation of Turbulent Convection over Flat and Wavy Surfaces

## ULRICH SCHUMANN

## 18.1 Introduction

This chapter reviews the principles of large eddy simulation (LES) and describes some results from LES of turbulent convection in a boundary layer heated from a flat or wavy lower surface and topped by an adiabatic frictionless flat top surface.

Much is known about thermal convection over ideally homogeneous horizontal surfaces (Busse 1978; Stull 1988). Land surfaces are, however, rarely homogeneous. They are often undulated and form hilly terrain. Even when the amplitude of such hilly surfaces stays below the mean height of the atmospheric boundary layer, one might expect that the topography influences the flow structure considerably. Here, we describe the principles of LES and investigate the effect of a wavy surface on turbulent convection within a boundary layer of finite depth for zero mean horizontal motion. The model describes turbulent motions of a Boussinesq fluid in a layer confined between two infinite horizontal walls for zero horizontal mean motion. The lower surface height varies sinusoidally in one horizontal direction while remaining constant in the other. Several cases are considered with amplitude $\delta$ up to $0.15H$ and wavelength $\lambda$ of 1 to $8H$, where $H$ is the mean fluid-layer height.

## 18.2 The LES Method

The first successful LES was performed by Deardorff (1970a). The term "large eddy simulation" was introduced by Leonard (1974) and Ferziger (1977). Earlier papers referred to the same method as direct numerical simulation (Schumann 1975), but this term is now restricted to solutions of the full Navier–Stokes equations that compute the evolution of all significant scales of motion without any turbulence models (Reynolds 1990).

An LES method computes the three-dimensional time-dependent details of the large eddies using a simple subgrid-scale (SGS) model for the effects of the small eddies on the large eddies. Here, the large eddies are those motion elements in a turbulent flow which carry most of the kinetic energy and most of the turbulent fluxes. These motions are simulated using a three-dimensional time-dependent numerical integration scheme which numerically resolves scales in between a lower limit of order $h$ as given by the grid scale or any equivalent resolution limit of the numerical integration scheme, and an upper limit as given by the size of the computational domain. We require the size of the domain to be large in comparison with the scale of the most energetic motion elements, i.e., the "large eddies." This is necessary not only to get the correct solutions but also to get sufficient data for reliable statistics. The scale $h$ has to be smaller than the scale of the large eddies. If the grid scale $h$ is close to the scale of the most energetic motions, such a simulation is called a "very large eddy simulation" (VLES) (Reynolds 1990). A VLES demands more accurate SGS models than an LES in which the SGS fluxes and variances are small. Such a VLES will still be more accurate than a turbulence model which tries to describe all scales of motion.

The numerical scheme approximates the basic equations which describe the motions in the flow under consideration together with proper boundary and initial conditions. Here we rely on the assumptions that, for flows which approach an asymptotic state, the statistics of the resultant flow are independent of details in the initial conditions. The boundary conditions have to be suitably selected to describe the mean forcing from larger scales. They should not constrain the internal dynamics of turbulent motions, which is particularly demanding at inflow and outflow boundaries (Friedrich and Arnal 1990).

The SGS model is necessary because of the nonlinearities of the basic equations. This is formally shown by averaging or filtering the basic equations with respect to motion scales smaller than $h$. The filtering results in equations for the large eddies that need to be solved. Various filter methods have been proposed (Aldama 1990), including volume mean averages (Lilly 1967; Deardorff 1970a), a volume balance method which integrates where possible and identifies SGS contributions as surface mean values (Schumann 1975; Grötzbach 1986; Friedrich and Arnal 1990), and a Gaussian filter within a convolution integral which corresponds to a sharp cutoff in wave number space (Leonard 1974). When the filter width is narrow, the effective filter is dictated by the approximation properties of the numerical scheme. Germano (1991) shows that the details of such filters are unimportant. What counts are the scales which are resolved in contrast to the subgrid scales which are to be modeled. On the other hand, Mason and Callen (1986) advocated the use of a filter which has a width considerably larger than the grid scale in order to obtain smooth fields which can be numerically approximated without approximation errors.

I believe that this is not necessary. On the contrary, one should try to simulate as large a fraction of the turbulent motion energy as is possible even if part of the simulations are affected by some second-order finite-difference error. This is demanding enough, as the following estimate shows: If the spectrum of kinetic energy follows the well-known von Kármán shape with arbitrary amplitude $C$, $E(k) = C(k/k_0)^4[1 + (k/k_0)^2]^{-17/6}$, where $k_0 \cong k_{max}/1.6$ is a wave number close to that of maximum kinetic energy $k_{max}$, then one finds by simple integration that all scales up to $k/k_0 \cong 55.4$ (19.5) need to be resolved in order to resolve 90% (80%) of all energy. Hence, an order of 50 (20) grid points is the minimum which one should have available to resolve most of the energy. In three dimensions this gives quite a large number of grid points and, therefore, one cannot afford to provide further grid points (say a factor of 3 more in each dimension) in order to resolve the large scales without any appreciable finite-difference errors.

The SGS fluxes have to be modeled. The model must simulate the transfer of kinetic energy and scalar variances from large to small eddies, where the dissipation by molecular diffusion takes place. For high Reynolds numbers (Peclet numbers) and remote from rigid boundaries, these models are usually constructed assuming that the turbulence at subgrid scales corresponds to Kolmogorov's inertial range of turbulence (Lilly 1967). This requires also that the scale $h$ is small in comparison with the buoyancy scale (Deardorff 1980; Schumann 1991a).

Accurate modeling of the fluxes itself is necessary only where the SGS fluxes get large in comparison with the resolved fluxes. This is typically the case at rigid surfaces where the scales of the most energetic and flux-carrying motions tend to zero. At such boundaries an LES actually becomes a VLES. Moreover, for some applications it might be important that the SGS model mimics the stochastic forcing from the small-scale motion onto the large eddies (Mason and Thomson 1991). Like any turbulence model, the SGS model must satisfy the usual requirements such as the correct dimensional and tensorial properties, invariance with respect to Galilean transformations of the coordinate system (Speziale 1985), and realizability (Schumann 1977, 1991a).

Common model variants are the Smagorinsky–Lilly model (Lilly 1967; Deardorff 1970a), using turbulent diffusivities which depend on the square of the filter scale and the deformation magnitude of the resolved velocity field, the first-order closure model of Prandtl–Kolmogorov type in which the diffusivity is computed as the product of filter scale times square root of the turbulent kinetic energy of the SGS motions (Schumann 1975, 1991a; Deardorff 1980; Wyngaard and Brost 1984; Moeng 1984; Haren and Nieuwstadt 1989), and second-order closure models which apply the transport equations of the fluxes (Deardorff 1973) or algebraic approximations (Sommeria 1976; Schemm and Lipps 1976; Schmidt and Schumann 1989). Mason (1989) and Mason and Derbyshire (1990) apply the Smagorinsky model with a damping function depending on the local Richardson number, similarly to that proposed by Lilly (1962). I prefer the first-order closure, which makes use of an energy transport equation because this equation accounts explicitly for the buoyant forcing and for local deviations from equilibrium, is easy to implement, is free of realizability problems, does not require explicitly specification of the critical Richardson number, and apparently provides as accurate simulations as second-order closures (Schumann 1991a). If the LES is successfully simulating the basic flow, then a factor of 2 increase in resolution certainly gives better results than any improvement in the SGS model.

Near the surface, the SGS model must be consistent with standard properties of surface layers. For instance, in the Prandtl layer, for high Reynolds number flows, we assume that the shear stress is related to the flow velocity in the first grid point. In neutrally stratified flows, the logarithmic law of the wall describes that relation. The logarithmic profile depends on the surface roughness and, hence, the effects of surface roughness get included. For stably or unstably stratified flows, Monin–Obukhov similarity is used as, e.g., in Schmidt and Schumann (1989). Such rather simple boundary conditions assume that the fluxes at the surface are in phase with the local velocity or other fields in the first grid cell (Schumann 1975). This approach gives results which are very close to various alternative schemes, at least for high Reynolds number flows (Piomelli et al. 1989).

The numerical scheme must simulate correctly the dynamics of the large eddies (Ferziger 1977, 1987). In particular it should be able to account for the tendency to local isotropy (which requires about isotropic resolution), and its approximation errors should be small in comparison with SGS effects (which requires at least second-order accuracy because the SGS diffusivities scale with $h^{4/3}$ in the inertial range of turbulence). The numerical scheme should satisfy integral conservation of mass, momentum, and second-order moments (kinetic energy and variances) to simulate the dynamics of the large eddies correctly. Such schemes exist for simulations of momentum and scalar fluctuations in incompressible fluids (e.g., Lilly 1965; Piacsek and Williams 1970; Ferziger 1987; Schumann 1975, 1985). The pressure field has to be determined such that the flow satisfies the continuity equation, a task which is easy to accomplish using fast elliptic solvers on regular Cartesian grids (Schumann 1980; Schumann and Sweet 1988) but computationally more demanding in curvilinear coordinates and grids with variable grid spacing in more than just one coordinate (Clark 1977; Schumann and Volkert 1984; Krettenauer 1991). However, for scalar quantities, like water concentrations in cloud models (Deardorff 1980; Moeng 1986), species concentrations in models with chemical reactions (Schumann 1989), or kinetic energy as used for SGS modeling, it is essential that the model guarantee positivity of the scalars. A prominent example of such a scheme is that given by Smolarkiewicz (1984). A refined method is described in Smolarkiewicz and Grabowski (1990) which avoids not only the appearance of negative values but also that of nonphysical overshootings. Unfortunately, a scheme which guarantees positiveness and, at the same time, conservation of variances is not known, so that compromises are unavoidable in this respect. The Smolarkiewicz scheme conserves variances to a higher degree when applied to a scalar with small fluctuations relative to a large mean value (Smolarkiewicz and Clark 1986; Schumann et al. 1987). Hence, one can often optimize the simulations with respect to the conflicting demands of positivity and variance conservation by proper selection of the mean value of the scalar when the mean value itself is irrelevant. This is typically the case for temperature in a Boussinesq fluid and for inert tracers (Ebert et al. 1989). Often, large finite-difference errors occur if the flow velocity possesses a large mean value relative to the fluctuating components. For such cases, if possible, a Galilean transformation should be used, i.e., a grid which moves with the mean velocity (Deardorff 1970a). Otherwise, the time step has to be taken very small and the discretization errors get large. This may have strong effects on the computed spectrum of turbulence and the energy transfer as measured by velocity derivative skewness (Schumann and Friedrich 1987). Alternatively, the mean advection should be treated by a higher order numerical scheme, e.g., by a pseudospectral method (Rogallo and Moin 1984; Moeng 1984; Gerz et al. 1989).

## 18.3 Description of the Problem

We use LES to investigate the effects of variable surface heat flux and variable surface height on the amplitude and scales of motion and on the related turbulent transports in the dry convective boundary layer (CBL); see Stull (1988). We assume zero or very weak mean wind and assume that the CBL is capped by a strong inversion so that vertical fluxes are vanishing at the top of the mixed layer. Moreover this chapter concentrates on the fully turbulent case of infinite Reynolds or Rayleigh number. The simulations for a flat surface will be compared with the laboratory experiments of Adrian et al. (1986) in a water tank.

Field observations of the structure of the atmospheric CBL by Kaimal et al. (1982), Druilhet et al. (1983), Jochum (1988), and Huynh et al. (1990) found that "gently rolling terrain" affects the turbulence only a little. These experimental findings are fairly consistent, but the experimental studies did not explain why the effects of terrain appear to be small in most respects.

Most previous LES studies considered flows over plane and homogeneous surfaces (Mason 1989; Moeng and Wyngaard 1989; Sykes and Henn 1989; Nieuwstadt 1990; Schumann and Moeng 1991a). The effects of inhomogeneous surface heating on the turbulent CBL over a flat surface has been investigated using LES by a few authors. Hadfield (1988) observed a general increase of velocity fluctuations due to an idealized two-dimensional surface heat-flux perturbation. Schmidt (1988) found considerable increase of horizontal velocity fluctuations with increasing inhomogeneity but a small reduction in the vertical velocity component. Graf and Schumann (1991) supported this result from simulations in comparison with field observations including a weak mean wind. Hechtel et al. (1990) simulated an observed case of the CBL with weak mean wind, including moisture effects and random surface properties; they found little influence of inhomogeneity on the turbulence statistics. The case of infinitely extended slope layers has been simulated by Schumann (1990). He showed that the turbulent motions dominate relatively to the upslope flow velocity for weak inclination.

In an earlier study (Krettenauer and Schumann 1989), we investigated the present problem for various finite Reynolds numbers using direct numerical simulation. We showed that for isothermal no-slip boundaries, two-dimensional convection sets in at subcritical Rayleigh numbers, in close quantitative agreement with the linear theory of Kelly and Pal (1978). These simulations were limited in the Reynolds number and in the domain size; the wavelength was $\lambda = 2H$ and equal to the domain size in all cases. Direct numerical simulations for larger domains (Moeng and Rotunno 1990; Krettenauer 1991) show a more turbulent structure.

For the present study, we incorporated an SGS model into the numerical method which is reported in Krettenauer (1991). It uses a "terrain-following" coordinate system (Clark 1977) in which the vertical coordinate $z$ is transformed by

$$\eta = H \frac{z - h}{H - h}. \tag{1}$$

Here $h(x) = \delta \cos(2\pi x/\lambda)$ is the height of the wavy surface and $H$ the depth of the domain. The SGS model is similar to that used by Schmidt (1988), Schmidt and Schumann (1989), Ebert et al. (1989), and Schumann (1989). It is simpler than that of Schmidt and Schumann (1989) in that buoyancy enters the SGS model only through the transport equation of kinetic energy and not through a second-order closure for the SGS fluxes. The same simplification was also used with success by Schumann (1990, 1991a). On the other hand, the terrain-following coordinates complicate the SGS model but allow for accurate representation of wavy surfaces.

Seven cases are treated. The cases are denoted by a string "LddlL," where the first L stands for LES, dd $\in$ 00, 10, 15 identifies the wave amplitude $\delta/H$ in percent, $l \in U$, 1, 2, 4, 8 the wavelength $\lambda/H$ (U for undefined), and the final L $\in$ 4, 8 denotes the domain size $L/H$. Figure 1 depicts the computational domain and the grid on the wavy surface for case L1014. In order to point out that the undulation is strong, we note that the maximum slope $2\pi\delta/\lambda$ reaches 0.942 with an inclination of 43° for $\delta/H = 0.15$. The surface roughness height (which enters the Monin–Obukhov relationships for momentum and heat transfer) is specified as $z_0 = 10^{-4}H$. As shown in Schmidt and Schumann (1989), the effect of the surface roughness is weak for prescribed heat flux in the convective cases. The numerical scheme uses a grid with $64 \times 64 \times 16$ grid points

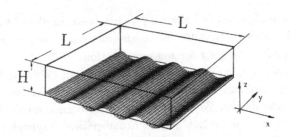

Fig. 1 Perspective sketch of the computational domain in three dimensions showing the sinusoidal surface wave in the $x$-direction; the surface height is constant in the $y$-direction. In the example, the wavelength is $\lambda = H$, the wave amplitude is $\delta = 0.1H$, and the lateral domain size is $L = 4H$. Case L1014.

for $L = 4H$ and $128 \times 128 \times 16$ grid points for $L = 8H$. Hence, the grid cells are close to isotropic with grid spacings of the order $H/16$.

The results are normalized by the characteristic velocity, height, temperature, and time scales,

$$w_* = (\beta g Q_S H)^{1/3}, \quad H, \quad T_* = Q_S/w_*, \quad t_* = H/w_*, \quad (2)$$

which are the "convective" scales proposed by Deardorff (1970b). Here, $\beta$ is the volumetric expansion coefficient, $g$ is gravity, $Q_s$ is the prescribed vertical temperature flux at the surface, and $H$ is the mean height of the flow domain. In the present simulations, the Reynolds number $Re = w_* H/\nu = \infty$, where $\nu$ is the viscosity of the fluid.

The initial conditions prescribe zero mean velocities, constant mean temperature, and constant SGS kinetic energy. Random perturbations are added to the temperature and velocity field to initiate turbulent convective motions. The mean value of the initial temperature is set to a large value in comparison with the fluctuations in order to reduce numerical diffusion in the (nonlinear) Smolarkiewicz scheme.

The time step of the integrations is $0.01t_*$. The computations are run until a maximum time of $35t_*$. This is a rather large time. For an atmosphere of 1000 m thickness and a heat flux of about 100 W m$^{-2}$ (temperature flux of 0.1 K s$^{-1}$) this corresponds to about 6 hours. Schmidt and Schumann (1989) ended their simulations of the atmospheric CBL at $t = 7t_*$. Hence, we can expect that our final results are close to steady state. Because of the finite domain size, the statistics do not reach steady state in a strict sense. Therefore, we will present results which are averaged over the last five time units of the simulations. The computation time on a Cray-YMP amounts to 4400 seconds for the smaller ($L/H = 4$) and 14,700 seconds for the larger domain ($L/H = 8$).

## 18.4 Results for Convection over a Flat Surface

The present method has been validated by comparison with atmospheric and laboratory measurements for the CBL which is topped by a stable fluid layer (Schmidt and Schumann 1989). These comparisons have shown that the results, for high grid resolution ($160 \times 160 \times 48$), are at least as accurate as available measurements. Graf and Schumann (1991) have compared LES results with recent field observations for a CBL with nonzero mean wind; they found good comparisons for much coarser grids. Further parameter studies and comparisons of results from four LES methods (Nieuwstadt et al. 1991) have shown that mean profiles of first-, second-, and third-order moments of turbulence quantities in the mixed layer of the CBL can be accurately represented even when using only $40 \times 40 \times 15$ grid cells in the CBL covering a domain of $4 \times 4 \times 1.5$ boundary layer depths. It still remains for us to validate the present code, because it contains many coding changes in order to treat curvilinear coordinates, contains a simpler SGS model than that of Schmidt and Schumann (1989), and is applied with a lid at the top of the mixed layer instead of a stable fluid layer.

For this purpose, the LES results will be compared with the laboratory results obtained by Adrian et al. (1986) for nonpenetrative convection between a heated lower and an adiabatic rigid upper surface. The same configuration is studied here except that we use a free-slip top surface. The experimental data are obtained for $Re$ between 552 and 1736. The experimental results, scaled by the convective scales, are virtually independent of $Re$ and are therefore assumed to be representative of a case with infinite Reynolds number. Hence, we compare these results with our LES case L00U4. Subsequently we compare mean profiles which are obtained by averaging over horizontal surfaces and over the time period from $t/t_* = 30$ to 35, as a function of the vertical coordinate $z$ with the experimental data.

The mean kinetic energy amounts to about $0.6w_*^2$. The SGS part amounts to less than 20% of the total energy and, hence, the results can be classified as LES in spite of the still rather coarse grid resolution. The mean temperature increases from its initial value because of constant heating in the integration period, as required for heat conservation. The mean vertical temperature profile increases in the upper part of the layer with increasing height. This countergradient heat transport is caused, as explained, e.g., in Schumann (1987) and Krettenauer and Schumann (1989), by narrow and fast-rising thermals which transport heat from the lower surface directly to the upper part of the fluid layer where the remainder of the fluid within the mixed layer is heated by turbulent diffusion and sinking warm fluid from above.

In Figure 2 we compare root-mean-square (rms) values of vertical and horizontal velocity fluctuations and of temperature fluctuations as computed from the LES with the measured results. The scatter of the experimental data characterizes approximately the standard deviation of these measurements. The vertical velocity fluctuations are largest in the mid-channel, as expected, and of a magnitude which is only a little larger than the values observed in the CBL. Schmidt and Schumann (1989) pointed out that $w'$ increases slightly with increasing stability; in this sense, the rigid top wall represents the infinite stability case. The measurements are close to the computed results. The systematic deviation to lower measured values in the upper part is caused in the experiment by friction at the top surface. Similar comments apply to the horizontal velocity fluctuations. The temperature fluctuations are largest near the heated wall. Most of these fluctuations originate from very small-scale turbulent eddies, as is known for atmospheric boundary layers. With respect to the vertical temperature flux, which is largest at the lower boundary, the large temperature fluctuations balance the reduced vertical velocity fluctuations. Overall, Figure 2 shows excellent agreement between LES and experiment. Sykes and Henn (1989) used the same data as shown in Figure 2 to test their LES for various grid discretizations including variable grid spacing in the vertical direction. However, the agreement with data was the same regardless of using variable or constant grid spacing.

The vertical temperature flux and the correlation coefficient of vertical velocity and temperature fluctuations are shown in Figure 3a and b. The measured data underestimate the flux. This is most obvious near the lower surface because there the normalized flux is unity by definition. This fact suggests that the LES gives higher accuracy than the measurements. The flux profile decreases linearly with height as required for steady state with constant heating rate according to the

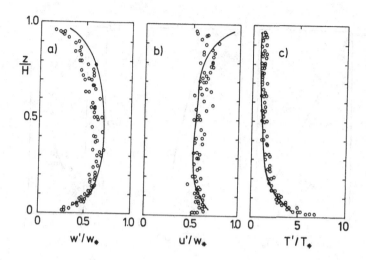

Fig. 2 Root-mean-square fluctuations of (a) vertical velocity variance $w'$, (b) horizontal velocity variance $u'$, (c) temperature variance $T'$ versus height for flat surface (Krettenauer and Schumann 1992). The circles represent the experimental results of Adrian et al. (1986).

vertical divergence of the flux. The entrainment flux, which usually acts in atmospheric cases, is absent for the present case because of the rigid lid. The correlation coefficient is quite high, which reveals that most of the motions are effective in transporting heat.

The third-order moment of the vertical velocity, shown in Figure 3c, is of importance with respect to the vertical transport of kinetic energy (in the vertical velocity component). Its value is positive throughout the mixed layer, and this indicates that the flow structure is composed of narrow updrafts with large upward velocity surrounded by wide and slow downdrafts. The magnitude of this quantity is only a few percent smaller than that found for the CBL. In the CBL, we expect a larger energy transport in order to balance the increased energy sink from entrainment at the inversion. The LES shows a small but negative value in the lowest grid cell. This result should not be considered realistic and is probably an effect of a too large dissipation rate near the surface, as discussed by Schmidt and Schumann (1989). Above the surface layer, the experiment seems to underestimate the energy flux. Our LES results are also very close to the direct numerical simulation results reported by Moeng and Rotunno (1990) for the same boundary conditions. For a Reynolds number $Re = 139$, they find about the same profile of variances

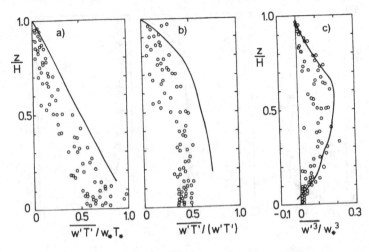

Fig. 3  Turbulent temperature flux and correlation coefficient versus height for a flat surface (Krettenauer and Schumann 1992). (a) Flux normalized by surface temperature flux, (b) flux normalized by the rms fluctuations of vertical velocity and temperature, (c) third-order moment of vertical velocity versus height for a flat surface. The circles represent the experimental results of Adrian et al. (1986).

and third-order moment of vertical velocity as we showed in Figure 3. This corroborates the independence of these statistics on the Reynolds number within the fully turbulent regime.

## 18.5 Results for Convection over a Wavy Surface

In this section we investigate the influence of the surface undulation on the turbulent convection in terms of mean profiles which are averages at constant transformed coordinates $\eta = $ const [see Eq. (1)] and averaged in addition over the last five time units. Figure 4 shows a few results from the LES. The undulation effect is largest in the $x$-component of the velocity variance. The horizontal variances increase with wavelength in the $x$-direction while that in the $y$-direction gets reduced. This is consistent with an increased excitation of rolls with axis parallel to the surface crests. However, the reduction in $v$ is less systematic than the increase in $u$ variance. The $w$ variance generally changes very little.

Inspection of the instantaneous flow fields shows that the motions are dominated by random motion components which hide the coherent

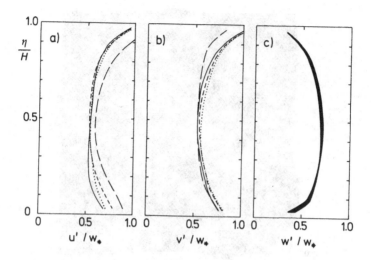

Fig. 4 Mean velocity fluctuations over wavy surfaces versus vertical coordinate. (a) in the $x$-direction, (b) $y$-direction, (c) $z$-direction.
········ L1014, ---- L1024, — — L1044, —— L00U4.

parts induced by the boundary forcing. In order to make the coherent
motion parts more visible, we will present results averaged over a finite
time period within the final part of the simulation period where the
time average filters out the short-lived small-scale random motions and
emphasizes the large-scale and long-lived parts of the turbulent motions.
For this purpose, we average over the time period $30 \leq t/t_* \leq 35$. The
limits of this interval are arbitrarily selected; the interval length of 5
convective time units is large enough to detect persistent structures but
still small enough to show up turbulent motions. If one would average
over an infinite time period, all motions parallel to the crests of the
surface waves should average out.

In Figure 5, we show the results from our "largest" (in terms of domain
size and computer effort) simulation. The wavelength is $\lambda/H = 1$, and
the amplitude is $\delta/H = 0.1$. We see that the flow is dominated by the
few strong updrafts that reach the top boundary. They cause the flow to
diverge horizontally at the upper boundary. Several such divergent flows
collide along lines where the fluid sinks back to the mixed layer. The
characteristic horizontal scale can be estimated to be 3 to $4H$. Although
these results are obtained for an undulated lower boundary, the results
look isotropic and do not show this forcing. The vertical cross section

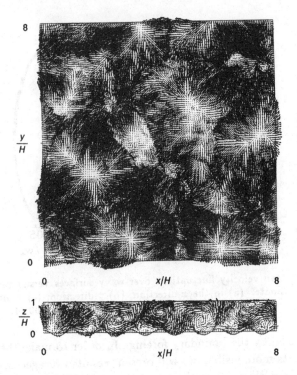

Fig. 5   Velocity field at $z = H$ (top) and at $y = 0$ (bottom), averaged
from $t/t_* = 30$ to $35$ when $L/H = 8$, $\lambda = H$, $\delta = 0.1H$ (case L1018).
Maximum velocity $1.72 w_*$ (Krettenauer and Schumann 1992).

shows clearly that the dominant turbulent scales are larger than the wavelength of the surface.

It appears as if short wavelengths excited by the surface get destroyed by the turbulent motions, but larger wavelengths might persist. In Figure 6 we investigate the effects of various surface undulations on the time and $y$-averaged flow fields in a vertical plane. We observe that a regular convection pattern, as would be expected from the forcing by the wavy surface, arises only for $\lambda/H \geq 4$. The averaged motion amplitude is strongest for $\lambda/H = 4$. It increases slightly with increasing wave amplitude.

This behavior can be explained approximately using a simple model as described in Schumann (1991b). The model assumes that the flow can be described (in the sense of VLES) by just two grid cells in the vertical and two grid cells in the horizontal per half-wavelength. The resultant

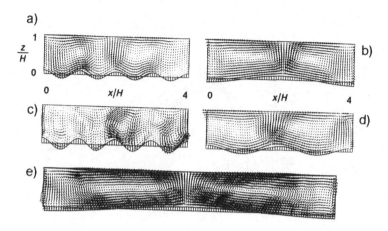

Fig. 6  Influence of wavelength of orography on the velocity field in a vertical plane for various surface undulations. These plots represent time and $y$-averaged results. (a) L1014, (b) L1044, (c) L1514, (d) L1024, (e) L1088. Maximum normalized velocity vectors (a) 1.03, (b) 1.34, (c) 1.32, (d) 0.62, (e) 1.0 (Krettenauer and Schumann 1992).

flow can be determined quasi-analytically. Figure 7 shows some results for illustration. We see that the model is not far off from the LES results.

From horizontal cross sections it has been found that the undulation for $\lambda/H = 2$ causes a flow pattern which is dominated by rolls with a horizontal axis. But the axes of these rolls are, surprisingly, not parallel but perpendicular to the waves, and these rolls get stronger by increasing wave amplitude. Qualitatively, the same effect was shown by Pal and Kelly (1979) and Krettenauer and Schumann (1989). Shorter surface waves enhance the rolls perpendicular to the crests while longer surface waves drive rolls parallel to the surface wave crests.

Several further analyses are presented elsewhere (Krettenauer and Schumann 1992). These include phase-averaged flow fields, co-spectra in the $x$- and in $y$-directions, characteristic length scales, time correlations, and heat transfer at the surface. In particular it is shown that the heat transfer over the wavy surfaces considered in this study is only very slightly sensitive to the undulation and follows closely the results deduced for flat surfaces (Schumann and Schmidt 1989).

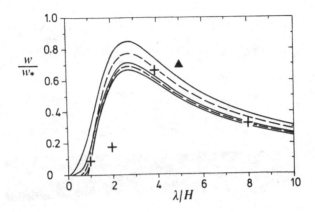

Fig. 7   Vertical convection velocity $w/w_*$ versus wavelength $\lambda/H$ for various values of terrain amplitude $\delta/H = 0$, 0.1, and 0.5, with homogeneous surface heating, $q/Q = 0$ (full curves), and for various values of surface heating $q/Q = 0$, 0.1, and 0.5, with $\delta/H = 0$ (dashed curves), from Schumann (1991b). The crosses denote the present LES results (average of maximum mean updraft and maximum mean downdraft velocity) for $\delta/H = 0.1$. The triangle denotes the result from a case with flat surface but variable surface heating $(q/Q = \frac{2}{3})$.

## 18.6 Conclusions

The method of LES has become a reliable tool for investigating turbulent convection. It appears that the large-scale thermal convection makes it particularly simple to simulate such flows (Nieuwstadt 1990). Rather coarse grids are sufficient to resolve these dominant thermal properties. This has become clear from the success of a very simple convection model which uses just four grid cells and, hence, represents a form of VLES (Schumann 1991b), and from comparisons of various LES methods for the CBL which show rather small differences in spite of essentially different numerical methods and SGS models (Nieuwstadt et al. 1991). Such studies are therefore very successful in investigations of the statistical properties of updrafts and downdrafts in such layers (Schumann and Moeng 1991a,b) and the spatial structure of such motion elements (Schmidt and Schumann 1989, Moeng and Schumann 1991).

In summary, the comparison between the LES results for the flat surface with the experimental results from Adrian et al. (1986) shows that the LES results can be taken to be as reliable as these laboratory measurements for high Rayleigh number flows. The experiments and results from further studies, including direct numerical simulations, have shown that the convection is fully turbulent for Reynolds numbers $Re \geq 100$. The convection pattern is composed of large-scale components which persist over more than five convective time units plus small-scale random turbulent motions. The motion structures are persistent over longer time periods in the presence of surface waves which seem to "lock" the motions to fixed positions. Otherwise, the effect of the surface wave is weak within the limits $\delta/H \leq 0.15$, $1 \leq \lambda/H \leq 8$, discussed in this chapter. In view of the fact that the maximum surface slope reaches $43°$ in one of the cases, this is a noteworthy result. Terrain-induced coherent structures can be expected only if the surface wavelength is large in comparison with the boundary layer depth. Shorter waves seem preferably to trigger rolls perpendicular to the crests, whereas long surface waves drive rolls parallel to the surface wave crests. In any case, the terrain-induced motion parts are rather small in comparison with the random turbulent components. The amplitudes of the coherent components are maximum at $\lambda/H = 4$. At small wavelengths the turbulence causes horizontal mixing, which destroys the flow structure imposed by the surface. At larger wavelengths local vertical mixing reduces the amplitude of large-scale horizontal motions. Also, large-scale motions require long times to get established, so that such regular structures will rarely appear in atmospheric boundary layers.

Unfortunately, laboratory data are not available for turbulent convection over wavy surfaces. However, our results agree basically with field observations (Kaimal et al. 1982; Druilhet et al. 1983; Jochum 1988; Huynh et al. 1990) which reveal few differences with respect to observations over homogeneous surfaces. In particular, we have found little systematic variation of length scales for vertical velocity with respect to terrain.

As a next step, we have started to investigate the effects of terrain on the turbulent boundary layer for cases with mean wind (Dörnbrack et al. 1991).

## Acknowledgment

This work was supported by the Deutsche Forschungsgemeinschaft.

# References

ADRIAN, R.J., FERREIRA, R.T.D.S. AND BOBERG, T. (1986) Turbulent thermal convection in wide horizontal fluid layers. *Exp. Fluids* 4, 121–141.

ALDAMA, A.A. (1990) *Filtering Techniques for Turbulent Flow Simulation.* Lecture Notes in Engineering, Vol. 56. Springer-Verlag.

BUSSE, F.H. (1978) Non-linear properties of thermal convection. *Rep. Prog. Phys.* 41, 1930–1967.

CLARK, T. (1977) A small-scale dynamical model using a terrain-following coordinate transformation. *J. Comput. Phys.* 24, 186–215.

DEARDORFF, J.W. (1970a) A numerical study of three-dimensional turbulent channel flow at large Reynolds number. *J. Fluid Mech.* 41, 453–480.

DEARDORFF, J.W. (1970b) Convective velocity and temperature scales for the unstable planetary boundary layer and for Rayleigh convection. *J. Atmos. Sci.* 27, 1211–1213.

DEARDORFF, J.W. (1973) The use of subgrid transport equations in a three-dimensional model of atmospheric turbulence. *J. Fluids Eng.* 95, 429–438.

DEARDORFF, J.W. (1980) Stratocumulus-capped mixed layers derived from a three-dimensional model. *Bound.-Layer Meteorol.* 18, 495–527.

DÖRNBRACK, A., KRETTENAUER, K. AND SCHUMANN, U. (1991) Numerical simulation of turbulent convective shear flows over wavy terrain. In Proc. 8th Turbulent Shear Flow Symp., Munich, 9–11 Sept. 1991, pp. 19.5.1–19.5.6.

DRUILHET, A., NOILHAN, J., BENECH, B., DUBOSCLARD, G., GUEDALIA, D. AND FRANGI, J. (1983) Étude expérimentale de la couche limite au-dessus d'un relief modéré proche d'une chaîne de montagne. *Bound.-Layer Meteorol.* 25, 3–16.

EBERT, E.E., SCHUMANN, U. AND STULL, R.B. (1989) Nonlocal turbulent mixing in the convective boundary layer evaluated from large-eddy simulation. *J. Atmos. Sci.* 46, 2178–2207.

FERZIGER, J.H. (1977) Large eddy simulations of turbulent flows. *AIAA J.* 15, 1261–1267.

FERZIGER, J.H. (1987) Simulation of incompressible turbulent flows. *J. Comput. Phys.* 69, 1–48.

FRIEDRICH, R. AND ARNAL, M. (1990) Analyzing turbulent backward-facing step flow with the lowpass-filtered Navier–Stokes equations. *J. Wind Eng. Ind. Aerodyn.* **35**, 101–128.

GERMANO, M. (1991) An algebraic property of the turbulent stress and its possible use in subgrid modeling. In Proc. 8th Turbulent Shear Flow Symp., Munich, 9–11 Sept. 1991, pp. 19.1.1–19.1.6.

GERZ, T., SCHUMANN, U. AND ELGHOBASHI, S.E. (1989) Direct numerical simulation of stratified homogeneous turbulent shear flows. *J. Fluid Mech.* **200**, 563–594.

GRAF, J. AND SCHUMANN, U. (1991) Simulation der konvektiven Grenzschicht im Vergleich mit Flugzeugmessungen beim LOTREX-Experiment. *Meteorol. Rdsch.* **43**, 140–148.

GRÖTZBACH, G. (1986) Direct numerical and large eddy simulation of turbulent channel flows. In *Encyclopedia of Fluid Mechanics*, Vol. 6. Ed. N. P. Cheremisinoff, pp. 1337–1391. Gulf Publ. Co.

HADFIELD, M.G. (1988) The response of the atmospheric convective boundary layer to surface inhomogeneities. Atmospheric Science Paper No. 433, Colorado State University.

HAREN, L. VAN AND NIEUWSTADT, F.T.M. (1989) The behavior of passive and buoyant plumes in a convective boundary layer, as simulated with a large eddy model. *J. Appl. Meteorol.* **28**, 818–830.

HECHTEL, L. M., MOENG, C.-H. AND STULL, R. B. (1990) The effects of nonhomogeneous surface fluxes on the convective boundary layer: A case study using large-eddy simulation. *J. Atmos. Sci.* **47**, 1721–1741.

HUYNH, B.P., COULMAN, C.E. AND TURNER, T.R. (1990) Some turbulence characteristics of convectively mixed layers over rugged and homogeneous terrain. *Bound.-Layer Meteorol.* **51**, 229–254.

JOCHUM, A. (1988) Turbulent transport in the convective boundary layer over complex terrain. In *Proceedings of the Eighth Symposium on Turbulence and Diffusion*, pp. 417–420. Amer. Meteorol. Soc.

KAIMAL, J.C., EVERSOLE, R.A., LENSCHOW, D.H., STANKOW, B.B., KAHN, P.H. AND BUSINGER, J.A. (1982) Spectral characteristics of the convective boundary layer over uneven terrain. *J. Atmos. Sci.* **39**, 1098–1114.

KELLY, R.E. AND PAL, D. (1978) Thermal convection with spatially periodic boundary conditions: Resonant wavelength excitation. *J. Fluid Mech.* **86**, 433–456.

KRETTENAUER, K. (1991) Numerische Simulation turbulenter Konvektion über gewellten Flächen. Dissertation, Report DLR-FB 91-12, DLR Oberpfaffenhofen.

KRETTENAUER, K. AND SCHUMANN, U. (1989) Direct numerical simulation of thermal convection over a wavy surface. *Meteorol. Atmos. Phys.* **41**, 165–179.

KRETTENAUER, K. AND SCHUMANN, U. (1992) Numerical simulation of turbulent convection over wavy terrain. *J. Fluid Mech.* **237**, 261–299.

LEONARD, A. (1974) Energy cascade in large-eddy simulations of turbulent fluid flows. *Adv. Geophys.* **18A**, 237–248.

LILLY, D.K. (1962) On the numerical simulation of buoyant convection. *Tellus* **14**, 148–172.

LILLY, D.K. (1965) On the computational stability of numerical solutions of time-dependent non-linear geophysical fluid dynamics problems. *Mon. Wea. Rev.* **93**, 11–26.

LILLY, D.K. (1967) The representation of small-scale turbulence in numerical simulation experiments. In *Proceedings of IBM Scientific Symposium on Environmental Sciences*. IBM Form No. 320-1951, pp. 195–210.

MASON, P.J. (1989) Large eddy simulation of the convective atmospheric boundary layer. *J. Atmos. Sci.* **46**, 1492–1516.

MASON, P.J. AND CALLEN, N.S. (1986) On the magnitude of the subgrid-scale eddy coefficient in large-eddy simulations of turbulent channel flow. *J. Fluid Mech.* **162**, 439–462.

MASON, P.J. AND DERBYSHIRE, S.H. (1990) Large-eddy simulation of the stably-stratified atmospheric boundary layer. *Bound.-Layer Meteorol.* **53**, 117–162.

MASON, P.J. AND THOMSON, D.J. (1991) Stochastic backscatter in the near wall region of large-eddy simulations. In Proc. 8th Turbulent Shear Flow Symp., Munich, 9–11 Sept. 1991, pp. 19.2.1–19.2.5.

MOENG, C.-H. (1984) A large-eddy-simulation model for the study of planetary boundary-layer turbulence. *J. Atmos. Sci.* **41**, 2052–2062.

MOENG, C.-H. (1986) Large-eddy simulation model of a stratus-topped boundary layer. Part I: Structure and budgets. *J. Atmos. Sci.* **43**, 2886–2900.

MOENG, C.-H. AND WYNGAARD, J.C. (1989) Evaluation of turbulent transport and dissipation closures in second-order modeling. *J. Atmos. Sci.* **46**, 2311–2330.

MOENG, C.-H. AND ROTUNNO, R. (1990) Vertical-velocity skewness in the buoyancy-driven boundary layer. *J. Atmos. Sci.* **47**, 1149–1162.

MOENG, C.-H., AND SCHUMANN, U. (1991) Composite structure of plumes in stratus-topped boundary layers. *J. Atmos. Sci.* **48**, 2280–2291.

NIEUWSTADT, F.T.M. (1990) Direct and large-eddy simulation of free convection. In *Proceedings of the 9th International Heat Transfer Conference*, Vol. I, pp. 37–47. Amer. Soc. Mech. Eng.

NIEUWSTADT, F.T.M., MASON, P.J., MOENG, C.-H. AND SCHUMANN, U. (1991) Large-eddy simulation of the convective boundary layer: A comparison of four computer codes. In Proc. 8th Turbulent Shear Flow Symp., Munich, 9–11 Sept. 1991, pp. 1.4.1–1.4.6. Extended version to appear in *Turbulent Shear Flows 8*, Ed. F. Durst et al. Springer-Verlag.

PAL, D. AND KELLY, R.E. (1979) Three-dimensional thermal convection produced by two-dimensional thermal forcing. Paper ASME 79-HT-109, Amer. Soc. of Mech. Eng.

PIACSEK, S.A. AND WILLIAMS, G.P. (1970) Conservation properties of convection difference schemes. *J. Comput. Phys.* **6**, 392–405.

PIOMELLI, U., FERZIGER, J., MOIN, P. AND KIM, J. (1989) New approximate boundary conditions for large-eddy simulations of wall-bounded flow. *Phys. Fluids A* **1**, 1061–1068.

REYNOLDS, W. C. (1990) The potential and limitations of direct and large eddy simulations. In *Whither Turbulence? Turbulence at the Crossroads*. Ed. J.L. Lumley, pp. 313–343. Springer-Verlag.

ROGALLO, R.S. AND MOIN, P. (1984) Numerical simulation of turbulent flows. *Ann. Rev. Fluid Mech.* **16**, 99–137.

SCHEMM, C.E. AND LIPPS, F.B. (1976) Some results of a simplified three-dimensional numerical model of atmospheric turbulence. *J. Atmos. Sci.* **33**, 1021–1041.

SCHMIDT, H. (1988) Grobstruktur-Simulation konvektiver Grenzschichten. Thesis, University of Munich, Report DFVLR-FB 88-30, DLR Oberpfaffenhofen.

SCHMIDT, H. AND SCHUMANN, U. (1989) Coherent structure of the convective boundary layer deduced from large-eddy simulation. *J. Fluid Mech.* **200**, 511–562.

SCHUMANN, U. (1975) Subgrid scale model for finite difference simulations of turbulent flows in plane channels and annuli. *J. Comput. Phys.* **18**, 376–404.

SCHUMANN, U. (1977) Realizability of Reynolds-stress turbulence models. *Phys. Fluids* **20**, 721–725.

SCHUMANN, U. (1980) Fast elliptic solvers and their application in fluid dynamics. *Computational Fluid Dynamics.* Ed. W. Kollmann, pp. 401–430. Hemisphere.

SCHUMANN, U. (1985) Conservation properties of finite difference Euler equations. *ZAMM* **65**, 243–245.

SCHUMANN, U. (1987) The countergradient heat flux in stratified turbulent flows. *Nucl. Eng. Des.* **100**, 255–262.

SCHUMANN, U. (1989) Large-eddy simulation of turbulent diffusion with chemical reactions in the convective boundary layer. *Atmos. Environ.* **23**, 1713–1727.

SCHUMANN, U. (1990) Large-eddy simulation of the upslope boundary layer. *Q. J. R. Meteorol. Soc.* **116**, 637–670.

SCHUMANN, U. (1991a) Subgrid length-scales for large-eddy simulation of stratified turbulence. *Theoret. Comput. Fluid Dynam.* **2**, 279–290.

SCHUMANN, U. (1991b) A simple model of the convective boundary layer over wavy terrain with variable heat flux. *Beitr. Phys. Atmos.*, **64**, 169-184.

SCHUMANN, U. AND VOLKERT, H. (1984) Three-dimensional mass- and momentum-consistent Helmholtz-equation in terrain-following coordinates. In *Notes on Numerical Fluid Mechanics*, Vol. 10. Ed. W. Hackbusch. Vieweg, pp. 109–131.

SCHUMANN, U. AND FRIEDRICH, R. (1987) On direct and large eddy simulation of turbulence. In *Advances in Turbulence.* Ed. G. Comte-Bellot and J. Mathieu, pp. 88–104. Springer-Verlag.

SCHUMANN, U., HAUF, T., HÖLLER, H., SCHMIDT, H. AND VOLKERT, H. (1987) A mesoscale model for the simulation of turbulence, clouds and flow over mountains: Formulation and validation examples. *Beitr. Phys. Atmos.* **60**, 413–446.

SCHUMANN, U. AND SWEET, R.A. (1988) Fast Fourier transforms for direct solution of Poisson's equation with staggered boundary conditions. *J. Comput. Phys.* **75**, 123–137.

SCHUMANN, U. AND SCHMIDT, H. (1989) Heat transfer by thermals in the convective boundary layer. In *Advances in Turbulence, Vol. 2.* Ed. H.-H. Fernholz and H.E. Fiedler, pp. 210–215. Springer-Verlag.

SCHUMANN, U. AND MOENG, C.-H. (1991a) Plume fluxes in clear and cloudy convective boundary layers. *J. Atmos. Sci.* **48**, 1746–1757.

SCHUMANN, U. AND MOENG, C.-H. (1991b) Plume budgets in clear and cloudy convective boundary layers. *J. Atmos. Sci.* **48**, 1758–1770.

SMOLARKIEWICZ, P.K. (1984) A fully multidimensional positive definite advection transport algorithm with small implicit diffusion. *J. Comput. Phys.* **54**, 325–362.

SMOLARKIEWICZ, P.K. AND CLARK, T.L. (1986) The multidimensional positive definite advection transport algorithm: Further development and applications. *J. Comput. Phys.* **67**, 396–438.

SMOLARKIEWICZ, P.K. AND GRABOWSKI, W.W. (1990) The multidimensional positive definite advection transport algorithm: Nonoscillatory option. *J. Comput. Phys.* **86**, 355–375.

SOMMERIA, G. (1976) Three-dimensional simulation of turbulent processes in an undisturbed trade wind boundary layer. *J. Atmos. Sci.* **33**, 216–241.

SPEZIALE, C.G. (1985) Galilean invariance of subgrid-scale stress models in large eddy simulation of turbulence. *J. Fluid Mech.* **156**, 55–62.

STULL, R.B. (1988) *An Introduction to Boundary Layer Meteorology.* Kluwer, 664pp.

SYKES, R.I. AND HENN, D.S. (1989) Large-eddy simulation of turbulent sheared convection. *J. Atmos. Sci.* **46**, 1106–1118.

WYNGAARD, J.C. AND BROST, R.A. (1984) Top-down and bottom-up diffusion of a scalar in the convective boundary layer. *J. Atmos. Sci.* **41**, 102–112.

# PHYSICAL OCEANOGRAPHY

# 19

---

# The Role of Oceans in Climate Change: A Challenge to Large Eddy Simulation

### GREG HOLLOWAY

## 19.1 Introduction

Modeling ocean circulation on space scales from ocean basin through global domain and over time scales from years to centuries poses an enormous computational challenge. This may be an extreme of large eddy simulation (LES), where the "large eddies" are already the oceanic gyres. Especially, to assess the significance of oceans for global climate change, this challenge must be addressed. The climate problem is particularly demanding because the domain is global while the time scales of interests are decades and longer. It is therefore paramount that ocean models successfully represent ("parameterize") as much of the smaller scale variability as possible. However, just as the desired grid spacing is as coarse as possible for reasons of computational efficiency, it is also required that the subgrid-scale (SGS) representation be quite skillful – lest errors at the SGS collect systematically over the long time scales of integration.

The range of scales involved in ocean modeling is huge, from planetary scales of 10,000 km down through the Kolmogorov dissipation scale of order 1 cm. Such a wide range might appear to be an invitation to LES, especially if an outer scale of a turbulent inertial range were well separated from dissipation scales. Unfortunately, this happy circumstance does not appear to occur. Instead, a host of processes are expressed differently at different scales, as cartooned in Fig. 1. In particular, the gravitational stability of the density stratification is so strong that overturning in the ocean interior rarely extends over about 1 m in the

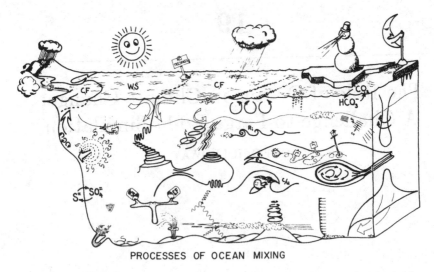

PROCESSES OF OCEAN MIXING

Fig. 1  A daunting variety of processes occur on scales between the basin
scale and dissipation.

vertical, leaving little more than one decade for any semblance of inertial subrange. A number of processes that affect vertical mixing and
transport are discussed with great insight by Müller (Chapter 21, this
volume). For the purposes of the present chapter, we restrict our attention to oceanic synoptic eddy scales, ranging in the horizontal from a
few kilometers to O(100 km).

Computational strategies for ocean climate modeling can be grouped
into two approaches: (1) eddy-resolving and (2) non-eddy-resolving,
where "eddy" refers to some portion of the oceanic synoptic scales. In
practice, the distinction depends upon whether the horizontal grid spacing is sufficient to resolve the first baroclinic radius of deformation, $R_1$.
While details of $R_1$ depend upon solving an eigenvalue problem, roughly
$R_1$ can be estimated by $NH/f$, where $H$ is the depth of fluid, $N$ is a
representative "stability frequency," measuring the density stratification
of the water column, and $f$ is Coriolis parameter (twice the vertical component of the earth's rotation). At low latitudes (small $f$), where the
ocean is fairly well stratified (modest to large $N$), $R_1$ is reasonably resolvable. However, at higher latitudes (large $f$) where the stratification
is weak (small $N$), as well as in shallow seas and on continental margins
(small $H$), $R_1$ may be as small as only a few kilometers. Thus, even
supercomputer models (e.g., Semtner and Chervin, 1988) which may be
"eddy-resolving" at subtropical latitudes will be non-eddy-resolving (or

only marginally resolving) at higher latitudes and in shallower water. Moreover, "eddy-resolving" has been put in quotes to draw attention to the ambiguity of what it means to be "eddy-resolving." Although model resolution may be adequate to allow the largest oceanic synoptic eddies to exist (just on the edge of viscous extinction!), it is by no means assured that such "eddies" are dynamically faithful. Especially, the transport properties of such eddies may be severely corrupted by the absence of degrees of freedom at smaller scales.

For the balance of this chapter, we turn to non-eddy-resolving strategies. We suppose only that we resolve the largest eddies, i.e., the main ocean gyres, but that we do not attempt to resolve the synoptic-scale eddies near $R_1$. Thus, we are denied a more common LES approach in which smaller scale eddies might be successively eliminated or filtered. Instead, we explore just one example of applying theoretical aspects of the statistical dynamics of eddies to larger scale circulation modeling.

## 19.2 Statistical Dynamics of the Synoptic Ocean Eddies

Efforts to model the global ocean circulation without resolving synoptic eddies have been made for decades (see Bryan, 1969), rooted historically in times when computational resources did not permit eddy-resolving global experiments. Today, despite enormous advances in computing power, a deliberate decision not to attempt to resolve eddies continues to offer opportunities to explore a range of ocean/climate possibilities evolving over significant time scales. However, the entire burden of representing eddy effects is then relegated to the SGS. For the most part, this has continued to be addressed in the form of Fickian gradient transport representations with constant coefficients. Attention has been given to adaptive (nonlinear) eddy diffusion coefficients, such that the structure of the large-scale fields determines those coefficients. However, attempts to verify these gradient transport relationships, either from direct observation or from process-oriented, high-resolution numerical experiments, have enjoyed mixed success at best. Although attempts to refine the basis for gradient transport kinds of representations will continue, it is clear that other sorts of ideas must be explored. This chapter illustrates a different sort of idea based upon the interaction of ocean currents with the shape of the seafloor topography.

### 19.2.1 Topographic Stress: The "Neptune Effect"

The interaction of SGS eddies with SGS variations of bottom topography can, in principle, exert powerful forces on the larger scale flow. This is accomplished through correlations between eddy pressure anomaly and topographic slopes. An estimate of the magnitude of this force can be made from the product of a topographic height fluctuation (say 100 m) times a horizontal pressure gradient which can be estimated from geostrophy at $fu$, where $f$ is the Coriolis parameter (say $10^{-4}$ s$^{-1}$) and $u$ is an eddy velocity (say $10^{-2}$ to $10^{-1}$ m s$^{-1}$). The product ($10^{-4}$ to $10^{-3}$ m$^2$ s$^{-2}$) corresponds to an effective stress (per unit area) of 1 to 10 dyn cm$^{-2}$. This is as large as or larger than typical wind stresses, which are thought to be the dominant driving mechanism for the oceans. Of course, neither the magnitude nor even the sign of a correlation coefficient between pressure gradients and topographic fluctuations is known, so the effective stress may be much smaller than the magnitudes estimated here. What may be alarming, though, is that the amplitudes of topographic fluctuation and of eddy speeds used in this estimation are small enough that they may be thought to be "comfortably ignored" in many large-scale ocean models. The challenge for dynamical theory is to estimate those correlation coefficients!

Intuitively one might suppose that the effects of topographic "roughness" would be to exert an enhanced drag against large-scale overlying currents. However, it has also been recognized for some time that large-scale gradients of $f/H$ lead to propagating pressure disturbances called topographic Rossby waves. These waves propagate with the sense of keeping shallow water to the right in the Northern Hemisphere, and to the left in the Southern Hemisphere. Then, if mean currents happen to be of a sense which opposes the intrinsic wave propagation, the phases of waves can become arrested in relation to topography. A persistent pressure–topography correlation can then occur, supporting a large drag against the mean flow while transferring energy into the wave. On the other hand, when the mean current has the same sense as intrinsic wave propagation, rapid phase change is expected and no persistent pressure–topography correlation is expected. The topographic drag is thus quite one-sided, resisting mean flows of one sense while being absent for mean flows of the opposite sense. The consequences of such an asymmetric drag have been considered, e.g., by Haidvogel and Brink (1986).

The story goes on. Topographic wave drag is a linear mechanism, neglecting effects of wave–wave interaction. At the opposite extreme, one may consider strongly nonlinear eddy fields, idealized as two-dimensional turbulence, overlying random topography. This was treated, e.g., by Herring (1977) and Holloway (1978). Among the results that emerged from these turbulence-theoretical analyses was the finding of a strong pressure–topography correlation. The sense of correlation is that positive pressure tends to occur above topographic elevations, hence exerting no net horizontal stress. Bringing together methods from turbulence theory and large-scale wave propagation, Holloway (1987) obtained a seemingly bizarre result: the drag due to pressure–topography correlations was not just asymmetrical (as in the case of linear wave drag), it was not even a "drag" at all. Correlations were foreseen to develop so strongly that the pressure force would *drive* the mean flow in the sense of intrinsic wave propagation.

Leaving details to the original literature, a brief outline of the theory follows. Consider motion of a homogeneous fluid under the effects of the earth's rotation overlying a bottom topography whose height varies by only a small fraction of the mean depth of fluid. Approximately, the low frequency motion (periods long compared with rotation period) is given by the evolution of the vertical component of vorticity, $\zeta = \nabla^2 \psi = \mathbf{z} \cdot \nabla \times \mathbf{u}$:

$$\partial_t \zeta + J(\psi, \zeta + \beta y + h) = F - D, \qquad (1)$$

where $\psi$ is the stream function defined by $\mathbf{u} = \mathbf{z} \times \nabla \psi$, $J(\ )$ denotes the Jacobian determinant with respect to horizontal coordinates $(x, y)$, and the total depth of fluid is written $H = H_0[1 - (h + \beta y)/f]$. The Coriolis parameter is $f$. External forcing $F$ and dissipation $D$ are listed only symbolically on the right side. The bottom topography includes a random part $h(x, y)$ and a mean slope $\beta y$. If one assumes that $u$ and $h$ satisfy periodic boundary conditions over some $x$ and $y$ length scales $L_x$ and $L_y$, then $u$ and $h$ may be spatially Fourier transformed to yield coefficients $\mathbf{u}_k$ and $h_k$, where $\mathbf{k}$ is a horizontal wavevector. In addition to the spatially periodic part of $\mathbf{u}$, there will occur a spatially uniform $x$-directed translation $U$ (with corresponding stream function contribution $-Uy$) with evolution given by

$$\partial_t U = -\int dA\, \psi \partial_x h + F_U - D_U, \qquad (2)$$

where $\int dA$ extends over the area of the flow domain, and $F_U$ and $D_U$ are mean external forcing (if present) and whatever frictional parameterization that one applies against $U$. Of special interest in (2) is the net pressure force $\psi \partial_x h$, where, by geostrophy, $\psi$ is proportional to pressure anomaly. How can we find $\psi \partial_x h$? The solution from Holloway (1987) is given in terms of expectations for second moments $Z_k = < \zeta_k \zeta_{-k} >$ and $C_k < \zeta_k h_{-k} >$, expressed in terms of the topographic variance spectrum $H_k = h_k h_{-k}$ and the specification of forcing and dissipation operators $F$, $D$, $F_U$, an $D_U$. Angle brackets $< >$ denote averaging over an ensemble of realizations of the flow, each proceeding under a randomly chosen realization of stochastic wind torques $F$. Following second-moment closure theory, the result for the expected force acting on $U$ is

$$-\int dA \psi \partial_x h = -\mathrm{Im} \sum_k \frac{k_x}{k^2} C_k = \sum_k \left[ -\frac{\eta_k + \nu_k}{\omega_k^2 + (\eta_k + \nu_k)^2} \frac{k_x^2}{k^2} U H_k \right.$$
$$\left. + \frac{\omega_k k_x}{\omega_k^2 + (\eta_k + \nu_k)^2} \frac{\gamma_k}{k^2} H_k \right], \tag{3}$$

where $\eta_k$ and $\gamma_k$ are given by expressions (see Holloway, 1987) involving sums of $Z_p$ and $H_p$ over all wavevectors $\mathbf{p}$, $\nu_k$ is the Fourier representation of $D$, and $\omega_k = k_x(U - \beta/k^2)$ is the frequency of propagation of a linear wave at wavevector $\mathbf{k}$, including Doppler shift by the mean flow $U$. Much can be gleaned from simple inspection of (3) given only the information that $\eta_k$ and $\gamma_k$ tend to be positive expressions of comparable amplitude. One observes that the first term in square brackets is indeed a drag (opposed in sign to $U$) which is asymmetric with respect to the sign of $U$ because of the role of $U$ in $\omega_k$. When $U < 0$, the mean flow is in the same sense as intrinsic wave propagation, $|U - \beta/k^2|$ is large and drag is small. When $U > 0$, there will be a value of $k$ for which $|U - \beta/k^2|$ is small and the drag contribution will be relatively large. Thus, we recover the classical view of an asymmetric drag, here modified only slightly to recognize that some drag still occurs for $U < 0$.

The fascinating aspect of this solution comes from the second term in square brackets. Here is a force contribution which takes the sign of $\omega_k$ and thus may not even oppose $U$ at all. In particular, if we considered a flow which was at rest in the mean ($U = 0$) while exhibiting a spectrum of eddy activity, we should predict that the eddies would latch onto the topography so as to force the flow into mean motion. We could further imagine a situation in which a mean force is applied, tending to drive $U$ positively, while further random torques excite an eddy field, and wind up with the net result that the mean flow would systematically oppose the mean forcing. When this seemingly bizarre result was foreseen, but

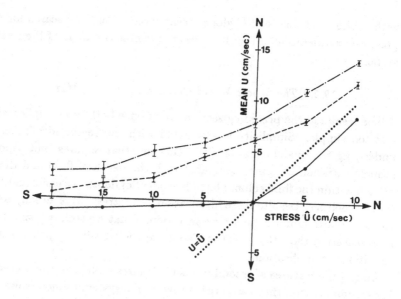

Fig. 2  In this numerical experiment, forcing and dissipation in (2) have been given simply as $F_U - D_U = E(\hat{U} - U)$. Apropos of an eastern ocean margin, mean depth increases toward the west. A mean stress, characterized by $\hat{U}$, may be directed northward (N) or southward (S). In the absence of topographic roughness, the mean frictional response is $U = \hat{U}$, shown as a dotted line. When topographic roughness of rms amplitude 30 m in a reference depth of 300 m is included, solutions are shown connected by a solid line, illustrating asymmetric drag. When random torques are included in (1), results are shown connected by a dashed line (low-amplitude torques) and by a dash–dot line (higher-amplitude torques). The bias toward northward mean $U$ even in the presence of southward mean stress s illustrates the nondrag aspect of topographic stress.

not yet tested numerically, it was termed the "Neptune effect" – being driven unaccountably and mysteriously from the seafloor. Numerical experimental confirmation soon followed, an example of which is shown in Fig. 2. See also Treguier (1989).

The "Neptune effect" shows the power of second-moment closure theory to *anticipate* quite unexpected phenomena that otherwise one might chance to discover by numerical experiment. In the case of "Neptune," only several lines of algebra were needed in lieu of substantial numerical experimentation.

The idealized circumstances underlying the derivation of "Neptune" as well as the complexity of its formula (3) prevent direct implementation

within LES modeling for the global ocean circulation. The search for a
*practical representation* in an LES ocean model is the topic of the next
section.

### 19.2.2 The Whole World Ocean According to Max

The results in the preceding section are at once both too complicated
and too simple. Simplicity is associated with the severe idealization
underlying the model equation (1). This equation requires that topographic variation be minor compared with the depth of fluid and that
motion within the fluid column be independent of depth. Oceanic reality
satisfies neither. Despite these idealizations, the stress (3) is already too
complicated to admit simple representation. What we seek is a simpler
representation than (3) which is also more robust with respect to its
conditions of application.

Among the features of second-moment closures such as in the preceding section are that they cause the statistics of a system to move toward
a maximum entropy configuration – apart from direct effects of external
forcing and dissipation. Entropy is defined as $S = -\int p(\mathbf{y}) \ln p(\mathbf{y}) \, dy$
where $p(\mathbf{y}) \, dy$ is the probability of finding the state of a flow (denoted by
phase space vector $\mathbf{y}$) in a neighborhood $dy$ of $\mathbf{y}$. Carnevale et al. (1981)
demonstrate that second moment closures have the property $dS/dt \geq 0$.
The stress given by (3) arises in order to permit the system to increase
$S$. Let us calculate directly the flow state $\mathbf{y}$ which maximizes $S$ subject
to the integrals of the motion from (1) and (2). Without loss of generality we can absorb (2) into (1), also omitting $\beta y$ while permitting $\psi$
and $h$ to vary in any way as functions of $x, y$. In an enclosed domain,
or under various special boundary conditions such as the periodicity assumption (Holloway, 1987), integrals of the motion (omitting $F$ and $D$)
are total energy $E = \frac{1}{2} \int dA |\nabla \psi|^2$ and a continuous family of function-
als $G = \int dA g(\zeta + h)$, where $g$ is any function. However, when (1) is
represented discretely for computation (as a finite-difference equation,
truncated spectral expansion, or whatever), almost all of the $G$ are no
longer conserved. Depending upon the scheme, it may be that only the
quadratic "potential enstrophy" $Q = \frac{1}{2} \int dA (\zeta + h)^2$ is conserved. Max-
imizing $S$ subject to constraints $E$ and $Q$ yields the expected circulation

$$(\alpha_1/\alpha_2 - \nabla^2) < \psi > = h, \tag{4}$$

where $\alpha_1$ and $\alpha_2$ are Lagrange multipliers which serve to impose the
constraints due to $E$ and $Q$. Result (4) was obtained by Salmon et al.

(1976). Second-moment closure calculations such as that of Holloway (1987) have the tendency to drive the system toward (4), although competing effects from $F$ and $D$ in actuality may keep the system far from (4).

Recall that our motivation in considering maximum entropy statistical equilibria was to find a simpler characterization of the stress obtained at (3). Let us first examine further simplification of (4). Note that $\alpha_1/\alpha_2$ has dimensions of an inverse square length, say $1/\lambda^2$. For geophysically interesting values of $E$ and $Q$, $\lambda$ is real. Appropriate values for $\lambda$ are not so narrowly determined. However, plausible choices for $\lambda$ of O(10 km) appear to yield plausible currents. Since our interest in ocean/climate studies is at scales much larger than $\lambda$, we consider only the larger scale character of (4). Then, the $\nabla^2$ drops out and we are left with

$$< \psi >= \lambda^2 h. \tag{5}$$

The suggestion is that this marvelously simple result may serve skillfully to guide LES ocean modeling. The influence of unresolved, small-scale eddies should be to cause larger scale flow to tend toward (5).

Let us first observe that (5) is consistent with the previous calculation leading to (3). Explicitly taking account of $\beta y$ as a contribution within $h$, one of the contributions to $< \psi >$ will be from $\lambda^2 \beta y$, corresponding to a mean translational flow $< U >= -\lambda^2 \beta$. The Neptune effect is seen only as a *disequilibrium* mechanism by which the system tends to move toward a maximum entropy configuration. It may be surprising that nonzero mean flows are anticipated *on account of entropy maximization*. What this makes clear is the failure of any sort of "eddy viscosity" which would have the character of causing the resolved motion to tend toward a state of rest. We see instead, given the shape of ocean basins as expressed by $h$, that a state of rest (in the mean) is extraordinarily improbable, being very far from maximum entropy. Experiments illustrating basin scale evolution toward equilibrium configurations can be seen in Griffa and Salmon (1989).

Second, let us consider some limitations of the theoretical underpinnings of (5). The problem as posed at (1) is barotropic; hence, motion is assumed to be independent of depth. In reality, motion in the ocean is intensified near the upper surface – in response to surface forcing. Thus, we note that the baroclinic (depth-dependent) motion is in response to external forcing. The case of maximum entropy equilibria which included provision for baroclinic motion was already addressed by Salmon

et al., with the result that motion on scales larger than the first deformation radius $R_1$ are barotropic while smaller-scale eddies should retain
more baroclinic shear. What is important to recognize is that we do not
attempt to idealize the ocean by pretending that its reality is barotropic.
Simply, we take account that the statistical dynamical *tendency* due to
eddies is *toward* barotropic motion on scales larger than $R_1$. This eddy
tendency competes with the nature of oceanic forcing. Significantly from
the view of ocean/climate modeling, the dominant scales of oceanic forcing are large enough that these can be treated explicitly in numerical
models.

The theory at (1) is also quasi-geostrophic. However, the maximum
entropy mean flow (5), assuming modest values of $\lambda$ and considered only
on scales larger than $R_1$, incurs only small ageostrophic correction. The
equator ($f = 0$) causes a singularity for quasi-geostrophy. Recalling
from its definition [following (1)] that $h$ is proportional to $f$, we observe
that the mean flow from (5) becomes very weak at small $f$. (Velocity,
given by $\mathbf{u} = \mathbf{z} \times \nabla\psi$, does not strictly vanish at $f = 0$.)

There is a further ambiguity arising from quasi-geostrophy. Because
the motion field is taken to be nearly horizontally nondivergent, $\psi$ is
usually taken to be a *velocity* stream function. To the same order, the
depth variation must be small. Defining a *transport* stream function $\phi$
from $H\mathbf{u} = \mathbf{z} \times \nabla\phi$, results from quasi-geostrophic theory lead as well to
$<\phi> = \lambda_2 h H_0$ or to $<\phi> = \lambda_2 h H$, with no distinction between actual
depth $H$ and some reference $H_0$. So, are we ruined? No. Requiring
that velocities remain nonsingular, we insist upon the velocity stream
function interpretation of (5). Since the motion from (5) is very nearly
along isobaths with only slight deviation due to latitudinal dependence
of $f$, a *transport* stream function can be given very nearly by $\phi = \psi H$
or, for suggested application in reality, $\phi = -\mu f H^2$, where $\mu = \lambda_2/2H_0$
is a length scale of $O(10 \text{ km})$.

## 19.3 Illustration

To take (5) as a theory of actual ocean circulation would be a huge,
indefensible leap – ignoring sun, wind, rain, etc. Of course the sun
does shine, the wind blows, the earth turns in the gravitational fields of
the sun and moon, etc. Oceans are a forced, dissipative, open system.
Perhaps even more fundamentally, we appreciate that the oceans participate with the rest of the planet as an entropy generator, receiving

solar radiation at low entropy and re-radiating in the infrared at higher entropy.

Inclusion of the gross forcing of the oceans is a manageable problem from the side of numerical modeling. For more than two decades, numerical models (Bryan, 1969) have recovered principal features of the upper ocean circulation. With modern computing resources, it has been possible to calculate the evolution of the global ocean with resolution of the larger transient eddies (Semtner and Chervin, 1988). However, certain systematic defects persist in these models. Among the apparent defects are that western boundary currents such as the Gulf Stream and Kuroshio tend to flow too far poleward, carrying warm waters into areas where such surface temperatures are not usually encountered. Coupled to atmospheric circulation models for global climate study, such defects drive unrealistic heat exchanges. Another defect occurs near eastern boundaries where models underpredict a poleward tendency in subsurface circulation (Neshyba et al., 1989).

Are these defects due to eddy-supported entropy-increasing tendencies within the actual oceans which are insufficiently resolved within model oceans? Can the models be "corrected" by taking account of theoretical maximum entropy solutions such as (5), making a hybron just such syntheses is in progress but not ready to report at this time. Nonetheless we can simply look at maps of theoretical flow $\mathbf{u}^* = \lambda \mathbf{z} \times \nabla h$ to see if maximum entropy mean flows are at all "realistic."

Figure 3 shows one example of $\mathbf{u}^*$, here drawn from the well-studied North Atlantic. $\nabla h$ is dominated by the strong depth changes along the continental margins. (A much weaker contribution from $\nabla f$ is not discernible on this figure.) A scale for the amplitude of $\mathbf{u}^*$ is not given, since this is rescalable according to $\lambda$. Typical speeds from a few centimeters per second through a few tens of centimeters per second may be obtained.

Along the western margin, $\mathbf{u}^*$ is strongly southward whereas the actual Gulf Stream flows northward. The Gulf Stream is usually understood to be a surface-intensified manifestation of the wind-driven circulation, whereas $\mathbf{u}^*$ (on scales larger than $R_1$) would be a depth-independent motion. Below the Gulf Stream, a southward flow of water of high latitude origin hugs the continental slope. A characteristic model defect, mentioned above, is that the Gulf Stream flows too far northward while remaining close to the coastline. A cold water mass, the "Slope Water," which would move southward along the shelf break, does not develop even in high resolution models (F. Bryan, personal commununication).

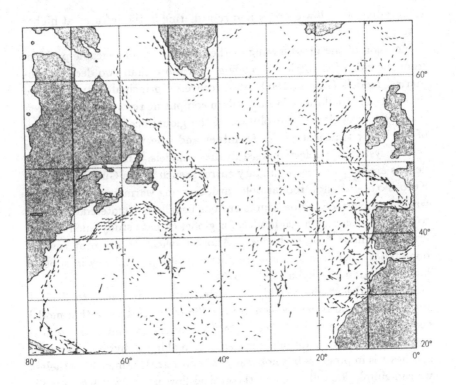

Fig. 3  Maximum entropy mean velocity u* is shown in the North Atlantic.

Efforts to "fix" limited area numerical models have sometimes turned to imposing a strong, southward undercurrent. We see from Fig. 3 that a tendency toward u* would appear to have the sense of improving these model aspects without such need to impose part of the answer.

On the eastern margin, u* is poleward, opposed to the surface circulation off northwest Africa while strengthening a poleward tendency from Portugal northward past Norway. What is not so apparent in Fig. 3 is the role of the Mid-Atlantic Ridge. Simply, the strength of flow is so dominated by continental margins that the gentler (on average) slopes on the Mid-Atlantic Ridge do not lead to such strong u*. There is a mean circulation, though, which is northward over the western ridge flank and southward over the eastern flank. The northward sense over the western flank may contribute to the path by which remnant Antarctic bottom water invades the North Atlantic.

It is easy to continue around the world with such qualitative indications that an eddy-driven entropy-maximizing tendency might improve the quality of ocean models. A more noteworthy comment is that **u**\* will oppose the Antarctic Circumpolar Current, whose transport is sometimes overpredicted by models, while **u**\* will supply a cyclonic tendency around the periphery of the Arctic Ocean.

## 19.4 Outlook

Practical value – if any – awaits testing by the incorporation of **u**\* into actual ocean circulation models. The bottom line question will be: Does it "work"? While that question remains to be answered, let us conclude here with some suggestions toward practical implementation.

It is often advantageous from the view of computational efficiency to separate a modeled velocity field $\mathbf{u}(\phi, \lambda, r, t)$, where $(\phi, \lambda, r)$ are spherical coordinates, into a baroclinic part $\mathbf{u}'$ and an external part given by a transport stream function, say $\Psi$. The definition is such that the depth integral of $\mathbf{u}'$ vanishes, leaving a depth-independent part given by $\Psi$. Thus, incorporation of **u**\* can be made with respect to $\Psi$. A question occurs: How is this $\Psi$ related to $\psi$ at (5)? Because the derivation leading to (5) is quasi-geostrophic, it is ambiguous whether $\psi$ is a velocity stream function or a transport stream function – the ambiguity due to restricting $h$ to vary by only a small fraction of $H$. In actuality $H$ varies from zero at the shoreline to great depth in the abyss. If we attempt to bring forward $\psi$ from (5) as a transport stream function, we clearly encounter a contradiction in practice since velocity singularities will result as $H \rightarrow O$. There is a further ambiguity in (5) due to quasi-geostrophy: one scales $h$ according to a reference depth $H_0$. Finally, the Coriolis parameter $f$, as it would enter (5), is constant, variations of Coriolis having been isolated in $\beta y$. The leap from quasi-geostrophic theory to practical global application is unclear. [One might feel a little better – or worse! – about this leap by recalling the severe idealizations (unforced, inviscid, truncated) underlying (5).] Given these uncertainties, I think that only the simplest plausible extension from (5) is warranted at this point. The simplest extension that suggests itself is a Neptune stream function given by

$$\Psi^* = -\mu f H^2,$$

where $f$ and $H$ are the actual Coriolis parameter and depth of fluid. The negative sign in (6) is due to the definition of $h$ in (5). Extension from (5) to (6) has also supposed that the spatial scale for variation of

$f$ is large compared with the scale for variation of $H$, as will be the case where $H$ is changing strongly. Where the change of $H$ is weaker, $\Psi^*$ itself will not matter very much. Finally, the practical suggestion is that, as a numerical model solves for the evolution of $\Psi$, one may blend in some of $\Psi^*$, with $\mu$ considered a "fudge factor" length scale which might be of O(10 km).

Two further extensions may be mentioned. The preceding discussion has been directed to scales larger than $R_1$ for which the "correction" $\mathbf{u}^*$ is barotropic. For global climate studies, in which only the coarsest possible resolution is desired, this large-scale $\mathbf{u}^*$ may be appropriate. However, modelers will also address questions at smaller scales, perhaps near a specific seamount or at high resolution along a specific coastline. The barotropic tendency is then inappropriate. Baroclinic effects were already treated by Salmon et al. (1976) within two-layer quasi-geostrophic idealization, showing that a barotropic tendency is expressed on scales larger than $R_1$, with baroclinic equilibria on smaller scales. One anticipates that smaller-scale topography will tend to induce mean flows that are more depth-intensified. As one moves upward in the water column, one should see $\mathbf{u}^*$ given by a smoothed version of the topography. [Near the actual benthic boundary, ageostrophic forces tend to retard $\mathbf{u}$ so that the overall tendency should be to observe a mid-depth maximum flow (in lower water column) directed along $\mathbf{u}^*$.] There are further difficulties here, inviting yet bolder leaps. Not only is the finite range of topographic variation a more severe problem, but also the vertical shear in a baroclinic $\mathbf{u}^*$ should demand compensating "thermal wind" tendencies in water property equations.

The second extension is with respect to "diagnostic" or "inverse" modeling. Rather than "simply" time-stepping prognostic equations for $\mathbf{u}$ or $\Psi$, as might be suggested in the preceding discussion, the goal is to utilize direct observations to find a feasible solution which is a "best fit" of observations and of dynamics. To the extent that one might express confidence in a tendency for actual oceans to approach $\mathbf{u}^*$, inverse calculations can attach a penalty to the distance $\mathbf{u} - \mathbf{u}^*$ (under some norm). A particular interest might arise in the context of "adjoint" models such as described by Tziperman and Thacker (1989) in which a tendency toward $\mathbf{u}^*$ would be included in the model dynamics with the option to treat $\lambda$, say, as a control parameter to be evaluated within the global optimization.

## Acknowledgments

Over a span of many years, I have benefited from numerous conversations with Rick Salmon. This work has been spurred along by valued conversations with Ken Brink, Dale Haidvogel, and Anne Marie Treguier, and especially in consequence of one "Red Herring" award. The term "Neptune effect" was suggested by John Patkau. Research has been supported in part by the Office of Naval Research (N00014-87-J-1262) and by the National Science Foundation (OCE-88-16366). This is contribution number 1833 of the School of Oceanography of the University of Washington.

## References

BRYAN, K. (1969) A numerical method for the study of the circulation of the world ocean. *J. Comput. Phys.* 4, 347–376.

CARNEVALE, G.F., FRISCH, U. AND SALMON, R. (1981) H-theorems in statistical fluid dynamics. *J. Phys. A* 14, 1701–1718.

GRIFFA, A. AND SALMON, R. (1989) Wind-driven ocean circulation and equilibrium statistical mechanics. *J. Marine Res.* 47, 457–492.

HAIDVOGEL, D.B. AND BRINK, K.H. (1986) Mean currents driven by topographic drag over the continental shelf and slope. *J. Phys. Oceanogr.* 16, 2159–2171.

HERRING, J.R. (1977) Two-dimensional topographic turbulence. *J. Atmos. Sci.* 34, 1731–1750.

HOLLOWAY, G. (1978) A spectral theory of nonlinear barotropic motion above irregular topography. *J. Phys. Oceanogr.* 8, 414–427.

HOLLOWAY, G. (1987) Systematic forcing of large-scale geophysical flows by eddy–topography interaction. *J. Fluid Mech.* 184, 463–476.

NESHYBA, S.J., MOOERS, C.N.K., SMITH, C.R.L. AND BARBER, R.T., EDS. (1989) *Poleward Flows Along Eastern Ocean Boundaries.* Springer-Verlag, 374 pp.

SALMON, R., HOLLOWAY, G. AND HENDERSHOTT, M.C. (1976) The equilibrium statistical mechanics of simple quasi-geostrophic models. *J. Fluid Mech.* 75, 691–703.

SEMTNER, A.J. AND CHERVIN, R.M. (1988) A simulation of the global ocean circulation with resolved eddies. *J. Geophys. Res.* 93, 15502–15522.

TREGUIER, A.M. (1989) Topographically generated steady currents in barotropic turbulence. *Geophys. Astrophys. Fluid Dynam.* **47**, 43–68.

TZIPERMAN, E. AND THACKER, W.C. (1989) An optimal-control/adjoint-equations approach to studying the oceanic general circulation. *J. Phys. Oceanogr.* **19**, 1471–1485.

# 20

# Modeling the Oceanic Planetary Boundary Layer

JAMES C. MCWILLIAMS, PATRICK C. GALLACHER,

CHIN-HOH MOENG, AND JOHN C. WYNGAARD

## 20.1 Introduction

Large eddy simulation (LES) of turbulent flows is a potent technique of numerical calculation. Where measurements are difficult to obtain with adequate sampling, as in oceanic and atmospheric planetary boundary layers (PBLs), LES solutions provide more complete information than measurements, and, eventually, we can hope to develop LES techniques that provide sufficiently accurate information that they become the primary standard against which concepts and parameterizations can be tested.

LES techniques have a substantial history of successful applications in engineering and meteorology, as can be seen from many of the chapters in this volume. In oceanography, however, their application is just beginning. Therefore, we are presently at the early stages of borrowing and beginning to adapt LES techniques and of discovering the uniquely oceanographic issues. In this chapter, we will briefly survey some of these issues and discuss some recent solutions for the rotating, stably stratified, stress-driven PBL.

## 20.2 Oceanic PBL Issues
### 20.2.1 Surface Gravity Waves

Surface gravity waves are a possibly large but mostly unknown influence on the PBL turbulence. They are energetic motions on time

scales of seconds and spatial scales up to tens of meters, agitating the
turbulent layer whose dominant space scales are only slightly larger and
whose time scales are tens of minutes. Gravity waves and their complex deformations of the air–sea interface are intimately involved with
the microphysical nature of exchanges across the interface (e.g., breaking crests, spray). They modify the upper-surface boundary condition
for the flow (a moving boundary). There are believed to be regimes
where surface waves drive boundary layer circulations (e.g., Langmuir
roll cells), although it is not yet known how frequently this occurs. Very
little has been done about modeling the wave influence on the PBL, and
the technical difficulties in doing so are daunting.

### 20.2.2 Thermodynamics of Buoyancy

Salinity and temperature are the controlling state variables near the
sea surface in a highly nonlinear equation of state, rather than just temperature or temperature and moisture as in the atmosphere. On the
other hand, the absence of phase changes (except across the interface)
and cloud radiation are helpful simplifications compared with the atmosphere. However, clouds in the atmosphere play a very substantial role
in determining the incident radiation that provides important buoyancy
forcing for the oceanic PBL, so we do not entirely escape the challenges
of treating them. The complications of the equation of state more often
add to geographical diversity of the PBL buoyancy structure rather than
contribute to local dynamical complexity, but there are exceptions, such
as cabeling and thermobaric instability.

### 20.2.3 Nature of Buoyancy Forcing

Heating and cooling of the oceanic PBL occurs through sensible exchange by conduction across the air–sea interface, through latent heat of
evaporation cooling of the ocean, through infrared radiation that causes
near-surface heating or cooling, and through solar radiation that causes
penetrative heating over tens of meters. Salinity forcing occurs through
evaporation and precipitation; the latter is often poorly measured and
poorly calculated in atmospheric models. Sea ice adds the complexities
of removing fresh water, or increasing salinity, as it freezes but, oddly
enough, does not yield much of the latent heat of fusion to the ocean;
rather, it goes to the atmosphere above. It adds fresh water when it
melts, and this is also a cooling process for the ocean. Perhaps most
interestingly, sea ice provides horizontal inhomogeneity to the surface

buoyancy forcing due to leads (ice cracks) on scales comparable to those of the PBL turbulence. It also provides a rough boundary for near-ice currents.

### 20.2.4 Coriolis Force

The Coriolis force due to the rotation of the earth is, of course, what makes a PBL distinct from an ordinary boundary layer. Its influences are similar in the atmosphere and ocean. Besides providing the primary balance to turbulent stresses (vertical momentum fluxes), as expressed in the classical Ekman layer solutions, Coriolis force adds the interesting complexity of energetic, low-frequency inertial waves (Foucault pendulum motions). Inertial waves are much better documented in the ocean than in the atmosphere, and probably they also are more abundant and important in the ocean due to the difference in the stress forcing mechanism (see below). Finally, there are some delightful subtleties due to the nonparallelism of the earth's gravitation and rotation vectors except at the poles (also see below).

### 20.2.5 Surface Stress Forcing

Surface wind stress is the primary mechanism for the production of PBL turbulence by velocity shear instability. This is in contrast to the atmosphere or laboratory where a horizontal pressure gradient in the fluid interior creates a shear to accommodate the no-slip surface velocity condition. In the upper-ocean PBL, of course, the surface velocity is not zero because it does not abut a solid surface. There is, however, an approximate correspondence between an idealized oceanic configuration – i.e., a shear layer with specified surface stress at a flat boundary and zero flow in the far interior – and the analogous atmospheric configuration – a shear layer with specified geostrophic flow in the interior and zero velocity at the flat surface. The latter is just a depth-independent, and dynamically inconsequent, Doppler shifting of the former by its mean surface velocity, insofar as the vertical momentum flux (or "eddy viscosity") has the same functional dependence upon the vertical profiles of mean shear and buoyancy. This correspondence overlooks any differences in near-surface turbulence structure: What is the "roughness length" in the ocean? What about the surface gravity waves? Are the Monin–Obukhov similarity profiles valid? Also, the correspondence depends on the oceanic surface stress being uninfluenced by the ocean's surface velocity, which is approximately true because oceanic currents are typically much slower than atmospheric winds. If one accepts this approximate correspondence, perhaps the most important distinction with the atmosphere is that the surface stress due to the near-surface wind is much more variable in space and time than is the low-level geostrophic flow in the atmosphere; thus, in this sense, the oceanic shear layer has the more complex, more variable stress forcing, and the more energetic inertial-wave field is likely a consequence of this.

### 20.2.6 The Convective PBL

In the absence of a strong wind stress, the surface gravity wave field usually is weak, and the dynamics of the oceanic PBL with unstable surface buoyancy forcing and entrainment mixing at the edge of the interior stable layer are similar to free (dry) convection in the atmosphere. The latter has been modeled extensively with LES techniques. In the subpolar oceans equatorward of the ice caps, the interior stratification can be quite weak, the buoyancy forcing can be strong, and the convection can penetrate very deep, even through thousands of meters to the bottom.

### 20.2.7 Bottom Topography and the Bottom PBL

It greatly simplifies the modeling of the upper-ocean PBL that there is no boundary topography (except that provided, in a sense, by the surface gravity waves). However, the bottom PBL in the ocean, like its atmospheric counterpart, does have dynamically significant topographic influences (although they have not yet been extensively examined even for the atmosphere). Over all, the oceanic bottom PBL is much like the atmospheric PBL in the absence of buoyancy forcing: a stably stratified shear layer, driven by the interior geostrophic flow.

## 20.3 LES of a Stably Stratified Shear PBL

Within the context described above, we focus on the particular problem of an idealized upper-oceanic PBL with specified surface stress, zero velocity in the interior, and a pre-existing stable stratification. We neglect surface gravity waves and buoyancy forcing. Also we assume a linear, single-component equation of state. Thus, given the approximate correspondence described above, we might just as well be addressing either the oceanic or atmospheric bottom PBL without topography.

### 20.3.1 Classical Ekman Layer

A PBL regime with which we can compare our stably stratified solutions is the classical Ekman layer in a uniform density fluid, with the planetary rotation vector perpendicular to the boundary; i.e., in a Cartesian coordinate frame,

$$\mathbf{f} = (0, 0, f_z). \tag{1}$$

With a uniform (i.e., spatially constant and isotropic) eddy viscosity, there exist steady boundary layer solutions. In them, the maximum velocity is at the surface, and the currents decay and rotate with distance from the boundary. The "surface angle" between the surface stress and velocity vectors is 45°. In the correspondence above, this is also the "ageostrophic angle" between the interior geostrophic velocity and either the near-surface velocity or minus the surface stress (which are parallel in the pressure-driven PBL). The transport (depth-integrated horizontal velocity) is zero in the direction of the surface stress: it lies in the perpendicular direction.

This problem has been solved with LES by Deardorff (1972) and Mason and Thomson (1987) and, quite recently, with direct numerical simulation (DNS, without any augmentation of molecular diffusivities by so-called subgrid-scale [SGS] parameterizations of transport processes associated with smaller scale fluid motions) by Coleman et al. (1990) for Reynolds numbers, $Re$, up to about 500. In these solutions the mean profiles are unstable and turbulence develops. These solutions are more similar than different: the PBL depth at which the vertical momentum flux drops to 5% of the surface stress is about $0.7u_*/f_z$, where $u_*$ is the square root of the kinematic surface wind stress; the turbulent eddy viscosity has a parabolic shape with depth, peaked in the middle of the turbulent layer; and the surface angle is much reduced from the Ekman solution value, to about 20° at largish $Re$. Collectively, these features indicate appreciable changes in the mean velocity profile compared with the steady solution described above. However, the transport remains perpendicular to the surface stress; this feature depends only upon momentum flux vanishing in the far field, which is a general property of a stress-driven PBL. The general agreement among these solutions, with their different SGS forms, broaches the question of how sensitive these solutions are to the SGS; we do not yet have a full answer.

### 20.3.2 Nonparallel Rotation Vector

The typical situation on a rotating planet is that there is an angle between the local vertical direction (i.e., parallel to the gravitational force) and the rotation vector, viz.,

$$\mathbf{f} = (0, f_y, f_z) \, . \tag{2}$$

This situation does not even admit a steady, horizontally homogeneous solution with uniform eddy viscosity. There is a heuristic argument by Garwood et al. (1985) that $f_y$ should have an important influence on the momentum balance by inducing anisotropy in the second moments of velocity, and a heuristic argument by Wyngaard (unpublished) suggests that it should be even more influential in the transport of passive scalars. There have been some recent demonstrations by Mason and Thomson (1987) and Coleman et al. (1990) that $f_y$ is indeed important: the surface angle changes appreciably with the surface stress direction – by $\pm 5°$ at $45°$ latitude, and even more as one moves equatorward. Probably the inclusion of the nonparallel component of $\mathbf{f}$ needs to become a standard element in studies of shear PBL, although it will not be in the following solutions.

### 20.3.3 Stably Stratified PBL

There are two distinct ways of posing this problem besides the choice of surface stress vs. geostrophic pressure gradient discussed above (see Fig. 1). One way is with a stabilizing surface buoyancy flux, acting on an otherwise unstratified fluid, in addition to the surface stress forcing. This problem has been dealt with recently by Mason and Derbyshire (1990) using LES and by Coleman et al. (1990) using DNS. The other natural way to pose the problem is to have a pre-existing stratification such that there is stable stratification within a distance much less than $0.7u_*/f_z$ from the surface. This is the problem we have solutions for. In the oceanic context, because of the large heat capacity of seawater, it requires quite a long time interval to develop a significant buoyancy gradient near the surface, during which windy events are likely to induce mixing that diminishes this gradient. Thus, it seemed to us more typical to have a stably stratified region well away from the surface as the context within which a stress-driven PBL develops. Specifically, we use an imposed surface stress (not a geostrophic pressure gradient) and an initial temperature profile that has a well-mixed layer above a stable thermocline. An essential difference between the alternatively

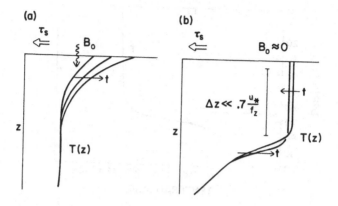

Fig. 1 The stably stratified, shear-driven, upper-ocean PBL: (a) stabilizing surface buoyancy flux ($B_0$) and an initially unstratified fluid; (b)
no surface buoyancy flux and an initially stably stratified fluid. The
sense of the slow evolution of the temperature profiles is indicated
by the temporal arrows labeled $t$.

posed problems is that the second one permits interactions between the
strongly turbulent layer and the stably stratified interior; this involves
entrainment dynamics and radiating gravity waves that are absent in
the first problem. Note that in both problems no stationary equilibrium
is possible, at least not within a time period of many days; however, on
a time scale of $O(f_z^{-1})$, a quite slowly evolving quasi-equilibrium can
arise if the buoyancy forcing or the initial stable stratification is not too
strong. This quasi-equilibrium phase will be our focus.

Specifically, we consider a particular solution with mid-latitude $f_z$
($= 0.7 \times 10^{-4}$ s$^{-1}$), strong wind stress ($\tau_s = 6.25 \times 10^{-4}$ m$^2$ s$^{-2}$), and
zero surface buoyancy flux; for more information on this and similar
solutions, see Gallacher et al. (1991). The temperature field is initially
well mixed to a depth of nearly 60 m, and the stable thermocline is
strongest at about 70 m depth; see Fig. 2. The domain dimensions are
150m × 150 m × 100 m. The horizontal boundary conditions are periodic.

We have used an SGS formulation devised by Deardorff during the
early 1970s, as it has been implemented by Moeng (1984) and has been
extensively explored by Moeng and Wyngaard (e.g., 1988) in the context
of the convective PBL. In addition to altering the surface boundary condition for specified stress and the corresponding surface layer structure
used for the shallowest model grid level, the most notable adaptation for
the present calculations is the implementation of the radiation boundary
condition of Klemp and Durran (1983) at the interior boundary, to allow

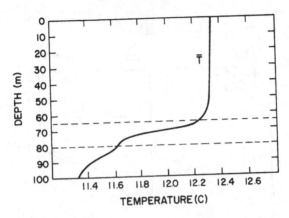

Fig. 2  Horizontally averaged temperature, $\overline{T}(z)$, during the quasi-equilibrium phase. The dashed lines (here and in the following figures) separate the well-mixed layer, the transition layer, and the gravity wave layer.

the Poincaré waves (i.e., inertial and internal gravity waves) generated in the stably stratified outer portion of the boundary layer to leave the domain.

This type of LES has essentially isotropic SGS dynamics, but it does include in the SGS energy equation representations of energy tendency, vertical transport, turbulent production, or dissipation. On the other hand, a diagnosis of this SGS equation in our solutions shows that there is principally a local balance between production and dissipation. Thus, the net result is very much like the simpler SGS formulation used by Mason and Thomson (1987) and Mason and Darbyshire (1990), which is a descendant of the Smagorinsky form of nonlinear viscosity. Thus, there is again an *a posteriori* indication of LES insensitivity with respect to the complexity of the SGS form.

We have examined solutions in the resolution range from $32^3$ to $64^3$ grid points, and a solution with $75^3$ is being calculated now. The results shown here are from the $64^3$ solution. Overall, we are struck by the weak resolution dependence of first- and second-moment quantities in these solutions, which on the face of it seems puzzling. On the other hand, there is only a slow decrease in the ratio of SGS to resolved-scale energies with resolution, suggesting that the effective $Re$ of the resolved flow (i.e., the ratio of $E^{1/2} \times \ell$ for the resolved and SGS components) increases only slightly faster than the linear resolution dimension, which is not very rapid. Experience with highly anisotropic regimes of turbulence, where there are important changes with $Re$ due to the development of dynamically significant coherent structures (McWilliams, 1984), would suggest that the present LES solutions are still exceedingly poorly resolved in this regard. Yet the LESs are also sufficiently expensive to compute that it is difficult to remedy this for these PBL regimes with nearly isotropic turbulence.

By these techniques, we have found quasi-equilibrium solutions that are physically plausible (see below). However, we have also found that the present LES techniques cannot arrive at this state from general velocity initial conditions, at least not for modest spatial resolution. Similar LES difficulties have been reported by others. We doubt that there are physically sensible multiple equilibria here, so we interpret this nonuniqueness as a technical failure of the LES with its particular SGS form.

In particular, excessive mean shear leads to excessive SGS energy and eddy viscosity, and the resolved-scale turbulence is damped to extinction; this precludes a successful spin-up from a state of rest, whose first stage is acceleration of a horizontally uniform flow confined to the shallowest model level. On the other hand, insufficient mean shear, as in an initially too deep Ekman profile, has sufficiently weak resolved-scale instabilities that the flow decays toward a spin-up state, and it again is unsuccessful in reaching quasi-equilibrium. Our present solutions were found by trying several initial conditions and then starting subsequent calculations by interpolation from the more successful of the previous ones. In the process, inertial oscillations have been suppressed, but they would be quite vigorous in a spin-up from rest. Clearly, there is a strong indication of insufficient generality for the present SGS formulation.

In the solution we will focus on, the horizontally averaged temperature profile, $\overline{T}(z)$ (Fig. 2) has only a slow rate of evolution in the sense indicated in Fig. 1. Compared with the uniform-density solutions (Section 2.1), the horizontally averaged velocity profile, $\overline{u}(z)$ (Fig. 3), is substantially different due to the stable stratification, even in the well-mixed layer (note the vertical partitioning in Fig. 2 et seq.): $\overline{u}(z)$ is compressed in $z$, most strongly where the stratification is most stable (i.e., in the transition layer); $\overline{u}(0)$ is increased for a given surface stress; the surface angle is increased (e.g., it is 37° in our solution, quite close to the value of Mason and Derbyshire, 1990); but, of course, the transport remains perpendicular to the surface stress. The turbulent momentum flux profiles (Fig. 4) are in balance with the mean velocity profiles as

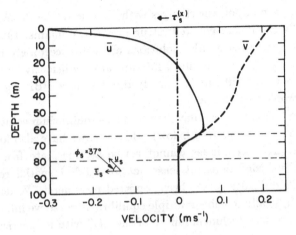

Fig. 3   $\overline{u}(z) = (\overline{u}(z), \overline{v}(z))$.

expected from Ekman dynamics:

$$-f\overline{v} = -\frac{\partial}{\partial z}\overline{u'w'}, \quad f\overline{u} = -\frac{\partial}{\partial z}\overline{v'w'}. \tag{3}$$

Outside the actively turbulent layers (i.e., in the gravity-wave layer), this balance is trivial: the mean velocities are zero, and the momentum flux, which is nonzero in the same component as the PBL transport, has no vertical gradient, hence no divergence.

The turbulent heat flux (Fig. 5) has primarily an entrainment profile (i.e., one-signed), as in, say, a convective PBL, but some countergradient flux also occurs in the lower portion of the transition layer, leading to a local intensification of the thermocline.

The turbulent velocities are appreciably anisotropic (Fig. 6), except in the gravity-wave layer, where they are nearly isotropic. The turbulent diffusivities (not shown) have a parabolic vertical structure within the well-mixed layer, but exhibit a sign reversal (i.e., locally countergradient flux) in the transition layer. The $y$-component of the eddy viscosity is quite large (and largely irrelevant) in the gravity-wave zone, where the mean flow is quite weak but the momentum flux remains finite. The resolved-scale Prandtl number, $Pr$, is somewhat smaller than Mason and Derbyshire (1990) and Coleman et al. (1990) have found (here, $Pr = 0.4$ in the vicinity of the boundary-layer maxima in the eddy diffusivities), but our SGS $Pr$ ($= 0.3$) is also somewhat smaller than theirs and this may have some influence.

The temperature variance (Fig. 7) is greatest in the thermocline in

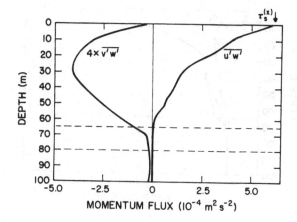

Fig. 4 $\overline{u'w'}(z)$ (solid line) and $\overline{v'w'}(z)$ (dashed line).

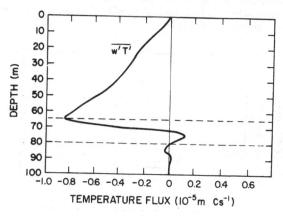

Fig. 5 $\overline{T'w'}$.

the transition layer. The gradient Richardson number,

$$Ri_g = -\frac{g\alpha\, d\overline{T}/dz}{(d\overline{u}/dz)^2}, \qquad (4)$$

where $\alpha$ is the thermal expansion coefficient, increases from zero at the surface to a value near the critical value (i.e., 0.25) between the well-mixed and transition layers. A bulk Richardson number across the well-mixed layer, $Ri_b$, is also near the critical value, but this is not a strongly discriminating statement since the buoyancy field, hence $Ri_b$, is a rapidly increasing function of depth at the base of the well-mixed layer.

The gravity waves are successfully transmitted out of the domain, and

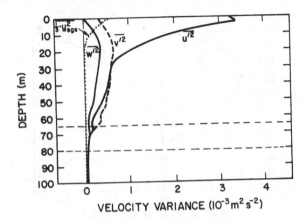

Fig. 6   $\overline{u'^2}(z)$ (solid line), $\overline{v'^2}(z)$ (dashed line), $\overline{w'^2}(z)$ (solid line), and, for comparison, the velocity variance for a single component of the isotropic SGS field (i.e., two-thirds of the horizontally averaged SGS kinetic energy; dotted line).

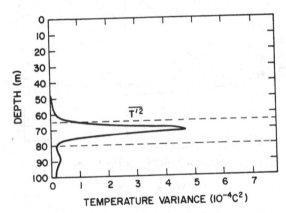

Fig. 7   $\overline{T'^2}(z)$.

their statistical properties are nearly independent of depth outside the PBL.

## 20.4 Prospects

Several issues associated with the stable, stress-driven PBL should be further investigated. Among those that particularly concern us are the role of inertial waves, the influences of a nonparallel component of the rotation vector, evolution from general velocity initial conditions, scalar transport properties, and general stratification profiles.

Better SGS formulations are, of course, the principal concern of this volume. Based upon the experience reported here, we believe that present LES solutions are reasonably insensitive to resolution and such variations as have yet been tried in the SGS forms. However, we are concerned that the latter do exert a sufficiently strong control on the solutions such that the effective $Re$, rather than being approximately infinite as declared in the LES conception – or manifesto – is really rather modest. Thus, the resolved-scale turbulence is dynamically constrained to a substantial degree, and the physical correctness of the solutions is likely to be strongly dependent upon the correctness of the SGS forms in ways we do not yet adequately understand.

## Acknowledgments

The National Center for Atmospheric Research is sponsored by the National Science Foundation.

## References

COLEMAN, G., FERZIGER, J. AND SPALART, P. (1990) A numerical study of the stratified turbulent Ekman layer. Report No. TF-48, Stanford University, Department of Mechanical Engineering.

DEARDORFF, J. (1972) Numerical investigation of neutral and unstable planetary boundary layers. *J. Atmos. Sci.* 29, 91–1150.

GALLACHER, P., McWILLIAMS, J., MOENG, C.-H. AND WYNGAARD, J. (1991) Large-eddy simulations of the upper-ocean planetary boundary layer. Preprint.

GARWOOD, R., GALLACHER, P. AND MULLER, P. (1985) Wind direction and equilibrium mixed-layer depth: General theory. *J. Phys. Oceanogr.* 15, 1325–1331.

KLEMP, J. AND DURRAN, D. (1983) An upper boundary condition permitting internal gravity wave radiation in numerical mesoscale models. *Mon. Wea. Rev.* 111, 430–440.

MASON, P. AND THOMSON, D. (1987) Large-eddy simulations of the neutral-static-stability planetary boundary layer. *Q. J. R. Meteorol. Soc.* 113, 413–443.

MASON, P. AND DERBYSHIRE, S. (1990) Large-eddy simulation of the stably-stratified atmospheric boundary layer. *Bound.-Layer Meteorol.* **53**, 117–162.

McWILLIAMS, J. (1984) The emergence of isolated, coherent vortices in turbulent flow. *J. Fluid Mech.* **146**, 21–43.

MOENG, C.-H. (1984) A large-eddy-simulation model for the study of planetary boundary-layer turbulence. *J. Atmos. Sci.* **41**, 2052–2062.

MOENG, C.-H. AND WYNGAARD, J. (1988) Spectral analysis of large-eddy simulations of the convective boundary layer. *J. Atmos. Sci.* **45**, 3573–3587.

# Diapycnal Mixing in the Ocean:
# A Review

PETER MÜLLER

## 21.1 Introduction

With the exception of the approximately 100 m deep, well-mixed upper layer, the oceanic interior can be thought of as consisting of complex surfaces of constant density (isopycnals). Particles moving along such isopycnals meet with little resistance, thus participating in the large-scale, quasi-two-dimensional oceanic circulation. Mixing across isopycnals (the so-called diapycnal mixing) is caused by small-scale processes that are not resolved in models of the large-scale circulation. It has, however, been realized that such diapycnal mixing plays an important role in stirring the ocean interior and in energy transformations between available potential and kinetic forms. Indeed, diapycnal mixing may be one of the unresolved, small-scale processes an understanding and proper quantification of which are crucial for successful modeling. This chapter provides an overview of the existing observational and theoretical information on diapycnal mixing of buoyancy in the ocean interior to guide the development and verification of subgrid-scale parameterizations designed for use in oceanic general circulation models including large eddy simulation (LES).

The conventional wisdom is that
- diapycnal mixing of buoyancy in the ocean interior is due to intermittent patches of turbulence with a vertical extent of a few meters;
- mesoscale eddies and internal waves cannot mix buoyancy across isopycnals;
- the turbulent patches are caused by breaking internal gravity waves;

• internal waves break by either shear or convective instabilities that are caused either by encounters of critical levels or by chance superpositions.

While these statements may be correct, this review will stress that all of our knowledge about the mixing process and mixing rates is indirect. Present-day instruments cannot measure diapycnal fluxes in the ocean. Theories and numerical models cannot calculate these fluxes for realistic ocean conditions. Thus, major basic issues are unresolved. Better understanding of diapycnal mixing is necessary since, as we will demonstrate in the first section, oceanic general circulation models, prognostic and diagnostic, are extremely sensitive to the value and functional dependence of mixing coefficients.

On the observational side, our knowledge about diapycnal mixing comes from diagnosing large-scale hydrographic data, from tracer release experiments, and from microstructure measurements. On the theoretical side, the statistics of the Richardson number and the energy flux through the internal wave field to high wave numbers have been calculated.

Among all these methods, the tracer release experiments provide perhaps the most direct evidence of diapycnal mixing. The interpretations of microstructure measurements, Richardson number statistics, and energy flux calculations all rely on the notion that breaking internal waves convert fixed fractions of the available kinetic energy into mixing and dissipation. As already pointed out, the crucial quantity in this scenario, the diapycnal buoyancy flux, can presently neither be measured nor calculated with confidence. This point will constitute our first major issue.

A second and related issue is the orientation of the diffusion tensor. It is usually assumed, but without any real physical basis, that it is diagonal in a horizontal/vertical coordinate system. The orientation is an issue since horizontal mixing across sloping isopycnals becomes diapycnal mixing.

Inferences from hydrographic data depend on the specific assumptions made. More importantly, the inferred diffusion coefficients are the ones that explain the large-scale hydrography and that are required in oceanic general circulation models that aim at prognosticating this hydrography. As our third issue we discuss whether diapycnal mixing induced by internal-wave breaking is relevant to coarse-resolution oceanic general circulation models.

Finally we ask whether diapycnal mixing in the ocean interior is relevant at all. All of the indirect inferences of internal wave-induced mixing in the ocean interior point to a value $O(10^{-5}\,\mathrm{m^2\,s^{-1}})$, an order of magnitude smaller than the value required to satisfy large-scale balances. One possible explanation is that most of the mixing happens near the boundaries of the oceans.

This review is by no means complete. Only a few examples are given and only some of the issues are discussed. A more complete discussion can be found in the proceedings of the 'Aha Huliko'a Workshop (Müller and Henderson, 1989), which is summarized in Müller and Holloway (1989) and in a review article by Gregg (1987). The difficulties and uncertainties of subgrid-scale representation are discussed with great insight by Holloway (1990).

A further limitation of this review is that we treat seawater as a one-component system and equate buoyancy and temperature. We therefore neglect the effects of double diffusion and the nonlinear equation of state on mixing. We also assume that passive chemical tracers behave like buoyancy. Finally, we do not distinguish between diapycnal and vertical mixing until this issue is discussed in Section 10.

## 21.2 Sensitivity Studies

Oceanic general circulation models (OGCMs) generally parameterize the subgrid-scale buoyancy fluxes by a Fickian diffusion term with a diffusion tensor that is diagonal in the horizontal/vertical coordinate system. The vertical diffusion coefficient $K_v$ (supposed to model the diapycnal diffusion) is much smaller than the horizontal coefficient and is usually assumed to be constant. The value of $K_v$ determines essential aspects of the thermohaline circulation. This has been most clearly demonstrated in a sensitivity study carried out by Bryan (1987) with a simplified OGCM. Figure 1 shows the calculated poleward heat flux (a quantity of importance for climate studies) as a function of latitude for various values of $K_v$. The heat transport increases approximately proportionally to the diffusion coefficient to the two-thirds power. Following Bryan (1987), this power law dependence can be rationalized as follows: the thermal wind relation implies a horizontal velocity scale

$$u = \frac{\Delta \rho \, g H}{\rho_o \, 2\Omega R},\tag{1}$$

where $g$ is the gravitational acceleration, $\Omega$ the earth's rotation rate, $R$ the earth's radius, $\Delta \rho$ the imposed surface density difference, $\rho_o$ a

*Peter Müller*

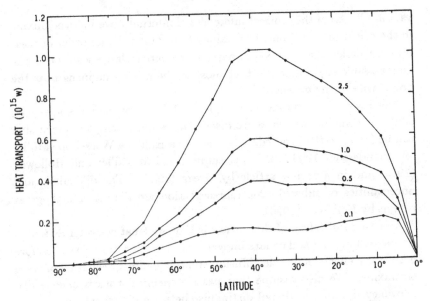

Fig. 1  Poleward heat transport in a coarse-resolution oceanic general circulation model as a function of latitude for various values of the vertical diffusivity ranging from 0.1 to $2.5 \cdot 10^{-4} \, \text{m}^2 \, \text{s}^{-1}$. From Bryan (1987).

reference density, and $H$ the vertical scale of the circulation. If a vertical advective–diffusive balance is assumed for the density field,

$$w \frac{\partial}{\partial z} \rho = \frac{\partial}{\partial z} K_v \frac{\partial}{\partial z} \rho, \qquad (2)$$

one finds

$$H = \frac{K_v}{w} = \frac{K_v R}{u H}, \qquad (3)$$

where $w$ is the vertical velocity scale and the incompressibility condition has been used. Solving (1) and (3) for $u$ and $H$ results in

$$u = \left( \frac{K_v \, \Delta\rho^2 \, g^2}{R \, 4\Omega^2 \, \rho_o^2} \right)^{1/3}, \qquad (4a)$$

$$H = \left( \frac{K_v \, R^2 \, 2\Omega \, \rho_o}{\Delta\rho \, g} \right)^{1/3}. \qquad (4b)$$

The heat transport, which is proportional to the product of $u$ and $\bar{H}$, then becomes

$$u \cdot H \sim K_v^{2/3}, \qquad (5)$$

as observed in the model calculation. This argument does not imply that the vertical advective–diffusive balance (2) holds locally, which it does not in the model, but only that it holds in some basin-wide integrated sense. Indeed, the horizontal advection terms in the density equation vanish upon integration over the area of the ocean, except for eddy-like terms.

While OGCMs with constant eddy viscosity and diffusivity coefficients of reasonable magnitude are capable of reproducing the broad features of the mid-latitude circulation, the same values do not allow reproduction of the complex structures of the equatorial circulation system, which includes a narrow, intense equatorial undercurrent and a sharp thermocline. Modeling of these features becomes more realistic when the vertical viscosity $A_v$ and mixing coefficient $K_v$ are assumed to depend on the Richardson number $Ri$ in a manner proposed by Pacanowski and Philander (1981),

$$A_v = \frac{A_o}{(1 + \alpha\,Ri)^n} + \nu,$$
$$K_v = \frac{A_v}{(1 + \alpha\,Ri)} + \kappa. \tag{6}$$

This form was motivated by the analysis of Munk and Anderson (1948), and is supposed to model the additional mixing caused by the strong undercurrent shear. In (6), $A_o$, $\alpha$, and $n$ are adjustable parameters, and $\nu$ and $\kappa$ are background values, in the absence of any shear-induced mixing. Note that the physical basis of (6) is unclear since the Richardson number $Ri$ appearing in (6) is the explicit model Richardson number, which is usually very different from and unrelated to the actual Richardson number.

Bryan's (1987) study demonstrates the sensitivity of prognostic calculations with respect to the value of the vertical diffusion coefficient. Inverse models show an even greater sensitivity. An argument advanced by Gargett (1984) demonstrates that the direction of the deep meridional circulation depends on such a subtle feature as the depth dependence of the diffusion coefficient. We repeat the argument here, for it clearly demonstrates possible pitfalls in our inference of the circulation from hydrographic data.

The observed density profile in most parts of the world's abyssal ocean is reasonably well approximated by

$$\rho(z) = \rho_o - \Delta\rho\exp(z/b), \tag{7}$$

where $\Delta\rho$ is the difference between the "surface" and bottom values of the density and $b$ is a scale depth, usually found to be $O(1.3\,\text{km})$. The associated Brunt-Väisälä frequency $N$ is

$$N^2(z) = -\frac{g}{\rho_o}\frac{\partial\rho}{\partial z} = \frac{g\,\Delta\rho}{b\,\rho_o}\,exp\left(\frac{z}{b}\right). \tag{8}$$

If this profile is interpreted as the result of the vertical advective–diffusive balance (2), we infer a vertical velocity

$$w = \frac{K_v}{b} + \frac{\partial}{\partial z}K_v. \tag{9}$$

If we assume

$$K_v(z) \sim N^{2q}(z), \tag{10}$$

the inferred vertical velocity becomes

$$w = \frac{K_v}{b}(1+q). \tag{11}$$

If we further use the planetary geostrophic potential vorticity balance

$$\beta v = f\frac{\partial w}{\partial z}, \tag{12}$$

where $f$ is the Coriolis frequency and $\beta$ is the beta parameter, we find a meridional velocity

$$v = \frac{f}{\beta}\frac{(1+q)q}{b^2}K_v, \tag{13}$$

which is equatorward for $-1 < q < 0$ and poleward otherwise. Observations discussed below have been interpreted as supporting values in the range from $q = 0$ to $q = -\frac{1}{2}$. Hence, inverse calculation might not even determine the sense of the meridional circulation.

A recent study by Cummins et al. (1990) shows that this dramatic effect does not occur in a prognostic calculation, partly because the vertical advective–diffusive balance (2) does not hold in that calculation and partly because explicit diapycnal mixing is overridden in some areas by implicit diapycnal mixing due to horizontal diffusion across sloping isopycnals.

## 21.3 Large-Scale Balances

Values of the diapycnal or vertical diffusion coefficients can be inferred from large-scale mass and heat balances. The original argument is due to Munk (1966), who applied it to the abyssal ocean in the following manner. Convection at high latitudes forms bottom water at a rate of $M = 50 \cdot 10^6 \, \text{m}^3 \, \text{s}^{-1}$. Conservation of mass requires that this water be upwelled with a velocity

$$w = \frac{M}{A} \sim 1.2 \, \text{cm} \, \text{d}^{-1}, \tag{14}$$

where $A = 3.6 \cdot 10^{14}\,\mathrm{m}^2$ is the area of the ocean. If the vertical advective–diffusive balance (2) holds on average, the scale height of the density field is given by

$$H = \frac{K_v}{w}. \tag{15}$$

The observed value of $H \sim 1.3\,\mathrm{km}$ then leads to

$$K_v = 1.6\,\mathrm{cm}^2\,\mathrm{s}^{-1}. \tag{16}$$

More recent estimates of the rate of bottom water formation are $O(20 \cdot 10^6\,\mathrm{m}^3\,\mathrm{s}^{-1})$ and would halve the values of the upwelling velocity and diapycnal mixing coefficient. Nevertheless, $w = O(1\,\mathrm{cm\,d}^{-1})$ and $K_v = O(1\,\mathrm{cm}^2\,\mathrm{s}^{-1})$ are still the benchmark values today. The same type of argument has also been applied to abyssal basins where the inflow over a sill has been measured. Hogg et al. (1982) estimated $K_v \sim 3$–$4\,\mathrm{cm}^2\,\mathrm{s}^{-1}$ for the Brazil Basin and Saunders (1987) found $K_v \sim 1.5$–$4\,\mathrm{cm}^2\,\mathrm{s}^{-1}$ for the Iberian Abyssal Basin.

## 21.4 Diagnostic Ocean Models

Any local estimate of the diffusion coefficient from the density equation must include the horizontal advection terms since $u/L \sim w/H$ because of incompressibility. The starting point of inverse estimates is hence the density equation in the form

$$u\partial_x \rho + v\partial_y \rho + w\partial_z \rho = D[\rho], \tag{17}$$

where $D[\rho]$ is the diffusion operator. Estimation of the diffusion coefficients from this equation requires knowledge of the three-dimensional velocity field and the gradient of the density field. Hydrographic data provide the density field and, by virtue of the thermal wind relation, the vertical shear of horizontal velocity,

$$\partial_z u = \frac{g}{\rho_o f}\,\partial_y \rho,$$

$$\partial_z v = -\frac{g}{\rho_o f}\,\partial_x \rho. \tag{18}$$

If, additionally, validity of the planetary potential vorticity balance is assumed,

$$\partial_z w = \frac{\beta}{f}\,v, \tag{19}$$

one can calculate the complete three-dimensional velocity field up to three constants of integration, the reference velocities $u_o$, $v_o$, and $w_o$. This expresses the classical indetermination of the "level of no motion."

*Peter Müller*

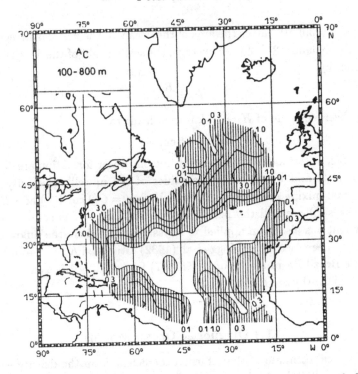

Fig. 2  Map of the diapycnal diffusion coefficient in the North Atlantic for the depth range 100 to 800 m, as inferred from beta-spiral calculations. Units are $10^{-4}\,\mathrm{m^2\,s^{-1}}$. Contours are logarithmically spaced with interval 0.5. Areas with values larger than $10^{-5}\,\mathrm{m^2\,s^{-1}}$ are shaded. From Olbers et al. (1985).

If this velocity field is substituted into the density equation (17), one obtains an equation containing the three unknown reference velocities and the unknown diffusion coefficients. Since this equation holds for each vertical level, one obtains a formally overdetermined system for the unknowns, which can be solved by appropriate mathematical methods. This is the essence of the beta-spiral method first put forward by Stommel and Schott (1977). There are other diagnostic methods that differ in the dynamical principles and the type of data used.

The beta-spiral method has been applied by Olbers et al. (1985) to the North Atlantic using the Levitus (1982) atlas. Figure 2 shows a map of the estimated diapycnal diffusion coefficient for the upper 800 m. The pattern of the diffusivities follows the pattern of the North Atlantic current system. The diffusivities are large where the currents are strong,

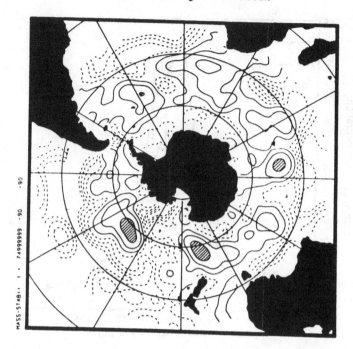

Fig. 3 Diapycnal diffusivity in the Antarctic Circumpolar Current for the depth range from 100 to 800 m, as inferred from beta-spiral calculations. Contours are logarithmically spaced with interval 0.5. Contours larger than $10^{-4}\,\mathrm{m^2\,s^{-1}}$ are full. Areas with values larger than $10^{-3}\,\mathrm{m^2\,s^{-1}}$ are shaded. From Olbers and Wenzel (1990).

and small (or indistinguishable from zero) in the center of the gyre. Overall the values are small, less than $10^{-4}\,\mathrm{m^2\,s^{-1}}$ in most areas.

Much higher values of the diapycnal diffusivity have been found by Olbers and Wenzel (1990) for the Antarctic Circumpolar Current, based on the Gordon et al. (1982) atlas. As seen in Fig. 3, values larger than $10^{-4}\,\mathrm{m^2\,s^{-1}}$ are found in most parts of the current.

## 21.5 Tracer Release Experiments

Another, perhaps more direct, estimate of the diapycnal diffusion coefficient can be obtained from tracer release experiments. In the deep ocean these experiments have been pioneered by Ledwell et al. (1986). They inject sulfur hexafluoride ($SF_6$) on an isopycnal and follow the vertical spread of the tracer about the initial surface in time (see Fig. 4).

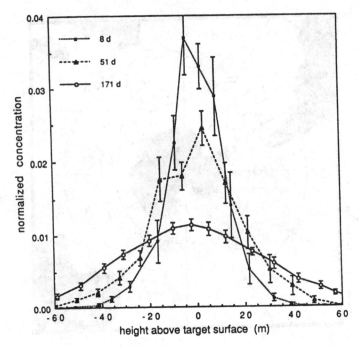

Fig. 4  Vertical spreading of the tracer during the Santa Monica Basin Experiment. Each profile is an average of 10 or so individual profiles, with the error bars indicating the variance in shape of the individual profiles. The target surface of the injection was the 5.085°C potential temperature surface. The height is really a transformed coordinate based on the potential temperature profiles and the mean height versus temperature profile for the middle (51 d) survey. The concentration shown is normalized so that the area under each curve is unity. The days after injection given in the key are nominal; in reality each survey cruise was about 10 days long. The diapycnal diffusivity inferred from the spreading is about $0.3\,\mathrm{cm^2\,s^{-1}}$. From Ledwell (1989).

A vertical diffusivity is then calculated from the formula

$$\sigma^2(t) = \sigma^2(t=0) + 2K_v t, \qquad (20)$$

where $\sigma^2$ is the variance of the vertical concentration distribution.

Their first experiment (in the Santa Monica Basin, release depth 800 m) yielded a vertical diffusivity of about $0.25\,\mathrm{cm^2\,s^{-1}}$. Their second experiment (in the Santa Cruz Basin, release depth 1500 m) yielded values larger than $1\,\mathrm{cm^2\,s^{-1}}$ (Ledwell, 1989).

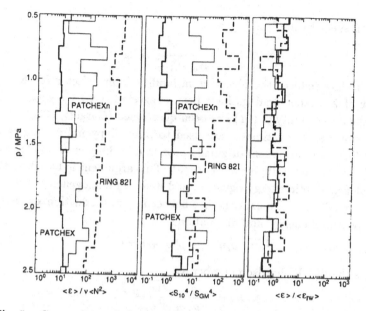

Fig. 5  Comparison of three different mid-latitude sites: PATCHEX, PATCHEXn, RING82I. The three panels show, as a function of pressure, the dissipation rate $\epsilon$ normalized by the buoyancy frequency $N$, the 10-m vertical shear $S$ normalized by the 10-m shear of the Garrett and Munk model, and the dissipation rate normalized by buoyancy frequency and shear. Note that the shear scaling greatly reduces the large differences when $\epsilon$ is scaled only by $N$. From Gregg (1989).

## 21.6 Microstructure Measurements

Though the vertical fluxes of buoyancy (and momentum) cannot be measured directly, microstructure profiling instruments are now capable of resolving the fluctuations of velocity and temperature in the centimeter range. From these microstructure profiles the dissipation rates of kinetic energy $\epsilon$ and of potential energy $\chi$ can be inferred if assumptions are made about the isotropy and statistics of the fluctuations. The dissipation rate $\chi$ is proportional to the dissipation rate of temperature or density variance.

Figure 5 (from Gregg, 1989) shows the kinetic energy dissipation $\epsilon$ as a function of depth for three experiments: PATCHEX, PATCHEXn, and RING82I. These profiles suggest that the kinetic energy dissipation rate is proportional to the buoyancy frequency squared and to the 10-m vertical shear to the fourth power:

$$\epsilon = 7 \cdot 10^{-10} \frac{N^2}{N_o^2} \frac{S_{10}^4}{S_G^4 M} \, \mathrm{W\,kg^{-1}}. \tag{21}$$

This scaling reduces the observed variability from a factor of 58 to a factor of 2. Other scalings have been suggested (e.g., Gargett, 1984) and the applicability of (21) has been questioned (Gargett, 1990).

Dissipation rate estimates are used to infer diapycnal mixing rates according to an argument originally proposed by Lilly et al. (1974) and applied to the ocean by Osborn (1980). This argument is based on the turbulence kinetic energy equation and assumes homogeneity and stationarity. The major balance is then between shear production, exchange with potential energy, and dissipation,

$$\overline{u'w'} \, \partial_z \bar{u} + \frac{g}{\rho_o} \overline{\rho'w'} + \epsilon = 0. \tag{22}$$

Define the flux Richardson number as

$$R_f = -\frac{g}{\rho_o} \frac{\overline{\rho'w'}}{\overline{u'w'} \, \partial_z \bar{u}}. \tag{23}$$

The vertical buoyancy flux is then given by

$$\frac{g}{\rho_o} \overline{\rho'w'} = \gamma\epsilon, \tag{24}$$

with an efficiency factor

$$\gamma = \frac{R_f}{1 - R_f}. \tag{25}$$

Since $\overline{\rho'w'} = -K_v \, \partial_z \bar{\rho}$, we find a relation between the diapycnal diffusion coefficient $K_v$ and the dissipation rate $\epsilon$,

$$K_v = \gamma\epsilon N^{-2}, \tag{26}$$

that involves only $\gamma$ as an unknown. Laboratory experiments and oceanic measurements (Thorpe, 1973; Oakey, 1982) seem to suggest a value of $\gamma = 0.2$. If this value is used, Gregg's measurements shown in Fig. 5 imply

$$K_v = 5 \cdot 10^{-6} \frac{S_{10}^4}{S_{GM}^4} \, \mathrm{m^2\,s^{-1}}, \tag{27}$$

independent of depth.

The dissipation rate $\chi$ of potential energy has been used to infer the diapycnal diffusion coefficient by assuming a balance between the conversion and dissipation terms (Osborn and Cox, 1972),

$$\frac{g}{\rho_o} \overline{\rho'w'} = \chi, \tag{28}$$

in the turbulence potential energy equation. This leads to

$$K_v = \chi N^{-2}. \tag{29}$$

Note that the assumed balances within the kinetic and potential energy equations are based on a scenario in which turbulence kinetic energy is generated by shear production. A fraction $R_f$ of this kinetic energy is converted to potential energy and then dissipated. The other fraction $(1 - R_f)$ is dissipated directly. This scenario does not hold in the numerical experiments of Ramsden and Holloway (1991), who studied inhomogeneous, remotely forced, stratified turbulent flows and found the divergence of the pressure work to be important in the turbulence kinetic energy equation.

## 21.7 Kinematic Estimates

If internal wave breaking is primarily shear driven, the Richardson number,

$$Ri = \frac{N^2}{(\partial_z u)^2 + (\partial_z v)^2}, \tag{30}$$

must be a key parameter. The rms Richardson number of the internal wave field is observed to be of the order of 2 (Munk, 1981), larger than the critical value $Ri = 0.25$. It was first recognized by Bretherton (1969) that a random superposition of internal waves would give a finite probability that $Ri < \frac{1}{4}$ locally. This statistical approach was pursued by Garrett and Munk (1972) and Garrett (1979b). The distribution function of the Richardson number has been derived by Desaubies and Smith (1982) for a Gaussian internal wave field. The distribution function depends on the rms strain $\lambda$ of the wave field, which is about $\lambda = 0.5$ for the ocean. Calculated and observed distributions agree fairly well, as can be seen in Fig. 6 for a data set from Evans (1982).

Using numerical simulations, Desaubies and Smith (1982) were also able to derive the statistics of the vertical distributions of regions where $Ri < \frac{1}{4}$. However, estimates of a diapycnal mixing coefficient require two additional parameters: (i) the amount of mixing that occurs in these regions and (ii) the frequency in time with which these events occur. Both these parameters are poorly established, although reasonable assumptions, such as complete mixing and a frequency of occurrence proportional to $N$, lead to reasonable values of $K_v$.

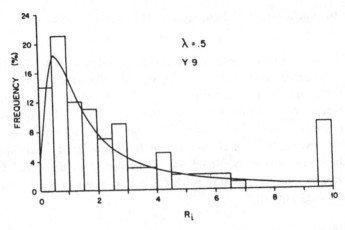

Fig. 6   Comparison of calculated probability density function of Richardson number with measured frequency function for a data set obtained by Evans (1982). From Desaubies and Smith (1982).

## 21.8 Dynamical Calculations

Nonlinear interactions among internal waves are a principal part of their dynamics and provide an important link in the overall energy cascade from large to small scales. Four approaches to their analysis are reviewed in Müller et al. (1986): (i) the evaluation of the transfer integral describing weakly and resonantly interacting waves, (ii) the application of closure hypotheses from turbulence theories to more strongly interacting waves, (iii) the integration of the eikonal equations describing the propagation of small-scale waves in a background of large-scale waves, and (iv) the direct numerical simulation of the basic hydrodynamic equations of motion. The weak interaction and eikonal calculations have provided most of the current wisdom about interactions within the oceanic internal wave field; notably, they have been used to calculate explicitly the energy flux $F$ to high wave numbers. If one assumes again that a fraction $R_f$ of this flux is used for mixing, then the diapycnal diffusivity can be estimated according to

$$K_v = R_f F N^{-2}. \tag{31}$$

The equation describing the nonlinear transfer of action or energy within an internal wave field due to weak resonant wave–wave interactions is given by

$$\frac{\partial}{\partial t} A(\mathbf{k}) = \int d\mathbf{k}' \, d\mathbf{k}'' \{ T^+ \delta(\mathbf{k} - \mathbf{k}' - \mathbf{k}'') \, \delta(\omega - \omega' - \omega'')$$
$$[A(\mathbf{k}')A(\mathbf{k}'') - A(\mathbf{k})A(\mathbf{k}') - A(\mathbf{k})A(\mathbf{k}'')]$$
$$+ 2T^- \delta(k - k' + k'') \, \delta(\omega - \omega' + \omega'')$$
$$[A(\mathbf{k}') A(\mathbf{k}'') + A(\mathbf{k}) A(\mathbf{k}') - A(\mathbf{k}) A(\mathbf{k}'')] \}, \tag{32}$$

where $A(\mathbf{k}) = E(\mathbf{k})/\omega(\mathbf{k})$ is the action density spectrum, $E(\mathbf{k})$ the energy density spectrum, $\mathbf{k}$ the wave number vector, $\omega = \omega(\mathbf{k})$ the frequency (given by the dispersion relation), and $T^+$ and $T^-$ transfer functions depending on $\mathbf{k}$, $\mathbf{k}'$, and $\mathbf{k}''$. Explicit expressions for $T^+$ and $T^-$ can be found in Müller and Olbers (1975) and Olbers (1976). The transfer equation (32) is a closed equation for the spectrum $A(\mathbf{k})$. The basic statistical closure hypothesis in its derivation is the assumption that the correlation time of the wave field is smaller than the interaction time, so that interacting wave modes can always be treated as statistically independent (Hasselmann, 1966, 1967; Benney and Saffmann, 1966). Waves interact only if the resonance conditions,

$$\mathbf{k}' \pm \mathbf{k}'' = \mathbf{k}$$

and
$$\omega' \pm \omega'' = \omega \tag{33}$$

are satisfied.

In analyzing the transfer equation (32), McComas and Müller (1981) assumed that internal wave energy is generated at low vertical wave numbers $\beta < \beta_*$ and dissipated at high vertical wave numbers $\beta > \beta_c$ (see Fig. 7). They then proved that an inertial range exists between $\beta_*$ and $\beta_c$, in which resonant interactions provide a constant (independent of vertical wave number) downscale energy flux $F$ from the generation to the dissipation region. At high frequencies the flux is provided by an induced diffusion mechanism, at low frequencies by a parametric subharmonic instability mechanism. The inertial range has a vertical wave number spectrum

$$E(\beta) \sim \beta^{-2} \tag{34}$$

as observed. The downscale energy flux is given by

$$F = \left[ 1 + \frac{27}{32\sqrt{10}} \right] \frac{\pi f}{N^2} E^2 \beta_*^2, \tag{35}$$

where $E$ is the total energy, and $f$ and $N$ are the Coriolis and Brunt-Väisälä frequencies, respectively. If the internal wave quantities scale

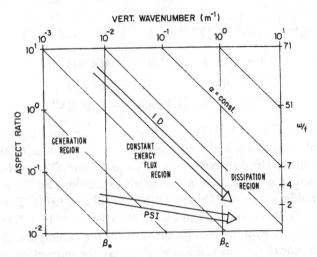

Fig. 7   Schematic representation of the dynamical balance of McComas and Müller (1981). Energy is generated at low vertical wave number $|k_z| < \beta_*$. Between $\beta_*$ and $\beta_c$ there is an inertial range where the induced diffusion (ID) mechanism at high frequencies and the parametric subharmonic instability (PSI) mechanism at low frequencies provide a constant energy flux to high wave numbers $|k_z| > \beta_c$, where energy is dissipated. The wave number $\beta_c$ is determined as the wave number where the spectrum must roll off because the nonlinear transfer can no longer keep up with dissipation. From McComas and Müller (1981).

like the Garrett and Munk model (Cairns and Williams, 1976) then

$$F = 2 \cdot 10^{-9} \frac{N^2}{N_o^2} \frac{E^2}{E_{GM}^2} \, \mathrm{W\,kg^{-1}}, \tag{36}$$

which has the same functional form as the estimate (21) from microstructure data and is a factor of 3 larger (if $S^2 \sim E$, which follows if one assumes $\beta_c$ = constant in the Garrett and Munk model [Gargett, 1990]). Energy flux calculations thus seem to be consistent with dissipation measurements.

A similar result was obtained by Henyey et al. (1986), who integrated the eikonal equations for small-scale waves propagating in a background of large-scale waves. In the geometric optics or WKB approximation, the frequency of a small-scale wave of wave number $\mathbf{k}$ propagating in a large-scale background flow $\bar{\mathbf{u}}$ is given by

$$\omega = \omega_o(\mathbf{k}) + \mathbf{k} \cdot \bar{\mathbf{u}}, \tag{37}$$

where $\omega_o(\mathbf{k})$ is the intrinsic frequency and $\mathbf{k} \cdot \bar{\mathbf{u}}$ is the Doppler shift. The eikonal equations then state that the vertical wave number of the small-scale wave changes according to

$$\dot{k}_z = \frac{d}{dt} k_z = -\frac{\partial \omega}{\partial z} = -\frac{\partial \bar{\mathbf{u}}}{\partial z} \cdot \mathbf{k}. \tag{38}$$

The energy flux past a certain wave number is given by

$$F = \dot{k}_z E(k). \tag{39}$$

If it is assumed that waves whose vertical wave number exceeds $\beta_c = 2\pi/5$ m break and are annihilated, then the flux past $\beta_c$ is made up only of waves whose vertical wave number increases. This flux was evaluated by Henyey et al. (1986) using a Monte Carlo simulation of wave trajectories in a current field $\bar{\mathbf{u}}$ selected randomly from an ensemble with a Garrett and Munk spectrum. They found

$$F = 6.4 \cdot 10^{-11} \frac{N^2}{N_o^2} \cosh^{-1} (N/f) \, \mathrm{W \, kg^{-1}} \tag{40}$$

from the simulations and a heuristic model. Their result has the same functional dependence, except for the $\cosh^{-1}$ factor, as McComas and Müller's result and is a factor of 6 smaller. A more important difference is, perhaps, that Henyey et al. predict a flux to higher frequencies, whereas McComas and Müller calculate a flux to lower frequencies (see Fig. 7). Note that both the resonant interaction and eikonal theory are used to calculate the flux of total, i.e., kinetic plus potential, energy to high wave numbers. Separate calculations of the transfers of kinetic and potential energy and the conversion from kinetic to potential energy have not yet been attempted.

These two dynamical studies assume that wave breaking is due to chance superposition within a random internal wave field. Wave breaking might also occur as the result of internal waves approaching a critical layer in an ambient geostrophic shear. Vertical critical layers are produced when changes in the geostrophic flow with depth force the intrinsic frequency toward the lower bound of the internal wave band, or when this lower bound is increased with depth. Horizontal critical layers occur either when the intrinsic frequency is Doppler shifted toward the buoyancy frequency $N$ by lateral changes in the geostrophic flow or when $N$ changes laterally. The specific case of vertical critical layers due to geostrophic shear was considered by Kunze and Müller (1989).

The intrinsic frequency following the mean flow is $\omega_o = \omega - \mathbf{k} \cdot \bar{\mathbf{u}}$, where the Eulerian frequency $\omega$ is invariant in a time-independent mean flow. As a wave propagates down from the surface, the change in geostrophic velocity $\Delta \bar{u}$ will result in a corresponding change of $k_z \Delta \bar{u}$ to its intrinsic frequency. A critical layer is reached at a depth where the intrinsic

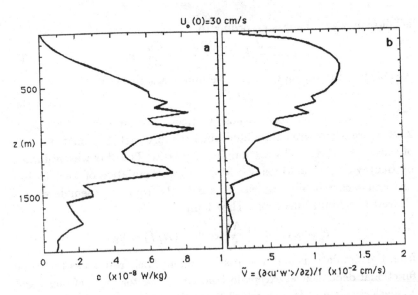

$U_o(0) = 30$ cm/s

$c$ $(\times 10^{-8}$ W/kg)

$\tilde{V} = (\partial\langle u'w'\rangle/\partial z)/f$ $(\times 10^{-2}$ cm/s)

Fig. 8   (a) Energy-flux divergence and (b) momentum-flux divergence due to critical layer absorption of internal waves in an ambient geostrophic shear with a 30-cm s$^{-1}$ surface velocity. The energy-flux divergence has maximum values of $\sim 5 \times 10^{-9}$ W kg$^{-1}$, which is comparable to the estimate of the energy flux through the internal wave field to high wave numbers. The momentum-flux divergence drives an "Ekman" mean flow $\tilde{v} = (1/f)\,\partial\langle u'w'\rangle/\partial z$ of up to 0.01 cm s$^{-1}$ (10 m d$^{-1}$), which is insignificant. From Kunze and Müller (1989).

frequency becomes $f = \omega - k_x \Delta\bar{u}$. As the wave approaches its critical layer, it slows down, steepens, breaks, and may convert part of its energy to mixing.

A quantitative estimate of the energy available for mixing was obtained by Kunze and Müller (1989). Following the work of Ruddick (1980) they considered an internal wave field of Garrett and Munk spectral intensity that propagates downward from the surface into a geostrophic flow that is a first baroclinic mode with a surface velocity of 30 cm s$^{-1}$. As the spectrum propagates downward those waves that encounter a critical level are eliminated. The resulting energy flux divergence is shown in Fig. 8 as a function of depth. A maximum value of about $5 \cdot 10^{-9}$ W kg$^{-1}$ is reached around a depth of 700 m, comparable to the estimate (36) of the energy flux to high wave numbers by wave–wave interactions.

## 21.9 Wave Breaking

At the heart of our understanding of diapycnal mixing is the concept that mixing is caused by the turbulence resulting from wave breaking either by random chance or in critical layers. Quantitative estimates assume that wave–wave interactions (weak or strong) cascade kinetic energy to small scales; a small fraction $R_f$ of this energy is converted to potential energy in mixing events and then dissipated. The other part $(1 - R_f)$ is dissipated directly. Much of our intuition about the wave breaking process itself has come from the analysis of simple situations where single waves break by shear or gravitational instability. What happens in a more complex environment consisting of many waves is by no means clear. At one extreme are the phenomenological theories of buoyant turbulence. The classical concepts based on the dominance of the buoyancy term in the turbulence kinetic energy equation lead to a kinetic energy spectrum (Lumley, 1964)

$$E(k) = A\epsilon_o^{2/3} [(1 + k_b/k)^{4/3}]k^{-5/3}, \tag{41}$$

where $A$ is the empirical Kolmogorov constant, $\epsilon_o$ is the kinetic energy dissipation rate, and $k_b = (N^3/\epsilon_o)^{1/2}$ is the buoyancy wave number. Weinstock (1985) calculates the same form for the temperature variance spectrum. The predicted spectral forms are roughly consistent with observations of shear spectra (Gargett et al., 1981) and of temperature gradient spectra (Gregg, 1977). However, the buoyancy flux co-spectrum calculated by Lumley (1964),

$$B(k) \sim N^2 \epsilon_o^{1/3} [1 + (k_b/k)^{4/3}]^{1/2} k^{-7/3}, \tag{42}$$

decreases with increasing wave numbers. Also, Weinstock's analysis implies a transfer of temperature variance toward lower wave numbers for $k < k_b$. These features are certainly contrary to the conventional wisdom of wave breaking.

Starting with the wave–wave interaction theory, Holloway (1983) also derives the expression (41) for the temperature and velocity variance spectra, but under the assumptions that there is little buoyancy flux and that nonlinearities cascade kinetic and potential energy independently to high wave numbers. As pointed out by Holloway (1989, 1990), there exist dynamical theories that are consistent with observed velocity and temperature spectra but that allow for quite different pathways of energy.

Direct numerical simulations of freely decaying buoyant turbulence have been carried out in three dimensions by Riley et al. (1981) and

Métais and Herring (1989). Siegel (1991) performed an LES of decaying buoyant turbulence at Reynolds numbers characteristic of the ocean in a box of about $(10\,\mathrm{m})^3$.

Forced-dissipative experiments have been carried out in two dimensions by Shen and Holloway (1986) with the following surprising energy balance. If external forcing of kinetic energy (and in some runs, potential energy) is applied at low wave numbers, then kinetic energy is converted to potential energy at low wave numbers. The potential energy then cascades to high wave numbers, where some of it is converted back to kinetic energy, which is then scattered to low and high wave numbers (Fig. 9). The buoyancy flux at the actual overturning events is upward; gravitational energy is released! It seems that this result is not an artifact of the two-dimensional geometry (Holloway and Ramsden, 1988; Ramsden and Holloway, 1991). Note that the scales of the buoyancy flux are crucial to any attempts to measure the buoyancy flux.

Inhomogeneous, remotely forced calculations have been carried out in two dimensions by Winters and D'Asaro (1989) and Ramsden and Holloway (1991) and in three dimensions by Winters and D'Asaro (1991). Winters and D'Asaro simulate the breakdown of an incoming finite-amplitude internal wave in a critical layer. In the two-dimensional simulation the breakdown is shear-driven, although unstable density gradients persist for many buoyancy periods. As shown by Winters and Riley (1992), modes driven by shear instability are oriented in the vertical plane of the background shear, whereas convectively unstable modes are oriented in the transverse plane and are hence eliminated from two-dimensional calculations. Ramsden and Holloway consider remotely forced turbulent flows in an environment with mean stratification but without mean shear. Ramsden and Holloway as well as Winters and D'Asaro find that the spatial separation of sources and sinks affects the dynamic balances. Furthermore, the buoyancy flux does not uniquely define mixing. A careful analysis of changes in available potential and base state potential energy is required to separate reversible wave effects from irreversible mixing effects (Winters and D'Asaro, 1991).

All these results certainly demonstrate that our basic notion about the buoyancy flux in wave breaking events is not as solidly founded as we sometimes presume.

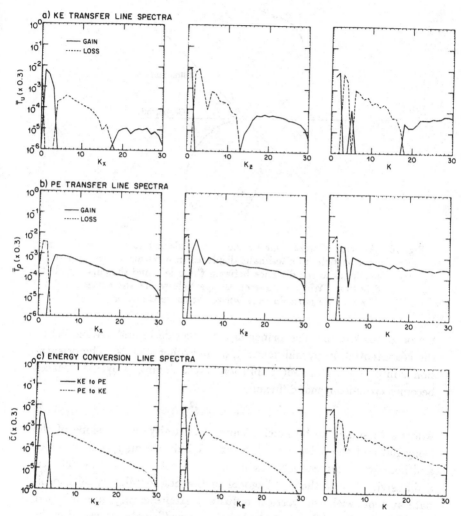

Fig. 9  Time-averaged transfer of kinetic energy and potential energy and conversion rate from kinetic to potential energy as a function of $k_x$, $k_z$, and total $k$ (from Shen and Holloway, 1986). Note that potential energy is converted to kinetic energy at the higher (overturning) wave numbers.

## 21.10 Horizontal Versus Isopycnal Mixing

A second major issue to be discussed here is that of horizontal versus isopycnal mixing. So far we have not stressed this issue and have used the terms "vertical" and "diapycnal" as synonyms. Diffusion in physical space requires specification of a diffusion tensor. The orientation of this

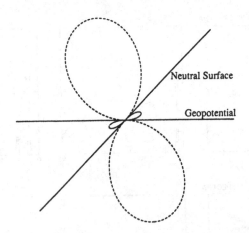

Fig. 10    Amount of work done by the Archimedean force. The work done
on a particle infinitesimally displaced in a certain direction is proportional to the distance between the origin and the curve in that
direction. Work is done on the particle where the curve is solid.
The particle must do work where the curve is dashed.

tensor is not known. The principal axes may lie in and orthogonal to
the geopotential, isopycnal, neutral, or any other surface. The orientation is important. Horizontal diffusion across sloping isopycnal surfaces
becomes cross-isopycnal diffusion,

$$K_d = K_v + K_h s^2, \qquad (43)$$

where $s$ (assumed to be much smaller than unity) is the slope of the
isopycnal and the subscripts $d$, $v$, and $h$ indicate diapycnal, vertical,
and horizontal, respectively (Redi, 1982). Most numerical models use
a diffusivity tensor that is diagonal in the horizontal/vertical coordinate system, with the vertical diffusivity (representing internal wave
breaking) much smaller than the horizontal coefficient (representing the
stirring by mesoscale eddies). Since mesoscale eddies mix or stir along
isopycnal surfaces, some researchers argue that the mixing tensor should
be diagonal in the isopycnal/diapycnal system.

Additional support for this representation seems to come from the
argument that the exchange of water particles on isopycnals does not
require any work against gravity. However, neither of these arguments is
fully convincing. Mesoscale eddies can mix properties across *mean* isopycnals as resolved in general circulation models (discussion below), and
exchange of water parcels on horizontal, i.e., geopotential, surfaces also
requires no work. Furthermore, potential energy is released when parcels

are exchanged within the wedge between the horizontal and isopycnal surfaces, the wedge of baroclinic instability. Indeed, the Archimedean work that is done on a parcel displaced by an infinitesimal amount is given by (Olbers and Wenzel, 1990)

$$dA = \rho[(\alpha \nabla \Theta - \beta \nabla S) \cdot d\mathbf{x}][\nabla \phi \cdot d\mathbf{x}], \qquad (44)$$

where $\phi$ is the geopotential, $\Theta$ is the potential temperature, $S$ is the salinity, $\alpha$ is the thermal expansion coefficient, and $\beta$ is the haline contraction coefficient. This form led Olbers and Wenzel to suggest a diffusion tensor (Fig. 10) with three principal components: a large value along the axis halfway between the isopycnal and geopotential surfaces where mixing would release the maximum amount of energy, a medium value along the intersection of the isopycnal and geopotential surfaces where mixing results in no release of potential energy, and a small value along the axis perpendicular to these two where mixing requires the maximum amount of work.

## 21.11 Baroclinic Instability

Another part of the conventional wisdom is that small-scale turbulence does the diapycnal mixing and that mesoscale eddies and internal waves cannot support any diapycnal mixing. Ultimately, all diapycnal mixing is done by molecular diffusion. Only molecular diffusion can transport buoyancy across instantaneous, actual density surfaces; mesoscale eddies, internal waves, and turbulence cannot. However, all these motions can transport tracers across mean isopycnals and it is the flux across such mean isopycnals that is relevant to general circulation models. An example is the diapycnal flux across the mean position of the Gulf Stream (Fig. 11), which is clearly supported by rings and meanders. The large diapycnal diffusivities seen in Olbers and Wenzel's (1990) inverse calculation might be another example. Water parcels carried across mean isopycnals by mesoscale eddy motions might become modified by mixing or air–sea interaction such that upon their return crossing they cause a net flux of buoyancy across the mean isopycnal. The question is, what controls the flux or rate of diffusion? It may well be (Garrett, 1989) that this flux is determined by the mesoscale eddies and that the smaller-scale motions just do what needs to be done, similar to the classical example of three-dimensional turbulence in a homogeneous, nonrotating fluid where energy is cascaded down to the dissipation scales without any back-effect on the larger scales.

478     *Peter Müller*

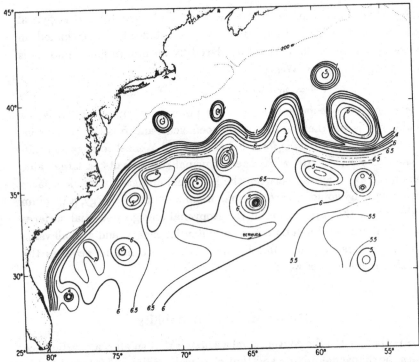

Fig. 11  Chart of the topography (hectometers) of the 15° isothermal surface showing the Gulf Stream, nine cold-core and three warm-core
rings. Contours are based on XBT, CTD, hydrographic, and satellite infrared data from the period March 16 to July 9, 1975. From
Richardson et al. (1978).

If this is indeed true, parameterization schemes could be based on
the theory of baroclinic instability. Take, e.g., Eady's (1949) classical
problem of baroclinic instability in a uniform vertical shear $\partial \bar{u}/\partial z$ over
depth $H$. The stream function of the fastest growing wave is given by

$$\varphi' = \phi(z) \cos(kx + \alpha(z)) \exp(\omega_i t), \qquad (45)$$

where the depth dependence of the amplitude $\phi(z)$ and the phase $\alpha(z)$
are sketched in Fig. 12. Since the transverse velocity is $v' \sim \partial_x \varphi'$ and
the temperature is $T' \sim \partial_z \varphi'$, the transverse heat flux becomes

$$\overline{v'T'} = \tfrac{1}{2}\phi^2 \, k \, \partial\alpha/\partial z \, \exp(2\omega_i t)$$

$$= O\left(H^2 \frac{N}{f} \frac{\partial \bar{u}}{\partial z} \frac{\partial \bar{T}}{\partial y}\right) \qquad (46)$$

if the transverse velocity is assumed to be limited by $v' = O(\bar{u})$ (Stone,
1974; Bryden, 1979). This transverse heat flux leads to a flux across the

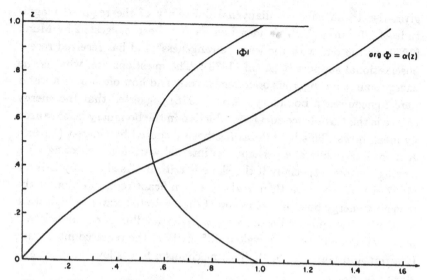

Fig. 12  The amplitude $|\phi(z)|$ and the phase $\alpha(z)$ as a function of normalized height $z$ for the most unstable wave in the Eady (1949) model of baroclinic instability. Note that the increase of $\alpha$ with height implies a transverse heat flux. From Pedlosky (1979).

sloping isopycnal surface with a diapycnal diffusivity (Garrett, 1989)

$$K_d \sim H^2 f Ri^{-3/2}. \tag{47}$$

To understand what determines diapycnal fluxes across mean isopycnals, eddy-resolving models need to be run with different "internal-wave" diffusivities. One could then see to what extent the fluxes across mean isopycnals depend on these "internal-wave" diffusivities or whether parameterizations like (47) are appropriate.

## 21.12 Boundary Mixing

The final point is the idea of boundary mixing. Diapycnal mixing coefficients in the main pycnocline have been inferred from dissipation measurements and calculations of the internal-wave energy flux to high wave numbers. All these estimates are of the order of $5 \cdot 10^{-6}$ to $5 \cdot 10^{-5}$ m$^2$ s$^{-1}$, an order of magnitude smaller than Munk's (1966) "abyssal recipes" value and the values required to satisfy abyssal mass and heat balances by a vertical advective–diffusive balance. One possible resolution of this discrepancy is that vigorous diapycnal mixing at the side wall boundaries of the ocean, in combination with along–isopycnal mixing or stirring, gives rise to an effective diapycnal diffusivity of the required magnitude of $10^{-4}\,\mathrm{m^2\,s^{-1}}$. This idea had already been suggested by Munk (1966) as the one with the least "strangeness" and has received recent observational support by Armi (1978). The questions are, what is the energy source for vigorous boundary mixing and how often does a water parcel encounter a boundary? Armi (1978) suggested that the energy source is the turbulence energy production in the boundary layer caused by mean flows. This hypothesis has been disputed by Garrett (1979a). A more likely source is perhaps the internal-wave field reflecting off a sloping bottom, especially if the slope is critical. As shown by Eriksen (1982, 1985) reflection then leads to a significant redistribution of the incoming energy flux. If it is assumed that reflected waves of high wave number break and produce mixing, a buoyancy flux is calculated that may sustain an abyssal equivalent diffusivity of the required magnitude (Garrett and Gilbert, 1988). This mechanism favors low latitudes and steep slopes.

Similarly, Müller and Xu (1992) show that the scattering of internal waves at random bottom topography leads to a transfer of the incoming flux to higher wave numbers. Although redistribution is smaller than for the reflection problem it is directed toward much higher wave numbers, so that both processes might be of equal efficiency (Fig. 13).

Though interior mixing and boundary mixing may have similar effects on the distribution of passive tracers, the dynamics of the flow are profoundly altered. If all the mixing were happening near the boundaries, then all the bottom water would be upwelled in boundary layers. The interior of the ocean would be quiescent except for the need to transfer fluid along isopycnals from one boundary layer to another (McDougall, 1989).

## 21.13 Summary and Conclusions

Diapycnal mixing coefficients are used in OGCMs to parameterize the effect of the subgrid-scale, cross-isopycnal buoyancy fluxes. Important features of the simulated circulation are sensitive to the value and functional dependence of the mixing coefficients. We have briefly reviewed the conventional wisdom that diapycnal mixing is induced by breaking internal gravity waves. All the inferences about the mixing processes and mixing rates are indirect. Diapycnal buoyancy fluxes have not yet been measured in the ocean interior. Major issues are unresolved. The

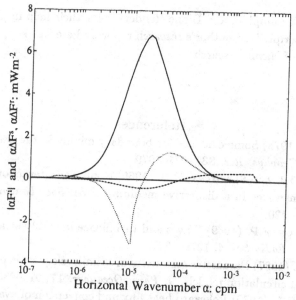

Fig. 13 Comparison of scattering and reflection for a typical oceanic wave spectrum, bottom spectrum, and bottom slope. The incident energy flux (solid line), the energy flux redistributed by scattering (dashed line), and the energy flux redistributed by reflection (dotted line) are shown as a function of horizontal wave number in a variance preserving representation. From Müller and Xu (1992).

basic physical problem is that we do not know what constitutes a typical breaking event and what the energy conversions are in such a typical event. The other issues are ones of relevance. Is internal-wave-induced mixing relevant to coarse-resolution oceanic general circulation models or must the mixing in these models be ascribed to other processes, such as baroclinic instability? Is mixing in the ocean interior of importance or is most of the mixing done near the boundaries of the ocean?

Our knowledge of ocean mixing is not as solid as often presumed; thus considerable research still needs to be done. The many surprises ahead make this a truly exciting field.

## Acknowledgments

I am grateful to all my colleagues on whose research efforts I have drawn in this review and who have generously shared their ideas about mixing with me. It is a pleasure to thank Crystal Miles, Naomi Yanag-

ishita, Rick Lumpkin, and Diane Henderson for their help in preparing
the manuscript. The author's research reported here was supported by
the Office of Naval Research.

# References

ARMI, L. (1978) Some evidence for boundary mixing in the deep ocean.
*J. Geophys. Res.* **83**, 1971–1979.

BENNEY, D.J. AND SAFFMANN, P.G. (1966) Nonlinear interactions of random
waves in a dispersive medium. *Proc. Soc. London A* **289**,
301–320.

BRETHERTON, F.P. (1969) Waves and turbulence in stably stratified fluids.
*Radio Sci.* **4**, 1279–1287.

BRYAN, F. (1987) Parameter sensitivity of primitive equation ocean general
circulation models. *J. Phys. Oceanogr.* **17**, 970–985.

BRYDEN, H.L. (1979) Poleward heat flux and conversion of available potential
energy in Drake Passage. *J. Mar. Res.* **37**, 1–22.

CAIRNS, J.L. AND WILLIAMS, G.O. (1976) Internal wave observations
from a midwater float, Part II. *J. Geophys. Res.* **81**, 1943–1950.

CUMMINS, P.F., HOLLOWAY, G. AND GARGETT, A.E. (1990) Sensitivity of
the GFDL ocean general circulation model to a parameterization
of vertical diffusion. *J. Phys. Oceanogr.* **20**, 817–830.

DESAUBIES, Y.J.F. AND SMITH, W.K. (1982) Statistics of Richardson
number and instability in oceanic internal waves. *J. Phys.
Oceanogr.* **12**, 1245–1259.

EADY, E.T. (1949) Long waves and cyclone waves. *Tellus* **1** (3), 33–52.

ERIKSEN, C.C. (1982) Observations of internal wave reflection off sloping
bottoms. *J. Geophys. Res.* **87**, 525–538.

ERIKSEN, C.C. (1985) Implications of ocean bottom reflection for internal
wave spectra and mixing. *J. Phys. Oceanogr.* **15**, 1145–1156.

EVANS, D.L. (1982) Observations of small-scale shear and density structure
in the ocean. *Deep-Sea Res.* **29**, 581–595.

GARGETT, A.E. (1984) Vertical eddy diffusivity in the ocean interior.
*J. Mar. Res.* **42**, 359–393.

GARGETT, A.E. (1990) Do we really know how to scale the turbulent
kinetic energy dissipation rate $\epsilon$ due to breaking of oceanic internal
waves? *J. Geophys. Res.* **95**, 15971–15974.

GARGETT, A.E., HENDRICKS, P.J., SANFORD, T.B., OSBORN, T.R. AND

WILLIAMS, A.J. III (1981) A composite spectrum of vertical shear in the upper ocean. *J. Phys. Oceanogr.* **11**, 1258–1271.

GARRETT, C.J.R. (1979a) Comment on "Some evidence for boundary mixing in the deep ocean" by Lawrence Armi. *J. Geophys. Res.* **84**, 5095.

GARRETT, C.J.R. (1979b) Mixing in the ocean interior. *Dynam. Atmos. Oceans* **3**, 239–265.

GARRETT, C. (1989) Are diapycnal fluxes linked to lateral stirring rates? In *Parameterization of Small-Scale Processes*. Proceedings, 'Aha Huliko'a Hawaiian Winter Workshop. Ed. P. Müller and D. Henderson, pp. 317–327. Hawaii Institute of Geophysics, University of Hawaii.

GARRETT, C.J.R. AND MUNK, W.H. (1972) Oceanic mixing by breaking internal waves. *Deep-Sea Res.* **19**, 823–832.

GARRETT, C. AND GILBERT, D. (1988) Estimates of vertical mixing by internal waves reflected off a sloping bottom. In *Small-Scale Turbulence and Mixing in the Ocean*. Ed. J. C. J. Nihoul and B. M. Jamart, pp. 405–423. Elsevier.

GORDON, A.L., MOLINELLI, E.J. AND BAKER, T.N. (1982) *Southern Ocean Atlas*. Columbia University Press, 250 pp.

GREGG, M.C. (1977) A comparison of finestructure spectra from the main thermocline. *J. Phys. Oceanogr.* **7**, 33–40.

GREGG, M.C. (1987) Diapycnal mixing in the thermocline: A review. *J. Geophys. Res.* **92**, 5249–5286.

GREGG, M.C. (1989) Scaling turbulent dissipation in the thermocline. *J. Geophys. Res.* **94**, 9686–9698.

HASSELMANN, K.F. (1966) Feynman diagrams and interaction rules of wave–wave scattering processes. *Rev. Geophys. Space Phys.* **4**, 1–32.

HASSELMANN, K.F. (1967) Nonlinear interactions treated by methods of theoretical physics (with applications to the generation of waves by wind). *Proc. Soc. London* **299**, 77–100.

HENYEY, F.S., WRIGHT, J. AND FLATTÉ, S.M. (1986) Energy and action flow through the internal wave field: An eikonal approach. *J. Geophys. Res.* **91**, 8487–8495.

HOGG, N.G., BISCAYE, P., GARDNER, W. AND SCHMITZ, W.J., JR. (1982) On the transport and modification of Antarctic bottom water in the Velma Channel. *J. Mar. Res.* **40**, 231–263.

HOLLOWAY, G. (1983) A conjecture relating oceanic internal waves and small-scale processes. *Atmosphere; Atmos.-Ocean* **21**, 107–122.

HOLLOWAY, G. (1989) Relating turbulence dissipation measurements to ocean mixing. In *Parameterization of Small-Scale Processes.* Proceedings, 'Aha Huliko'a Hawaiian Winter Workshop. Ed. P. Müller and D. Henderson, pp. 329–339. Hawaii Institute of Geophysics, University of Hawaii.

HOLLOWAY, G. (1990) Subgrid-scale representation. In: *Oceanic Circulation Models: Combining Data and Dynamics.* Ed. D.L.T. Anderson and J. Willibrand, pp. 513–583. Kluwer.

HOLLOWAY, G. AND RAMSDEN, D. (1988) Theories of internal wave interaction and stably stratified turbulence: Testing against direct numerical experimentation. In *Small-Scale Turbulence and Mixing in the Ocean.* Proceedings of the 19th International Liège Colloquium on Ocean Hydrodynamics. Ed. J.C.J. Nihoul and B.M. Jamart, pp. 363–377. Elsevier Oceanography Series, Vol. 46.

KUNZE, E. AND MÜLLER, P. (1989) The effect of internal waves on vertical geostrophic shear. In *Parameterization of Small-Scale Processes.* Proceedings, 'Aha Huliko'a Hawaiian Winter Workshop. Ed. by P. Müller and D. Henderson, pp. 271–285. Hawaii Institute of Geophysics, University of Hawaii.

LEDWELL, J.R. (1989) A strategy for open ocean mixing experiments. In *Parameterization of Small-Scale Processes.* Proceedings, 'Aha Huliko'a Hawaiian Winter Workshop. Ed. P. Müller and D. Henderson, pp. 157–163. Hawaii Institute of Geophysics, University of Hawaii.

LEDWELL, J.R., WATSON, A.J. AND BROECKER, W.S. (1986) A deliberate tracer experiment in Santa Monica Basin. *Nature* **323**, 322–342.

LEVITUS, S. (1982) *Climatological Atlas of the World Ocean.* NOAA Res. Rep. No. 13, 173 pp.

LILLY, D.K., WACO, D.E. AND ADELFANG, S.I. (1974) Stratospheric mixing estimated from high-altitude turbulence measurements. *J. Appl. Meteorol.* **13**, 488–493.

LUMLEY, J.L. (1964) The spectrum of nearly inertial turbulence in a stably stratified fluid. *J. Atmos. Sci.* **21**, 99–102.

MCCOMAS, C. H. AND MÜLLER, P. (1981) The dynamic balance of internal waves. *J. Phys. Oceanogr.* **11**, 970–986.

MCDOUGALL, T.J. (1989) Dianeutral Advection. In *Parameterization of Small-Scale Processes.* Proceedings, 'Aha Huliko'a Hawaiian Winter Workshop. Ed. P. Müller and D. Henderson, pp. 289–315. Hawaii Institute of Geophysics, University of Hawaii.

MÉTAIS, O. AND HERRING, J.R. (1989) Numerical simulations of freely evolving turbulence in stably stratified fluids. *J. Fluid Mech.* **202**, 117–148.

MÜLLER, P. AND HENDERSON, D. (EDS.) (1989) *Parameterization of Small-Scale Processes.* Proceedings, 'Aha Huliko'a Hawaiian Winter Workshop. Hawaii Institute of Geophysics, Special Publication, University of Hawaii, 365 pp.

MÜLLER, P. AND HOLLOWAY, G (1989) Parameterization of small-scale processes. *Trans. Amer. Geophys. Union* **70** (36), 818–820, 830.

MÜLLER, P., HOLLOWAY, G., HENYEY, F. AND POMPHREY, N. (1986) Nonlinear interactions among internal gravity waves. *Rev. Geophys.* **24**, 493–536.

MÜLLER, P. AND OLBERS, D.J. (1975) On the dynamics of internal waves in the deep ocean. *J. Geophys. Res.* **80**, 3848–3860.

MÜLLER, P. AND XU, N. (1992) Scattering of oceanic internal gravity waves off random bottom topography. *J. Phys. Oceanogr.* **22**, 474–488.

MUNK, W.H. (1966) Abyssal Recipes. *Deep-Sea Res.* **13**, 707–730.

MUNK, W.H. (1981) Internal waves and small-scale processes. In *Evolution of Physical Oceanography.* Ed. B. A. Warren and C. Wunsch, pp. 264–291. MIT Press.

MUNK, W.H. AND ANDERSON, E.R. (1948) Notes on a theory of the thermocline. *J. Mar. Res.* **7**, 276–295.

OAKEY, N.S. (1982) Determination of the rate of dissipation of turbulent energy from simultaneous temperature and velocity shear microstructure measurements. *J. Phys. Oceanogr.* **12**, 256–271.

OLBERS, D.J. (1976) Nonlinear energy transfer and the energy balance of the internal wave field in the deep ocean. *J. Fluid Mech.* **74**, 375–399.

OLBERS, D.J., WENZEL, M. AND WILLEBRAND, J. (1985) The inference of North Atlantic circulation parameters from climatological hydrographic data. *Rev. Geophys.* **23**, 313–356.

OLBERS, D.J. AND WENZEL, M. (1990) Determining diffusivities from hydrodynamic data by inverse methods with applications to the Circumpolar Current. In *Oceanic Circulation Models: Combining Data and Dynamics.* Ed. D.L.T. Anderson and J. Willibrand, pp. 95–139. Kluwer.

OSBORN, T.R. (1980) Estimates of the local rate of vertical diffusion from dissipation measurements. *J. Phys. Oceanogr.* **10**, 83–89.

OSBORN, T.R. AND COX, C.S. (1972) Oceanic fine structure. *Geophys. Fluid Dynam.* **3**, 321–345.

PACANOWSKI, R.C. AND PHILANDER, S.G.H. (1981) Parameterization of vertical mixing in numerical models of tropical oceans. *J. Phys. Oceanogr.* 11, 1443–1451.

PEDLOSKY, J. (1979) *Geophysical Fluid Dynamics*, 2nd ed. Springer-Verlag, 710 pp.

RAMSDEN, D. AND HOLLOWAY, G. (1991) Energy transfer across an internal wave/vortical mode spectrum. In *The Dynamics of Oceanic Internal Gravity Waves*. Proceedings, 'Aha Huliko'a Hawaiian Winter Workshop. Ed. P. Müller and D. Henderson, pp. 295–314. School of Ocean and Earth Science and Technology, University of Hawaii.

REDI, M.H. (1982) Oceanic isopycnal mixing by coordinate rotation. *J. Phys. Oceanogr.* 12, 1154–1158.

RICHARDSON, P.L. CHENEY, R.E. AND WORTHINGTON, L.V. (1978) A census of Gulf Stream rings, Spring 1975. *J. Geophys. Res.* 83 (C12), 6136–6143.

RILEY, J.J., METCALFE, R.W. AND WEISSMAN M.A. (1981) Direct numerical simulations of homogeneous turbulence in density-stratified fluids. In *Nonlinear Properties of Internal Waves*, Vol. 76. Ed. B.J. West, pp. 79–112. American Institute of Physics.

RUDDICK, B. (1980) Critical layers and the Garrett–Munk spectrum. *J. Mar. Res.* 38, 135–145.

SAUNDERS, P.M. (1987) Flow through discovery gap. *J. Phys. Oceanogr.* 17, 631–643.

SHEN, C.Y. AND HOLLOWAY, G. (1986) A numerical study of the frequency and the energetics of nonlinear internal gravity waves. *J. Geophys. Res.* 91 (C1), 953–973.

SIEGEL, D.A. (1991) Large-eddy simulation of internal wave motions. In *The Dynamics of Oceanic Internal Gravity Waves*. Proceedings, 'Aha Huliko'a Hawaiian Winter Workshop. Ed. P. Müller and D. Henderson, pp. 315–340. School of Ocean and Earth Science and Technology, University of Hawaii.

STOMMEL, H. AND SCHOTT, F. (1977) The beta spiral and the determination of the absolute velocity field from hydrographic station data. *Deep-Sea Res.* 24, 325–329.

STONE, P.H. (1974) The meridional variation of the eddy heat fluxes and their parameterization. *J. Atmos. Sci.* 31, 444–456.

THORPE, S.A. (1973) Turbulence in stably stratified fluids: A review of laboratory experiments. *Bound. Layer Meteorol.* 5, 95–119.

WEINSTOCK, J. (1985) On the theory of temperature spectra in a stably stratified fluid. *J. Phys. Oceanogr.* **15**, 475–477.

WINTERS, K.B. AND D'ASARO, E.A. (1989) Two-dimensional instability of finite amplitude internal gravity wave packets near a critical level. *J. Geophys. Res.* **94**, 12709–12719.

WINTERS, K. AND D'ASARO, E. (1991) Diagnosing diapycnal mixing. In *The Dynamics of Oceanic Internal Gravity Waves*. Proceedings, 'Aha Huliko'a Hawaiian Winter Workshop. Ed. P. Müller and D. Henderson, pp.279–294. School of Ocean and Earth Science and Technology, University of Hawaii.

WINTERS, K.B. AND RILEY, J.J. (1991) Instability of internal waves near a critical level. *Dynam. Atmos. Oceans* **16**, 249–278.

# 22

---

# Near-Surface Mixing
# and the Ocean's Role in Climate

## MARK A. CANE

## 22.1 Introduction

Of the many aspects of the ocean that inspire our interest, we are here concerned with the ocean's role in the climate system. From this viewpoint the most important consideration is the ocean's influence on the atmosphere. The primary direct influence is sea surface temperature (SST), which exerts a strong control on the surface heat exchange and hence on atmospheric circulation features. A secondary influence is gas exchange, which also is strongly temperature dependent. Hence in modeling the ocean for climate, the central task is to simulate global SST. It will be our exclusive concern here: while other aspects must enter in the ultimate ocean component of a climate model, a model which cannot simulate SST is surely inadequate.

The ocean is almost everywhere topped by a well-mixed surface layer. Figure 1 illustrates the near uniformity of density, temperature and salinity within this layer. This ocean mixed layer (OML) is the oceanic counterpart to the atmospheric planetary boundary layer (PBL). It can be seen that simulating SST is equivalent to finding the temperature of the OML. (However, there can be a small difference between the OML temperature and the skin temperature, especially in calm conditions.)

For a given heat flux convergence into the OML, the temperature change will be inversely proportional to the layer depth, $h$. This depth, along with the mixing which makes the layer, is determined by near-surface turbulent processes. Not only are these processes poorly understood, they are also very difficult to measure in the field: the open

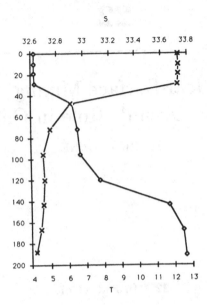

Fig. 1 Temperature $(T)$ and salinity $(S)$ profiles at Ocean Weather Ship P ($145°$W, $50°$N) on October 9, 1977, as a function of depth (m). A well-mixed surface layer is found almost everywhere in the ocean. (Courtesy of D. Archer.)

ocean is not an easy platform to work from. As will be discussed in the concluding section, these difficulties create an opportunity for the application of large eddy simulation (LES) techniques. However, the body of the chapter, which is not about LES, is an essay on the impediments to applying insights from LES to the OML. Figure 2, which is derived from the most comprehensive and sophisticated model yet used to simulate SST, is an illustration of this cautionary tale. The errors are comparable to the signal. The next sections suggest some possible reasons for this, and some of their broader implications. In particular, it will be difficult to establish that complex mixing models perform better than the current simple ones. It may even prove difficult to establish that, in this context, their basis is more physically sound.

## 22.2 Processes Affecting SST

Changes in SST are governed by the temperature equation

$$\frac{\partial T}{\partial t} = -\mathbf{u} \cdot \nabla T - \frac{w_e}{h}(T - T_e) + Q. \tag{1}$$

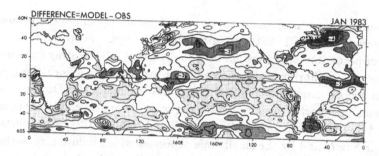

Fig. 2   SST differences between the simulation with the ocean GCM of Rosati and Miyakoda (1988) and observations.  The model uses Mellor–Yamada Level 2.5 vertical mixing and Smagorinsky (1963) non-linear eddy viscosity. Contour interval is 1°C; dark shading in areas with differences $< -1$°C; light shading in areas with differences $> +1$°C. From Rosati and Miyakoda (1988).

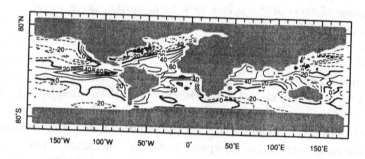

Fig. 3   Annually averaged flux of heat into the ocean $(\mathrm{W}\,m^{-2})$ according to the atlas of Esbensen and Kushnir (1981).

Here **u** is the horizontal velocity in the mixed layer, $w_e$ is the entrainment velocity at the base of the mixed layer $h$ is the mixed layer depth, $T_e$ is the temperature of water entrained into the mixed layer and $Q$ is the surface heat flux, i.e., the heat exchanged with the atmosphere. $Q$ has four components: the solar (shortwave) radiation absorbed by the ocean; the net flux of infrared (longwave) radiation into the ocean; and the turbulent transfers of latent and sensible heat. The entrainment velocity is the rate at which a volume of fluid per unit area crosses into the mixed layer, so the time rate of change of mixed layer depth, $h$, is the difference between this and the vertical (upwelling) velocity.

Figure 3 is a map of annually averaged heat flux at the ocean surface. In only a few regions (e.g., the Gulf Stream, the equator, eastern boundary upwelling zones) are large-scale ocean dynamics the primary determinant of SST. In these regions the ocean dynamics (i.e., horizontal transports; upwelling) are effective in driving the OML away from near thermal equilibrium with the overlying atmosphere. Hence these are the regions marked by relatively large heat exchange, $Q$. On the other hand, over most of the ocean the large-scale dynamical effects are small, and SST is determined locally (cf. Gill and Niiler, 1973) by the surface heat flux $Q$ and by the small scale processes which induce near-surface mixing. In particular, the mixing processes determine $h$.

## 22.3 The Surface Heat Flux

The surface heat flux is not well known and standard atlases differ enormously. Figure 4 is an example. Note that a (not uncommon) difference of $20\,\mathrm{W}\,m^{-2}$ applied to a 50-m surface layer would give a temperature difference of almost 3°C in a year. For most of the world ocean this is a substantial part of the annual variation. Weare et al. (1981) estimated mean errors of $30\,\mathrm{W}\,m^{-2}$ (4°C year$^{-1}$). Clearly such large uncertainties render these fields useless as forcing for ocean models.

It should be emphasized that neither of the atlases used for Fig. 4 – or any of the many other possible choices not shown here – are wrong in any obvious way. The formulas used to compute $Q$ are highly uncertain and the data they require are sparse and often inaccurate. These issues are discussed broadly by Seager et al. (1988). Blanc (1987) provides a trenchant account of the problems with the parameterizations of latent and sensible heat transfer. He concludes that the paltry experimental basis for these formulas leaves room for systematic errors as large as 40%. Added to the common problems in relating such boundary layer turbulent transfers to a few bulk variables is the variability of the underlying surface. The surface roughness length, for example, depends on the surface wave field, which in turn depends on the near-surface wind profile.

The radiative components also present severe problems. It may be that the problem of radiative transfer through the clear atmosphere is solved, but the radiative effects of clouds are highly uncertain. For example, Cess et al. (1989) show large discrepancies among calculations of cloud effects in the radiative transfer codes in state of the art general circulation models (GCMs) of the atmosphere. These codes work with a representation of the temperature and cloud profiles through the at-

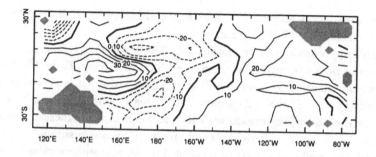

Fig. 4 Annually averaged heat flux W $m^{-2}$: difference between the values of Esbensen and Kushnir (1981) and Weare et al. (1980)

mospheric column. For many applications, such as climatological and decades-long calculations, all that is available are the variables observable from a merchant ship: SST; some approximation of the surface wind, air temperature and humidity; and a visual estimate of cloud cover. Obviously, the bulk formulas for radiative heating devised to use only this limited information are necessarily imprecise. It is not surprising that the many formulas available in the literature disagree considerably (cf. Fung et al. 1984).

In addition to the parameterization issues, the data collected by volunteer observers on merchant ships are not research quality. A wind measurement on a moving ship is clearly problematical, and one need only look upward and attempt to determine the percentage of the sky covered by clouds – let alone their ability to block sunlight or trap heat – to appreciate how uncertain such estimates must be. The problems have been compounded by the way the data are handled by scientific and operational centers; for example, the World Meteorological Organization (WMO) scale routinely used to convert Beaufort (sea state) estimates to wind speed is incorrect (Cardone et al., 1990). And, of course, merchant ships do not cover the globe: there are gaping holes in the tropical and South Pacific and the Southern Ocean. The hope for the future is a combination of satellite measurements and the assimilation of data into numerical weather prediction models (e.g., Reynolds et al., 1989).

Since no available heat flux compilation is sufficiently accurate to yield acceptable answers if used to drive an ocean model, it is preferable to calculate $Q$ on the basis of model SST and climatological data. This approach inherently incorporates the negative feedback between SST and $Q$: excessively high SST induces the ocean to give off more heat by

increasing latent and sensible heat flux and increasing radiative losses.
As a result the SST is reduced.

This approach can be too successful. The surface air temperature $T_a$
over the ocean is rarely far from the SST. Specifying $T_a$ in the bulk
formulas will strongly tend to pull the SST toward it – it is very close to
specifying SST. In the real climate system this air temperature is largely
set by the underlying ocean, with its enormous thermal inertia. Hence
while specifying $T_a$ will yield a good answer in an ocean simulation, the
model may well fail as part of a coupled ocean-atmosphere calculation.
Recognizing this, Seager et al. (1988) devised "bulk formulas" which
use only SST, surface wind speed and cloud cover, in effect modeling $T_a$
and the surface humidity.

In view of the uncertainties in the bulk formulas, Seager et al. modified the coefficients in these formulas within the bounds suggested by the
literature. Blumenthal and Cane (1989) took this procedure further, to
a formal inverse calculation in which the heat flux parameters are determined by the requirement that the ocean model give the best agreement
with observed SST, with the parameters still being constrained to lie
within acceptable limits. In addition to being dependent on the data
and form of the heat flux equations the resultant parameters and heat
fluxes will not be independent of the ocean model used. The model is described in detail by Seager et al. and by Blumenthal and Cane. Briefly,
the model uses filtered equations and linear dynamics, and parameter-
izes the temperature at the base of the mixed layer, $T_e$, as a function
of the thermocline depth. Most important for what follows, the mixed
layer is taken to have a constant depth of 50 m. Grandiosely restated,
the mixing physics in this model are such as always to adjust the mixed
layer depth to 50 m.

Figure 5 shows the results for the difficult transition month of April.
In this calculation six parameters were effectively tuned to yield the best
simulation of SST over the climatological seasonal cycle in the tropical
Pacific. The main effect of the systematic tuning, as compared with the
more subjective approach of Seager et al., was to reduce systematic errors which gave rise to biases in the annual mean. The overall pattern of
the SST simulation compares well with observations, and the computed
heat fluxes look reasonable. The lack of a definitive heat flux atlas makes
it impossible to say more. The SST errors, which are almost everywhere
less than 2°C, are considerably smaller than those of the more physically
and numerically elaborate and, presumably, correct model used to compute Fig. 2. (Seager, 1989, gives results for individual months, which

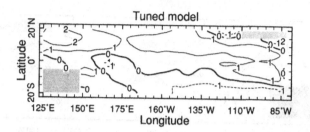

Fig. 5  Data-model difference of Pacific climatological SST for April. The model and heat flux parameterization are those of Blumenthal and Cane (1989).

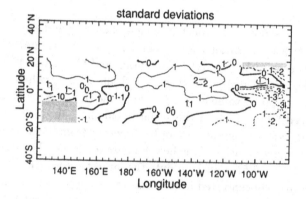

Fig. 6  April Pacific heat flux residuals in units of heat flux data error $(35\text{W }m^{-2})$. Residuals are the extra heat fluxes required to make the model SST agree exactly with observations. From Blumenthal and Cane (1989).

are more directly comparable to the case illustrated in Fig. 2; the errors in his calculation are approximately as in Fig. 5.)

Though the simulated SST is an improvement over the other calculation, the errors are still substantial. Are these errors unequivocally attributable to flaws in the ocean model, with its overly simple treatment of the effects of near-surface mixing physics, or could they still be due to shortcomings of the surface heating? The tuning can only minimize the latter effect, not eliminate it.

Figure 6 is an attempt to answer this question. It shows the additional corrective heat flux which would be needed for the model simulation to match exactly the observed SST. It is expressed in units of standard

496 Mark A. Cane

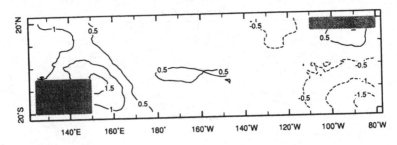

Fig. 7  The difference in simulated SST due to a difference in cloud climatology: clouds from Weare et al. (1980) or Esbensen and Kushnir
(1980). Shown for July, but other months are similar. Calculated
with the model of Blumenthal and Cane (1989).

error, i.e., the error to be expected in view of missing and erroneous data
and the intrinsic limitations of the bulk formulas. This standard error is
estimated to be 35 W $m^{-2}$ (cf. Blumenthal and Cane, 1989). Since the
error in Fig. 6 is usually less than 1, and almost everywhere less than
2, one would appear to be entitled to conclude that the remaining error
could indeed be the fault of the surface heat flux. In other words, the
ocean model's shortcomings cannot be discerned from the data used here
because the uncertainties in the heat flux are too large. The structure of
the error in Fig. 6 may offer an out: the largest errors are not random,
but are clearly concentrated in the cold upwelling tongue at the eastern
end of the equator. However, this unusual region of cold SST is also
marked by atypical cloud regimes, so we should expect the errors in
heat flux to be larger there. Finally, the estimate of standard error is
admittedly uncertain, and may be overstated. Figure 7, which shows
that the use of two different cloud data sets can change the simulated
SST by $O(1°C)$ over substantial areas, suggests that this estimate is
more likely conservative.

It has proved possible to improve on standard formulas for surface
heat flux to some extent by an inverse calculation with an ocean model.
An important by-product of such a calculation is the conclusion that
it is very difficult to test mixing parameterizations against SST data
because of the large uncertainties in the surface heat flux. *A priori*, one
would have expected large errors in the higher latitudes (say, poleward
of 15N,S), where there is a significant seasonal variation in the depth of
the mixed layer. Yet the assumption of a constant mixed layer depth
of 50 m does not cause any evident problems. One must turn to other

variables, such as mixed layer depth, to discriminate among different formulations of mixing physics. Obviously, this new test will show a different mixing formulation to be better than the simple $h = 50$ m, but it is not clear whether it will succeed in showing anything to be better than the next simplest formulation (see below). The desire for physical correctness aside, there are reasons to seek improvement. Though a better parameterization may not give a significantly better simulation of SST, it should yield a better estimate of the heat exchange with the atmosphere.

## 22.4 Models for Near-Surface Mixing

A very long list of mechanisms has been suggested as contributors to near-surface mixing in the ocean, including free convection, surface waves, Langmuir circulations, mean flow instabilities, shear instabilities associated with internal inertia-gravity waves and fish. The usual fluid mechanical investigations of boundary layer behavior, computational or otherwise, do not include most of these effects. The atmospheric PBL is in many respects similar to the OML, but the surface waves are absent and the surface boundary condition is different. While the importance of the uniquely oceanographic features for mixing is uncertain, their mere presence raises questions about the applicability of more standard results from the study of the PBL and other boundary layers.

By now it is well established that much of the mixing in the ocean is highly intermittent in space and time. The passage of a storm greatly enhances wind stirring and generates inertial motions with strong vertical shears (Large et al., 1986). A still less predictable source of mixing events is strong shears resulting from the random superposition of internal waves. Even a statistical description of such mixing events may have to wait for advances in our understanding of internal-wave generation.

A detailed description of the physics of near-surface mixing is essentially complex. Looking at the issue from the perspective of climate, we take the point of view that we need not be concerned with the precise timing of mixing events, only their long-term effects. While all turbulence closures presume it is possible to average over short time or space scales, the time scale (though not the space scale) separation here is especially strong. The mixing times are order of minutes at most, while the climatically important variations are over months or more. This point of view permits the hope that the complexities will average out, and that even if we are unable to capture the mechanisms of turbulent mixing in detail, we will be able to account for their adjustment of the climatic time scale circulation.

In oceanography, the most common instance of treating turbulence as an adjustment process is "convective adjustment," The hydrostatic approximation used in large-scale ocean models eliminates the possibility of explicit simulation of free convection. If a static instability is found to be present after a model time step, then the vertical profile is immediately adjusted by mixing the unstable profile to neutral stability while conserving heat and momentum.

It is straightforward to extend this approach to a "dynamic adjustment" in which the profile is partially adjusted if some dynamic instability criterion, such as one based on a Richardson number ($Ri$), is violated. Relying on shear instability could be sufficient even though it ignores surface waves and Langmuir circulations, among other things. Certainly, if shear instabilities are present we would expect mixing. Perhaps, in the real ocean, some other cause of mixing would have acted sooner and prevented the shear unstable situation from arising, but if the timing and other details are not important on climate scales, then it may be that we will adjust to the correct state regardless.

Numerically, such a dynamical adjustment process may be regarded as a time splitting scheme with a fully implicit realization of an eddy viscosity formulation. For example, write the temperature equation as

$$\partial T / \partial t = A + \partial q / \partial z,$$

where $q$ is the (vertical component of) the turbulent heat flux and $A$ represents everything else. Discretize in time and space: $T_j^n = T(j\Delta z, n\Delta t)$. Suppose we first account for $A$ in advancing from time $n - 1$ to time $n$ and obtain a new value $T^*$. Write

$$q_{j-1/2} = \kappa_{j-1/2}(T_{j-1} - T_j),$$

where $\kappa$ may be a function of $Ri$, etc. With a second order in space, fully implicit scheme one obtains

$$-a_j^- T_{j-1}^n + T_j^n - a_j^+ T_{j+1}^n = (1 - a_j^- - a_j^+)T_j^*$$

with

$$a_j^\pm = \frac{\kappa_{j\pm 1/2}}{1 + \kappa_{j+1/2} + \kappa_{j-1/2}}$$

and $\kappa$ is now nondimensionalized by the grid spacing and time step. All of the information about the adjustment of the large scale by turbulent mixing is contained in the $a's$ and the corresponding coefficients for momentum (with the common assumption that all scalars mix like $T$). For $\kappa > 0$, $0 \le a \le 1$; as $\kappa \to 0$ then $a \to 0$, and as both relevant values of $\kappa \to \infty$, then a $\to 0.5$. More generally, if $\kappa$ is smooth then $a$ cannot be much greater than 0.5. Taking $a = 0.5$ recovers a form of the convective adjustment algorithm.

Parameterizations of the OML can be divided into two classes: (i) bulk or integral parameterizations, which posit, *a priori*, the well-mixed structure of Fig. 1; and (ii) those which take mixing to be a function of depth with the expectation of recovering the OML structure where appropriate. In the purest versions of (i) all the momentum and heat put in at the surface are assumed to go into the mixed layer, leaving the deeper ocean unchanged. Then, given the conservation laws for heat and momentum, the only variable to be determined is the OML depth, $h$.

The simplest formulation is to take $h = $ constant, as was done above. This may allow SST to be determined, but it is obviously physically inadequate. The first of the commonly used bulk parameterizations works from the turbulence kinetic energy (TKE) balance. In its original and simplest form (Kraus and Turner, 1967), it is assumed that the net surface input of TKE, $mu_*^3$, where $u_*$ is the surface friction velocity and $m$ a constant, all goes into raising the potential energy of the ocean. This is accomplished by mixing warm water down, i.e., by deepening the OML. Subsequent versions have elaborated the TKE balance to include other effects such as shear-generated entrainment at the base of the OML (see especially Niiler and Kraus, 1977). Each of these introduces new uncertain parameters; $m$, the fraction of the surface energy available to raise the TKE, is one such constant (though it is not obvious that it should be constant).

A second bulk approach begins from the mean kinetic energy equation. Define a bulk mixed layer Richardson number as

$$Ri = \frac{\Delta b h}{\Delta u^2},$$

where $\Delta b$ and $\Delta u$ are the jumps in buoyancy and momentum at the base of the mixed layer. Then $h$ is determined by the condition that $Ri = Ri_c$, a critical Richardson number. In the original version of this formulation, Pollard et al. (1973) took $Ri_c = 1$. Price (1979) found $Ri_c \approx .65$ by fitting to data from laboratory entrainment experiments.

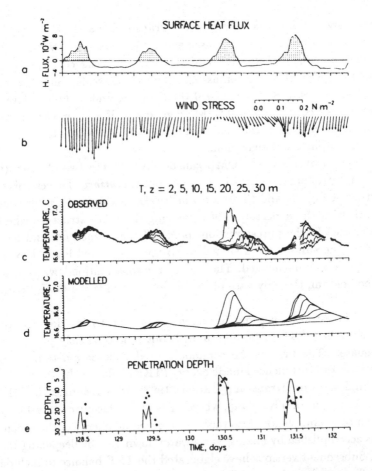

Fig. 8 Simulation of observations of mixed layer evolution at approximately 30°N, 124°W in May 1980. (a) Surface heat flux; (b) wind stress; (c) CTD-measured temperatures at depths of 2, 5, 10, 25 and 30n m; (d) model-computed temperatures; (e) mixed layer penetration depth calculated from the CTD data (heavy dots) and from the model (solid curves). Model and figure are from Price et al. (1986).

This has been shown to work well in a number of applications to the ocean; Figure 8, from Price et al. (1986), is an example.

It is not clear why this simple scheme works as well as it does; Pollard et al. (1973) offer three different and not altogether consistent justifications, none of which is completely satisfying. Two kinds of evidence suggest that it cannot be entirely adequate. Figure 9 shows temperature profiles from two locations before and after the passage of a storm. It is

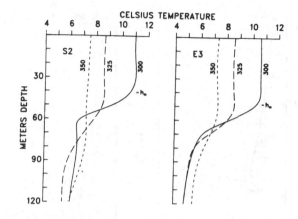

Fig. 9 Smoothed temperature profiles from the beginning (day 300), middle (day 325) and end (day 350) of the STREX observations at two sites near OWS-P (50°N, 145°W). Late autumn periods in 1980 (site S2) and 1981 (site E3) both show warming of the ocean beneath the mixed layer. From Large et al. (1986).

clear that temperatures below the mixed layer have increased: heat has been mixed down below the final OML depth. The most likely cause is the action of inertial waves generated by the storm (Large et al., 1986). Capturing such an effect requires adding some mixing below the OML to the bulk approach. This can be done (e.g., Price et al., 1986), but it does not relate in any apparent way to the bulk $Ri$. A new and distinct justification is required. A second problem is presented by those places where there is no clear mixed layer. The most important of these is the equator, where the strong shears associated with the subsurface equatorial undercurrent appear to disallow an a well-defined OML.

It is more common to use the non bulk formulations in ocean GCMs. If the simplest of these, $\kappa$ = constant, is tuned to mixed layer values, it creates too much mixing in the deeper ocean. The next simplest idea, making the mixing a function of a gradient Richardson number, was first applied to the ocean more than 40 years ago by Munk and Anderson (1948). The version most often used in GCMs is due to Pacanowski and Philander (1981), who tuned their parameterization so the GCM simulation fit observations of equatorial currents in the Indian Ocean. As compared with turbulence measurements in the equatorial Pacific (Peters et al., 1988), their parameterization gives too little mixing at low $Ri$ and too much at high $Ri$. It is also smoother than the data, lacking the abrupt transition at a critical $Ri$ (Figure 10). These differences can

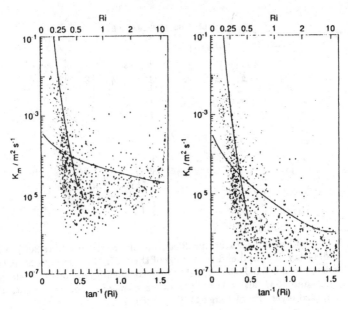

Fig. 10    Eddy coefficients $\kappa_m$ and $\kappa_H$ versus $Ri$ from hourly average equatorial measurements in the Pacific. From Peters et al. (1988).

of course be eliminated by deriving the $Ri$ dependence directly from a curve fit to the Peters et al. (1988) data. This has been done by P. Schopf (private communication) with some success.

A more serious difficulty must be confronted if the data of Fig. 10 are correctly interpreted as saying that mixing does not depend on $Ri$ alone. This seems a viable interpretation of Fig. 10, especially when $Ri$ is near critical. It is usually expected that $Ri$ alone is insufficient; a dimensional argument indicates that $\kappa$ should depend as well on a turbulent length and velocity scale. A counterargument might be essayed from the adjustment viewpoint, since the $a's$ are nondimensional. Nonetheless, it is more plausible that mixing depends on some (perhaps not entirely local) measure of turbulent intensity as well as on $Ri$.

In addition there is the dependence of $Ri$ on scale. Taking data smoothing and the measurement process into account, Peters et al. (1988) estimate that the $Ri$'s of Fig. 10 are based on an effective scale of 17 m. It is possible that a gradient $Ri$ model might be adequate, and that the scatter of Fig. 10 is a consequence of coarse vertical sampling.

The most elaborate turbulence closure scheme yet used in an ocean GCM is the Level 2.5 scheme of Mellor and Yamada (1982). A variant of this scheme was used in the simulation of Rosati and Miyakoda (1988);

illustrated in Fig. 2. It is a second-moment closure, employing plausible assumptions systematically to obtain a closure with a number of empirical constants. Values for these are fit to data from various neutral flow experiments. A hierarchy of models is obtained by an ordering in a small parameter that is a measure of the anisotropy of the Reynolds stress tensor. Of interest here are the level 2 and the level 2.5 models appropriate to the boundary layer. In this case define

$$-(\overline{u'w'}, \overline{v'w'}) = \kappa_m \left(\frac{\partial u}{\partial z}, \frac{\partial v}{\partial z}\right), \qquad -\overline{w'T'} = \kappa_H \frac{\partial T}{\partial z}; \qquad (2)$$

$$\kappa = lqS_M, \quad \kappa_H = lqS_H; \qquad (3)$$

where $l$ is a length scale, and the $S$'s are stability functions. The TKE, $q^2/2$, is determined from the equation

$$\frac{d}{dt}\left(\frac{q^2}{2}\right) - \frac{\partial}{\partial z}\left[lqS_q \frac{\partial}{\partial z}\left(\frac{q^2}{2}\right)\right] = P_s + P_b - \epsilon, \qquad (4)$$

where $P_s$ and $P_b$ are shear and buoyancy production and the dissipation $\epsilon$ is modeled as

$$\epsilon = \frac{q^3}{B_1 l}, \qquad (5)$$

with $B_1$ being another empirical constant. In the level 2 model the left hand side of (4) is absent: there is a local balance between production and dissipation. In this case the $S$'s depend only on local $Ri$ (and $q$ is determined by local shear and local $Ri$). The $S$'s $= 0$ for $Ri \geq Ri_c \approx .2$ (see Fig. 4 of Mellor and Yamada, 1982) and rise steeply as $Ri \to 0$. This behavior is much like the gradient $Ri$ parameterization of Price et al. (1986). For the level 2.5 model, one would anticipate that in oceanographic modeling applications the $d/dt$ term would be small compared with the diffusion term; the results of Rosati and Miyakoda (1988) are in accord. Diffusion tends to smooth the mixing profile, making the mixing at any level dependent on the level of production in its turbulent neighborhood.

The basis for determining $l$ is not firm; as Mellor and Yamada(1982), (p. 852) put it: "The major weakness of all the models probably relates to the turbulent master length scale." Some have used the algebraic formula

$$l = l_o \frac{\kappa z}{l_o + \kappa z}; \quad l_o \equiv \quad \text{const} \frac{\int_o^\infty zq \, dz}{\int_o^\infty q \, dz}, \tag{6}$$

which varies from $\kappa z$ at small $z$ to $l_0$ at large $z$. The asymptotic length $l_0$ may be adjusted via the arbitrary constant in (6). For the ocean it may do just as well to dispense with the logarithmic layer and take $l = l_0$. The most complex approach, used by Rosati and Miyakoda (1988), adds an equation like (4) for $lq^2$ (Eq. (48) of Mellor and Yamada, 1982). They adopted this approach in the hope of treating the separate region of strong mixing associated with the equatorial undercurrent. However, it turned out to be necessary to add a restriction in the presence of stable regions, as suggested by Galperin et al. (1988). This restriction was also found to be useful in the successive runs of the model including data assimilation (Derber and Rosati, 1989).

Comparisons of various OML models by Martin (1985, 1986) show that the performance of the Mellor and Yamada model is quite similar to the much simpler, far less expensive model of Price et al. (1986): it behaves like an $Ri$-dependent model. The discussion above anticipates this result. The MY models tend to allow more shear and buoyancy gradient in the near-surface layer than the observations indicate (viz., Price et al. 1986). This discrepancy could be reduced by increasing the mixing length. It may be a symptom of a missing process, such as Langmuir cells (which would have the scale of the OML). Or it may be that fitting the logarithmic layer, while appropriate to the atmospheric PBL, results in an underestimate of the near-surface mixing in the vicinity of the ocean's nonrigid upper boundary.

We have already seen in Fig. 2 that Rosati and Miyakoda's global simulation using the Mellor and Yamada level 2.5 model was less than satisfactory. This calculation was carried out with the ocean GCM developed at Geophysics Fluid Dynamics Laboratory over several decades, but with the Mellor and Yamada vertical mixing and the Smagorinsky (1963) nonlinear lateral eddy viscosity as subgrid-scale closures. It is a global model with horizontal resolution of 1° of latitude and longitude; its 12 vertical levels are concentrated so that 6 points lie in the upper 70 m. The model is driven by wind stress and heat flux derived from the twice daily National Meteorological Center operational analyses.

This is the most comprehensive model yet used to simulate the global ocean, with the most sophisticated turbulence closure. It is not clear why the performance is as disappointing as it is. (The prediction of

OML depth is no more accurate than that for SST; cf. Fig. 11 of Rosati and Miyakoda, 1988.) Errors in the forcing data and errors in the model's large-scale dynamics are possibilities, along with the subgrid-scale closures.

Regardless of its flaws, the simulation has a number of novel and interesting features. It is the first to generate fields of TKE. In the model, as in nature, these fields are highly intermittent in time and space. Comparing simulated TKE with observations will clearly be a challenging task.

## 22.5 Discussion

Among the many facets of the ocean's role in the climate system, its influence on SST is most essential. Therefore, the first task of an ocean model intended for climate studies is to simulate SST accurately. SST is determined by (i) surface heat flux (heat exchanged with the atmosphere); (ii) near-surface mixing physics and (iii) large-scale dynamics via horizontal advection and upwelling.

There are uncertainties in all of these. We are most confident of our ability to model numerically the large scale dynamics. But even if the dynamical model is impeccable, the ocean circulation is forced, primarily by the surface wind stress, and there are surely errors in the forcing data fields.

The errors in the standard surface heat flux compilations are too large to be used as forcing for ocean models: the consequent errors in SST simulations can be as large as the signal. Both observational data and the heat flux parameterizations are significant sources of error. Global cloud data are inadequate for the accurate calculation of the radiative components of the surface heat balance; in addition, the parameterization of cloud-radiative effects is uncertain. It may not be fanciful to say that clouds are the most serious problem in modeling the ocean for climate.

The parameterization of sensible and latent heat exchange is uncertain for the full range of oceanographic conditions. (There is also the thorny issue of the relationship between observed variables, e.g., the wind observed from a merchant ship, and the variables needed in these formulas, e.g., the 10 m wind.) Because measurements at sea are so difficult to obtain, an observational approach alone is unlikely to provide a remedy. LES simulations of the atmospheric PBL could make a critical contribution, but it will be necessary to account for the peculiar condition of a wavy lower boundary which is itself determined by the state of the PBL.

In view of the difficulties with surface heat fluxes, it is better to use the SST the model predicts together with atmospheric data to calculate heat fluxes. With this approach the SST errors are somewhat self-limiting because of the negative feedback between SST and heat flux. One may go on to tune the uncertain parameters in the surface heat flux formulas in order to minimize the error in the model simulation of SST. Any remaining errors must than be due either to irreducible errors in the heat flux, including those attributable to data errors or to shortcomings of the ocean model such as an inadequate parameterization of mixing physics. In the example given above, it was plausible that all the SST error was due to heat flux problems.

It is then not possible to use such a model-data SST comparison to identify flaws in the model. In this example the OML depth is fixed at 50 m, the simplest "mixing model." Since the OML depth is observed to vary in space and time, we know the model is incorrect. But we do not know whether a better one would, in the face of the other errors, yield a better simulation of SST. The more elaborate model with the more physically correct turbulence parameterization used to produce Fig. 2 did not do as well on SST as the simple model. If the comparison were between a more adequate but still simple parameterization such as the bulk OML model of Price, and a more elaborate one such as Mellor and Yamada Level 2.5 or the still more ideal product of a future program of LES, then it would no longer be obvious that all the ocean data available are up to the task of picking a clear winner.

The choice of parameterization could be based on careful comparison with boundary layer turbulence in a laboratory setting or direct numerical simulation, but one must be mindful of the complexities of the oceanic setting. What fits the laboratory may be undone in the ocean by the action of surface and internal waves, Langmuir cells, etc. This caveat will be with us until LES or some other technique is extended to a true proxy for the ocean.

Physical verisimilitude is not the only consideration. In view of the size of the overall computational task in global ocean modeling, the parameterization of turbulence cannot ignore the issue of computational efficiency. Even more than the time limitations, relatively limited vertical resolution must be accommodated. This limits the input to a parameterization, but also limits what is needed as output. One would expect parameterizations to depend on temporal and vertical (and perhaps horizontal) resolution; however, this has not been true for ocean models.

The computational constraints make it undesirable to complicate parameterizations unless the additions can be demonstrated to be an improvement. We have argued that such verifications are likely to be difficult in view of the complexity of the oceanic setting and the poor quality of the data. However, the strategy we applied to the surface heat flux parameterization will apply here as well. We will always want to begin with what we believe to be a physically correct turbulence closure, albeit a somewhat incomplete one. For example, there are presently no OML models which account for variations in sea state. Some parameterization of the closure may be needed to adapt it to ocean modeling e.g., to account for coarse vertical resolutions. This also acknowledges that the parameterization does not stand alone, but is part of a model system. The full model can then be exercised and verified against as many data as possible, tuning any uncertain parameters in the process. If the number of tunable parameters or the uncertainties in them is small, then the verification will still allow a meaningful evaluation of model performance. If some verifiable aspect of the model is sensitive to the parameterization, then this very sensitivity can be exploited to determine a best parameterization (or parameter value). If nothing is sensitive, then it doesn't matter what is chosen, so the least expensive parameterization recommends itself.

## Acknowledgments

I am grateful for valuable discussions with Yeuchen Zhao, Tony Rosati and Boris Galperin. My thanks go to Virginia DiBlasi-Morris for preparing the manuscript for publication. This work was supported by NASA Grant NAGW-916 and NSF Grant OCE-9000127.

## References

BLANC, T. V. (1987) Accuracy of bulk method determined flux, stability, and sea surface roughness. *J. Geophys. Res.*, **80**, 3867–3876.

BLUMENTHAL, M. B. AND CANE, M. A. (1989) Accounting for parameter uncertainties in model verification: An illustration with tropical sea surface temperature. *J. Phys. Oceanogr.*, **19**, 815–830.

CARDONE, V. J., GREENWOOD, J. G., AND CANE, M. A. (1990) On trends in historical marine wind data. *J. Climate*, **3**, 113–127.

CESS, R., POTTER, G., BLANCHET, J., BOER, G., GHAN, S., KIHEL, J., LE-TREUT, H., LI, Z.-X., LIANG, X.-Z., MITCHELL, J., MORCRETTE, J.-J., RANDALL, D., RICHES, M., ROECKNER, E., U.SCHLESE, SLINGO, A., K.E, T., WASHINGTON, W., WETHERALD, R., AND YAGAI, I. (1989) Interpretation of cloud–climate feedback as produced by 14 atmospheric general circulation models. *Science*, **245**, 513–516.

ESBENSEN, S. AND KUSHNIR, Y. (1981) The heat budget of the global ocean: An atlas based on estimates from surface mairne observations, Climatic Researach Institute, Report No. 29, Oregon State University, Corvallis, ORegon, 27 pp., 188 charts.

FUNG, I., HARRISON, D., AND LACIS, A. (1984) On the variability of the net longwave radiation at the ocean surface. *Rev. Geophys. Space Phys.*, **22**, 177–193.

GALPERIN, B., KANTHA, L., HASSID, S., AND ROSATI, A. (1988) A quasi-equilibrium turbulent energy model for geophysical flows. *J. Atmos. Sci.*, **45**, 55–62.

GILL, A. AND NIILER, P. (1973) The theory of seasonal variability in the ocean. *Deep-Sea Res.*, **20**, 141–177.

KRAUS, E. AND TURNER, J. (1967) A one-dimensional model of the seasonal thermocline. Part II: The general theory and its consequences. *Tellus*, **19**, 98–106.

LARGE, W., McWILLIAMS, J., AND NIILER, P. (1986) Upper ocean thermal response to strong autumnal forcing of the northeast Pacific. *J. Phys. Oceanogr.*, **16**, 1524–1550.

MARTIN, P. (1985) Simulation of the mixed layer at OWS November and Papa with several models. *J. Geophys. Res.*, **90**(C1), 903–916.

MARTIN, P. (1986) Testing and comparison of several mixed-layer models. Naval Ocean Research and Development Activity Tech. Rep., 143 pp.

MELLOR, G. AND YAMADA, T. (1982) Development of turbulence closure model for geophysical fluid problems. *Rev. Geophys. Space Phys.*, **20**(4), 851–875.

MUNK, W. AND ANDERSON, J. (1948) Notes on a theory of the thermocline. *J. Mar. Res.*, **7**, 276–295.

NIILER, P. AND KRAUS, E. (1977) One-dimensional models of the upper ocean. In *Modelling and Prediction of the Upper Layers of the Ocean.*, Ed. E. Kraus, pp. 143–172. Pergamon Press.

PACANOWSKI, R. AND PHILANDER, S. (1981) Parameterization of vertical mixing in numerical models of tropical oceans. *J. Phys. Oceanogr.*, **11**, 1443–1451.

PETERS, H., GREGG, M., AND TOOLE, J. (1988) On the parameterization of equatorial turbulence. *J. Geophys. Res.*, **93**, 1199–1218.

POLLARD, R., RHINES, P., AND THOMPSON, R. (1973) The deepening of the wind-mixed layer. *Geophys. Fluid Dynam.*, **3**, 381–404.

PRICE, J. (1979) On the scaling of stress-driven entrainment experiments. *J. Fluid Mech.*, **90**, 509–529.

PRICE, J., WELLER, R., AND PINKEL, R. (1986) Diurnal cycling: Observations and models of the upper ocean response to diurnal heating, cooling, and wind mixing. *J. Geo. Res.*, **91**, 8411–8427.

REYNOLDS, R., ARPE, K., GORDON, C., HAYES, S., LEETMAA, A., AND MCPHADEN, M. (1989) A comparison of tropical Pacific surface wind analyses. *J. Climate*, **2**, 105–111.

ROSATI, A. AND MIYAKODA, K. (1988) A general circulation model for upper ocean simulation. *J. Phys. Oceanogr.*, **18**, 1601–1626.

SEAGER, R. (1989) Modeling tropical Pacific sea surface temperature: 1970–1987. *J. Phys. Oceanogr.*, **19**, 419–434.

SEAGER, R., ZEBIAK, S., AND CANE, M. A. (1988) A model of the tropical Pacific sea surface temperature climatology. *J. Geophys. Res.*, **93**, 1265–1280.

SMAGORINSKY, J. (1963) General circulation experiments with the primitive equations. Part I: The basic experiment. *Mon. Wea. Rev.*, **91**, 99–164.

WEARE, B., STRUB, P., AND SAMUEL, M. (1980) *Marine Climate Atlas of the Tropical Pacific Ocean.* University of California, Davis, Dept. of Land, Air and Water Resources, 147 pp.

WEARE, B., STRUB, P., AND SAMUEL, M. (1981) Annual mean surface heat fluxes in the tropical pacific ocean. *J. Phys. Oceanogr.*, **11**, 705–717.

# ENVIRONMENTAL FLOWS

# Conjunctive Filtering Procedures in Surface Water Flow and Transport

KEITH W. BEDFORD AND WOON K. YEO

## 23.1 Introduction

The modeling and prediction of turbulent flow and transport in surface waters is a particularly challenging problem for a number of reasons. First, the flows are highly time varying in both the mean and fluctuating components and are the result of a variety of interacting fluid phenomena. Second, these interacting nonlinear fluid mechanisms often do not separate into well-defined spectral regions; i.e., they overlap. Third, stratification affects the macroscale level through internal waves, blocking, etc. Fourth, the planform geometry and bottom bathymetry are irregular with both large-scale and small-scale variability, which, in the case of lakes, reservoirs and estuaries, is of the same scale as the flow physics. Finally and of significance is that the depth scale is two to three orders of magnitude smaller than the horizontal length scale.

No better example of the type of water body being considered with this class of models exists than the Great Lakes. Certainly, any estuary or other large lake or reservoir is equally complex. Figure 1 (Boyce 1974; Bedford and Abdelrhman 1987) presents a schematic of the basic coherent fluid mechanisms and their typical length and time scales. The dynamic response of the Lakes is particularly complex (Dingman and Bedford 1984; Libicki and Bedford 1990), as resonant or nearly resonant interactions result in strong storm surges which develop in a matter of hours ($\sim 6 - 7$) and result in internal waves, near-shore Kelvin waves and jets and upwelling and downwelling events lasting a matter of days. Persistent wind-driven currents continue against a backdrop of turbulence.

Traditional model formulations of these processes depend upon average equations prepared with Reynolds definition of the average. Field data collected in a number of surface water sites at time and space intervals which resolve turbulence (e.g., Gross and Nowell 1983, 1985;

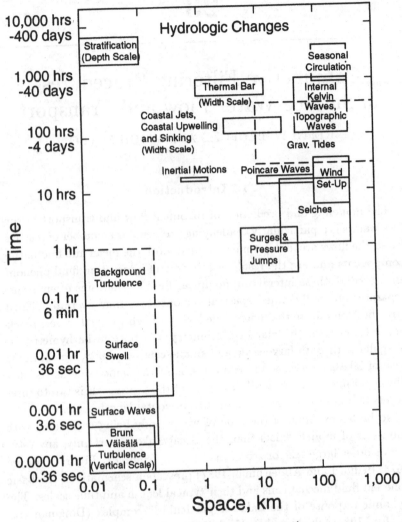

Fig. 1   Space—time process schematic.

Bedford et al. 1987b) clearly indicate that the stationarity requirement is met at the 90% confidence interval for averaging periods of approximately 10–15 minutes. Occasionally these periods might extend to an hour during quiescent periods with no tide or seiche activity. Consequently there is no steady Reynolds mean at periods greater than 10–15 minutes.

In 1978 our research group began undertaking an examination of the structural aspects of the three-dimensional models used for these calculations. Attention was first drawn to examining the effect of the compressed vertical scale by analysis of time and space requirements imposed by the traditionally used coarse horizontal grid on the interpretation of data calculated at scales below the "horizontal scale" and its associated time scale. Babajimopoulos and Bedford (1980) and Bedford and Babajimopoulos (1980) examined a rigid lid rectangular lake by means of a model prepared via Leonard's (1974) higher order averaging procedure. The length was assumed to be three orders of magnitude larger than the depth, and a constant (in the mean) spectrally correct wind shear was imposed at the surface. Sampling at the basin center at mid-depth was used for the reported results. A free surface version of this model was also developed and tested.

In examining the results (Bedford 1981) a number of differences between the rigid and free surface model were explored, but two items continue to be relevant here. The first is that in both formulations there was a two-sloped spectrum during all stationary portions of the simulation. At wave numbers, $k_d$, corresponding to the local water depth and higher, the spectra were minus 5/3 sloped, while at $k_d$ and lower, the slopes were proportional to $k^{-3}$. Essentially, the models propagated energy inputs according to two-dimensional (2D) turbulence theory and it was concluded that the use of the term "3D" for these models was misleading.

The resolution of the minus 5/3 region in the spectra, while at first glance an indication of possible 3D turbulence activity, occurred in a wave number and frequency region which was well below the minimal horizontal grid scale. Therefore, a second result of this work spoke to the very anisotropic grid structure and quite small time steps required for stability (free surface) purposes. In essence there are two fixed spatial scales in these models below which model-resolved results should either be interpreted with great care or eliminated with proper averaging. These are the scale of the horizontal grid $\Delta x$ and its associated time of travel, $\Delta x/u = t_l$, and the depth, $d$, and its associated time scale, $t_d$. In most model grids $\Delta x >> h$ as, e.g., in Lake Erie, where the horizontal grid = 500 meters and the average depth is 20 meters. The corresponding time scales based upon the vertical depth, $t_d$, and time step for the model, $\Delta t$, are quite small in contrast to the grid-based horizontal time scale, $t_l$. Therefore, in using $\Delta x$ and $t_l$ as the minimum resolution scales it was argued that very small time steps resolved portions of the spectra that were not fully supported by 3D physics. That is, $\Delta t$ inadvertently would resolve improper physics. It was argued that time resolution of the order of $t_l$ was the minimally justified temporal resolution of the model.

To address this consistency issue, combined space–time filtering was suggested (Bedford and Dakhoul 1982) and tested (Dakhoul and Bedford 1986a,b) and a consistency requirement between space–time averaging scales was examined. The methodology was initially tested on a Burgers equation solution and compared with results calculated with a variety of other averaging methods of recent (large eddy simulation, LES) and historical (Reynolds) relevance. The theory and tests were limited to isotropic filters. The acute problem of anisotropic grids and filters generated by shallow flows was not addressed.

Recognizing that the use of moving averages gives rise to a different and reduced set of averaging operations, Bedford et al. (1987a) and Dingman (1986) explored the requirement of consistent higher order analog averaging (suggested by LES) on the linear differential terms in the governing equations. In so doing it became possible to unify a wide variety of seemingly disparate numerical methods and view them as digital versions of either a Gaussian or uniform weight function analog filter with each digital method distinguished only by what integer coefficient was used in the analog representation. The key operation permitting this unification was at the time interpreted as a necessity for double or cascade averaging of the linear terms to be consistent with the perception that the nonlinear terms were also "double averaged." The requirement for analyzing the linear terms consistently with the nonlinear terms is not often (if at all) discussed in the literature.

In a thorough treatment, by Aldama (1990; see also Aldama 1985), the space–time filter was invoked as part of a three-scale approach to the LES method. Space–time consistency was placed upon a sound analytical footing, and Burgers flow calculations further validated the space–time filter approach. Of significance was the extremely thorough asymptotic expansion analysis of the inertia and cross terms resulting from space–time filtering. The analysis of convergence and scale effects led to the conclusion that the subgrid-scale terms were so unimportant that they could almost be ignored. This essential result, exploited by Rosman (1987), led to a turbulent flow model that was essentially closure-free, a conclusion rather at odds with most published turbulence modeling techniques. In testing this formulation, Rosman applied it to the calculation of flow over a backward step and performed detailed comparisons with laboratory data. It was unsettling that a small amount of dissipation via a Smagorinsky-like term was required.

While the above "closure-free" approximation was argued primarily from the effects of scale, Yeo (1987) and Yeo and Bedford (1988, in press a,b) provided a full theoretical derivation of these results via a conjunctive filtering operation based upon the combined use of a low pass and a high pass filter. Based upon this filter an analysis of the resulting terms revealed the presence of mean flow representations for shear-, velocity- and vortex-stretching-based energy transfer mechanisms.

When coupled with the requirement for double or cascade averaging the presence of these formulations raises a number of questions, not the least of which is what happened to the subgrid-scale terms.

The purpose of this chapter is to review this conjunctive averaging technique and to demonstrate that, by reformulating the derivation of the governing equations, the role of double averaging is now seen as an act of separation followed by averaging. The conjunctive averaging is also seen to provide resolution sufficient to achieve fully and robustly the low pass average equations. Finally, with the subgrid terms being so unimportant in the average equations, the issue is raised as to whether we are really using the average equations or whether in fact we are using equations that are merely separated. The current practice of analyzing only the nonlinear portion of the equations is seen as an incomplete implementation of either the averaging or the separation approach and thereby continues to require empirical augmentation.

## 23.2 Filtering Procedures
### 23.2.1 Low Pass Filtering/Averaging

Following the seminal work of Leonard (1974) a total field variable $(f)$ is decomposed into its low-pass averaged component $(\overline{f})$ and its deviation $(f')$ with the low-pass filtered variable defined as

$$\overline{f}(\underline{x},t) = \int\int\int\int\limits_{-\infty}^{+\infty} G(\underline{x}-\underline{x}',t-t')\,f(\underline{x}',t')\,d\underline{x}'\,dt' \qquad (1)$$

Here $G(\underline{x},t)$ is the weight function constrained such that

$$\int\!\!\int\!\!\int\!\!\int_{-\infty}^{+\infty} G(\underline{x},t)\, d\underline{x}\, dt \;=\; 1.0 \qquad\qquad (2)$$

The uniform or Gaussian filters (Leonard 1974; Kwak et al. 1975; Clark et al. 1979) are the most widely used weight functions with the space−time filter (Dakhoul and Bedford 1986a,b) being composed as

$$G(\underline{x},t) = \left(\frac{\gamma_t}{\pi}\right)^{1/2}\frac{1}{\Delta_t}\exp\left(\frac{-\gamma_t t^2}{\Delta_t^2}\right)\prod_{i=1}^{3}\left(\frac{\gamma_i}{\pi}\right)^{1/2}\frac{1}{\Delta_i}\exp\left(\frac{-\gamma_i x_i^2}{\Delta_i^2}\right) \qquad (3)$$

In Eq. (3) $\gamma_t$ and $\gamma_i$ are the standard filter coefficients and $\Delta_t$ and $\Delta_i$ are related to the second moments as

$$\frac{\Delta_t}{2\gamma_t} = \int_{-\infty}^{+\infty} t^2 G(\underline{x},t)\, dt \qquad\qquad (4)$$

$$\frac{\Delta_i}{2\gamma_i} = \int\!\!\int\!\!\int_{-\infty}^{+\infty} |x_i| G(\underline{x},t)\, d\underline{x} \qquad\qquad (5)$$

The response function for the low pass filter is found from taking the Fourier transform, $F[\ ]$, of the filtering operation such that

$$R(\underline{k},\omega) \;=\; \frac{F[\bar{f}]}{F[f]} \;=\; F[G] \;=\; L(\underline{k},\omega) \qquad\qquad (6)$$

In Eq. (6) $\underline{k}$ is the wave number vector and $\omega$ is the radian frequency.

### 23.2.2 High Pass Filter

From Yeo (1987) and Yeo and Bedford (1988, in press a,b) the complementarity principle from the signal processing literature is adopted to derive a high pass filter response function to resolve the high frequency portions of the signal. With regard to Figure 2 the response functions, summing to 1.0, give

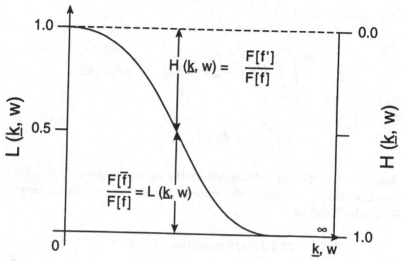

Fig. 2  Schematic of high pass and low pass filter response function.

$$1 = \frac{F[\overline{f}]}{F[f]} + \frac{F[f']}{F[f]} \qquad (7)$$

The first term on the right hand side of Eq. (7) is the response function of the low pass filter, $L(\underline{k},\omega)$, and the second term is the response function of the high pass filter, $H(\underline{k},\omega)$, i.e.,

$$1 = L(\underline{k},\omega) + H(\underline{k},\omega) \qquad (8)$$

Therefore the high pass response functions for uniform and Gaussian space filters become

$$H(\underline{k}) = 1 - \frac{\sin(\underline{k}\Delta_i/2)}{\underline{k}\Delta_i/2}; \quad H(\underline{k}) = 1 - \exp\left(-\frac{\Delta_i^2 \underline{k}^2}{4\gamma_i}\right) \qquad (9)$$

The inverse Fourier transform gives

$$H(\underline{x}) = F^{-1}[H(\underline{k})] = \delta(\underline{x}) - G(\underline{x}) \qquad (10)$$

where $\delta(\underline{x})$ is the Dirac delta function. Therefore, an equation for the fluctuating components can be found by applying the high pass filter to the governing equations

$$f'(\underline{x},t) = \int\limits_{-\infty}^{+\infty} \int \int [\delta(\underline{x}-\underline{x}') - G(\underline{x}-\underline{x}')]f(\underline{x}')\,d\underline{x}'$$

$$= c\Delta_i^2 \frac{\partial^2 f}{\partial \underline{x}^2} + O(\Delta_i^4) \tag{11}$$

where $c$ is a coefficient determined by the second moment of the filter. Note that this is essentially the same result but in generalized form as derived by Kwak et al. (1975).

### 23.2.3 Implementation of Filters

If, for example, the Gaussian space filter is used, then the response function $L(\underline{k})$ is written as

$$L(\underline{k}) = \exp\left(-\sum_{i=1}^{3} a_i k_i^2\right)$$

$$= 1 - (a_1 k_1^2 + a_2 k_2^2 + a_3 k_3^2)$$
$$+ \frac{1}{2!}(a_1 k_1^2 + a_2 k_2^2 + a_3 k_3^2)^2 - \cdots \tag{12}$$

where $a_i = \Delta_i^2/4\gamma_i$. Using a general formula from Fourier analysis (e.g.; Hildebrand 1976; Blinchikoff and Zverev 1976),

$$F[f'] = F[f] - F[\overline{f}]$$

$$= H(\underline{k})F[f]$$

$$= [(a_1 k_1^2 + a_2 k_2^2 + a_3 k_3^2)$$
$$- \frac{1}{2!}(a_1 k_1^2 + a_2 k_2^2 + a_3 k_3^2)^2 + \cdots ]F[f]$$

$$= -F\left[\left(a_1 \frac{\partial^2}{\partial x^2} + a_2 \frac{\partial^2}{\partial y^2} + a_3 \frac{\partial^2}{\partial z^2}\right)f\right.$$

$$+ \frac{1}{2!}\left(a_1^2\frac{\partial^4}{\partial x^4} + a_2^2\frac{\partial^4}{\partial y^4} + a_3^2\frac{\partial^4}{\partial z^4} + 2a_1a_2\frac{\partial^4}{\partial x^2\partial y^2}\right.$$

$$\left. + 2a_2a_3\frac{\partial^4}{\partial y^2\partial z^2} + 2a_3a_1\frac{\partial^4}{\partial z^2\partial x^2}\right)f + \cdots\Bigg]$$

$$= -F\left[\Phi(f) + \frac{1}{2!}\Phi^2(f) + \frac{1}{3!}\Phi^3(f) + \cdots\right] \qquad (13)$$

With the assumption of a symmetric filter, $\Delta = \Delta_1 = \Delta_2 = \Delta_3$ and $\gamma = \gamma_1 = \gamma_2 = \gamma_3$, and

$$f'(\underline{x},t) = f - \overline{f} = -\alpha\nabla^2 f - \frac{1}{2!}\alpha^2\nabla^4 f - \frac{1}{3!}\alpha^3\nabla^6 f + \cdots \qquad (14)$$

where $\alpha = \Delta^2/4\gamma$. This series to the second term is identical to the Leonard–Clark formulation (Leonard 1974; Clark et al. 1979). While this is a more generalized form of the Leonard–Clark relationship, the fact remains that the fluctuating quantity is related to the total field variable. It is desirable to relate $f'(\underline{x},t)$ to the low pass average variable $\overline{f}(\underline{x},t)$.

Using the complementarity principle, we can write the Fourier transform, $F[f']$, as

$$F[f'] = H(\underline{k})F[f] = \frac{H(\underline{k})}{L(\underline{k})}F[\overline{f}]$$

$$= F\left[-\Phi(\overline{f}) + \frac{1}{2!}\Phi^2(\overline{f}) - \frac{1}{3!}\Phi^3(\overline{f}) + \cdots\right] \qquad (15)$$

The final form for Eq. (15) becomes

$$f'(\underline{x},t) = -\Phi(\overline{f}) + \frac{1}{2!}\Phi^2(\overline{f}) - \frac{1}{3!}\Phi^3(\overline{f}) + \cdots \qquad (16)$$

and is a series expansion for the fluctuating component, which is based on the low-pass averaged variable. For convenience the series expansion for $f'$ as a function of the total field variable $f$ will be called the YB-I series and the series for $f'$ in terms of $\overline{f}$ will be called the YB-II series (Yeo and Bedford 1988).

### 23.2.4 The Nonlinear Terms

The inertia, $u_iu_j$, or advection terms, $u_i\phi$, can be analyzed with the

YB-II series in the following manner. Two functions $f$ and $g$ are written with the YB-II series as

$$f = \overline{f} - \Phi(\overline{f}) + \frac{1}{2!}\Phi^2(\overline{f}) - \frac{1}{3!}\Phi^3(\overline{f}) + \cdots$$

$$\tag{17}$$

$$g = \overline{g} - \Phi(\overline{g}) + \frac{1}{2!}\Phi^2(\overline{g}) - \frac{1}{3!}\Phi^3(\overline{g}) + \cdots$$

The product of $f$ and $g$ in Eq. (17) becomes

$$fg = \overline{f}\,\overline{g} - \overline{f}\Phi(\overline{g}) - \overline{g}\Phi(\overline{f}) + \frac{1}{2!}\overline{f}\Phi^2(\overline{g})$$

$$+ \Phi(\overline{f})\Phi(\overline{g}) + \frac{1}{2!}\overline{g}\Phi^2(\overline{f}) + \cdots \tag{18}$$

Now the average $\overline{fg}$ in terms of the product of average variables is therefore required from the YB-I series,

$$\overline{fg} = fg + \Phi(fg) + \frac{1}{2!}\Phi^2(fg) + \frac{1}{3!}\Phi^3(fg) + \cdots \tag{19}$$

After substitution of Eq. (18) into Eq. (19) and considerable algebra (Yeo 1987) there results the following power series for $\overline{fg}$:

$$\overline{fg} = \overline{f}\,\overline{g} + 2\alpha\overline{f}_{,k}\,\overline{g}_{,k} + \frac{1}{2!}(2\alpha)^2\overline{f}_{,kl}\,\overline{g}_{,kl} + \frac{1}{3!}(2\alpha)^3\overline{f}_{,klm}\,\overline{g}_{,klm} + \cdots$$

$$\tag{20}$$

Here a symmetric filter has again been assumed for clarity's sake; i.e., $\alpha = \Delta^2/4\gamma$. This series is called the YB-III series.

Via application to the inertia and advection terms in the governing equations, the YB-III series gives

$$R_{ij}^l = \overline{u_i u_j} - \overline{u}_i\,\overline{u}_j = 2\alpha\overline{u}_{i,k}\overline{u}_{j,k} + \frac{1}{2!}(2\alpha)^2\overline{u}_{i,kl}\overline{u}_{j,kl} + \cdots \tag{21}$$

and

$$Q_j^l = \overline{u_j\phi} - \overline{u}_j\,\overline{\phi} = 2\alpha\overline{u}_{j,k}\overline{\phi}_{,k} + \frac{1}{2!}(2\alpha)^2\overline{u}_{j,kl}\overline{\phi}_{,kl} + \cdots \tag{22}$$

## 23.3 Equation Preparation

It is customary even in the LES method simply to start with the average equations and then apply the filtering operations only to the inertial or advection terms. This analysis is carried forward via a decomposition of the nonlinear terms and subsequent analysis of the product of the mean field variables and the cross terms via the low pass filter operations. The subgrid-scale terms are then empirically represented by mean field variables. One of the issues this chapter raises is that if attention is focused on just the nonlinear terms several conceptual issues are ignored which, if properly addressed, might provide a more consistent approach to the question of averaging. Therefore, this section reviews the governing equation and the various possible forms from separation and averaging.

### 23.3.1 Total Field Equations

The starting point is the total field equations for velocity, $u_i$, and scalar $\phi$:

$$\frac{\partial u_i}{\partial t} + (u_i u_j)_{,j} + \frac{1}{\rho} P_{,i} - \nu u_{i,jj} = L(\underline{x}, t) = 0 \tag{23}$$

$$\frac{\partial \phi}{\partial t} + (u_j \phi)_{,j} - \lambda \phi_{,jj} = M(\underline{x}, t) = 0 \tag{24}$$

In Eqs. (23) and (24) $\nu$ and $\lambda$ are the kinematic viscosity and molecular diffusivity respectively.

### 23.3.2 Separated Total Field Equations

Before proceeding to the average equation, let us reverse the order of the operations usually applied and separate the total field variable $f$ into an average or slowly evolving term $\overline{f}$ and the rapidly evolving or fluctuating component $f'$ and reformulate Eqs. (23) and (24) without any loss of information. Equations. (23) and (24) become

$$\frac{\partial \overline{u}_i}{\partial t} + (\overline{u}_i \overline{u}_j)_{,j} + \frac{1}{\rho} \overline{P}_{,i} - \nu \overline{u}_{i,jj} = A_i$$

$$= -\left\{ \frac{\partial u_i'}{\partial t} + (u_i' u_j' + u_i' \overline{u}_j + \overline{u}_i u_j')_{,j} + \frac{1}{\rho} P_{,i}' - \nu u_{i,jj}' \right\} \tag{25}$$

$$\frac{\partial \overline{\phi}}{\partial t} + (\overline{u}_j \overline{\phi})_{,j} - \lambda \overline{\phi}_{,jj} = B$$

$$= - \left\{ \frac{\partial \phi'}{\partial t} + (u'_j \phi' + \overline{u}_j \phi' + u'_j \overline{\phi})_{,j} - \lambda \phi'_{,jj} \right\} \qquad (26)$$

At first glance these are equations that are expressed in terms of averaged variables and do not contain the problems with the nonlinear terms that are so often encountered. It is, of course, recognized that no variability has been eliminated from the these equations either. Finally, the scale of the separation is set by the filter coefficient which, for the equation preparation phase, does not necessarily have to be connected to the grid size at all.

### 23.3.3 Averaged Field Equations

To derive the averaged equations, Eqs. (23) and (24) are filtered via Eq. (1). The operation is straightforward and results in the following equations:

$$\frac{\partial \overline{u}_i}{\partial t} + (\overline{u_i u_j})_{,j} + \frac{1}{\rho}\overline{P}_{,i} - \nu \overline{u}_{i,jj} = \overline{L}(\underline{x}, t) = 0 \qquad (27)$$

$$\frac{\partial \overline{\phi}}{\partial t} + (\overline{u_j \phi})_{,j} - \lambda \overline{\phi}_{,jj} = \overline{M}(\underline{x}, t) = 0 \qquad (28)$$

The resolution of these equations is defined by the response function of the filter, which implies (Figure 3a) that no variability is resolved past $k_c$, the cutoff wave number, nor above the response function in the region with wave numbers less than $k_c$. With the average defined as in Eq. (1) the resolution of the filter functions is defined for the continuum equations all the way to $k = \infty$ (Figure 3a) even though $L(\underline{k}, \omega)$ might equal zero at $k_c$ well below $k = \infty$. With this definition of the analog filter, zone II in Figure (3a) defines the fluctuations theoretically resolved by the analog high pass filter. Zone I represents the low-pass filtered portion of the response function.

The implementation of the analog filter, while straightforward, requires examination. The traditional analysis step has been to decompose or separate the nonlinear term and proceed with averaging. However, as suggested in Eqs. (25) and (26) the act of separation, if applied to one portion of the equation, should for consistency be applied to all the

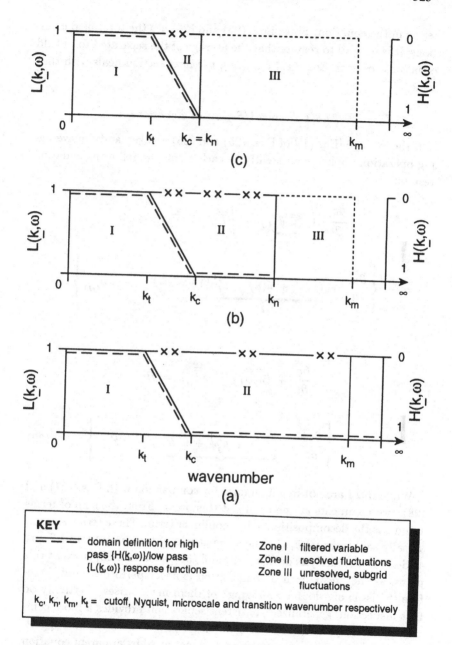

KEY

- - - - -  domain definition for high    Zone I    filtered variable
pass {H($\underline{k}$,ω)}/low pass         Zone II   resolved fluctuations
{L($\underline{k}$,ω)} response functions    Zone III  unresolved, subgrid
                                                       fluctuations

$k_c$, $k_n$, $k_m$, $k_t$ = cutoff, Nyquist, microscale and transition wavenumber respectively

Fig. 3  High pass/low pass response function domain for (a) analog,
(b) digital ($k_c < k_n$) and (c) digital ($k_c = k_n$) filters.

terms in the equation. Since Eqs. (25) and (26) are the total field equations; it is natural to require that the average of the separated total field equations result in Eqs. (27) and (28); the next section deals with that issue.

### 23.3.4 Averaged/Separated Equations

If the average [Eq. (1)] of Eqs. (25) and (26) is taken and the averaging operation applied consistently to each term, the following equation results:

$$\frac{\partial \bar{\bar{u}}_i}{\partial t} + \underbrace{(\overline{\bar{u}_i \bar{u}_j})_{,j}}_{1} + \frac{1}{\rho}\bar{\bar{P}}_{,i} - \nu \bar{\bar{u}}_{i,jj} = \bar{A}_i$$

$$= -\left\{ \frac{\partial \overline{u'_i}}{\partial t} + \underbrace{(\overline{\bar{u}_i u'_j} + \overline{u'_i \bar{u}_j} + \overline{u'_i u'_j})_{,j}}_{2} + \frac{1}{\rho}\overline{P'}_{,i} - \nu \overline{u'_i}_{,jj} \right\} \quad (29)$$

and

$$\frac{\partial \bar{\bar{\phi}}}{\partial t} + \underbrace{(\overline{\bar{u}_j \bar{\phi}})_{,j}}_{3} - \lambda \bar{\bar{\phi}}_{,jj} = \bar{B}$$

$$= -\left\{ \frac{\partial \overline{\phi'}}{\partial t} + \underbrace{(\overline{\bar{u}_j \phi'} + \overline{u'_j \bar{\phi}} + \overline{u'_j \phi'})_{,j}}_{4} - \lambda \overline{\phi'}_{,jj} \right\} \quad (30)$$

While these are not in a final form for comparison with Eqs. (27) and (28), two comments are necessary at this point. First, the sum of terms 1 and 2 is the decomposition of the nonlinear term. These terms are universally said to equal $\overline{u_i u_j}$. The same is said of the sum of terms 3 and 4. Second, the analysis of Eqs. (29) and (30), even in LES, stops at this point, as the linear term decomposition is never performed and therefore the issue of "double averaging" of them never arises. Certainly on first inspection, generalized consistent separation provides a more complex equation system when averaged than traditional treatments suggest. While seemingly not an improvement as measured by apparent equation simplification, the terms in the fully prepared Eqs. (29) and (30) will provide the heretofore missing terms necessary to explain several of the questions raised earlier.

## 23.4 Equation Analysis

It is desirable to compare the averaged equations in (27) and (28) to the filtered separated equations in (29) and (30), as they should be equal. To make this comparison Eqs. (27) and (28) are first analyzed.

### 23.4.1 Analysis of the Average Field Equations

By use of the YB-III series expansion in Eqs. (21) and (22), Eqs. (27) and (28) become

$$\frac{\partial \overline{u}_i}{\partial t} + (\overline{u}_i \overline{u}_j)_{,j} + \frac{1}{\rho}\overline{P}_{,i} - \nu \overline{u}_{i,jj} + R^l_{ij,j} = \overline{L}(\underline{x},t) = 0 \quad (31)$$

$$\frac{\partial \overline{\phi}}{\partial t} + (\overline{u}_j \overline{\phi})_{,j} - \lambda \overline{\phi}_{,jj} + Q^l_{j,j} = \overline{M}(\underline{x},t) = 0 \quad (32)$$

These are the essential equations resulting from the "closure-free" formulation referred to in the introduction. The closure-free label results from the fact that no turbulent correlations are present which must be empirically specified.

A further comparison points to yet another matter requiring explanation. As noted, the YB-III expansion for symmetric filters results in Eq. (21), which is rewritten as

$$\overline{u_i u_j} = \overline{u}_i \overline{u}_j + R_{ij} = \overline{u}_i \overline{u}_j + 2\alpha(\overline{u}_{i,k}\overline{u}_{j,k}) + \cdots \quad (33)$$

Equation (33) may be compared with the Leonard–Clark reduction for $\overline{u_i u_j}$ as follows:

$$\overline{u_i u_j} = \overline{\overline{u}_i \overline{u}_j} + \overline{\overline{u}_i u'_j} + \overline{u'_i \overline{u}_j} + \overline{u'_i u'_j}$$

$$= \overline{u}_i \overline{u}_j + [2\alpha(\overline{u}_{i,k}\overline{u}_{j,k}) + \cdots] + \overline{u'_i u'_j} \quad (34)$$

It is clearly seen that at least through the second order filter length terms, these expansions are identical except for the presence of the "closure" terms in the Leonard–Clark approximation. An explanation of this difference will follow, but first the comparison of the average equations in (31) and (32) with Eqs. (29) and (30) will be finished.

### 23.4.2 Analysis of the Generalized Averaged/Separated Equation

It is necessary to look a bit more at the linear terms. From Dingman (1986) or Bedford et al. (1987a) it is a very simple matter to show that two important averaging rules used in Reynolds-averaged equations do not adapt to the case of the moving average definitions in use for the LES procedure. For any function $f(\underline{x},t)$, where $f(\underline{x},t) = \overline{f}(\underline{x},t) + f'(\underline{x},t)$,

$$\overline{f'}(\underline{x},t) \neq 0 \tag{35}$$

and

$$\overline{\overline{f}}(\underline{x},t) \neq \overline{f}(\underline{x},t) \tag{36}$$

Concentrating on the impact of Eq. (36) the linear terms in Eqs. (29) and (30) are double averaged and can be further simplified by applying the higher order averaging definition. Therefore, for symmetric filters

$$\overline{\overline{f}} = \overline{f} + \alpha \overline{f}_{,kk} + (\text{HOT}) \tag{37}$$

and Eqs. (29) and (30) become

$$\frac{\partial \overline{u}_i}{\partial t} + \left[ \overline{\overline{u}_i \overline{u}_j} + \overline{\overline{u}_i u'_j} + \overline{u'_i \overline{u}_j} + \overline{u'_i u'_j} \right]_{,j} + \frac{1}{\rho} \overline{P}_{,i} - \nu \overline{u}_{i,jj} = C_i$$

$$= -\alpha \left[ \underbrace{\frac{\partial}{\partial t}(\overline{u}_{i,kk})}_{a} + \underbrace{\frac{1}{\rho}\overline{P}_{,ikk}}_{b} - \underbrace{\nu \overline{u}_{i,jjkk}}_{c} \right]$$

$$- \left[ \overbrace{\frac{\partial}{\partial t}(\overline{u'_i})}^{a} + \overbrace{\frac{1}{\rho}\overline{P'}_{,i}}^{b} - \overbrace{\nu \overline{u'}_{i,jj}}^{c} + \cdots \right] \tag{38}$$

and

$$\frac{\partial \overline{\phi}}{\partial t} + \left[ \overline{\overline{u}_j \overline{\phi}} + \overline{\overline{u}_j \phi'} + \overline{u'_j \overline{\phi}} + \overline{u'_j \phi'} \right]_{,j} - \lambda \overline{\phi}_{,jj} = D$$

$$= -\alpha \left[ \underbrace{\frac{\partial}{\partial t}(\overline{\phi}_{,kk})}_{d} - \underbrace{\lambda \overline{\phi}_{,jjkk}}_{e} + \cdots \right]$$

$$- \left[ \overbrace{\frac{\partial}{\partial t}(\overline{\phi'})}^{d} - \overbrace{\lambda \overline{\phi'}_{,jj}}^{e} \right] \tag{39}$$

The relationship of the conjunctive and high pass filter is seen as follows. If Eq. (16) is used for the fluctuating term, then the terms $a - e$ cancel out respectively. The remaining portion of Eqs. (38) and (39) are equal to Eqs. (27) and (28) and (31) and (32) under the following condition:

$$\overline{u_i u_j} = \overline{u}_i \overline{u}_j + R^l_{ij} = \overline{\overline{u}_i \overline{u}_j} + \overline{\overline{u}_i u'_j} + \overline{u'_i \overline{u}_j} + \overline{u'_i u'_j} \tag{40}$$

$$\overline{u_j \phi} = \overline{u}_j \overline{\phi} + Q^l_j = \overline{\overline{u}_j \overline{\phi}} + \overline{\overline{u}_j \phi'} + \overline{u'_j \overline{\phi}} + \overline{u'_j \phi'} \tag{41}$$

As noted in Yeo and Bedford (1990a) the first term of $R_{ij}$ ($R^{l=1}_{ij}$) and the first term of the Leonard–Clark analyses for $\overline{\overline{u}_i u'_j} + \overline{u'_i \overline{u}_j}$ are equivalent. This equivalence is then established through terms second order in the filter length. The term $\overline{u'_i u'_j}$ and the higher order terms from expansion of the cross terms are fourth order and higher and equivalent respectively to $R^{\geq 2}_{ij}$ and $Q^{\geq 2}_j$. Therefore, the equivalence to Eqs. (31) and (32) is established for

$$\overline{u'_i u'_j} + \sum_{l=2}^{\infty} LC^u_{ij}(\alpha^l) = R^{\geq 2}_{ij} = \frac{1}{2!}(2\alpha)^2 \overline{u}_{i,kl} \overline{u}_{j,kl} + \text{HOT} \tag{42}$$

and

$$\overline{u'_j \phi'} + \sum_{l=2}^{\infty} LC^\phi_{ij}(\alpha^l) = Q^{\geq 2}_j = \frac{1}{2!}(2\alpha)^2 \overline{u}_{j,kl} \overline{\phi}_{,kl} + \text{HOT} \tag{43}$$

In Eqs. (42) and (43) $LC$ represents the power series expansion for the respective Leonard–Clark terms in terms of the filter term $\alpha$ ($= \Delta^2/4\gamma$).

Before leaving this section it is important to point out that the act of separation followed by averaging is not and should not be confined to the nonlinear terms. The linear terms resulting from separation also require "closure," and one measure of a satisfactory representation for the low pass and high pass components is whether terms $a - e$ in Eqs. (38) and (39) cancel. Only a comprehensively derived equation for $\overline{\alpha}$ and $\alpha'$ will allow this to occur. Therefore, these cancellations become a constraint that must be adhered to in the derivation of a closure for $\alpha'$.

### 23.4.3 Digitization/Discretization: Interpretive Constraints

The act of discretization of a continuous domain is a digitization, and the signal processing literature (e.g., Hamming 1990) provides considerable material on the constraints in adapting and interpreting analog and digital filters. One of the most elementary is the Nyquist wave number, $k_n$ (or frequency). If for the time being we assume that $\underline{k}_c = \underline{k}_n$ and $\omega_c = \omega_n$ (Figure 3c) then from the signal processing literature it is well known that all variability is removed via the low pass filter at $\underline{k}_c = \underline{k}_n$ and that any information from the spectrum for $\underline{k}_c > \underline{k}_n$ is not usable, as it is aliased data. Therefore, the low pass filter domain does not extend beyond $\underline{k}_c$. If we are again faithful to the signal processing literature, the high pass filter defined from $L(\underline{k}, \omega)$ is also subject to the same constraint (Hamming 1990). Therefore, the fluctuations are also defined only in a domain from $0 \longrightarrow \underline{k}_n = \underline{k}_c$ and not beyond as in zone III (Figures 3b,c). The implication then is that the turbulence ($f'$) being defined and parameterized in the averaged equations is defined only at the resolved grid scales and lower (in wave number). Therefore, fluctuations at the grid scale or higher (zone III, Figures 3b,c) are not defined or included in the averaged turbulent flow equations. This is intuitive. In light of this constraint and its implication, the use of the term "subgrid-scale" to refer to the term $\overline{u_i' u_j'}$ is misleading because the term does not include zone III fluctuations from the grid wave number and higher, which are thought to require parameterization. Rather, only the portion of the domain in zone II is parameterized. It is possible (Figure 3b) to have $\underline{k}_c < k_n$, and a much larger region of fluctuation response function domain (zone II) obtains. However, with the perfectly flat low pass response between $k_c$ and $k_n$, $\overline{u_i' u_j'}$ in zone II could still average to zero. Should the response function fluctuate about $R = 0$ in zone II ($k_c \longrightarrow k_n$) such as with the uniform filter response function, then the contribution to $\overline{u_i' u_j'}$ will be finite. In numerical models, therefore, Figure 3c represents the response function with quite small $\overline{u_i' u_j'}$ contributions.

The region between $k_t$ and $k_c$ is the transition region, and it is this region where the cross-product terms $\overline{u_i' \overline{u}_j}$ and $\overline{\overline{u}_i u_j'}$ are important.

What the YB-III expansion on the nonlinear terms $\overline{u_i u_j}$ and $\overline{u_j \phi}$ does is to specify fully and completely the actual average equations resolved to the permissible limit $\underline{k}_n$ or its numerical model form where $\underline{k}_c = \underline{k}_n$. It has also been argued that the limit of the filter imposed by the model grid is $\underline{k}_c = \underline{k}_n$, and that the $\overline{u_i' u_j'}$ (or $\overline{u_i' \phi'}$) terms are quite small and therefore *do not* represent contributions from the "subgrid scale" (zone III, Figures 3b,c) as they are beyond the Nyquist limit of the filter domain.

### 23.4.4 A Final Note on $R_{ij}$

The representation of turbulence in zone II in terms of the mean flow by $R_{ij}$ $(Q_j)$ is generalized and contains several interesting features. It is a simple matter to show the structure of $R_{ij}$ in terms of vorticity, $\Omega_{ij}$, and strain rate, $S_{ij}$. In terms of the averaged field variables,

$$S_{ij} = \tfrac{1}{2}(\overline{u}_{i,j} + \overline{u}_{j,i}) \tag{44}$$

and

$$\Omega_{ij} = \tfrac{1}{2}(\overline{u}_{i,j} - \overline{u}_{j,i}) \tag{45}$$

The first term in $R_{ij}^1$ can then be written as

$$R_{ij}^1 = (2\alpha)[\underbrace{S_{ik}S_{jk}}_{a} + \underbrace{S_{ik}\Omega_{jk} + \Omega_{ik}S_{jk}}_{b} + \underbrace{\Omega_{ik}\Omega_{jk}}_{c}] \tag{46}$$

Term $a$ is the deformation tensor, which is a basis for the Smagorinsky model. Term $c$ is the rotation tensor, which has also been used as the basis for empirical functions describing flows with high velocity. The terms $b$ are essentially vortex stretching terms.

It is interesting that all three of these terms result from full resolution of low pass average equations including the cross-term contribution. These terms represent activity between $k_t$ and $k_c$ (i.e., zone I and zone II) and do not represent subgrid-scale activity. Yet the fact remains that these terms in more simplified form are the basis for a variety of "closures" for the subgrid-scale activity. With these average equations containing so small a contribution from the $\overline{u_i'u_j'}$ $(\overline{u_i'\phi})$ terms and no subgrid-scale contribution, how might equations be formulated which include the effects of turbulence from all wave numbers up to the microscale wavenumber (zone III)? The next section explores this requirement.

### 23.5 The Separated Equations
#### 23.5.1 Analog Considerations: A Two-Scale Separation

The total field or continuous analog equations in Eqs. (23) and (24) are valid for all wave numbers up to the microscale, $k_m$, as are the separated equations in Eqs. (25) and (26). The separated and total field equations are quite appealing, as the decomposition of the nonlinear term is not a problem. Furthermore comparison of total equations (25) and (26) with the averaged equations (31) and (32) reveals considerable equivalence in the terms which are functions of the averaged variables. However, the essential difference between the two equations is that the full effect of the fluctuating physics is retained in Eqs. (25) and (26), and not in Eqs. (31) and (32). It is quite simple to derive an equation for the fluctuating components. Subtracting Eqs. (31) and (32) from Eqs. (25) and (26) gives

$$\frac{\partial u_i'}{\partial t} + (u_i'\overline{u}_j + \overline{u}_i u_j' + u_i'u_j')_{,j} - R_{ij,j} = \frac{-1}{\rho}P_{,i}' + \nu u_{i,jj}' \qquad (47)$$

$$\frac{\partial \phi'}{\partial t} + (u_i'\overline{\phi} + \overline{u}_i\phi' + u_i'\phi')_{,i} - Q_{i,i} = \lambda\phi_{,jj}' \qquad (48)$$

Notice that there is an explicit turbulence and mean flow interaction as contrasted with the mean flow equations. The two-scale separation employed here is permitted because of the analog nature of the filter wherein the filter function is essentially continuously defined to the mesoscale limit. Therefore, the turbulence represents all the domain in zone II of Figure 3a. The model/digital form is different, however.

### 23.5.2 The Digital Formulation: A Three-Scale Approach

As noted in the preceding section the introduction of the grid essentially digitizes the continuum and for all intents and purposes moves the Nyquist wave number from $\infty$ back to $k_c$, the cutoff wave number (Figure 3c). This reorganization results in the low pass/high pass filter domain extending only to $k_c$, $k_n$ and results in a total field equation separated into the low frequency terms (zone I in Figure 3c), the resolved turbulent fluctuations ($f'$, zone II in Figure 3c) and the subgrid scale turbulent fluctuations ($f^*$, zone III, Figure 3c). With this three-scale definition a total field variable is separated into three components such that

$$f(\underline{x},t) = \overline{f}(\underline{x},t) + f'(\underline{x},t) + f^*(\underline{x},t) \qquad (49)$$

This separation can be directly inserted into the total field equations;

however, it is more convenient to recognize that the resolved scale turbulence can be "closed" via the YB-II series valid to $k_c = k_n$, i.e.,

$$f(\underline{x},t) = \overline{f}(\underline{x},t) + (-\alpha \overline{f}_{,ii} + \text{HOT}) + f^*(\underline{x},t) \qquad (50)$$

Using only the first term of the $f'$ expansion, substitution of Eq. (50) into Eqs. (23) and (24) gives for isotropic filters

$$\frac{\partial \overline{u}_i}{\partial t} + (\overline{u}_i \overline{u}_j)_{,j} + \frac{1}{\rho}\overline{P}_{,i} - \nu \overline{u}_{i,jj}$$

$$- \alpha \left\{ \frac{\partial}{\partial t}(\overline{u}_{i,kk}) + (\overline{u}_i \overline{u}_{j,kk} + \overline{u}_{i,kk}\overline{u}_j + \overline{u}_{i,kk}\overline{u}_{j,ll})_{,j} \right.$$

$$\left. + \frac{1}{\rho}\overline{P}_{,ijj} - \nu \overline{u}_{i,jjkk} \right\}$$

$$= \frac{\partial u_i^*}{\partial t} + (u_i^* u_j^*)_{,j} + \frac{1}{\rho}P_{,i}^* - \nu u_{i,jj}^*$$

$$+ \underbrace{\left\{ (\overline{u}_i - \alpha \overline{u}_{i,kk})(u_j^*) + (u_i^*)(\overline{u}_j - \alpha \overline{u}_{j,kk}) \right\}_{,j}}_{a} \qquad (51)$$

$$\frac{\partial \overline{\phi}}{\partial t} + (\overline{u}_i \overline{\phi})_{,i} - \lambda \overline{\phi}_{,jj}$$

$$- \alpha \left\{ \frac{\partial}{\partial t}\overline{\phi}_{,jj} + (\overline{u}_{i,jj}\overline{\phi} + \overline{u}_i \overline{\phi}_{,jj} + \overline{u}_{i,jj}\overline{\phi}_{,kk})_{,i} - \lambda \overline{\phi}_{,jjkk} \right\}$$

$$= \frac{\partial \phi^*}{\partial t} + (u_i^* \phi^*)_{,i} - \lambda \phi_{,jj}^*$$

$$+ \underbrace{\left\{ (\overline{u}_i - \alpha \overline{u}_{i,kk})(\phi^*) + (u_i^*)(\overline{\phi} - \alpha \overline{\phi}_{,kk}) \right\}_{,i}}_{b} \qquad (52)$$

The terms on the left hand side of Eqs. (51) and (52) represent activity valid in the region $k < k_c = k_n$ expressed in terms of the filtered quantity, while the right hand side represents the actual subgrid-scale activity and its interaction with the resolved (not mean) flow through terms $a$ and $b$. It is noted that the first four terms in the Eq. (51) and the first three terms of Eq. (52) are identical to the first four terms in the average momentum equation (31) and the first three terms in the transport equation (32) respectively. Yet Eqs. (51) and (52) are wholly different in that unlike Eqs. (31) and (32) these equations include all the resolved and subgrid-scale activity. That the filtered lower order differential terms in the average and total equations are identical is a major source of misconception about whether the average equations are being used in turbulence models or whether in fact the total field equations separated into the three basic components should be used.

## 23.6 Concluding Remarks

An attempt has been made in this chapter to clarify some misinterpretations of the role and use of the averaged equations for turbulent flow and transport modeling. The intended result of these clarifications is to direct modelers' attention to the fact that many, if not most, modelers who employ the concept of subgrid-scale parameterization are not using (nor should they use) the averaged equations, since the concept of subgrid-scale contributions to properly low-pass filtered or averaged equations is inapplicable. If one wishes to robustly include the subgrid effects, then the total field equations must be used. The process of separation into filtered (averaged) and turbulent components yields a two-scale separation which can be formally applied to the continuous or analog equation. However, the digitization of the total equations via the numerical grid introduces a third separation resulting from the introduction of the Nyquist wave number and its equivalence to the cutoff wave number. In this way a three-scale separation is achieved, i.e., averaged variables, grid-resolved turbulent fluctuations which can be formally related to the averaged variables, and the subgrid-scale turbulence. By contrasting the average and the total/separated equations it is noticed that one of the possible sources of conceptual confusion exists because the fundamental terms in both sets of equations are identical, i.e. the filtered first order differential terms and the molecular terms. Several more specific comments apply.

First by use of high pass averaging techniques from the signal processing literature the fluctuation components can, for Gaussian (and uniform) filters, be expanded in a power series as a function of not only the unfiltered or total variable but more importantly the filtered variable. This series can be used to create a direct power series expansion for the average nonlinear advection and inertia terms.

Second, the use of this generalized series expansion results in a set of averaged governing differential equations which do not contain any formal turbulent correlation closure terms as they were not derived by a separation methodology applied to the nonlinear terms.

Third, the traditional preparation of the average equations by separation of the nonlinear portions of the equation followed by averaging is seen to be incomplete in that the entire equation must be formally separated and then averaged. These separated then filtered equations formally reduce to the directly averaged equations and in so doing it is shown that the Leonard–Clark terms are formally equal to the first term of the conjunctive filter derived power series expansion of the nonlinear terms.

Finally, the terms $\overline{u_i' u_j'}$ and $\overline{u_i' \phi'}$ are shown to be quite small in the averaged equations. This is intuitive because the domain of the filter extends to the Nyquist limit only when one is considering a discretization (digitization) of the continuous equations. Thus the concept of subgrid-scale contributions is not applicable to the correctly averaged equations.

## Acknowledgments

This chapter is a contribution from the Great Lakes Forecasting Program. Support from CRAY Research Inc., NOAA-Ohio Sea Grant, Ohio State University and the U.S. Geological Survey Water Resources Center is gratefully acknowledged. Earlier support for this work came from National Science Foundation Grant No. CEE 8410552 and is also appreciated. The authors benefited considerably from conversations with C.F. Kuan, D. Podber, G. Baker, T. Herbert and Y. Guzennac and their suggestions are appreciated.

## Rerefences

ALDAMA, A. (1985) Theory and applications of two and three scale filtering approaches for turbulent flow simulations. Ph.D. thesis, Massachusetts Institute of Technology.

ALDAMA, A. (1990) *Filtering Techniques for Turbulent Flow Simulations*, Lecture Notes in Engineering. Springer-Verlag.

BABAJIMOPOULOS, C. AND BEDFORD, K. (1980) Formulating lake models which preserve spectral statistics. *J. Hydraulic Eng.* **106**, 1.

BEDFORD, K. (1981) Spectra preservation capabilities of Great Lakes transport models. In *Transport Models for Island and Coastal Waters*. Ed. H. Fisher, pp. 172–221. Academic Press.

BEDFORD, K. (in press) Diffusion, dispersion and subgrid scale parameterization. In *Coastal, Estuarial and Harbour Engineers Reference Book*. Ed. M. Abbott and W. Price. Chapman and Hall.

BEDFORD, K. AND ABDELRHMAN, M. (1987) Analytical and experimental studies of the benthic boundary layer and their applicability to near bottom transport in Lake Erie. *J. Great Lakes Res.* **13**, 628.

BEDFORD, K. AND BABAJIMOPOULOS, C. (1980) Verifying lake transport models with spectral statistics. *J. Hydraulic Eng.* **106**, 21.

BEDFORD, K. AND DAKHOUL, Y. (1982) Applying LES turbulence modeling to open channel flow. In *Applying Research to Hydraulic Practice*. Ed. P. Smith, pp. 32–43. Amer. Soc. Civil Engineers.

BEDFORD, K., DINGMAN, S. AND YEO, W. (1987a) Preparation of estuary and marine model equations by generalized filtering methods. In *Three-Dimensional Models of Marine and Estuary Dynamics*. Ed. J. Nihoul and B. Jamart, pp. 113–125, Elsevier.

BEDFORD, K., WAI, O., LIBICKI, C. AND VAN EVRA, R III (1987b) Sediment entrainment and deposition measurements in Long Island Sound. *J. Hydraulic Eng.* **113**, 1325.

BLINCHIKOFF, H. AND ZVEREV, A. (1976) *Filtering in the Time and Frequency Domain*. Wiley.

BOYCE, F. (1974) Some aspects of Great Lakes physics of importance to biological and chemical processes. *J. Fish. Res. Board of Canada* **31**, 689.

CLARK, R., J. FERZIGER AND REYNOLDS, W. (1979) Evaluation of subgrid scale models using an accurately simulated turbulent flow. *J. Fluid Mech.* **91**, 1.

DAKHOUL, Y. AND BEDFORD, K. (1986a) Improved averaging method for turbulent flow simulation. Part I : Theoretical development and application to Burgers transport equation. *Int. J. Num. Meth. Fluids* **6**, 49.

DAKHOUL, Y. AND BEDFORD, K. (1986b) Improved averaging method for turbulent flow simulation. Part II : Calculations and verifications. *Int. J. Num. Meth. Fluids* **6**, 65.

DINGMAN, J. (1986) The development and application of generalized higher order filtering techniques to the continuum wave equations. Ph.D. thesis, Ohio State University.

DINGMAN, J. AND BEDFORD, K. (1984) The Lake Erie response to the January 1978 cyclone. *J. Geophys. Res.* **89**, 6427.

GROSS, T. AND NOWELL, A. (1983) Mean flow and turbulence scaling in a tidal boundary layer. *Cont. Shelf Res.* **2**, 109.

GROSS, T. AND NOWELL, A. (1985) Spectral scaling in a tidal boundary layer. *J. Phys. Oceanogr.* **15**, 496.

HAMMING, R. (1989) *Digital Filters*, 3rd ed. Prentice Hall.

HILDEBRAND, F. (1976) *Advanced Calculus for Applications*, 2nd ed. Prentice Hall.

KWAK, D., REYNOLDS, W. AND FERZIGER, J. (1975) Three-dimensional time dependent computation of turbulent flow. Dept. TF-5, Dept. of Mechanical Engineering, Stanford University.

LEONARD, A. (1974) Energy cascade in large eddy simulations of turbulent fluid flows. *Adv. Geophys.* **18A**, 237.

LIBICKI, C. AND BEDFORD, K. (1990) Sudden, extreme Lake Erie storm surges and the interaction of wind-stress, resonance and geometry. *J. Great Lakes Res.* **16**, 380.

ROSSMAN, P. (1987) Modeling shallow water bodies via filtering techniques. Ph.D. thesis, Massachusetts Institute of Technology.

YEO, W. (1987) A generalized high pass/low pass filtering procedure for deriving and solving turbulent flow equations. Ph.D. thesis, Ohio State University.

YEO, W. AND BEDFORD, K. (1988) Closure-free turbulence modeling based upon a conjunctive higher order averaging procedure. In *Computational Methods in Flow Analysis*. Ed. H. Niki and M. Kawahara, pp. 844–851. Okayama University of Science.

YEO, W. AND BEDFORD, K. (in press a) A generalized high pass/low pass averaging procedure for deriving turbulent flow and transport model equations.

YEO, W. AND BEDFORD, K. (in press b) A first analysis of the generalized high pass/low pass averaging procedure and resulting turbulent flow and transport model equations.

# 24

---

# Leonard and Cross-Term Approximations in the Anisotropically Filtered Equations of Motion

## ALVARO A. ALDAMA

## 24.1 Introduction

The large eddy simulation (LES) approach is a procedure aimed at making feasible the numerical simulation of high Reynolds number turbulent flows. The LES technique is based on evidence which indicates that, while the large-scale dynamics in turbulent flows are very sensitive to flow domain geometry and boundary conditions and, therefore, vary from flow to flow, the small-scale behavior tends to be quite universal. Accordingly, in LES calculation the large-scale motion is explicitly resolved and the effects of the unresolved scales are accounted for through the use of simple closures, such as the Smagorinsky–Lilly eddy viscosity model (Smagorinsky, 1963; Lilly, 1967). The explicit resolution of large scales is made possible by applying a filtering operation to the equations of motion, thus obtaining filtered equations which govern the dynamics of the large scales. These equations are then numerically solved employing mesh spacings which are of proper size for resolving the large scales.

The LES technique has been quite successfully used in the simulation of various flows. For instance, this procedure has been employed by Ferziger et al. (1977), Antonopoulos-Domis (1979), and Bardina et al. (1985) for the simulation of isotropic turbulence; by Deardorff (1970), Schumann (1975), Moin and Kim (1982), Mason and Callen (1986), and Piomelli et al. (1988) for the simulation of turbulent channel flow; by Mansour et al. (1978) for the simulation of turbulent mixing layers; by Eidson (1985) for the simulation of turbulent natural convection; by Deardorff (1973, 1974) for the simulation of the planetary boundary layer; and by Babajimopoulos and Bedford (1980), Bedford and Babajimopoulos (1980), and Bedford (1981) for the simulation of lake circulation. Different methods and models have been employed in those investigations to handle the nonlinear terms that appear as a result of filtering the Navier–Stokes equations, namely the Leonard terms (Leonard, 1974), the cross terms, and the subgrid-scale (SGS) terms. Among these, the cross terms represent the direct interaction between resolved and unresolved scales, which gives rise to backscatter or energy transfer from small scales to large scales. In most LES computations this interaction has not been properly accounted for, in that the cross terms are lumped with the SGS stresses and then modeled using some dissipative closure scheme, which is unable to model the aforementioned energy transfer mechanism.

Backscatter appears to be quite important in near-wall regions, transitional flows, and "thin flows" which tend to exhibit nearly two-dimensional behavior. Flows of the latter type are ubiquitous in environmental and geophysical applications in which there exists a disparity between characteristic horizontal and vertical scales in the flow domain (e.g., atmospheric boundary layer flows, lake circulation, near-shore ocean circulation). The proper representation of the cross terms is thus very important in flows in which backscatter is expected to play a significant role.

It is also well known that one can compute the Leonard terms exactly in LES calculation by treating them explicitly (in the time stepping sense). Nevertheless, this approach imposes stiff stability constraints which force the use of small time steps. These constraints may be undesirable in the simulation of environmental and geophysical flows where there exists a very wide range of dynamically significant scales. Hence, in such cases, the use of local (i.e., nonintegral) approximations for the Leonard terms seems to be attractive. One such approximation is that introduced by Leonard in 1974. The Leonard approximation was employed in some LES computations performed in the 1970s and early 1980s, but appears to have fallen into disuse since then. One possible explanation for this (besides the fact that the Leonard terms can be computed exactly) is that the convergence of the series that gives rise to the Leonard approximation cannot be proven, as pointed out by Love (1980).

Aldama (1990) and Aldama and Harleman (1991) have derived an approximation of the cross terms and have analyzed its mathematical nature and that of the Leonard approximation, in physical space and Fourier space, for the case of isotropically filtered equations. In many applications, strong anisotropy is imposed by the geometry of the flow domain, as in the case of thin environmental and geophysical flows, or by the action of buoyancy. It is thus of interest to consider the application of anisotropic filters to the equations of motion. The aforementioned analyses of the Leonard and cross-term approximations are extended in this chapter to the case of the anisotropically filtered Navier–Stokes equations.

## 24.2 Space-Filtered Equations of Motion and the Leonard Approximation

The LES procedure is based on the use of space filtering on the equations of motion. Let us denote all instantaneous flow variables by uppercase letters and assume that $U_i = U_i(\mathbf{x}, t)$ (where $\mathbf{x}$ is a Cartesian position vector and $t$ is time) represents the instantaneous velocity field. The corresponding filtered or large-scale (LS) velocity is defined by the convolution integral

$$\bar{U}_i(\mathbf{x}, t) = \int G(\mathbf{x} - \mathbf{x}') U_i(\mathbf{x}', t)\, dr'^3, \qquad (1)$$

where from hereon the integral sign should be interpreted as a Lebesgue integral taken over an unbounded domain (in the understanding that it is assumed that the definition of all flow variables is properly extended outside bounded domains), $dr^3 = dx\, dy\, dz$ and $G(\mathbf{x})$ is a mean-preserving filter function. As was mentioned before, the interest in this chapter is focused on the case of anisotropically filtered equations of motion. Thus, in what follows it will be assumed that $G$ is an anisotropic Gaussian filter, defined by the following expression:

$$G(\mathbf{x}) = \left(\frac{\gamma}{\pi}\right)^{3/2} \frac{\exp\{-\gamma[(x/\lambda_x)^2 + (y/\lambda_y)^2 + (z/\lambda_z)^2]\}}{\lambda^3}, \qquad (2)$$

where $\gamma$ is a constant parameter, normally taken equal to 6 (for a discussion on this choice, see Aldama, 1990); $\lambda_x$, $\lambda_y$, and $\lambda_z$ are the filter widths in the $x$, $y$, and $z$ directions, respectively; and $\lambda = (\lambda_x \lambda_y \lambda_z)^{1/3}$. The rationale for choosing a Gaussian filter will become apparent later.

Let us now introduce the following decomposition:

$$U_i = \bar{U}_i + u_i, \qquad (3)$$

where $u_i$ represents the small-scale or SGS component of the velocity field. Spatially filtering the incompressible Navier–Stokes equations, including Coriolis effects, and employing Eq. (3), the following result is obtained:

$$\frac{\partial \bar{U}_i}{\partial t} + \frac{\partial \bar{U}_i \bar{U}_j}{\partial x_j} + 2\epsilon_{ijk}\Omega_j\bar{U}_k$$

$$= -\frac{1}{\rho}\frac{\partial \bar{P}}{\partial x_i} + \frac{\partial}{\partial x_j}\left(\nu\frac{\partial \bar{U}_i}{\partial x_j} - L_{ij} - C_{ij} - T_{ij}\right), \qquad (4)$$

where $\Omega_i$ represents the earth's rotation angular velocity; $P$, the instantaneous dynamic pressure; $\rho$, the fluid density; $\nu$, the fluid kinematic viscosity; $L_{ij} = \overline{\bar{U}_i\bar{U}_j} - \bar{U}_i\bar{U}_j$, the Leonard terms; $C_{ij} = \overline{\bar{U}_i u_j + u_i\bar{U}_j}$, the cross terms; and $T_{ij} = \overline{u_i u_j}$, the SGS terms. For a general filtering operation, Reynolds postulates do not hold, i.e., $L_{ij} \neq 0$, $C_{ij} \neq 0$.

Among other things, the Gaussian filter defined by Eq. (2) is a good function, in that it is infinitely differentiable and all its derivatives vanish at infinity more rapidly than any algebraic power of $|\mathbf{x}|$. The instantaneous velocity field (and for that matter, all instantaneous flow variables) has finite energy, in the sense that the Lebesgue integral $\int U_i U_i \, dr^3$ exists finitely. Therefore, it is *very* reasonable to assume that $U_i$ is in a so-called class K of functions, which are Lebesgue-integrable when divided by a polynomial of a proper degree. There exists a theorem in harmonic analysis which states that the convolution of a function in class K and a good function is everywhere continuous and infinitely differentiable (Champeney, 1989). Whence, in view of Eq. (1), the LS component of the velocity field, $\bar{U}_i$, is everywhere continuous and infinitely differentiable. It is therefore reasonable to assume $\bar{U}_i$ to be regular or expandable in a Taylor series in the scale of the filter widths $\lambda_x$, $\lambda_y$, and $\lambda_z$. The Leonard approximation can be justified on these grounds (when the LS component of the velocity field is defined by convoluting the instantaneous velocity with a good function), and is derived for the case of the anisotropic Gaussian filter as follows.

The space-filtered LS advective terms are defined by

$$\overline{\bar{U}_i\bar{U}_j}(\mathbf{x}, t) = \int G(\mathbf{x} - \mathbf{x}')\bar{U}_i\bar{U}_j(\mathbf{x}', t) \, dr'^3. \qquad (5)$$

Formally expanding $\bar{U}_i\bar{U}_j(\mathbf{x}', t)$ in a Taylor series around $(\mathbf{x}, t)$ in Eq. (5) yields:

$$\overline{\bar{U}_i \bar{U}_j}(\mathbf{x}, t) = \int G(\mathbf{x} - \mathbf{x}') \sum_{k=0}^{\infty} \frac{1}{k!} \left[ (x'_l - x_l) \frac{\partial}{\partial x_l} \right]^k \bar{U}_i \bar{U}_j(\mathbf{x}, t) \, dr'^3. \quad (6)$$

Since $G(\mathbf{x})$ is even in $x$, $y$ and $z$, all its odd moments with respect to the coordinate directions vanish. Hence, Eq. (6) becomes:

$$\overline{\bar{U}_i \bar{U}_j} = \bar{U}_i \bar{U}_j + \frac{\sigma_{kl}}{2} \frac{\partial^2 \bar{U}_i \bar{U}_j}{\partial x_k \partial x_l} + \frac{\tau_{klmn}}{24} \frac{\partial^4 \bar{U}_i \bar{U}_j}{\partial x_k \partial x_l \partial x_m \partial x_n} + \cdots, \quad (7)$$

where $\sigma_{kl} = \int x_k x_l G(\mathbf{x}) \, dr^3$ is the tensor of second order moments of $G$, and $\tau_{klmn} = \int x_k x_l x_m x_n G(\mathbf{x}) \, dr^3$ is the tensor of fourth order moments of $G$. It should be observed that $\sigma_{kl}$ is a diagonal tensor and that its diagonal elements are given by

$$\sigma_{ii} = \frac{\lambda_i^2}{2\gamma} \quad \text{(no summation)}. \quad (8)$$

It can be shown, by a lengthy but straightforward calculation, that

$$\tau_{klmn} \frac{\partial^4 \bar{U}_i \bar{U}_j}{\partial x_k \partial x_l \partial x_m \partial x_n} = 3\sigma_{kl} \sigma_{mn} \frac{\partial^4 \bar{U}_i \bar{U}_j}{\partial x_k \partial x_l \partial x_m \partial x_n}. \quad (9)$$

Thus, Eq. (7) becomes:

$$\overline{\bar{U}_i \bar{U}_j} = \bar{U}_i \bar{U}_j + \frac{\sigma_{kl}}{2} \frac{\partial^2 \bar{U}_i \bar{U}_j}{\partial x_k \partial x_l} + \frac{\sigma_{kl} \sigma_{mn}}{8} \frac{\partial^4 \bar{U}_i \bar{U}_j}{\partial x_k \partial x_l \partial x_m \partial x_n}$$
$$+ O(\sigma_{kl} \sigma_{lm} \sigma_{mk}), \quad (10)$$

which constitutes the Leonard approximation for the case of an aniso-tropic Gaussian filter.

## 24.3 The Approximation of the Cross Terms

The development of a local approximation of the cross terms was first pursued by Clark et al. (1977), who proposed to expand both the LS velocity, $\bar{U}_i$, and the SGS velocity, $u_j$, in a Taylor series inside the integral that defines the cross terms, $\overline{\bar{U}_i u_j}$. As was mentioned before, it is mathematically justifiable to assume that $\bar{U}_i$ is regular in the scale of the filter widths, but this is definitely not the case for $u_j$, given its highly fluctuating nature. Therefore, Clark et al.'s procedure is mathematically unjustified. Rogallo and Moin (1984) suggested an alternative procedure for the derivation of a local approximation of the cross terms, consisting of expanding only $\bar{U}_i$ inside the convolution integral that defines the cross terms. However, these authors did not fully pursue their idea, probably because of the infeasibility of proving the convergence of the series that gives rise to approximations of this kind. As will be shown later, proving convergence is not necessary in order to justify the use of these approximations. Aldama and Harleman (1991) have derived an approximation of the cross terms in the realm of Rogallo and Moin's suggestion, for the case of an isotropic Gaussian filter. This result is generalized below for the case of an anisotropic Gaussian filter.

To the knowledge of the author, a local approximation of the cross terms can be developed only when a Gaussian filter is employed. This is another reason for choosing such a filter function. Now, the cross terms are defined by

$$\overline{U_i u_j}(\mathbf{x}, t) = \int G(\mathbf{x} - \mathbf{x}') \bar{U}_i(\mathbf{x}', t) u_i(\mathbf{x}', t) \, dr'^3. \tag{11}$$

Formally expanding the LS velocity field, $\bar{U}_i(\mathbf{x}', t)$, in a Taylor series around $(\mathbf{x}, t)$ in Eq. (11) yields

$$\overline{U_i u_j}(\mathbf{x}, t) = \int dr'^3 \, G(\mathbf{x} - \mathbf{x}')$$
$$\times \sum_{k=0}^{\infty} \frac{1}{k!} \left[ (x_l' - x_l) \frac{\partial}{\partial x_l} \right]^k \bar{U}_i(\mathbf{x}, t) u_j(\mathbf{x}', t), \tag{12}$$

or

$$\overline{U_i u_j}(\mathbf{x}, t) = \bar{U}_i \bar{u}_j + \frac{\partial \bar{U}_i}{\partial x_k} \int (x_k' - x_k) G(\mathbf{x} - \mathbf{x}') u_j(\mathbf{x}', t) \, dr'^3$$
$$+ \frac{1}{2} \frac{\partial^2 \bar{U}_i}{\partial x_k \partial x_l} \int (x_k' - x_k)(x_l' - x_l) G(\mathbf{x} - \mathbf{x}') u_j(\mathbf{x}', t) \, dr'^3$$
$$+ \frac{1}{6} \frac{\partial^3 \bar{U}_i}{\partial x_k \partial x_l \partial x_m} \int dr'^3 \, G(\mathbf{x} - \mathbf{x}') u_j(\mathbf{x}', t)$$
$$\times (x_k' - x_k)(x_l' - x_l)(x_m' - x_m) + \cdots. \tag{13}$$

An anisotropic Gaussian filter, as defined by Eq. (2), satisfies the following relation:

$$x_k G(\mathbf{x}) = -\sigma_{kl} \frac{\partial G(\mathbf{x})}{\partial x_l}. \tag{14}$$

Hence,

$$\int (x_k' - x_k) G(\mathbf{x} - \mathbf{x}') u_j(\mathbf{x}', t) \, dr'^3 = -\sigma_{kl} \int \frac{\partial G(\mathbf{x} - \mathbf{x}')}{\partial x_l'} u_j(\mathbf{x}', t) \, dr'^3$$

$$= \sigma_{kl} \int G(\mathbf{x} - \mathbf{x}') \frac{\partial u_j(\mathbf{x}', t)}{\partial x_l'} \, dr'^3 = \sigma_{kl} \frac{\partial \bar{u}_j}{\partial x_l}, \tag{15}$$

where Gauss' theorem and the fact that the anisotropic Gaussian filter is a good function have been used. A similar procedure yields the following results:

$$\int (x_k' - x_k)(x_l' - x_l) G(\mathbf{x} - \mathbf{x}') u_j(\mathbf{x}', t) \, dr'^3$$

$$= \sigma_{kl} \bar{u}_j + \sigma_{km}\sigma_{ln} \frac{\partial^2 \bar{u}_j}{\partial x_m \partial x_n}, \tag{16}$$

$$\int (x_k' - x_k)(x_l' - x_l)(x_m' - x_m) G(\mathbf{x} - \mathbf{x}') u_j(\mathbf{x}', t) \, dr'^3$$

$$= (\sigma_{kl}\sigma_{mn} + \sigma_{km}\sigma_{ln} + \sigma_{kn}\sigma_{lm}) \frac{\partial \bar{u}_j}{\partial x_n} + \sigma_{kn}\sigma_{lp}\sigma_{mq} \frac{\partial^3 \bar{u}_j}{\partial x_n \partial x_p \partial x_q}. \tag{17}$$

Substituting Eqs. (15)–(17) in Eq. (13) results in

$$\overline{U_i u_j} = \bar{U}_i \bar{u}_j + \sigma_{kl} \frac{\partial \bar{U}_i}{\partial x_k} \frac{\partial \bar{u}_j}{\partial x_l} + \frac{\sigma_{kl}}{2} \frac{\partial^2 \bar{U}_i}{\partial x_k \partial x_l} \bar{u}_j + \frac{\sigma_{km}\sigma_{ln}}{2} \frac{\partial^2 \bar{U}_i}{\partial x_k \partial x_l} \frac{\partial^2 \bar{u}_j}{\partial x_m \partial x_n}$$

$$+ \frac{1}{6}(\sigma_{kl}\sigma_{mn} + \sigma_{km}\sigma_{ln} + \sigma_{kn}\sigma_{lm}) \frac{\partial^3 \bar{U}_i}{\partial x_k \partial x_l \partial x_m} \frac{\partial \bar{u}_j}{\partial x_n}$$

$$+ \frac{1}{6}\sigma_{kn}\sigma_{lp}\sigma_{mq} \frac{\partial^3 \bar{U}_i}{\partial x_k \partial x_l \partial x_m} \frac{\partial^3 \bar{u}_j}{\partial x_n \partial x_p \partial x_q} + \cdots. \tag{18}$$

Now, by definition (cf. Eq. (3)): $\bar{u}_j = \bar{U}_j - \bar{\bar{U}}_j$. Moreover, employing a Leonard-type of procedure, it can be shown that

$$\bar{\bar{U}}_j = \bar{U}_j + \frac{\sigma_{kl}}{2} \frac{\partial^2 \bar{U}_j}{\partial x_k \partial x_l} + O(\sigma_{kl}\sigma_{lk}). \tag{19}$$

Therefore,

$$\bar{u}_j = -\frac{\sigma_{kl}}{2} \frac{\partial^2 \bar{U}_j}{\partial x_k \partial x_l} + O(\sigma_{kl}\sigma_{lk}). \tag{20}$$

Substituting Eq. (20) in Eq. (18) the following expression is obtained:

$$\overline{U_i u_j} = -\frac{\sigma_{kl}}{2} \bar{U}_i \frac{\partial^2 \bar{U}_j}{\partial x_k \partial x_l} - \frac{\sigma_{kl}\sigma_{mn}}{2} \left( \frac{\partial \bar{U}_i}{\partial x_k} \frac{\partial^3 \bar{U}_j}{\partial x_l \partial x_m \partial x_n} \right.$$

$$\left. + \frac{1}{2} \frac{\partial^2 \bar{U}_i}{\partial x_k \partial x_l} \frac{\partial^2 \bar{U}_j}{\partial x_m \partial x_n} \right) + O(\sigma_{kl}\sigma_{lm}\sigma_{mk}). \tag{21}$$

A similar procedure yields

$$\overline{u_i U_j} = -\frac{\sigma_{kl}}{2} \frac{\partial^2 \bar{U}_i}{\partial x_k \partial x_l} \bar{U}_j - \frac{\sigma_{kl}\sigma_{mn}}{2} \left( \frac{\partial^3 \bar{U}_i}{\partial x_l \partial x_m \partial x_n} \frac{\partial \bar{U}_j}{\partial x_k} \right.$$

$$+ \frac{1}{2}\frac{\partial^2 \bar{U}_i}{\partial x_m \partial x_n}\frac{\partial^2 \bar{U}_j}{\partial x_k \partial x_l}\right) + O(\sigma_{kl}\sigma_{lm}\sigma_{mk}). \tag{22}$$

Equations (21) and (22) constitute the approximation of the cross terms for the case of anisotropically filtered equations of motion.

Combining the Leonard approximation, Eq. (10), with the approximation of the cross terms, Eqs. (21) and (22), yields

$$\overline{\bar{U}_i\bar{U}_j} + \overline{\bar{U}_i u_j} + \overline{u_i\bar{U}_j} = \bar{U}_i\bar{U}_j + \sigma_{kl}\frac{\partial \bar{U}_i}{\partial x_k}\frac{\partial \bar{U}_j}{\partial x_k}$$

$$+ \frac{\sigma_{kl}\sigma_{mn}}{2}\left[\frac{1}{4}\left(\frac{\partial^4 \bar{U}_i}{\partial x_k \partial x_l \partial x_m \partial x_n}\bar{U}_j + \bar{U}_i\frac{\partial^4 \bar{U}_j}{\partial x_k \partial x_l \partial x_m \partial x_n}\right)\right.$$

$$\left.+ \frac{\partial^2 \bar{U}_i}{\partial x_k \partial x_m}\frac{\partial^2 \bar{U}_j}{\partial x_l \partial x_n} - \frac{1}{2}\frac{\partial^2 \bar{U}_i}{\partial x_k \partial x_l}\frac{\partial^2 \bar{U}_j}{\partial x_k \partial x_l}\right] + O(\sigma_{kl}\sigma_{lm}\sigma_{mk}). \tag{23}$$

As was mentioned before, the convergence of the series that gives rise to the Leonard and cross-term approximations cannot be proven. This underscores the importance of performing analyses of their nature in physical and Fourier spaces. Such analyses are presented in the following sections. In particular, the asymptotic nature of both approximations is demonstrated, a fact that allows a formal perturbation theory to be built around them.

## 24.4 Analysis of the Leonard Approximation in Physical Space

Equation (6) can be formally expressed as

$$\overline{\bar{U}_i\bar{U}_j}(\mathbf{x},t) = \sum_{k=0}^{\infty}\frac{1}{k!}\int G(\mathbf{x}-\mathbf{x}')\left[(x_l'-x_l)\frac{\partial}{\partial x_l}\right]^k \bar{U}_i\bar{U}_j(\mathbf{x},t)\, dr'^3. \tag{24}$$

Using the fact that $G(\mathbf{x},t)$ is even in $x$, $y$, and $z$, and employing the multinomial expansion, Eq. (24) can be written as

$$\overline{\bar{U}_i\bar{U}_j}(\mathbf{x},t) = \sum_{k=0}^{N-1}\sum_{l=0}^{k}\sum_{m=0}^{l}\frac{1}{[2(k-l)]![2(l-m)]!(2m)!}$$

$$\times \frac{\partial^{2k}\bar{U}_i\bar{U}_j}{\partial x^{2(k-l)}\partial y^{2(l-m)}\partial z^{2m}}(\mathbf{x},t)$$

$$\times \int (x'-x)^{2(k-l)}(y'-y)^{2(l-m)}(z'-z)^{2m} G(\mathbf{x}-\mathbf{x}')\, dr'^3$$

$$+ (R_{ij})_N, \tag{25}$$

where $(R_{ij})_N$ is the residual given by

$$(R_{ij})_N = \sum_{l=0}^{N} \sum_{m=0}^{l} \frac{1}{[2(k-l)]![2(l-m)]!(2m)!}$$

$$\times \int (x'-x)^{2(k-l)}(y'-y)^{2(l-m)}(z'-z)^{2m}$$

$$\times G(\mathbf{x}-\mathbf{x}') \frac{\partial^{2N}\bar{U}_i\bar{U}_j}{\partial x^{2(N-l)}\partial y^{2(l-m)}\partial z^{2m}}(\mathbf{x}_R,t)\, dr'^3, \quad (26)$$

and $\mathbf{x}_R$ lies somewhere on the line segment joining $\mathbf{x}$ and $\mathbf{x}'$.

Even moments of the anisotropic Gaussian filter are given by the following expression:

$$\int x^{2k}y^{2l}z^{2m}G(\mathbf{x})\, dr^3 = \frac{\lambda_x^{2k}\lambda_y^{2l}\lambda_z^{2m}}{\pi^{3/2}\gamma^{k+l+m}}\Gamma(k+\tfrac{1}{2})\Gamma(l+\tfrac{1}{2})\Gamma(m+\tfrac{1}{2}). \quad (27)$$

Substituting Eq. (27) in Eq. (25) and employing the duplication formula for the gamma function yields

$$\bar{U}_i\bar{U}_j(\mathbf{x},t) = \sum_{k=0}^{N-1}(a_{ij})_k(\mathbf{x},t) + (R_{ij})_N, \quad (28)$$

where

$$(a_{ij})_k(\mathbf{x},t) = \frac{1}{(4\gamma)^k}\sum_{l=0}^{k}\sum_{m=0}^{l}\left[\frac{\lambda_x^{2(k-l)}\lambda_y^{2(l-m)}\lambda_z^{2m}}{(k-l)!(l-m)!m!}\right.$$

$$\left.\times \frac{\partial^{2k}\bar{U}_i\bar{U}_j}{\partial x^{2(k-l)}\partial y^{2(l-m)}\partial z^{2m}}(\mathbf{x},t)\right]. \quad (29)$$

Now let

$$\frac{\partial^{k+l+m}\bar{U}_i\bar{U}_j}{\partial x^k\partial y^l\partial z^m} = \frac{U^2}{\Lambda_x^k\Lambda_y^l\Lambda_z^m}\frac{\partial^{k+l+m}\bar{U}_i^*\bar{U}_j^*}{\partial x^{*k}\partial y^{*l}\partial z^{*m}}, \quad (30)$$

where $U$ represents the characteristic scale of the LS velocity field; $\Lambda_x$, $\Lambda_y$, and $\Lambda_z$ are the corresponding characteristic length scales in the $x$, $y$, and $z$ directions, respectively; and the dimensionless derivative that appears on the right hand side of Eq. (30) is of $O(1)$. This implies that there exists a positive constant $\mathcal{M}$ such that (Erdélyi, 1956)

$$\left|\frac{\partial^{k+l+m}\bar{U}_i^*\bar{U}_j^*}{\partial x^{*k}\partial y^{*l}\partial z^{*m}}\right| \leq \mathcal{M}^2. \quad (31)$$

Furthermore, the filter widths $\lambda_x$, $\lambda_y$, and $\lambda_z$ give a measure of the smallest resolvable scales in the $x$, $y$, and $z$ directions, and therefore a small parameter $\epsilon$ may be introduced as follows:

$$\epsilon = \frac{\lambda_x}{\Lambda_x} = \frac{\lambda_y}{\Lambda_y} = \frac{\lambda_z}{\Lambda_z} \ll 1. \quad (32)$$

Making the ratio of filter widths in every coordinate direction to the corresponding length scale characteristic of the LS velocity field equal to the other two (cf. Eq. (32)) does not restrict the generality of the analysis, since any constant of $O(1)$ can be absorbed in the definition of $\Lambda_x$, $\Lambda_y$, and $\Lambda_z$ .

Employing Eqs. (27), (30)–(32), the triangle inequality, and the multi-nomial expansion formula in Eq. (26) yields

$$|(R_{ij})_N| \leq \frac{\mathcal{M}^2 \mathcal{U}^2}{(4\gamma)^N} \epsilon^{2N} \sum_{l=0}^{N} \sum_{m=0}^{l} \frac{1}{(N-l)!(l-m)!(m)!}$$

$$= \frac{\mathcal{M}^2 \mathcal{U}^2}{(\frac{4}{3}\gamma)^N N!} \epsilon^{2N}. \tag{33}$$

Also, let

$$(P_{ij})_{N-1} = \frac{(4\gamma)^{N-1}}{\mathcal{U}^2 \epsilon^{2(N-1)}} \left| \frac{(a_{ij})_{N-1}}{\sum_{l=0}^{N-1} \sum_{m=0}^{l} \frac{1}{(N-l-1)!(l-m)!(m)!}} \right|$$

$$= \frac{\gamma^{N-1}(N-1)!}{\mathcal{U}^2 \epsilon^{2(N-1)}} |(a_{ij})_{N-1}| = O(1) \qquad \forall N \tag{34}$$

by Eqs. (29), (30), (32), and the properties of the order relations.

Finally, let us compare the magnitude of the residual, $|(R_{ij})_N|$, with the magnitude of the $(N-1)$th term in Eq. (28):

$$\left| \frac{(R_{ij})_N}{(a_{ij})_{N-1}} \right| \leq \frac{3\mathcal{M}^2}{4\gamma N (P_{ij})_{N-1}} \epsilon^2 = o(1), \tag{35}$$

where Eqs. (33) and (34) have been employed. The inequality (35) is precisely the statement of the asymptotic nature of the series (28).

If only terms of first order in $\sigma_{kl}$ (second order in $\epsilon$) are kept in Eq. (10), as was originally proposed by Leonard (1974), the error made in employing the Leonard approximation is of $O(\epsilon^4)$, in the asymptotic sense.

## 24.5 Analysis of the Approximation of the Cross Terms in Physical Space

The formal series expansion of the cross terms can be written as

$$\overline{U_i u_j}(\mathbf{x}, t) = \sum_{k=0}^{\infty} \frac{1}{k!} \int dr'^3 G(\mathbf{x} - \mathbf{x}')$$

$$\times \left[ (x'_l - x_l) \frac{\partial}{\partial x_l} \right]^k \overline{U}_i(\mathbf{x}, t) u_j(\mathbf{x}', t). \tag{36}$$

Upon using the multinomial expansion in Eq. (36), the following result is obtained:

$$\overline{\bar{U}_i u_j}(\mathbf{x}, t) = \sum_{k=0}^{\infty} \sum_{l=0}^{k} \sum_{m=0}^{l} \frac{1}{(k-l)!(l-m)!m!} \frac{\partial^k \bar{U}_i}{\partial x^{k-l} \partial y^{l-m} \partial z^{m-n}}(\mathbf{x}, t)$$

$$\times \int (x'-x)^{k-l}(y'-y)^{l-m}(z'-z)^m G(\mathbf{x}-\mathbf{x}') u_j(\mathbf{x}', t) \, dr'^3. \quad (37)$$

It is convenient to separate odd and even derivatives in Eq. (37) in the following form:

$$\overline{\bar{U}_i u_j} = \bar{U}_i \bar{u}_j + \sum_{k=1}^{N-1} [(a'_{ij})_k + (b'_{ij})_k] + (R'_{ij})_N, \quad (38)$$

where

$$(a'_{ij})_k = \sum_{l=0}^{2k-1} \sum_{m=0}^{l} \frac{1}{(2k-l-1)!(l-m)!m!} \frac{\partial^{2k-1} \bar{U}_i}{\partial x^{2k-l-1} \partial y^{l-m} \partial z^{m-n}}(\mathbf{x}, t)$$

$$\times \int (x'-x)^{2k-l-1}(y'-y)^{l-m}(z'-z)^m G(\mathbf{x}-\mathbf{x}') u_j(\mathbf{x}', t) \, dr'^3, \quad (39)$$

$$(b'_{ij})_k = \sum_{l=0}^{2k} \sum_{m=0}^{l} \frac{1}{(2k-l)!(l-m)!m!} \frac{\partial^{2k} \bar{U}_i}{\partial x^{2k-l} \partial y^{l-m} \partial z^{m-n}}(\mathbf{x}, t)$$

$$\times \int (x'-x)^{2k-l}(y'-y)^{l-m}(z'-z)^{mn} G(\mathbf{x}-\mathbf{x}') u_j(\mathbf{x}', t) \, dr'^3, \quad (40)$$

and $(R'_{ij})_N$ represents the residual, given by the expression

$$(R'_{ij})_N = \sum_{l=0}^{2N-1} \sum_{m=0}^{l} \frac{1}{(2N-l-1)!(l-m)!m!} \frac{\partial^{2N-1} \bar{U}_i}{\partial x^{2N-l-1} \partial y^{l-m} \partial z^{m-n}}(\mathbf{x}, t)$$

$$\times \int (x'-x)^{2N-l-1}(y'-y)^{l-m}(z'-z)^m G(\mathbf{x}-\mathbf{x}') u_j(\mathbf{x}', t) \, dr'^3$$

$$+ \sum_{l=0}^{2N} \sum_{m=0}^{l} \frac{1}{(2N-l)!(l-m)!m!} \int (x'-x)^{2N-l}(y'-y)^{l-m}(z'-z)^m$$

$$\times G(\mathbf{x}-\mathbf{x}') \frac{\partial^{2N} \bar{U}_i}{\partial x^{2N-l} \partial y^{l-m} \partial z^m}(\mathbf{x}_R, t) u_j(\mathbf{x}', t) \, dr'^3, \quad (41)$$

where $\mathbf{x}_R$ lies somewhere on the line segment joining $\mathbf{x}$ and $\mathbf{x}'$.

Now let

$$\frac{\partial^{k+l+m} \bar{U}_i}{\partial x^k \partial y^l \partial z^m} = \frac{\mathcal{U}}{\Lambda_x^k \Lambda_y^l \Lambda_z^m} \frac{\partial^{k+l+m} \bar{U}_i^*}{\partial x^{*k} \partial y^{*l} \partial z^{*m}}, \quad (42)$$

$$\frac{\partial^{k+l+m} \bar{u}_j}{\partial x^k \partial y^l \partial z^m} = \frac{v}{\Lambda_x^k \Lambda_y^l \Lambda_z^m} \frac{\partial^{k+l+m} \bar{u}_j^*}{\partial x^{*k} \partial y^{*l} \partial z^{*m}}, \quad (43)$$

where $v$ represents the characteristic scale of the filtered SGS velocity field, and the dimensionless derivatives appearing on the right hand side of Eqs. (42) and (43) are of $O(1)$. This implies that positive constants $B$ and $\beta$ can be chosen such that

$$\left| \frac{\partial^{k+l+m} \bar{U}_i^*}{\partial x^{*k} \partial y^{*l} \partial z^{*m}} \right| \leq B, \tag{44}$$

$$\left| \frac{\partial^{k+l+m} \bar{u}_j^*}{\partial x^{*k} \partial y^{*l} \partial z^{*m}} \right| \leq \beta. \tag{45}$$

On the basis of Eqs. (32), (42), and (43), the following asymptotic estimates for $(a'_{ij})_k$ and $(b'_{ij})_k$ can be obtained by repeated integration by parts, for $\epsilon \to 0$ (for a similar, more detailed derivation for the isotropic case, see Aldama, 1990):

$$
(a'_{ij})_k = \frac{1}{2\pi\gamma^k} \left\{ \sum_{l=0}^{k-1} \sum_{m=0}^{l} \frac{\Gamma(k-l-\frac{1}{2})\Gamma(l-m+\frac{1}{2})\Gamma(m-n+1)}{[2(k-l-1)]![2(l-m)]!(2m+1)!} \right.
$$
$$
\times \lambda_x^{2(k-l-1)} \lambda_y^{2(l-m)} \lambda_z^{2(m+1)}
$$
$$
\times \frac{\partial^{2k-1}\bar{U}_i}{\partial x^{2(k-l-1)}\partial y^{2(l-m)}\partial z^{2m+1}} \frac{\partial \bar{u}_j}{\partial z}
$$
$$
+ \frac{\Gamma(k-l-\frac{1}{2})\Gamma(l-m+1)\Gamma(m-n+\frac{1}{2})}{[2(k-l-1)]![2(l-m)]!(2m+1)!} \lambda_x^{2(k-l-1)} \lambda_y^{2(l-m+1)} \lambda_z^{2m}
$$
$$
\times \frac{\partial^{2k-1}\bar{U}_i}{\partial x^{2(k-l-1)}\partial y^{2(l-m)+1}\partial z^{2m}} \frac{\partial \bar{u}_j}{\partial y}
$$
$$
+ \frac{\Gamma(k-l)\Gamma(l-m+\frac{1}{2})\Gamma(m-n+\frac{1}{2})}{[2(k-l-1)]![2(l-m)]!(2m)!} \lambda_x^{2(k-l)} \lambda_y^{2(l-m)} \lambda_z^{2m}
$$
$$
\left. \times \frac{\partial^{2k-1}\bar{U}_i}{\partial x^{2(k-l)-1}\partial y^{2(l-m)}\partial z^{2m}} \frac{\partial \bar{u}_j}{\partial x} \right\} [1 + o(1)], \tag{46}
$$

$$
(b'_{ij})_k = \frac{1}{\pi^{3/2}\gamma^k} \left\{ \sum_{l=0}^{k} \sum_{m=0}^{l} \frac{\Gamma(k-l+\frac{1}{2})\Gamma(l-m+\frac{1}{2})\Gamma(m-n+\frac{1}{2})}{[2(k-l)]![2(l-m)]!(2m)!} \right.
$$
$$
\left. \times \lambda_x^{2(k-l)} \lambda_y^{2(l-m)} \lambda_z^{2m} \bar{u}_j \right\} [1 + o(1)]. \tag{47}
$$

Now let the following quantity be defined:

$$
Q_{N-1} = \sum_{l=0}^{N} \sum_{m=0}^{l} \frac{\Gamma(N-l+\frac{1}{2})\Gamma(l-m+\frac{1}{2})\Gamma(m-n+\frac{1}{2})}{[2(N-l)]![2(l-m)]!(2m)!} \tag{48}
$$

$$
= \left( \frac{3}{4} \right)^{N-1} \frac{\pi^{3/2}}{(N-1)!},
$$

where the duplication formula for the gamma function and the multinomial expansion formula have been used in the last equality. Let us also consider the quantity $\pi^{1/2}\Gamma(p+1)/(2p+1)!$, where $p$ represents a nonnegative integer. By writing the coefficient in front of the bracketed terms in Eq. (46) as $1/(2\pi^{3/2}\gamma^k)$, that quantity appears several times there. It can be shown that $\pi^{1/2}\Gamma(p+1)/(2p+1)! \leq \Gamma(p+\frac{1}{2})/(2p)!$. Thus, the following quantity can be defined:

$$(S_{ij})_{N-1} = \frac{\pi^{3/2}\gamma^{N-1}}{\mathcal{U}v\epsilon^{2(N-1)}} \left| \frac{(a'_{ij})_{N-1} + (b'_{ij})_{N-1}}{Q_{N-1}} \right|$$

$$= \frac{(N-1)!\left(\frac{4}{3}\gamma\right)^{N-1}}{\mathcal{U}v\epsilon^{2(N-1)}} \left|(a'_{ij})_{N-1} + (b'_{ij})_{N-1}\right| = O(1) \quad (49)$$

for all fixed finite values of $N$ as $\epsilon \to 0$, by Eqs. (32), (42), (43), (46), (48) and the properties of the order relations.

Moreover, the use of Eqs. (32), (41)–(45), (46), (48), and the triangle inequality yields

$$|(R_{ij})_N| < \frac{3\mathcal{B}\beta\mathcal{U}v}{2\left(\frac{4}{3}\gamma\right)^N} \epsilon^{2N}\left[1 + o(1)\right] + (J_{ij})_N, \quad (50)$$

where

$$(J_{ij})_N = \left| \sum_{l=0}^{2N} \sum_{m=0}^{l} \frac{1}{[(2N-l)]![(l-m)]!m!} \int dr'^3\, G(\mathbf{x} - \mathbf{x}')u_j(\mathbf{x}',t) \right.$$

$$\left. \times (x'-x)^{2N-l}(y'-y)^{l-m}(z'-z)^m \frac{\partial^{2N}\bar{U}_i}{\partial x^{2N-l}\partial y^{l-m}\partial z^m}(\mathbf{x}_R,t) \right|. \quad (51)$$

An upper bound for $(J_{ij})_N$ can be found by defining the following functions:

$$v^{+}_{ij,2N,l,m}(\mathbf{x}',t) = \begin{cases} \xi_{i,2N,l,m}u_j(\mathbf{x}',t), & \text{for } \eta_{ij,2N,l,m} > 0, \\ 0, & \text{for } \eta_{ij,2N,l,m} \leq 0, \end{cases} \quad (52)$$

$$v^{-}_{ij,2N,l,m}(\mathbf{x}',t) = \begin{cases} 0, & \text{for } \eta_{ij,2N,l,m} \geq 0, \\ \xi_{i,2N,l,m}u_j(\mathbf{x}',t), & \text{for } \eta_{ij,2N,l,m} < 0, \end{cases} \quad (53)$$

$$\xi_{i,2N,l,m} = \text{sgn}\left[\frac{\partial^{2N}\bar{U}_i}{\partial x^{2N-l}\partial y^{l-m}\partial z^m}(\mathbf{x}_R,t)\right], \quad (54)$$

$$\eta_{ij,2N,l,m} = (x'-x)^{2N-l}(y'-y)^{l-m}(z'-z)^m$$

$$\times \frac{\partial^{2N}\bar{U}_i}{\partial x^{2N-l}\partial y^{l-m}\partial z^m}(\mathbf{x}_R,t)u_j(\mathbf{x}',t)\,dr'^3. \quad (55)$$

Hence,

$$(J_{ij})_N \leq (J_{ij})^{+}_N + (J_{ij})^{-}_N, \quad (56)$$

where

$$(J_{ij})_N^{\pm} = \left| \sum_{l=0}^{2N} \sum_{m=0}^{l} \frac{1}{[(2N-l)]![(l-m)]!m!} \right.$$

$$\times \int (x'-x)^{2N-l}(y'-y)^{l-m}(z'-z)^m$$

$$\left. \times \, G(\mathbf{x}-\mathbf{x}') \left| \frac{\partial^{2N}\bar{U}_i}{\partial x^{2N-l}\partial y^{l-m}\partial z^m}(\mathbf{x}_R,t) \right| v_{ij,2N,l,m}^{\pm}(\mathbf{x}',t)\, dr'^3 \right| \quad (57)$$

and where no summation is implied. Using Eqs. (42) and (44) in Eq. (57) results in

$$(J_{ij})_N^{\pm} \le BU \left| \sum_{l=0}^{2N} \sum_{m=0}^{l} \frac{\Lambda_x^{-(2N-l)}\Lambda_y^{-(l-m)}\Lambda_z^{-m}}{[(2N-l)]![(l-m)]!m!} \right.$$

$$\times \int (x'-x)^{2N-l}(y'-y)^{l-m}(z'-z)^m$$

$$\left. \times \, G(\mathbf{x}-\mathbf{x}')v_{ij,2N,l,m}^{\pm}(\mathbf{x}',t)\, dr'^3 \right|. \quad (58)$$

Now employing Eqs. (32), (42)–(45), (48), (56), (58), and a procedure analogous to the one used to derive Eq. (47) yields

$$(J_{ij})_N \le \frac{2B\beta U v}{\left(\frac{4}{3}\gamma\right)^N N!}\epsilon^{2N}\,[1+o(1)]. \quad (59)$$

Finally, using Eqs. (49), (50), and (59), the magnitudes of the residual, $(R_{ij})_N$, and the $(N-1)$th term in Eq. (38) may be compared as follows:

$$\left| \frac{(R_{ij})_N}{(a'_{ij})_{N-1}+(b'_{ij})_{N-1}} \right| < \frac{21B\beta}{8\gamma N(S_{ij})_{N-1}}\epsilon^2\,[1+o(1)] = o(1), \quad (60)$$

which proves that the series contained in Eq. (85) is asymptotic.

Once again, if only terms of first order in $\sigma_{kl}$ (second order in $\epsilon$) are kept in Eqs. (21) and (22), the error made in employing the cross-term approximation is of $O(\epsilon^4)$, in the asymptotic sense.

## 24.6 Analysis of the Leonard Approximation in Fourier Space

The space Fourier transform of the anisotropic Gaussian filter, defined by Eq. (2), is given by

$$\widehat{G}(\mathbf{k}) \equiv \mathcal{F}\{G(\mathbf{x})\} = \exp\left(-\frac{\sigma_{lm}}{2}k_l k_m\right), \quad (61)$$

where $\widehat{(\cdot)}$ and $\mathcal{F}\{\cdot\}$ are alternative notations for the Fourier transform and $\mathbf{k}$ is the wave number vector. It is instructive to note that $\widehat{G}(\mathbf{k})$ is also Gaussian. By the convolution theorem, the Fourier transform of the filtered LS advective terms, $\bar{U}_i \bar{U}_j$, (cf. Eq. (5)) is

$$\widehat{\overline{\bar{U}_i \bar{U}_j}}(\mathbf{k}, t) = \widehat{G}(\mathbf{k}) \widehat{\bar{U}_i \bar{U}_j}(\mathbf{k}, t). \tag{62}$$

Now let $\widehat{G}(\mathbf{k})$ be approximated by a truncated Taylor series as follows:

$$\widehat{G}(\mathbf{k}) \simeq 1 - \frac{\sigma_{lm}}{2} k_l k_m. \tag{63}$$

Evidently, Eq. (63) entails approximating the Gaussian filter by an osculating paraboloid in Fourier space. Thus, from hereon, Eq. (63) will be referred to as the parabolic approximation of the Gaussian filter in Fourier space.

Substituting Eq. (63) in Eq. (62) yields

$$\widehat{\overline{\bar{U}_i \bar{U}_j}}(\mathbf{k}, t) \simeq \widehat{\bar{U}_i \bar{U}_j}(\mathbf{k}, t) - \frac{\sigma_{lm}}{2} k_l k_m \widehat{\bar{U}_i \bar{U}_j}(\mathbf{k}, t)$$

$$= \mathcal{F}\left\{ \bar{U}_i \bar{U}_j + \frac{\sigma_{lm}}{2} \frac{\partial^2 \bar{U}_i \bar{U}_j}{\partial x_l \partial x_m} \right\}, \tag{64}$$

which shows that the Leonard approximation, keeping terms up to $O(\epsilon^2)$ times the leading order term in Eq. (10), entails employing a parabolic approximation of the Gaussian filter in Fourier space.

## 24.7 Analysis of the Approximation of the Cross Terms in Fourier Space

The Fourier transform of the cross terms, $\overline{\bar{U}_i u_j}$, defined by Eq. (11), is

$$\widehat{\overline{\bar{U}_i u_j}}(\mathbf{k}, t) = \frac{\widehat{G}(\mathbf{k})}{(2\pi)^3} \int \widehat{\bar{U}_i}(\mathbf{k}' - \mathbf{k}, t) \widehat{u_j}(\mathbf{k}', t) \, dk'^3, \tag{65}$$

where $dk^3 = dk_1 \, dk_2 \, dk_3$. In addition, from Eq. (1), the Fourier transform of the LS velocity is given by

$$\widehat{\bar{U}_i}(\mathbf{k}, t) = \widehat{G}(\mathbf{k}) \widehat{U_i}(\mathbf{k}, t). \tag{66}$$

Combining Eq. (66) with the Fourier transform of Eq. (3) results in:

$$\widehat{u_j}(\mathbf{k}, t) = \left[ \frac{1 - \widehat{G}(\mathbf{k})}{\widehat{G}(\mathbf{k})} \right] \widehat{\bar{U}_j}(\mathbf{k}, t). \tag{67}$$

Substituting now Eqs. (63) and (67) in Eq. (65), keeping only quadratic terms in the wave number vector components yields

$$\widehat{\overline{U_i u_j}}(\mathbf{k},t) \simeq \frac{\sigma_{lm}}{2(2\pi)^3} \int \widehat{\overline{U}}_i(\mathbf{k'}-\mathbf{k},t) k'_l k'_m \widehat{\overline{U}}_j(\mathbf{k'},t)\, dk'^3$$

$$= \mathcal{F}\left\{ -\frac{\sigma_{lm}}{2} \bar{U}_i \frac{\partial^2 \bar{U}_j}{\partial x_l \partial x_m} \right\}. \tag{68}$$

A similar derivation results in

$$\widehat{\overline{u_i U_j}}(\mathbf{k},t) \simeq \frac{\sigma_{lm}}{2(2\pi)^3} \int k'_l k'_m \widehat{\overline{U}}_i(\mathbf{k'},t)\widehat{\overline{U}}_j(\mathbf{k'}-\mathbf{k},t)\, dk'^3$$

$$= \mathcal{F}\left\{ -\frac{\sigma_{lm}}{2} \frac{\partial^2 \bar{U}_i}{\partial x_l \partial x_m}\, \bar{U}_j \right\}. \tag{69}$$

Equations (68) and (69) show that the approximation of the cross terms, when only terms up to $O(\epsilon^2)$ times the leading order advective terms in the equations of motion, $\bar{U}_i \bar{U}_j$, also entails the use of the parabolic approximation of the Gaussian filter in Fourier space.

## 24.8 Concluding Remarks

In environmental and geophysical applications the range of dynamically significant scales is very wide. Furthermore, it is often the case that in environmental and geophysical flow simulations the scales of interest are very large, not only in space, but also in time. This is partly related to considerations of computational economy, which in many instances force the use of large time steps in those simulations. Therefore, the use of at least partially implicit time stepping schemes is necessary. This makes the use of local approximations of the Leonard and cross terms very attractive in environmental and geophysical large eddy simulations. In addition, the occurrence of strongly anisotropic flows is very frequent in environmental and geophysical applications.

The aforementioned observations have motivated the study presented herein. Thus, the Leonard approximation has been extended to anisotropically filtered equations. Also, an approximation of the cross terms has been derived in the same context. The nature of both approximations has been analyzed in physical space, and it has been shown that both possess an asymptotic nature in a small parameter defined as the ratio of the filter widths in every coordinate direction to the corresponding length scale characteristic of the filtered velocity field. As a consequence, it was concluded that the (asymptotic) error made in using the Leonard and cross-term approximations is of fourth order in the referred small parameter. The Leonard and cross-term approximations were also analyzed in Fourier space and it was concluded that their use involves approximating the Gaussian filter by an osculating paraboloid in Fourier space.

The demonstration of the asymptotic character of both the Leonard and the cross-term approximations provides theoretical support for their use in the context of a perturbation theory. Moreover, since the cross terms have been expressed as a function of the filtered velocity field, the need of lumping them with the subgrid stresses and parameterizing the result via a closure model has been eliminated. It is believed that employing the approximation of the cross terms will better capture the physics of backscatter, decreasing the burden that has to be carried by the closure model, which would then have only to represent unresolved processes.

## Acknowledgments

The assistance of Roger Beckie in typesetting this chapter is greatly appreciated. This work was partially supported by the National Science Foundation under Grant No. CTS-8908877.

## References

ALDAMA, A.A. (1990) *Filtering techniques for turbulent flow simulation*, Lecture Notes in Engineering, Vol. 56 (Springer-Verlag, Heidelberg).

ALDAMA, A.A. AND HARLEMAN, D.R.F. (1991) The approximation of nonlinearities in the filtered Navier-Stokes equations. *Advances in Water Resources* **14**, 15.

ANTONOPOULOS-DOMIS, M. (1979) Aspects of large eddy simulation of homogeneous isotropic turbulence. QMCEP 6038, Department of Nuclear Engineering, Queen Mary College.

BABAJIMOPOULOS, C. AND BEDFORD, K. (1980) Formulating lake models which preserve spectral statistics. *J. Hydr. Div., ASCE* **106**, 1.

BARDINA, J., FERZIGER, J.H. AND ROGALLO, R.S. (1985) Effect of rotation on isotropic turbulence: Computation and modeling. *J. Fluid Mech.* **154**, 321.

BEDFORD, K. (1981) Spectra preservation capabilities of Great Lakes transport models. In *Transport models for inland and coastal waters*. Ed. H.B. Fischer (Academic Press, New York), p. 172.

BEDFORD, K. AND BABAJIMOPOULOS, C. (1980) Verifying lake models with spectral statistics. *J. Hydr. Div., ASCE* **106**, 21.

CLARK, R.A., FERZIGER, J.H. AND REYNOLDS, W.C. (1977) Evaluation of subgrid-scale turbulence models using a fully simulated turbulent flow. Report No. TF-9, Department of Mechanical Engineering, Stanford University.

CHAMPENEY, D.C. (1989) *A handbook of Fourier theorems* (Cambridge University Press, Cambridge).

DEARDORFF, J.W. (1970) A numerical study of three-dimensional turbulent channel flow at large Reynolds numbers. *J. Fluid Mech.* **41**, 453.

DEARDORFF, J.W. (1973) The use of subgrid transport equations in a three-dimensional model of atmospheric turbulence. *J. Fluids Eng., ASME* **95**, 429.

DEARDORFF, J.W. (1974) Three-dimensional numerical study of the height and mean structure of a heated planetary boundary layer. *Boundary Layer Meteorol.* **7**, 81.

EIDSON, T. (1985) Numerical simulation of the turbulent Rayleigh-Bénard problem using subgrid modeling. *J. Fluid Mech.* **158**, 245.

ERDÉLYI, A. (1956) *Asymptotic expansions* (Dover, New York).

FERZIGER, J.H., METHA, U.B. AND REYNOLDS, W.C. (1977) Large eddy simulation of homogeneous isotropic turbulence. Symposium on Turbulent Shear Flows, Penn State University, University Park.

LEONARD, A. (1974) Energy cascade in large-eddy simulations of turbulent flows. *Adv. in Geophys.* **18A**, 237.

LILLY, D. K. (1967) The representation of small-scale turbulence in numerical simulation experiments. Proc. IBM Sci. Comput. Symposium Environ. Sci., IBM Data Process Div., White Plains, N. Y., p. 195.

LOVE, M. D. (1980) Subgrid modeling studies with Burgers' equation. *J. Fluid Mech.* **100**, 87.

MASON, P. J. AND N. S. CALLEN (1986) On the magnitude of the subgrid-scale eddy coefficient in large eddy simulations of turbulent channel flow. *J. Fluid Mech.* **162**, 439.

MANSOUR, N.N., FERZIGER, J.H. AND REYNOLDS, W.C. (1978) Large eddy simulation of a turbulent mixing layer. Report TF-11, Department of Mechanical Engineering, Stanford University.

MOIN, P. AND KIM, J. (1982) Numerical investigation of turbulent channel flows. *J. Fluid Mech.* **118**, 323.

PIOMELLI, U., MOIN, P. AND FERZIGER, J.H. (1988) Model consistency in large eddy simulation of turbulent channel flows. *Phys. Fluids* **31**, 1884.

ROGALLO, R.S. AND MOIN, P. (1984) Numerical simulation of turbulent flows. *Ann. Rev. Fluid Mech.* **16**, 99.

SCHUMANN, U. (1975) Subgrid scale model for finite difference simulations of turbulent flows in plane channels and annuli. *J. Comp. Phys.* **18**, 376.

SMAGORINSKY, J. (1963) General circulation experiments with the primitive equations: I. The basic experiment. *Mon. Weather Rev.* **91**, 216.

# 25

## Large Eddy Simulation as a Tool in Engineering and Geophysics: Panel Discussion

### THOMAS A. ZANG

This panel was chaired by Spiro Lekoudis and Thomas Zang. The panel members were (in order of presentation) Spiro Lekoudis, Ugo Piomelli, Marcel Lesieur, Peyman Givi, Joel Ferziger, Ulrich Schumann, Paul Mason and John Wyngaard. The following is not a verbatim transcript but rather a summary of the points made during the prepared remarks by the panel members and during the discussion.

**Lekoudis:** The panel members, at least the first four, whose primary interest is the application of large eddy simulation (LES) to problems in hydrodynamics and aerodynamics, have been asked to focus their remarks on the following issues: (1) What is the current status/role of LES? (2) What might be the status/role of LES in 10 years (assuming that computers can run LES codes at, say, 100 Gflops)? (3) What is the relationship of LES to lower-level, i.e., Reynolds-averaged Navier–Stokes (RANS), models, particularly second-order closures? (4) What is the relationship of LES to experiments and how can this relationship be improved? (5) What efforts would accelerate progress in LES?

**Piomelli:** At present the problems amenable to LES are flows in simple geometry and at low Reynolds number, e.g., channel flow and flow past a backward-facing step. To become a useful engineering tool, LES should be applied to problems in which its cost is comparable to that of lower-level models, or to flows in which lower-level models fail; these include unsteady or separated flows, three-dimensional (3D) boundary layers and transitional flows; no practical purpose is served by applying LES to attached flow past an airfoil at cruise conditions, since this problem can be treated reasonably well with lower-level models. In the short term LES is limited to the range of moderate to moderately high Reynolds numbers (up to the order of $10^5$), but it can still be utilized for validation of lower-level models, particularly since it can be extended to Reynolds numbers much higher than those feasible by direct simulations. A successful LES should produce accurate mean flows, fairly accurate Reynolds stresses and pressure statistics (which are very hard to measure) and qualitative structural features. To make an impact within the engineering community, LES in 10 years needs to be feasible for routine runs on a desktop computer.

Further development is required in various areas. Of course, more accurate models are desirable. To be applied to the nonequilibrium flows mentioned before, such models should have the following features: (1) The models should be formulated in real rather than in wave number space, to facilitate their use in finite-difference codes. (2) The subgrid-scale (SGS) stresses should vanish at the wall (preferably with the correct asymptotic behavior). (3) The SGS stresses should vanish in laminar flow. (4) The models should be able to account for backscatter.

Both direct numerical simulation (DNS) and experiments fill an important role in aiding this development. Direct simulations can be used for *a priori* tests of models, and filtered DNS velocity fields can be used for *a posteriori* comparisons. The low Reynolds numbers attainable with DNS, and the limited application of DNS to flows in complex geometries, however, indicate that the final comparisons must be made with experimental results. Experiments, however, need to be well documented (especially regarding the initial and boundary conditions) and to provide detailed data. Filtering the experimental data, if possible, is desirable.

Improved formulations of the boundary conditions must also be developed. In the first place, the use of approximate boundary conditions is desirable if LES is to be used at very high Reynolds numbers. They should capture more of the physics of the wall layer than those presently in use, which typically require only that the mean velocity satisfy a logarithmic law. Alternatively, the use of unstructured grids in the near-wall region should be explored.

For space-developing problems, inflow/outflow conditions are necessary. While some progress has been made on the latter, inflow boundary conditions still hinder the application of LES to more complex flows. In spatially developing flows with a turbulent inflow, the time-dependent velocity must be specified at each point in the inflow plane. This can be done by using the results of a separate calculation of a spatially periodic flow, or by specifying a random velocity satisfying some given constraints (having a given mean profile or spectrum, for instance); the flow must then be allowed to develop for some distance before it becomes realistic. The first solution is feasible in some cases, but requires performing a separate calculation and storing a significant amount of data. With the second, which is perhaps more generally applicable, a significant part of the computational box may be wasted if the distance required for the inflow conditions to become realistic is large. Matching energy spectra and Reynolds stress profiles or introducing information on the structure of turbulence might be necessary to reduce this "numerical entry length."

*Schumann:* Let me note that some LESs of flow past a backstep have taken inflow conditions from a channel flow LES (see Friedrich and Arnal, 1990). This is a reasonable approach.

*Piomelli:* Reasonable, but costly, since one must run a channel flow LES in addition to the backstep LES.

*Yakhot:* Why do we need wall models? Couldn't we just resolve the wall region and use models which turn themselves on away from the wall?

*Piomelli:* The difficulty is that it is extremely expensive to resolve all the relevant scales in the spanwise and wall-normal directions in the wall region.

*Yakhot:* It is not a problem if we find the right set of coordinates.

*Piomelli:* That's what I meant by introducing structural information into the model.

*Moin:* There is no question that we need unstructured grids in LES algorithms. These would permit the use of embedded meshes near the wall.

*Lesieur:* I would like to relate my personal experience in applying LES to engineering problems. My remarks perhaps represent a perspective from Grenoble–France–Europe. I will mention three examples from our published work. These examples are LES of flow over a backstep, transition in a Mach 5 flat-plate boundary layer and a DNS of a Mach 1 jet. I wish to stress that my group got involved in these applications at the urging of French industry. The code used for the backstep computations was a conventional, industrial computational fluid dynamics code with the RANS models removed for the DNS calculations and replaced by eddy-damped quasi-normal Markovian SGS models for the LES. The computations were performed as both a DNS and an LES.

The LES results concerning statistical quantities such as mean velocity,
pressure coefficient and turbulent stresses are in much better agreement
with experimental data. LES also has the great advantage of providing
valuable information on the coherent structures in the flow. The supersonic boundary-layer transition problem is one for which DNS is not
feasible and LES is the only alternative.

I anticipate that progress in LES will arise from the increased computational capacity of forthcoming parallel computers and from rapid
image processing capability. We need to develop new models which account for compressibility effects. With a 100-Gflop computer, LES of
the boundary-layer transition problem would permit resolution into the
Kolmogorov range.

*Ferziger*: I would call the examples you showed building-block problems rather than industrial flows. The real flows are much more complicated.

*Lesieur*: My point is that these were studies sponsored by industry.
Such companies as Avions Marcel-Dassault, SNECMA and Renault are
interested in using our LES technology.

**Givi**: My remarks will concentrate not on modeling per se, but rather
on some of the critical issues pertaining to LES for turbulent combustion.
Combustion engineers have traditionally used the product of turbulence
modelers. Codes employing Reynolds-averaged models (up to the two-
equation level) have been applied to reacting flows since the 1970s. LES
has not seen the same level of effort. The presence of extra parameters such as the Damköhler number and the Zeldovich number has had
something to do with this. The flow regimes that prevail in combustion
problems are such that the Smagorinsky model is not applicable and
SGS models based on transport equations are desired.

DNS work on combustion problems has been for flows with very simple kinetics, such as equilibrium chemistry, and it is time for LES to be
applied to these cases. When flames are present it appears that stochastic models are needed. (Dynamic eddy viscosity models are unlikely to
work well since the effects of flames occur at high wave numbers.) In this
case the computations should be performed in the composition domain
rather than the physical domain.

This conference has illustrated the good coupling that has existed
for many years between the LES communities in fluid dynamics and in
atmospheric sciences. Now may be the time to make this a three-way
coupling, with the inclusion of the combustion community.

*Schumann*: I have found that adding chemical reactions is not too difficult because slow chemistry presents no problem and very fast chemistry limits the reaction zones to the interface between the reacting components. For this case special methods exist (see, e.g., Schumann, 1989).

*Givi*: The limits of zero or infinite Damköhler number are fairly easy, even for RANS. The real difficulties are that most of the kinetics occur at intermediate Damköhler numbers and that, in problems with complex chemistry, reactions occur over a wide range of Damköhler numbers.

*Ferziger*: An important consideration is the objective of the LES. For example, accurate predictions of the production of nitrogen oxide by a flame require a lot of resolution and simple models do not work.

**Ferziger:** Some of the flows of interest to me include airfoils; internal combustion engines; lakes (pollution and dispersion effects); turbomachinery (which includes effects of rotation, heat transfer, unsteadiness, three-dimensionality and corners); and wind engineering (for which the fluctuating forces on a body are desired).

I foresee continuation of LES of building-block flows. They should be flows containing physical effects which give conventional turbulence models trouble. LES can be used to understand the physics of these flows, thus assisting in devising effective flow control techniques and parameterizations (models) for the engineer who needs to make practical predictions.

For the next decade one important task will be programming LES codes on parallel computers. Early real engineering applications of LES are likely to be turbine blade flows and engine cylinder flows; these are particularly good candidates for early LES due to their complexity and low Reynolds numbers. LES should also be applied to more complicated building-block flows, including real chemistry, hypersonics and simple models of real combustors. The type of experiments that will need to be done will change. We will need fewer experiments but ones that are chosen and designed more carefully, are better documented and produce more detailed data. We should treat DNS and LES as equal partners with experiments, taking best advantage of the strengths of each.

From the theory we need improved models, especially ones which contain improved parameterizations of extra strains (rotation, compressibility, chemistry, etc.) and improved wall models.

From the mathematics community we need improved numerics, especially methods for complex geometries and ones which are time accurate. As always, we need faster algorithms.

*Wyngaard*: What do you mean by "engineering design tool"? In the atmospheric sciences there is the perception that modeling centers are required. Can we really expect an individual engineer to have the expertise to use LES on a personal computer?

*Lesieur:* In my experience engineers in industry genuinely want to apply LES to their problems.

*Vanka:* Is it feasible to use LES to develop second-order closure models and should we (those performing LES) do this particular job? After all, the LES community could act as many experimentalists do and simply furnish the data to the modeling community. I am also concerned that as we begin applying LES to fully nonperiodic flows the extra computational expense will be too great.

*Ferziger:* For 3D, time-dependent flows, one may just as well do LES as use a RANS model. In other flows, one can use LES occasionally to check the performance of RANS models in critical cases. I do not foresee a universal model for LES.

**Schumann:** I believe that LES is not restricted to moderate Reynolds numbers. On the contrary, LES is most suitable in the limit of infinite Reynolds numbers; consider the applications in atmospheric sciences. The regime with moderate Reynolds numbers appears to be the most difficult because it requires high resolution: asymptotic theories like the inertial range concept are not yet applicable.

I would also like to address the issue of promoting the application and success of LES, not just in the meteorological field but in the wider LES community. We should bring open questions which have to do with opinions of what is better to a point, e.g., must we use spectral methods or must we use second-order closures for the SGS model? We should have test cases to settle such questions instead of relying on the opinions of influential people.

I will give an example of an effort in this direction undertaken by four of us (Nieuwstadt, Mason, Moeng and Schumann 1991). Each of us has a convective boundary layer LES code. We agreed to apply these codes to the same test case. I present here some preliminary figures from an as yet unpublished paper. We all used the same modest resolution ($40^3$ grid) on a computational domain of moderate size. Despite the strong differences among the SGS models, numerical methods and grid distributions of the four codes, the principal results were remarkably similar, even to the level of third moments. The sharpest differences were in the spectra. My point is that, at least on this simple problem, none of the differences between the codes and the models really mattered.

I suggest that we do more such comparisons, especially for flow problems in which shear is the important ingredient because I think that differences between the LES codes are larger there. If anyone wants to participate in such a comparison, please let me know.

**Mason:** An important concern in meteorological LES simulations is the implementation of high Reynolds number boundary conditions. These boundary conditions are an empirical description based on observations. In contrast to low Reynolds number simulations it is inconceivable that an LES grid would ever resolve down to the scale of a laminar wall layer and avoid this empiricism. The boundary conditions which are used seek to relate the velocity in the first grid plane off the surface to the stress at the surface. A frequently used boundary condition is based on mean flow relations and gives a surface stress proportional to the square of the velocity gradient just above it. Another boundary condition asserts that the stress is proportional to the velocity gradient. To establish the correct boundary conditions, observations averaged over horizontal areas are needed. To date judgments have had to be based on point measurements. For the atmosphere these confirm that the usual logarithmic wall relations with stress proportional to velocity squared are better than proposed alternatives. High Reynolds number LES will always depend on empirical descriptions of the near-surface flow. There is a need to encourage experimentalists to make appropriate measurements of these regions.

*Schumann:* Let me take this opportunity to note that I have long since abandoned the boundary condition which appears in Schumann (1975) (and is often referred to as Schumann's model) in favor of the boundary condition just described. See the description in Schumann and Schmidt (1989). The old method required the knowledge of the mean shear stress, which was known from an integral balance in the Poiseuille flow case but is unknown in general. The assumption of a linear relationship between local stress and flow velocity in the first grid point was just a first-order approximation. At least for convective situations, the details of this boundary condition are rather unimportant.

**Wyngaard:** In geophysics we don't have the advantage of the rich history that engineering fluid dynamics has in experimental work. We just don't have comparable access to suitable experiments. The detailed experimental databases on turbulent flows with the important forcing effects that one finds in geophysics – buoyancy, stratification, and rotation – are not often possible to obtain. If we even had turbulence data from *laboratory models* of geophysical flows, say from a tank of water, I believe (1) we could demonstrate that typical approaches to geophysical flow modeling – e.g., most second-order closures – are seriously flawed because they do not properly accommodate the structural changes that these extra forcings cause; and (2) they would provide a better opportunity than we now have in geophysics to test LES techniques in a definitive way. I believe that generating experimental data bases against which we could make some definitive tests of models is the most important single thing that we can do to advance geophysical flow modeling.

*Kerr:* In the physics community a lot of stir has been caused by the recent experimental data of Joseph Libchaber of the University of Chicago that shows nonclassical scaling exponents for the Nusselt number. Some of his unpublished data for large aspect ratio are reproducible with my unpublished DNS. This seems to be an ideal situation for developing LES of convection. There is a hard number (analogous to the wall stress in shear flows) that we would like to reproduce as accurately as possible.

## References

FRIEDRICH, R. AND ARNAL, M. (1990) Analyzing turbulent backward-facing step flow with the lowpass-filtered Navier–Stokes equations. *J. Wind Eng. Ind. Aerodyn.* **35**, 101–128.

NIEUWSTADT, F.T.M., MASON, P.J., MOENG, C.H. AND SCHUMANN, U. (1991) Large-eddy simulation of the convective boundary layer: a comparison of four computer codes. In *Eighth Symposium on Turbulent Shear Flows*, September 9-11, 1991, Technical University of Munich, Munich, Germany.

SCHMIDT, H. AND SCHUMANN, U. (1989) Coherent structure of the convective boundary layer deduced from large-eddy simulation. *J. Fluid Mech.* **200**, 511–562.

SCHUMANN, U. (1975) Subgrid scale model for finite difference simulations of turbulent flows in plane channels and annuli. *J. Comput. Phys.* **18**, 376–404.

SCHUMANN, U. (1989) Large-eddy simulation of turbulent diffusion with chemical reactions in the convective boundary layer. *Atmos. Environ.* **23**, 1713–1727.

## PART FOUR

# LARGE EDDY SIMULATION
# AND MASSIVELY PARALLEL COMPUTING

# 26

---

# Parallel Computing for Large Eddy Simulation

## CECIL E. LEITH

## 26.1 Introduction

The final session of the workshop on large eddy simulation (LES) was a general discussion on the evolution of massively parallel computing hardware and techniques and its impact on the future of LES. The discussion was led by a panel consisting of Bill Dannevik (Lawrence Livermore National Laboratory), Gordon Erlebacher (Institute for Computer Applications in Science and Engineering), George Karniadakis (Princeton University), Chuck Leith (Lawrence Livermore National Laboratory), Lynn Lewis (Silicon Graphics), and Vijay Sonnad (IBM). There was general agreement that massively parallel computing could provide the increases by orders of magnitude in computing power needed in many LES applications, but that there would be some difficulties in developing the best software tools and in identifying the most appropriate hardware configurations. This report is an attempt to summarize in a somewhat undifferentiated way the many comments made by the panel members and other participants. It includes, however, many personal views based on developments at the Lawrence Livermore National Laboratory (LLNL) since the time of the workshop. It does not therefore necessarily reflect the current beliefs of the other panel members.

The development in recent years of mass-produced microprocessor chips for industrial computing is leading to a new approach to high speed scientific and engineering computing that differs radically from the conventional CRAY-class supercomputer. The large market and intense competition have driven the development of ever more powerful chips. Such chips, which cost about $1000, are already matching the processor speed of the CRAY-1 for scalar arithmetic and are expected within a year or two to exceed that of the CRAY-2. It seems fairly clear that within a few years the mass-produced microprocessor will replace the handcrafted processor, which has been at the heart of scientific computers for decades.

There are two natural consequences of this historic development. Sequential computations are being moved to the more cost-effective engineering workstations that use these new microprocessors. New codes are being written for highly parallel computers in which hundreds or thousands of these new chips are assembled as nodes in an array that is tightly coupled through fast communication switches and channels.

Efficient use of such parallel hardware has two simple requirements. The computing load should be distributed more or less uniformly over processor nodes to keep them all busy, and communication of necessary data between nodes should be fast enough to feed the node arithmetic. Of these two requirements, the first appears to be relatively easily met by many computational fluid dynamics (CFD) codes. The second requirement means that the hardware should provide a ratio of communication time to floating point operation time that is as small as possible and that the algorithm used should minimize the required data transfer between nodes.

The discussion can be divided into three main topics – hardware, programming models, and software tools – and will be summarized in the next three sections. In the final section, an attempt will be made to look ahead to the parallel computing center of the future.

## 26.2 Highly Parallel Hardware Configurations

Highly parallel computer architecture is usually divided into two main types: single instruction, multiple data (SIMD) and multiple instruction, multiple data (MIMD). Unfortunately there was little representation of the SIMD approach among the participants, in spite of the large number of SIMD machines and users in the CFD community. For this reason, the panel discussion focused primarily on MIMD machines.

In SIMD architecture, all nodes simultaneously carry out the same instruction sequence while operating, of course, on different data. In some ways such a machine is a large vector computer, and regular mesh explicit CFD algorithms requiring simple nearest neighbor communication can be and have been implemented with high efficiency.

In contrast, in MIMD architecture, each node is a complete computer with its own code and data stored in its own local memory. Clearly MIMD provides for far greater flexibility of use, but the consequent complexity can lead to far more difficult programming tasks.

A further distinction is often made between MIMD machines with shared memory and those with distributed local memory. Machines with shared memory are familiar from the widely available multiprocessor CRAYs. When the number of processors is greatly increased, it is no longer feasible to provide each processor with equal access to all the memory. Instead some sort of switching and communication mechanism is employed for remote access to distributed local memories. If the switch is fast enough the memory may be considered to be shared, but this is more a programming model than an inherent hardware property.

There is a general consensus that massively parallel computers will in coming years evolve toward the more general MIMD architecture, especially since an MIMD machine can always be used in an essentially SIMD mode. The microprocessor nodes are rapidly becoming more powerful in order to satisfy the needs of the engineering workstation market. More powerful arithmetic requires more and faster memory, and this too is coming but not as rapidly. The key problem for massively parallel computing in the future is likely to be whether the switching and communication technology will keep pace with that of the microprocessor chips.

## 26.3 Programming Models for LES

Three principal programming models appear to be emerging: data parallel, message passing, and shared memory. The data-parallel model deals directly with multidimensional data arrays and has been particularly useful on SIMD machines using the new FORTRAN 90 array syntax. The message-passing model, developed for MIMD machines, leaves to the programmer the task of exchanging the appropriate data between the local memories of the various nodes. The shared-memory model treats the total memory of the machine as equally accessible to each node and thus mimics on MIMD machines the multiprocessor approaches of CRAY-class machines. Each of these models has its strengths and weakness in terms of efficiency of hardware use and of ease of programming. It is likely that all will be needed to satisfy the widest range of future applications.

The data-parallel model is naturally suited to SIMD hardware. One deals directly with data arrays and acquires neighboring information through data shifts. Many CFD problems in which data are carried on a well-defined regular mesh easily fit this programming model. For them the compiler automatically handles the parallel aspects of the program, and thus this approach provides simple programming for simple problems.

Perhaps the easiest way to put a CFD code onto MIMD hardware is through the domain-decomposition message-passing (DDMP) model. This is trivial for explicit finite-difference schemes for integrating the equations of compressible fluid dynamics which involve only local space–time physics. By decomposing the total computational domain of the flow into subdomains, each attached to a processor node responsible for all arithmetic in it, only border data need be transmitted between subdomains, and communication is diminished relative to arithmetic by a surface to volume ratio.

For schemes requiring a global solve, such as for incompressible fluids, the problem is not so trivial, but an effort is made to reduce the algebraic problem to one involving only border data in order to maintain the surface to volume ratio benefit. The spectral element method for fluid flow is a prime example of this kind of DDMP with each element being a subdomain.

The simplicity of the DDMP programming model is largely a consequence of the separability of physics, algebra, and communication. The algebra, if needed, and the communication depend on the parallel architecture, but once worked out they can serve a range of physics applications without essential modification. Programming with message passing is similar in nature and complexity to the more familiar programming with writes to and reads from working storage needed for problems too large to fit in the core memory.

The DDMP model satisfies scaling laws such that to maintain efficiency it is necessary to increase the problem size as the number of processors increases in order to keep the same amount of mesh on each node. Massively parallel computers are thus not for solving present problems faster but rather for solving bigger problems in the same time.

The shared-memory model is a natural extension of multitasking on CRAY-class machines. It is not yet clear how well this model will scale to many more processors. Synchronization is the key problem here, and it requires explicit use of barriers and locks. By contrast, the message passing of the DDMP model imposes implicit synchronization on the process.

There has been much discussion of the need for new algorithms for parallel computing, but often an old algorithm is more appropriate. The choice of algorithm for a particular application has always involved compromise, and now suitability for parallel implementation has introduced a new consideration. Even the choice of physical model may be influenced by hardware. It is, for example, easier to put compressible hydrodynamics on parallel architecture than incompressible. It is therefore tempting to replace incompressible spectral transform LES codes by compressible grid-point codes run at moderately low Mach number.

## 26.4 Parallel Languages and System Software

The most controversial aspect of highly parallel computing is the development of appropriate software tools. This has lagged behind the development of hardware simply because no one would commit much effort without knowing how and how well the hardware worked. The potential impasse is being avoided by a large number of computational scientists who are putting their models onto parallel hardware without the benefit of the many proposed parallel programming tools.

Automatic parallelization of all existing FORTRAN codes now appears to be essentially impossible. Parallel extensions of the FORTRAN language have already been developed for multitasking on CRAY-class machines, and these are being further extended to highly parallel hardware. The usual kind of committee efforts at standardization are progressing at the usual rate.

Certain new languages, such as SISAL, are more suited to expressing parallel constructs than is FORTRAN, but efficient compilers for them on existing hardware take time to implement. Replacement of FORTRAN, the mother tongue of scientific computing, by some theoretically superior Esperanto of parallel computing may remain a forlorn hope.

Fine grain parallelism is fairly easy to implement in a compiler, but it may involve too much data flow between processor nodes to be efficient. Coarse grain parallelism, as in DDMP, is inherently more efficient, but it is better left to human ingenuity to work out the best data flow strategies. The greatest needs of those currently doing this are timing and debugging tools for tracking down inefficiency and errors.

A consequence of the great power of the microprocessor chips is the difficulty of feeding them data fast enough from their large, slow local memory. Small, fast cache memories are being interposed in some designs in order to alleviate this problem, but now a new one, that of planning the efficient use of the cache, arises. This local cache flow problem may pose a more important challenge to system software developers than the more traditional problems of parallel data flow.

## 26.5 Future Computer Centers

Existing computer centers are facing difficult times. On the one hand, they see their users fleeing to more cost-effective workstations; on the other, they are facing a wrenching shift to highly parallel computing. It is clear that the shift must eventually be made, but it is not clear when and how.

At LLNL a consensus is being reached on the Livermore model of the computing center of the future and how to get there. Although the details are still fuzzy, the outline is becoming clear on the basis of experience with the Massively Parallel Computing Initiative (MPCI) at LLNL and of knowledge of the kind of computing that has been carried out at LLNL for several decades.

The Livermore model is based on an MIMD architecture with thousands of nodes that can be partitioned in arbitrary ways. In particular, hundreds of the nodes are available for single-node use as effective replacements for workstations. In this way, capacity is provided for the large number of users with no need for parallel computing. The rest of the thousands of nodes are available for users who request clusters of various sizes for parallel applications. The center in the Livermore model then becomes a seamless UNIX cycle server with no particular distinction between single-node and highly parallel use. Resources are distributed by a gang scheduler that distributes nodes and time among users, as is currently being done for the 128-node BBN TC-2000 being shared by about 100 users at the LLNL MPCI. The first step to move to the Livermore model is likely to be the hardest, and that is to convert existing single-node codes to a standard compiler language for running on a single UNIX node. The later spreading out to parallel clusters of nodes can then evolve in a continuous way using any one or a mixture of programming models.

The Livermore model tends to counter the centrifugal move to individual workstations, which have often proven to be cost effective even when used only 5% of the time. In the Livermore model the user, with a standard X-window terminal that costs 10 times less than does a workstation, has immediate access to workstation computing power when needed. Workstations that have specialized hardware – for visualization, for example – would, however, continue to be useful in this environment.

## Acknowledgment

This work was performed under the auspices of the U.S. Department of Energy by the Lawrence Livermore National Laboratory under Contract No. W-7405-ENG-48.

# 27

## Geophysical Fluid Dynamics, Large Eddy Simulation, and Massively Parallel Computing

WILLIAM P. DANNEVIK

### 27.1 Introduction

The character of large-scale scientific computing is undergoing radical change. The rate of increase in throughput of individual vector super-computer processors is slowing due to fundamental physical constraints as well as to engineering limitations. At the same time, market forces in the high-performance workstation arena are rapidly driving the performance of RISC-based "commodity" microprocessors toward the level of recent-generation supercomputers. Mass-produced microprocessors operate near a performance/cost level about 10 times larger than that of a typical vector supercomputer processor.

These developments influence large-scale scientific computing in at least three ways. First, some problems which we typically execute today on vector supercomputers can be executed on dedicated workstations, particularly when it is recognized that many computational scientists really have access to only a small percentage of the vector supercomputer, since it is being shared with many other workers. Second, it is becoming easy to coerce several microprocessors to work cooperatively on a single computational problem to solve it in one-tenth the wall clock time or one-tenth the cost (but not both, as yet). Finally, and perhaps most importantly, it is possible to think of "many" (defined arbitrarily here as 100 or more) microprocessors cooperating to solve a single large problem in wall clock time of hours instead of months, or to solve a previously insoluble problem in a few days or weeks (assuming that the problem has been declared important enough to warrant the resulting expense).

The first two impacts have already been experienced in the production environment. That the third, of massively parallel *production* computing, is "just around the corner" has been claimed for several years. This is now a reality. (For example, in geophysical fluid dynamics applications, atmospheric and oceanic general circulation models used in climate simulation have been rewritten for several massively parallel architectures as part of the DOE CHAMMP Program.) Previously diverse trends in the evolution of basic massively parallel computer (MPC) design have begun to converge sufficiently that the design of software development environments essential to their productive use can be contemplated. It is fair to say that massively parallel computing has been "housebroken" (Smith, 1991), and that if one is not computing on an MPC, one is not "supercomputing" (Brooks, 1991).

What impact will these developments have in the area of LES? In what follows, we will offer some thoughts on this question in the context of geophysical fluid dynamics (GFD) applications. These observations are based primarily on our recent experiences at Lawrence Livermore National Laboratory (LLNL). In many cases, our work is concerned with the adaptation of *existing* models to MPC systems, without major change in algorithms. In addition, our work relates to finite-difference-based models as contrasted with spectral models. Much more will undoubtedly be accomplished in the future involving new algorithms. We hasten to mention that GFD applications are a small subset of the application-level MPC research topics addressed at LLNL, and are linked closely to particular MPC resources present at the laboratory. Nevertheless, the field is changing rapidly enough that current views can quickly become dated. The following can best be viewed as a snapshot of our evolving thinking.

## 27.2 The Earliest Viewpoint

Before going further, it is well to point out that people have been thinking about such issues for a long time. It was estimated some time ago that about 64,000 processors would be needed to complete a comprehensive global-scale computation in geophysical fluid dynamics, including a rather sophisticated approach to parameterization of many subgrid-scale physics processes in addition to "ordinary" fluid turbulence, such as cloud microphysics, atmospheric chemistry, and radiative transfer. When we consider that multipipe RISC microprocessors currently offer sustained performance levels of a few tens of Mflops, this estimate is seen to correspond to the holy grail of "Grand Challenge" scientific computation, i.e., 1 Tflops. Furthermore, this early estimate anticipated many of the fundamental challenges in achieving good efficiency in the utilization of massively parallel resources, e.g., the need to manage effectively interprocessor communications and computational load imbalance, and the question of the best granularity of parallelization.

What is particularly remarkable about these estimates is that they were first published in 1922 (Richardson, 1965). This may or may not be the first reference to the concepts of LES, but it is certainly the first mention of the role of massively parallel computing in LES. The type of computers contemplated by L. F. Richardson were, of course, the kind composed of a *human* computer equipped with pencil, paper, and adding machine. The application was numerical weather prediction. What Richardson could not anticipate was that the speed of individual processors in his proposed MPC would be accelerated to the point that a single *electronic* central processing unit (CPU) could eventually handle the load which he quite accurately estimated would be associated with global numerical weather prediction with a grid size of the order of a few hundred kilometers. For many years now, this has been accomplished on a routine basis with single-processor vector supercomputers.

Today, it is again being estimated that on the order of thousands of CPUs will be needed to meet the needs of geophysical fluid dynamics problems. For example, to improve the scientific basis for global change estimates based on increased greenhouse gas emission, next-generation climate simulation models must have increased spatial resolution, more complete subgrid physics process representations, and full coupling of major components such as ocean and atmospheric dynamics, atmospheric and marine biogeochemistry, and ecosystem dynamics. Also, there will be a flood of observational data connected with constellations of space-based instrument platforms designed for remote sensing of environmental variables relevant to climate monitoring and initialization of advanced mesoscale weather prediction models.

As in the past, GFD applications are presently one of the driving forces in the supercomputer marketplace. Since these applications also exhibit some of the highest effective Reynolds numbers of naturally occurring flows, it is certain that no foreseeable increases in throughput associated with MPC developments will replace the need for an important SGS parameterization component in these models. In the next few sections, we offer some observations on the complex interplay between the nature of SGS models and expected parallel performance of typical GFD models.

## 27.3 MPC Architectures and the MIMD Domain Decomposition Paradigm

MPC hardware can be divided loosely into two classes, depending on the degree of independence possessed by the individual computational nodes. The SIMD (single instruction, multiple data) class is characterized by all nodes executing an identical instruction stream in lockstep. Often, one may think of this class as simply a vector architecture with a very large number of vector registers. In the MIMD (multiple instruction, multiple data) architecture, each processor may be executing different instructions at a given instant. The simplest example of this class is a gaggle of workstations linked by a local network and occasionally exchanging data files among applications. It is possible to emulate SIMD mode on a MIMD computer, but not vice versa.

A further classification is possible. Multiprocessor versions of CRAY-type computers are known as *shared-memory* machines, since all processors must share access to common locations in a single central memory (usually facilitated via some form of direct memory access capability). It is not clear whether such access procedures scale well to thousands of processors, and many other MPC architectures feature some form of hierarchical *distributed memory*. Here, each computational unit has one or several forms of local memory. It is possible to dedicate a portion of each node's local memory to a common pool which can be used to *imitate* shared memory. However, a shared-memory machine cannot convincingly imitate a distributed-memory machine, since the multitude of connections possible in a distributed-memory architecture cannot be duplicated on the limited connectivity of the shared-memory bus structure.

The experiences related in this chapter were acquired on an MIMD distributed-memory developmental platform, the BBN TC2000 system. This machine has been useful as an experimental environment because it can simulate shared-memory and SIMD operational modes as well. However, we have found the direct MIMD distributed-memory mode straightforward and adequate for most of our purposes. The TC2000 version which we use has 128 computational nodes connected in a two-dimensional (2D) toroidal network implemented via 8 × 8 crossbar switches. Each node has about 16 mbytes of local memory, instruction and data cache units, and a sustained double-precision throughput of a few Mflops.

In many operational GFD models, the computational load can be separated into two main components. One is the direct load associated with the time advance of 3D hydrodynamic fields. In these equations are complex source terms representing the effect of additional "nonhydrodynamic" processes (e.g., heating due to radiative transfer) and parameterization of SGS processes such as turbulent transfer. The computation of these source terms represents a second major computational load. Some of the SGS parameterizations are turned on and off depending on local conditions in the resolved-scale flow field. For example, cumulus convection parameterization is turned on in regions where moisture and vertical velocity fields and thermal stratification are conducive to penetrative convection. Thus, the distribution of computational load within the model varies in space and time. These features suggest to us that MIMD capabilities may be important in achieving good parallel efficiency.

As mentioned in Section 27.1, our early work has centered on the porting of existing models to MPC systems, with little change in algorithm. The climate and numerical weather prediction (NWP) models with which we have worked are finite-difference-based and are characterized by explicit time-advance algorithms. The applications can be characterized as low Mach number density stratified flows, with a correspondingly finite signal speed throughout.

One parallelization strategy becomes obvious for such problems. If the computational domain is divided in some fashion into chunks, then the time advance in most of the interior of these subvolumes can be accomplished without knowledge of conditions outside the chunk. Thus, we can think of a processor being assigned to the work associated with each chunk. Only data near chunk boundaries need be communicated to the nearest neighboring processors. This *domain decomposition* approach is applicable to arbitrary local space–time physics problems solved with time-explicit algorithms. In addition, we may think of several processors being assigned to each chunk, each being responsible for the work connected with a given physics process. For example, some may advance hydrodynamic fields, while others compute the source terms due to radiative transfer or SGS processes. This coarse-grain *process parallelism* stems from the explicit and operator-split nature of the underlying algorithms.

One should note that spectral algorithms and Fourier transforms are generally not local in physical and wave number space, so that the domain decomposition approach would require much more than nearest-neighbor communication. On the other hand, alternative procedures with favorable scaling exist for such computations, but fall outside the scope of this chapter.

In the following section, we will consider some measures of parallel efficiency associated with these parallelization paradigms, as well as the impact on parallel efficiency of variations in the SGS parameterization approaches.

## 27.4 Parallel Performance of the Domain Decomposition/ Message-Passing Paradigm with "Local" SGS Modeling

Having an MPC with, say, 1,000 processors is one thing; getting performance equivalent to 1,000 times the single processor throughput can be quite another. There can be at least two culprits. One is that the existing algorithm resists parallelization, so that the solution method must be modified to achieve *any* parallel efficiency. If the parallel algorithm operates less effectively on a single processor than does the original serial algorithm, then aggregate parallel performance will suffer. We have not encountered this case in our early work. The second and more common culprit is the need for substantial interprocessor communication for all but the most "embarrassingly parallel" problems. (Some Monte Carlo problems, e.g., fall into this latter category, with little communication required until the problem is basically completed.) Especially for the MIMD distributed-memory method outlined above, parallel efficiency largely hinges on keeping the time spent in interprocessor communication small compared with the time spent doing useful arithmetic. The interprocessor communication is generally accomplished in the form of messages sent from one node to one or several other nodes.

The business of defining and measuring parallel performance has some subtle aspects. With due respect for those who have spent parts of a career on such matters, we will be content with the following simple characterizations. Suppose that we have an MPC which is dedicated to our use, so that our measurements will not be contaminated by extraneous uncontrollable events. We first run the parallel model for a number of time steps using only a single processor ($n_p = 1$). Next, we run the parallel model using some larger number $n_p$ of processors. The "parallel speedup" of the second experiment is defined as

$$S(n_p) = T(1)/T(n_p), \tag{1}$$

where $T(n_p)$ is the wall clock time measured for completion of the job using $n_p$ processors. If the match between machine architecture, model algorithms, and parallelization approach is perfect, then $S(n_p)$ will equal $n_p$, and we can say that the parallelization is perfect. (A number of factors are ignored here, such as the fact that some additional processor may be required to initialize even the single-processor job.) We may also define a "parallel efficiency" as $e(n_p) = S(n_p)/n_p$, so that $e(n_p) = 1$ for the ideal case.

Clearly, speedup and efficiency are crude measures of parallel performance, and it is easy to conceive of more elaborate figures of merit, such as some measure of the cost of a "sufficiently accurate" solution, with the side constraint of completion within a desired turnaround time. Our discussions here are too simple to require such sophistication.

The parallel efficiency will be determined by the time spent in communication $T_c$ versus that spent doing the time-advance algorithm $T_a$. If we assume blocking communication (a node processor can do no other work while messages are processed and sent), then $T(n_p) = T_c(n_p) + T_a(n_p)$, and $T(1) = T_a(1)$, where $()_c$ and $()_a$ refer to communication and arithmetic, respectively. It follows that

$$S(n_p) = n_p[1 + T_c(n_p)/T_a(n_p)]^{-1}, \tag{2}$$

showing the important role played by the communications "overhead" $O_c \equiv T_c/T_a$. To estimate the latter quantity, we must say more about the problem being solved and the decomposition strategy.

The simplest case is that of direct (closure-free) numerical solution of the Euler equations discretized uniformly in a regular domain of space dimension $d$, decomposed uniformly in $D \leq d$ dimensions. (Note that one result of this is that each subdomain requires the same amount of arithmetic, so that there are no load balancing issues involved.) Supposing that there are $n_g$ grid points distributed in the total solution domain, the time spent by a processor in advancing the fields in a subdomain can be written as

$$T_a(n_p, n_g) = (Fn_g)/(R_a n_p), \tag{3}$$

where $F$ is the number of flops required per grid point per time step and $R_a$ is a computation rate per node processor.

An estimate of the communications time may be developed in similar fashion, but is more complicated. The time depends on the "radius" $L$ of finite-difference stencil used for spatial differencing, on the volume $W$ of border data communicated per border grid point, and on communications hardware performance parameters and the software implementation of message-passing utilities on the machine. In the limit in which message lengths are sufficient to allow neglect of message latency and setup time, and in which the number of interior grid points in each space dimension is much larger than the differencing stencil radius, the communication time is given by ($d \leq 3$)

$$T_c(n_p, n_g) \cong \frac{\alpha \beta L W}{R_c} n_p^{(\gamma - D + 1)/d} n_g^{(D-1)/D}, \qquad (4)$$

where

$$\alpha = \begin{cases} 4 \text{ for } (D,d) = (2,2) \text{ or } (2,3) \\ 6 \text{ for } (D,d) = (3,3), \end{cases}$$

$$\gamma = \begin{cases} 1 \text{ for } (D,d) = (2,3) \\ 0 \text{ otherwise}, \end{cases}$$

$R_c$ is the communications rate, and $\beta$ is a machine-specific parameter related to message latency and network topography.

In this limit, the communications overhead is then

$$O_c(n_p, n_g) \cong \frac{\alpha \beta L W R_a}{F R_c} n_p^{(\gamma - D + 1 + d)/d} n_g^{-1/D}. \qquad (5)$$

Estimate (5) expresses the generalized surface-to-volume ratio of time spent communicating versus doing arithmetic for the DDMP approach for explicit time-marching problems (for closely related estimates, see Flatt, 1984; Flatt and Kennedy, 1989). For example, for a 2D problem decomposed in two dimensions, $O_c \sim (n_p/n_g)^{-1/2}$. One interesting case of estimates similar to (5) is that of a partitioning of, for example, a 3D domain into *thin* 2D slabs (Procassini and Dannevik, 1991a), for which it follows that $S \sim n_p/(1 + \lambda n_p)$. On the other hand, a 2D decomposition of the same domain into columns with bases of unit aspect ratio gives $S \sim n_p/(1 + \lambda n_p^{1/2})$. The difference is crucial, because in the first case, $S$ eventually *saturates* at $S \sim \lambda^{-1}$ as larger and larger processor counts $n_p$ are considered for a fixed problem size (fixed $n_g$). The second strategy yields $S \sim \lambda^{-1} n_p^{1/2}$ in the same limit (see Fig. 1).

Now, we are supposed to be examining the relation of all of this to

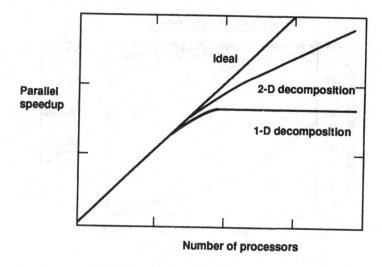

**Fig. 1**    Qualitative behavior of the scaling of speedup with processor count for a fixed problem size using one- and two-dimensional DDMP paradigms.

LES and SGS modeling. Some simple inferences can immediately be drawn from estimates such as (5). Consider a hierarchy of more and more sophisticated LES models. To begin with, let us recall the first "LES" proposal, that of solving the Navier–Stokes equations instead of the Euler equations, i.e., adding *molecular* dissipation terms. This immediately raises the spatial order of derivatives explicitly appearing in the equation set, which may elevate $L$ in (5). $O_c$ is directly proportional to $L$, so that, other factors being equal, this leads to reduced parallel efficiencies.

Of course, other factors do *not* often remain equal in such scenarios. The inclusion of explicit dissipation effects may result in the need for a higher-order spatial differencing scheme to improve treatment of internal or wall-induced boundary layers, which again will influence $L$. The existence of strong shear may tempt one to introduce static or dynamically adaptive nonuniform grids, which will lead to load balancing issues for parallelization schemes as simple as DDMP. In the static case, one may think of arranging subdomain volumes of unequal size in physical space but containing approximately the same computational requirements, so that processors will finish nearly synchronously. Parallel efficiency anal-

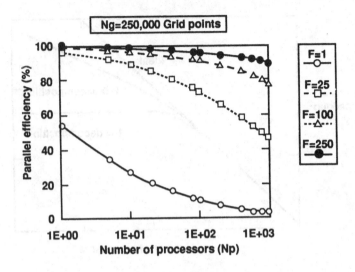

Fig. 2   Parallel efficiency as a function of processor count (fixed problem size) for various levels of grid point computational intensity.

ysis of such schemes will remain a largely empirical endeavor in the near term.

Some other effects of various SGS schemes are easier to analyze. Let us return again to the simple uniform problem and consider the step of moving from the Navier–Stokes equations to LES with a spatially varying nonlinear SGS viscosity. The prototype example for GFD is the Smagorinsky formulation. Computation of the eddy viscosity can be carried out in terms of resolved flow-field variables which are already available in each processor subdomain. This operation increases the computational "intensity" parameter $F$ in (3), and thus reduces the communication overhead (5). This leads to an apparent *increase* in parallel efficiency. This is illustrated in Figs. 2 and 3, which show the variation in parallel efficiency as a function of floating point intensity $F$ for a 2D decomposition of a 2D domain. Addition of a Smagorinsky viscosity to a Navier–Stokes-like code may not increase $F$ dramatically in most cases. However, one can see that more sophisticated eddy viscosity formulations for density-stratified flows in which mean temperature gradients and velocity shear are incorporated will begin to make a difference.

In large-scale GFD applications such as climate modeling or numerical weather prediction, other SGS parameterizations can become extremely

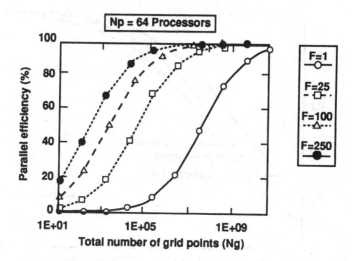

Fig. 3   Parallel efficiency as a function of problem size (fixed processor count) for various levels of grid point computational intensity.

demanding. Such computations are typically responsible for half of the CPU time consumed by global atmospheric general circulation models. Many of the parameterizations in these models are intrinsically 1D and column-oriented, with no horizontal data connectivity. An example is the parameterization of SGS cumulus convection processes. An obvious DDMP strategy for these codes is a 2D horizontal decomposition, with no decomposition of the vertical coordinate. Physics modules meeting these specifications have a favorable impact on parallel efficiency, because they increase the computational intensity without dramatically changing communications requirements.

We have verified some of these considerations with a simple prototype GFD model on the BBN TC2000 (Procassini and Dannevik, 1991b). Consider first the 2D DDMP strategy applied to the shallow water equations on the sphere. Suppose the addition of SGS computation of the type described above results in a computational intensity $F' = \lambda F$, where $F$ is the intensity for the equations without the SGS model. Then the resulting parallel efficiency according to (5) will be

$$e' = [1 + \lambda^{-1}(e^{-1} - 1)]^{-1}, \tag{6}$$

where $e$ is the efficiency without the SGS model. Figure 4 shows the

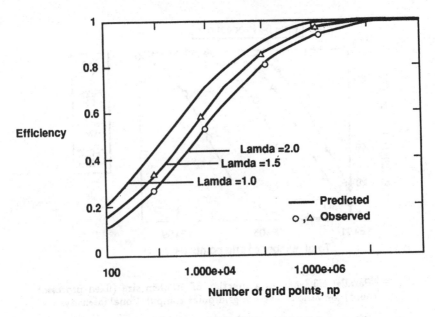

Fig. 4  Predicted and observed variation of parallel efficiency with computational intensity for the two-dimensional DDMP paradigm applied
to a time-explicit second-order accurate algorithm for solution of
the shallow water equations on a sphere.

variation of $e'$ with $n_g$ for a fixed processor count as predicted by (5)
and (6) and as observed experimentally for the cases $\lambda = 1.5$ and 2.0.
The additional arithmetic for these timing test cases was arbitrary.

These estimates survive even in a very complex GFD code such as a
3D global general circulation model. Figure 5 shows observed parallel
efficiency of the UCLA general circulation model using the 2D DDMP
scheme as a function of processor count for a fixed resolution of $4° \times
5° \times 9$ vertical levels [$n_g \sim O(10^4)$]. The two data sets are for the
hydrodynamics only and for hydrodynamics plus all additional column
physics and SGS parameterizations.

By now it should be clear what will be the impact of adding more
layers of sophistication to the SGS model such as additional evolution
equations for subgrid volume averaged prognostic variables like kinetic
energy and thermal variance. Each additional prognostic field will introduce additional communications overhead and computational intensity

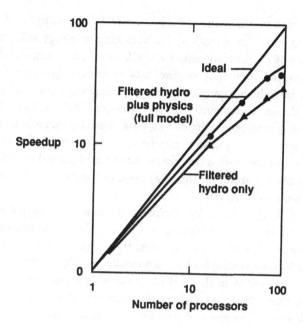

Fig. 5  Parallel speedup for the UCLA atmospheric general circulation model as a function of processor count: effect of column physics parameterizations.

which will affect efficiency according to (3) and (4). Whether the effect on parallel performance is favorable will depend on the details of the SGS formulation, and can be estimated readily. We leave this to the reader with regard to whatever SGS model may be of interest.

We have to this point considered only "local" SGS models, which depend on resolved-scale fields only in the near vicinity of a grid point. More elaborate parameterization schemes exist, of course, and are considered briefly in the next section.

## 27.5 Impact of "Nonlocal" SGS Parameterizations on Parallel Performance

Adding to a given resolved-scale equation set the types of SGS models discussed so far does not *qualitatively* change the computational or data complexity of the modified code (at least from the standpoint of the parallelization issues connected with the DDMP paradigm). For example, adding evolution equations for SGS kinetic energy will change the parameters of (5) but not usually the form of (5). Some SGS parameterizations, however, may require data from nonadjacent subdomains. This may introduce communications requirements with scaling properties that depart significantly from expressions such as (4). Examples would be any scaling law depending on a spatially variable integral turbulent length scale, or a wave number-dependent eddy viscosity formulation incorporated into a pseudospectral or grid point-based LES code. For the DDMP paradigm, this may require communications extending over all other subdomains.

Let us first consider a less extreme example. Baroclinic instability mechanisms act in the ocean at scales an order of magnitude smaller than those in the atmosphere, because the ocean density stratification is intrinsically more stable, leading to a smaller Rossby deformation radius. Baroclinic instability in the ocean leads to eddy formation on scales typically less than 50 km; explicit resolution of these eddies would require $O(10^6)$ horizontal degrees of freedom in a global 3D ocean circulation model. This is substantially beyond the present state of the art, so that SGS parameterization is needed to represent the effects of SGS baroclinic eddies. A crude treatment of this situation is simply to dissipate baroclinic eddies at the deformation radius and hope that the omitted dynamics has a negligible effect on the gross resolved-scale flow features. This has led to the common practice of adding ad hoc dissipation terms of the form $c_1 \nabla^2 u$ or a more "scale-selective" $c_2 \nabla^4 u$, where constants $c_1$, $c_2$ are chosen experimentally.

In the context of the DDMP paradigm, one may immediately see an issue with regard to the choice of harmonic vs. biharmonic diffusion, in that use of the latter may double the communication time estimate (4), *reducing* the parallel speedup (assuming the spatial differencing is second-order accurate throughout). One the other hand, simply increasing horizontal resolution will *increase* parallel speedup according to (5) and (2). Which is the better strategy depends on the parallel efficiency level of the code before modification.

This simple trade-off situation can become much more acute with more sophisticated SGS models which approach a "nonlocal" form. Consider the anticipated potential vorticity method (APVM) of Sadourney and Basdevant (1985). This method results in a physics-based lateral diffusion scheme with nonlinear coefficient framed in terms of the resolved-scale kinetic enstrophy. However, the combined requirements of kinetic energy conservation and potential enstrophy dissipation lead to a diffusion operator in the form of an iterated Laplacian, $\tau^{-1}c_n(-\nabla^2)^n u$, where $\tau$ is a representative eddy turnover time at the grid scale, and $n \sim 8$. This results in a very large finite-difference stencil and correspondingly large communications requirements for the DDMP parallelization approach. This would suggest that simply increasing horizontal resolution might be more effective or, more likely, that a completely different parallelization paradigm might be more suitable.

Finally, let us consider the impact of fully nonlocal SGS models. Such models can result from an approximate physical-space representation of a spectral turbulence closure, and may require for their pointwise computation the communication of data from *all* other subdomains. To illustrate the sorts of issues involved in this problem, consider the well-known "pole problem" in global-scale geophysical fluid dynamics modeling using spherical coordinates with explicit time advance algorithms. Here, convergence of meridians in the polar regions results in an onerous Courant–Friedrichs–Lewy (CFL) stability criterion, so that shorter time steps are required as compared with lower latitudes. This situation can be relieved by a method which removes high frequency excitation from the primitive prognostic variables in a physically meaningful way. (With some stretch of the imagination, this can be viewed as SGS modeling in the time domain.) Arakawa (1966) suggested that this can be accomplished by application of a low-pass filter in wave number space which removes linearly unstable modes which would otherwise violate the CFL criterion. This requires, of course, several Fourier transform operations (1D, since only the meridians are converging).

In the simplest implementation of these ideas within the DDMP paradigm, the resulting speedup estimate becomes ($D = 2 = d$)

$$S(n_p, n_g) = n_p \left[ 1 + \frac{\alpha\beta LW R_a}{FR_c} \left( \frac{n_p}{n_g} \right)^{1/2} \left( \frac{1 + \delta n_g^{1/2}(n_p - 1)^{1/2}}{1 + \mu n_g^{1/2}} \right) \right]^{-1}, \quad (7)$$

where $\delta \equiv L^f/(\alpha LW)$, $L^f$ is a filter stencil radius, $\mu \equiv F^f/F$, and $F^f$ is the number of flops per zone required for filter arithmetic.

In the limit in which an increasing number of processors is envisioned for a large but fixed-size problem, this speedup saturates at $S \sim \mu FR_c n_g^{1/2}/(\alpha\beta\delta LW R_a)$, whereas the speedup in the absence of filtering continues to grow at a rate $S \sim FR_c n_g^{1/2}/(\alpha\beta LW R_a)n_p^{1/2}$. The performance (7) can be improved, however, by using a tree-based global gather–scatter algorithm to accomplish the nonlocal communica-

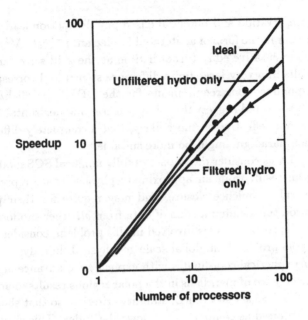

Fig. 6   Parallel speedup for filtered and unfiltered hydrodynamics module
of the UCLA atmospheric general circulation model.

tion, with the result (for a simple 2D mesh processor connectivity)

$$S(n_p, n_g) = n_p \left[1 + \frac{\alpha\beta LW R_a}{FR_c}\left(\frac{n_p}{n_g}\right)^{1/2}\left(\frac{1 + \delta n_g^{1/2}\log_2(n_p-1)^{1/2}}{1+\mu n_g^{1/2}}\right)\right]^{-1}. \quad (8)$$

This behaves as $S \sim \mu FR_c n_g^{1/2}/(\alpha\beta\delta LW R_a)n_p^{1/2}(\log_2 n_p)^{-1}$, a substantial improvement over (7). But in either case, parallel efficiency has been degraded by the filtering operation. Figure 6 shows this effect for the AGCM code mentioned previously. Unless the efficiency is already very high without filtering, the impact could be substantial. Depending on the parameters appearing in (7) or (8), it could be more economical simply to decrease the time step and bypass the filtering procedure altogether.

    The impact of multidimensional transforms would be even more severe, and would probably call into question the effectiveness of the DDMP paradigm for such problems.

## 27.6 Concluding Remarks

We first repeat that the preceding observations are colored strongly by the particular mix of numerical algorithm, parallelization paradigm, and machine architecture we have chosen. A change in any one of these three ingredients could lead to qualitatively different conclusions for any given SGS parameterization. Furthermore, the quantitative impact of any proposed trade-off between resolution and SGS model choice depends on the numerical values of parameters in formulas such as (5), which often must be determined by detailed experiment. [For example, whether 1D filtering degrades or enhances parallel performance in the limit of high resolution with a fixed number of processors depends on the value of $\delta/\mu$ in (7) or (8).] These parameters can vary substantially among different members within the same machine architecture class, and among different software implementations of generically similar communications utilities.

Our early experiences suggest that the advent of massively parallel computing resources will add a new level of complexity to the tasks of building effective SGS parameterizations appropriate for a given CFD algorithm and assessing the traditional resolution/SGS model trade-off. [We have not discussed here the fact that the basic algorithms used in both direct and large eddy simulation will be undergoing continual reappraisal (e.g., full spectral vs. grid-based) as MPC architectures and software environments evolve and mature.] In this sense, the transition to the MPC environment will be more painful than was the transition from serial to vector supercomputer technology.

The transition may become easier in the future with the advent of portable parallel software toolsets to aid parallelization. Also, it is likely that the next generation of MPC resources will possess nonblocking communication capability, so that the issue of communication overhead will be important only if the time to accomplish arithmetic is insufficient to cover the time required for communication.

In the meantime, however, the business of SGS model building and testing on advanced parallel supercomputers is likely to become more complex and challenging. It might be amusing to speculate on the nature of computing resources which might be needed to eliminate the need for LES approaches in geophysical fluid dynamics. With the rule of thumb that each doubling of (microscale) Reynolds number would require an order of magnitude increase in throughput, one can estimate that a $10^8$-fold increase over today's Gflops levels would begin to help. This could be accomplished on a parallel machine with 10,000 nodes, each with Tflops throughput. Tflops performance could be attained on the individual processor level if each node were itself a parallel machine (i.e., an *iterated* parallel architecture). Is this out of the question? Note that today's TMC CM-5 is rated at 128 Mflops per node peak performance because each node is in fact four processors which can operate in a data parallel (SIMD) mode. Finally, note that Richardson's 1922 estimate of 64,000 processors for numerical weather prediction also seemed naive and unworkable at the time. Today, that goal is in hand.

## Acknowledgments

This work has been supported by the U.S. Department of Energy at Lawrence Livermore National Laboratory under Contract No. W-7405-ENG-48.

## References

ARAKAWA, A. (1966) Computational design for long-term integration of the equations of atmospheric motion. *J. Comput. Phys.* 1, 119–143.

BROOKS, E. (1991) The massively parallel computing initiative. In *The 1991 MPCI Yearly Report: The Attack of the Killer Micros*. Ed. E. Brooks and K. White, pp. 1–5. Lawrence Livermore National Laboratory Report UCRL-ID-107022.

FLATT, H.P. (1984) A simple model for parallel processing. *Computer* 17, 95–108.

FLATT, H.P. AND KENNEDY, K. (1989) Performance of parallel processors. *Parallel Computing* 12, 1–18.

PROCASSINI, R. AND DANNEVIK, W. (1991a) A shared-memory implementation of a global ocean model on a MIMD parallel computer. In *The 1991 MPCI Yearly Report: The Attack of the Killer Micros*. Ed. E. Brooks and K.. White, pp. 119–126. Lawrence Livermore National Laboratory Report UCRL-ID-107022.

PROCASSINI, R. AND DANNEVIK, W. (1991b) Implementation of a shallow water model on a MIMD parallel computer. In *The 1991 MPCI Yearly Report: The Attack of the Killer Micros*. Ed. E. Brooks and K. White, pp. 111–118. Lawrence Livermore National Laboratory Report UCRL-ID-107022.

RICHARDSON, L.F. (1965) *Weather Prediction by Numerical Process.* Dover, 236 pp.

SADOURNEY, R. AND BASDEVANT, C. (1985) Parameterization of subgrid scale barotropic and baroclinic eddies in quasi-geostrophic models: Anticipated potential vorticity method. *J. Atmos. Sci.* **42**, 1353–1364.

SMITH, N.P. (1991) Supercomputing review: News and analysis. *Supercomput. Rev.* **4**, 12–15.

# Index

adjustment
  convective, 498
  dynamic, 498
anisotropic flows, 129
anticipated potential vorticity
    method (APVM), 590
approximate boundary
    condition, 121, 560
average
  cascade, 516–17
  conjunctive, 517
  double, 516–17, 526
  Favré, 244, 260, 316, 318
  low-pass, 517, 521, 531
  moving, 516, 528
  phase, 373, 377
  Reynolds, 38, 55, 129, 160,
    193, 209, 316, 322, 513,
    528, 559, 562

bathymetry, 513
beta density, 330, 334–42
bulk, 67, 390, 451, 491, 500
  formulas, 493–5, 500
  parameterizations, 499
    Kraus and Turner, 499
    Price, 507

Burgers equation, 516

cache flow, 574
climate simulation models, 578–9
communications overhead,
    583–4, 587
compressible hydrodynamics,
    105–13, 232, 573
computational intensity, 586–8
condensation processes, 6
convection
  cumulus, 581, 587
  free, 497–8, 444
  nonpenetrative, 407
  penetrative, 444, 581
  slant-wise, 7, 28
  two-dimensional, 405
Coriolis parameter, 426–9, 437
cross terms, 120–2, 262–3, 516,
    523, 529, 531, 540–55

deformation, tensor, 531
diffusion flames, 288, 292, 299
direct interaction approximation
    (DIA), 79–82, 92, 95, 123, 197,
    257, 288, 292, 299
distributed memory, 580, 582